Pitman Research Notes in Mathematics Series

Submission of proposals for consideration
Suggestions for publication, in the form of outlines and representative samples, are invited by the Editorial Board for assessment. Intending authors should approach one of the main editors or another member of the Editorial Board, citing the relevant AMS subject classifications. Alternatively, outlines may be sent directly to the publisher's offices. Refereeing is by members of the board and other mathematical authorities in the topic concerned, throughout the world.

Preparation of accepted manuscripts
On acceptance of a proposal, the publisher will supply full instructions for the preparation of manuscripts in a form suitable for direct photo-lithographic reproduction. Specially printed grid sheets are provided and a contribution is offered by the publisher towards the cost of typing. Word processor output, subject to the publisher's approval, is also acceptable.

Illustrations should be prepared by the authors, ready for direct reproduction without further improvement. The use of hand-drawn symbols should be avoided wherever possible, in order to maintain maximum clarity of the text.

The publisher will be pleased to give any guidance necessary during the preparation of a typescript, and will be happy to answer any queries.

Important note
In order to avoid later retyping, intending authors are strongly urged not to begin final preparation of a typescript before receiving the publisher's guidelines and special paper. In this way it is hoped to preserve the uniform appearance of the series.

Longman Scientific & Technical
Longman House
Burnt Mill
Harlow, Essex, UK
(tel (0279) 26721)

Longman Scientific & Technical
Churchill Livingstone Inc.
1560 Broadway
New York, NY 10036, USA
(tel (212) 819-5453)

Titles in this series

1 Improperly posed boundary value problems
A Carasso and A P Stone
2 Lie algebras generated by finite dimensional ideals
I N Stewart
3 Bifurcation problems in nonlinear elasticity
R W Dickey
4 Partial differential equations in the complex domain
D L Colton
5 Quasilinear hyperbolic systems and waves
A Jeffrey
6 Solution of boundary value problems by the method of integral operators
D L Colton
7 Taylor expansions and catastrophes
T Poston and I N Stewart
8 Function theoretic methods in differential equations
R P Gilbert and R J Weinacht
9 Differential topology with a view to applications
D R J Chillingworth
10 Characteristic classes of foliations
H V Pittie
11 Stochastic integration and generalized martingales
A U Kussmaul
12 Zeta-functions: An introduction to algebraic geometry
A D Thomas
13 Explicit *a priori* inequalities with applications to boundary value problems
V G Sigillito
14 Nonlinear diffusion
W E Fitzgibbon III and H F Walker
15 Unsolved problems concerning lattice points
J Hammer
16 Edge-colourings of graphs
S Fiorini and R J Wilson
17 Nonlinear analysis and mechanics: Heriot-Watt Symposium Volume I
R J Knops
18 Actions of fine abelian groups
C Kosniowski
19 Closed graph theorems and webbed spaces
M De Wilde
20 Singular perturbation techniques applied to integro-differential equations
H Grabmüller
21 Retarded functional differential equations: A global point of view
S E A Mohammed
22 Multiparameter spectral theory in Hilbert space
B D Sleeman
24 Mathematical modelling techniques
R Aris
25 Singular points of smooth mappings
C G Gibson
26 Nonlinear evolution equations solvable by the spectral transform
F Calogero
27 Nonlinear analysis and mechanics: Heriot-Watt Symposium Volume II
R J Knops
28 Constructive functional analysis
D S Bridges
29 Elongational flows: Aspects of the behaviour of model elasticoviscous fluids
C J S Petrie
30 Nonlinear analysis and mechanics: Heriot-Watt Symposium Volume III
R J Knops
31 Fractional calculus and integral transforms of generalized functions
A C McBride
32 Complex manifold techniques in theoretical physics
D E Lerner and P D Sommers
33 Hilbert's third problem: scissors congruence
C-H Sah
34 Graph theory and combinatorics
R J Wilson
35 The Tricomi equation with applications to the theory of plane transonic flow
A R Manwell
36 Abstract differential equations
S D Zaidman
37 Advances in twistor theory
L P Hughston and R S Ward
38 Operator theory and functional analysis
I Erdelyi
39 Nonlinear analysis and mechanics: Heriot-Watt Symposium Volume IV
R J Knops
40 Singular systems of differential equations
S L Campbell
41 N-dimensional crystallography
R L E Schwarzenberger
42 Nonlinear partial differential equations in physical problems
D Graffi
43 Shifts and periodicity for right invertible operators
D Przeworska-Rolewicz
44 Rings with chain conditions
A W Chatters and C R Hajarnavis
45 Moduli, deformations and classifications of compact complex manifolds
D Sundararaman
46 Nonlinear problems of analysis in geometry and mechanics
M Atteia, D Bancel and I Gumowski
47 Algorithmic methods in optimal control
W A Gruver and E Sachs
48 Abstract Cauchy problems and functional differential equations
F Kappel and W Schappacher
49 Sequence spaces
W H Ruckle
50 Recent contributions to nonlinear partial differential equations
H Berestycki and H Brezis
51 Subnormal operators
J B Conway

52 Wave propagation in viscoelastic media
F Mainardi
53 Nonlinear partial differential equations and their applications: Collège de France Seminar. Volume I
H Brezis and J L Lions
54 Geometry of Coxeter groups
H Hiller
55 Cusps of Gauss mappings
T Banchoff, T Gaffney and C McCrory
56 An approach to algebraic K-theory
A J Berrick
57 Convex analysis and optimization
J-P Aubin and R B Vintner
58 Convex analysis with applications in the differentiation of convex functions
J R Giles
59 Weak and variational methods for moving boundary problems
C M Elliott and J R Ockendon
60 Nonlinear partial differential equations and their applications: Collège de France Seminar. Volume II
H Brezis and J L Lions
61 Singular systems of differential equations II
S L Campbell
62 Rates of convergence in the central limit theorem
Peter Hall
63 Solution of differential equations by means of one-parameter groups
J M Hill
64 Hankel operators on Hilbert space
S C Power
65 Schrödinger-type operators with continuous spectra
M S P Eastham and H Kalf
66 Recent applications of generalized inverses
S L Campbell
67 Riesz and Fredholm theory in Banach algebra
B A Barnes, G J Murphy, M R F Smyth and T T West
68 Evolution equations and their applications
F Kappel and W Schappacher
69 Generalized solutions of Hamilton-Jacobi equations
P L Lions
70 Nonlinear partial differential equations and their applications: Collège de France Seminar. Volume III
H Brezis and J L Lions
71 Spectral theory and wave operators for the Schrödinger equation
A M Berthier
72 Approximation of Hilbert space operators I
D A Herrero
73 Vector valued Nevanlinna Theory
H J W Ziegler
74 Instability, nonexistence and weighted energy methods in fluid dynamics and related theories
B Straughan
75 Local bifurcation and symmetry
A Vanderbauwhede
76 Clifford analysis
F Brackx, R Delanghe and F Sommen
77 Nonlinear equivalence, reduction of PDEs to ODEs and fast convergent numerical methods
E E Rosinger
78 Free boundary problems, theory and applications. Volume I
A Fasano and M Primicerio
79 Free boundary problems, theory and applications. Volume II
A Fasano and M Primicerio
80 Symplectic geometry
A Crumeyrolle and J Grifone
81 An algorithmic analysis of a communication model with retransmission of flawed messages
D M Lucantoni
82 Geometric games and their applications
W H Ruckle
83 Additive groups of rings
S Feigelstock
84 Nonlinear partial differential equations and their applications: Collège de France Seminar. Volume IV
H Brezis and J L Lions
85 Multiplicative functionals on topological algebras
T Husain
86 Hamilton-Jacobi equations in Hilbert spaces
V Barbu and G Da Prato
87 Harmonic maps with symmetry, harmonic morphisms and deformations of metrics
P Baird
88 Similarity solutions of nonlinear partial differential equations
L Dresner
89 Contributions to nonlinear partial differential equations
C Bardos, A Damlamian, J I Díaz and J Hernández
90 Banach and Hilbert spaces of vector-valued functions
J Burbea and P Masani
91 Control and observation of neutral systems
D Salamon
92 Banach bundles, Banach modules and automorphisms of C*-algebras
M J Dupré and R M Gillette
93 Nonlinear partial differential equations and their applications: Collège de France Seminar. Volume V
H Brezis and J L Lions
94 Computer algebra in applied mathematics: an introduction to MACSYMA
R H Rand
95 Advances in nonlinear waves. Volume I
L Debnath
96 FC-groups
M J Tomkinson
97 Topics in relaxation and ellipsoidal methods
M Akgül
98 Analogue of the group algebra for topological semigroups
H Dzinotyiweyi
99 Stochastic functional differential equations
S E A Mohammed

100 Optimal control of variational inequalities
V Barbu
101 Partial differential equations and dynamical systems
W E Fitzgibbon III
102 Approximation of Hilbert space operators. Volume II
C Apostol, L A Fialkow, D A Herrero and D Voiculescu
103 Nondiscrete induction and iterative processes
V Ptak and F-A Potra
104 Analytic functions – growth aspects
O P Juneja and G P Kapoor
105 Theory of Tikhonov regularization for Fredholm equations of the first kind
C W Groetsch
106 Nonlinear partial differential equations and free boundaries. Volume I
J I Díaz
107 Tight and taut immersions of manifolds
T E Cecil and P J Ryan
108 A layering method for viscous, incompressible L_p flows occupying R^n
A Douglis and E B Fabes
109 Nonlinear partial differential equations and their applications: Collège de France Seminar. Volume VI
H Brezis and J L Lions
110 Finite generalized quadrangles
S E Payne and J A Thas
111 Advances in nonlinear waves. Volume II
L Debnath
112 Topics in several complex variables
E Ramírez de Arellano and D Sundararaman
113 Differential equations, flow invariance and applications
N H Pavel
114 Geometrical combinatorics
F C Holroyd and R J Wilson
115 Generators of strongly continuous semigroups
J A van Casteren
116 Growth of algebras and Gelfand–Kirillov dimension
G R Krause and T H Lenagan
117 Theory of bases and cones
P K Kamthan and M Gupta
118 Linear groups and permutations
A R Camina and E A Whelan
119 General Wiener–Hopf factorization methods
F-O Speck
120 Free boundary problems: applications and theory, Volume III
A Bossavit, A Damlamian and M Fremond
121 Free boundary problems: applications and theory, Volume IV
A Bossavit, A Damlamian and M Fremond
122 Nonlinear partial differential equations and their applications: Collège de France Seminar. Volume VII
H Brezis and J L Lions
123 Geometric methods in operator algebras
H Araki and E G Effros
124 Infinite dimensional analysis–stochastic processes
S Albeverio
125 Ennio de Giorgi Colloquium
P Krée
126 Almost-periodic functions in abstract spaces
S Zaidman
127 Nonlinear variational problems
A Marino, L Modica, S Spagnolo and M Degiovanni
128 Second-order systems of partial differential equations in the plane
L K Hua, W Lin and C-Q Wu
129 Asymptotics of high-order ordinary differential equations
R B Paris and A D Wood
130 Stochastic differential equations
R Wu
131 Differential geometry
L A Cordero
132 Nonlinear differential equations
J K Hale and P Martinez-Amores
133 Approximation theory and applications
S P Singh
134 Near-rings and their links with groups
J D P Meldrum
135 Estimating eigenvalues with *a posteriori/a priori* inequalities
J R Kuttler and V G Sigillito
136 Regular semigroups as extensions
F J Pastijn and M Petrich
137 Representations of rank one Lie groups
D H Collingwood
138 Fractional calculus
G F Roach and A C McBride
139 Hamilton's principle in continuum mechanics
A Bedford
140 Numerical analysis
D F Griffiths and G A Watson
141 Semigroups, theory and applications. Volume I
H Brezis, M G Crandall and F Kappel
142 Distribution theorems of L-functions
D Joyner
143 Recent developments in structured continua
D De Kee and P Kaloni
144 Functional analysis and two-point differential operators
J Locker
145 Numerical methods for partial differential equations
S I Hariharan and T H Moulden
146 Completely bounded maps and dilations
V I Paulsen
147 Harmonic analysis on the Heisenberg nilpotent Lie group
W Schempp
148 Contributions to modern calculus of variations
L Cesari
149 Nonlinear parabolic equations: qualitative properties of solutions
L Boccardo and A Tesei
150 From local times to global geometry, control an[d] physics
K D Elworthy

151 A stochastic maximum principle for optimal control of diffusions
U G Haussmann
152 Semigroups, theory and applications. Volume II
H Brezis, M G Crandall and F Kappel
153 A general theory of integration in function spaces
P Muldowney
154 Oakland Conference on partial differential equations and applied mathematics
L R Bragg and J W Dettman
155 Contributions to nonlinear partial differential equations. Volume II
J I Díaz and P L Lions
156 Semigroups of linear operators: an introduction
A C McBride
157 Ordinary and partial differential equations
B D Sleeman and R J Jarvis
158 Hyperbolic equations
F Colombini and M K V Murthy
159 Linear topologies on a ring: an overview
J S Golan
160 Dynamical systems and bifurcation theory
M I Camacho, M J Pacifico and F Takens
161 Branched coverings and algebraic functions
M Namba
162 Perturbation bounds for matrix eigenvalues
R Bhatia
163 Defect minimization in operator equations: theory and applications
R Reemtsen

Dynamical systems and bifurcation theory

M I Camacho & M J Pacifico

Federal University of Rio de Janeiro

F Takens

University of Groningen

Dynamical systems and bifurcation theory

Longman Scientific & Technical

Copublished in the United States with
John Wiley & Sons, Inc., New York

Longman Scientific & Technical
Longman Group UK Limited
Longman House, Burnt Mill, Harlow
Essex CM20 2JE, England
and Associated Companies throughout the world.

Copublished in the United States with
John Wiley & Sons, Inc., 605 Third Avenue, New York, NY 10158

First published 1987

AMS Subject Classifications: 34CXX, 58–XX, 53–XX

ISSN 0269–3674

British Library Cataloguing in Publication Data
Dynamical systems – bifurcation theory.—
(Pitman research notes in mathematics
series, ISSN 0269–3674; 160).
1. System analysis
I. Camacho, M.I.T. II. Pacifico, M.J.
III. Takens, F.
515.3′5 QA402

ISBN 0-582-44261-3

Library of Congress Cataloging-in-Publication Data
Dynamical systems—bifurcation theory.

(Pitman research notes in mathematics series,
ISSN 0269–3674; 160)
Proceedings of a meeting held in Aug. 1985 at the
Federal University of Rio de Janeiro.
Bibliography: p.
1. Differentiable dynamical systems—Congresses.
2. Bifurcation theory—Congresses. I. Camacho, M.I.T.
II. Pacifico, M.J. III.Takens, Floris. IV. Series.
QA614.8.D939 1987 515.3′52 87-4032
ISBN 0-470-20843-0 (USA only)

Printed and bound in Great Britain by
Biddles Ltd, Guildford and King's Lynn

Contents

Preface

This volume contains the proceedings of the meeting on Dynamical Systems and Bifurcation Theory held at the Federal University of Rio de Janeiro in August of 1985. The occasion was particularly rewarding to us because of the active participation of a number of specialists in this field. We hope that this first meeting on Dynamical Systems at the Federal University will be followed by others with the same enthusiasm.

We are indebted to several colleagues and friends for their contribution which made the meeting possible, especially Luis Carlos Guimarães, Mário C. de Oliveira and Jacob Palis Jr. The financial support was given by IBM do Brasil, IMPA-CNPq, Fundação Universitária José Bonifácio, the Brazilian Mathematical Society and above all we had full support of the Institute of Mathematics of the Federal University of Rio de Janeiro. We wish to express our gratitude to all of them.

We are grateful to the lectures and to the authors.

M.I. Camacho*, M.J. Pacifico* and F. Takens**

* Instituto de Matemática
Universidade Federal do Rio de Janeiro
Caixa Postal 68530, CEP 21.910
Rio de Janeiro, RJ, Brazil

** Mathematatisch Instituut
Rijksuniversiteit Groningen
G.P.O. 800
Groningen - The Netherlands

J A BELOQUI

A quasi transversal Hopf bifurcation

Introduction

Bifurcation of generic one parameter families of vector fields with simple recurrence are caused by, either local phenomena such as non-hyperbolicity of singularities or periodic orbits, or by semi-local phenomena, like the non transversal intersection of the invariant manifolds of two critical elements.

The persistent ones, i.e., those that remain after small perturbations, are the saddle node and the Andronov-Hopf bifurcation [1], [3], [11], [12] and [10], in case we lose hyperbolicity. When the family bifurcates by losing transversality, we get persistence by imposing quasi-transversality, that is, non transversality of the least degenerated kind [6], [7], [8].

There are different ways of defining equivalence between families X_μ and $\tilde{X}_\mu$. In the first place, for fields we have that a topological equivalence between two vector fields X, $\tilde{X}$ on M is a homeomorphism $h: M \to M$ such that h sends orbits of X onto orbits of $\tilde{X}$, preserving time orientation. If in addition h preserves time, that is $h X_t = \tilde{X}_t h$ holds, then h is called a conjugacy. A vector field X is called structurally stable if it is equivalent to any nearby vector field.

We say that a family X_μ at μ_o is mildly equivalent to

$\tilde{X}_\mu$ at $\tilde{\mu}_o$, if there exists a reparametrization (homeomorphism) $\lambda: (U \subset \mathbb{R}^n, \mu_o) \to (U, \tilde{\mu}_o)$ and a family of homeomorphisms $h_\mu : M \to M$ so that h_μ maps orbits of X_μ onto orbits of $\tilde{X}_{\lambda(\mu)}$ for μ near μ_o. If we can choose the homeomorphisms h_μ to depend continuously on μ, then we say that X_μ is continuously (or strongly) equivalent to $\tilde{X}_\mu$ (at μ_o and $\tilde{\mu}_o$).

The conditions under which two persistent families X_μ and $\tilde{X}_\mu$ are equivalent are well known. When the bifurcation is due to the loss of hyperbolicity of a singularity, the saddle-node and the Andronov Hopf bifurcation are locally strongly equivalent to any nearby family, or strongly stable.

On the other hand, some classes of families going through quasi transversal bifurcations are not strongly or even mildly stable. But Van Strien [9] characterized the conditions under which two families X_μ and $\tilde{X}_\mu$ are either mildly or strongly semi-locally equivalent, that is, in a neighborhood of the orbit of tangency. He did this in terms of moduli of stability [10],[11],[12],[13],[14].

This concept appears naturally. For example, let f be a diffeomorphism of a surface which exhibits two hyperbolic fixed points p_1, p_2, such that $W^u(p_1)$ meets $W^s(p_2)$ quasi transversally, that is, with parabolic contact. Take g near f. Then it shall be topologically equivalent to f if and only if it "looks like" f and $\lambda = \dfrac{\ln \beta^s(p_1)}{\ln \beta^u(p_2)} = \dfrac{\ln \beta^s(\tilde{p}_1)}{\ln \beta^u(\tilde{p}_2)} = \tilde{\lambda}$ where $\beta^\sigma(p_i)$ $(\beta^\sigma(\tilde{p}_i))$ is the eigenvalue associated to $W^\sigma(p_i)$ $(W^\sigma(\tilde{p}_i))$ $(\sigma = s,u;\ i=1,2)$ [9], [5]. Thus, besides f and g having the same shape, there exists a differentiable real invariant which is preserved under topological equivalence. In this context we call λ a modulus for equivalence.

Here we are interested in a class of 2-parameter families X_μ of vector fields with simple recurrences in M^3, which is persistent. It appears in a natural way: X_μ bifurcates by simultaneously losing hyperbolicity and transversality.

We will give necessary and sufficient conditions for the existence of a weak equivalence between members of this class. The notion of modulus of stability is essential for our characterization. We shall also exhibit some moduli of stability for strong equivalence.

More precisely,

Let $\mathfrak{X}_1(M^3)$ be the space of C^∞ vector fields endowed with the C^∞ Whitney topology and let $\mathfrak{X}_2(M)$ be the space of the mappings $X: I \times I \to \mathfrak{X}_1(M)$ with the usual C^∞ topology $(I = [-1,1])$. Suppose $X(\mu)$, is a 2-parameter family of vector fields such that X_μ at $\mu = (0,0)$, or shortly X_o is a field which exhibits:

a) a saddle type periodic orbit $\sigma_1 = \sigma_1(0)$ of period 1 hyperbolic, with associated eigenvalues $0 < |\beta_1| < 1 < \beta_2$, C^2 linearizable;

b) a quasi-hyperbolic singularity $p=p(0)$ of saddle type such that $\dim W^u(p(0)) = 1$, associated to the eigenvalue $\alpha_3 > 0$ and $\dim W^s(p(0)) = 2$, associated to eigenvalues $\alpha_1 \pm i\alpha_2$ $(\alpha_1 = 0)$. Consider $X/W^s(p(0))$ as a "vague attractor";

c) a unique orbit γ in $W^u(p(0)) \cap W^s(\sigma_1(0))$ that is, a quasi-transversal intersection.

Call $E \subset \mathfrak{X}_1(M^3)$ the set of fields like X_o and E' the set of families as the one described above.

Let us now give the semilocal versions of the notions of equivalence.

A semilocal equivalence between X and $\tilde{X}$ and $\tilde{X} \in E$ shall be an equivalence defined from a neighborhood of $\bar{\gamma}$ onto a neighborhood of $\bar{\bar{\gamma}}$.

We can now state the following

<u>Theorem A</u>. Let X, $\tilde{X}$ be two vector fields in E, $\varepsilon - C^r$ near.

Then they are semilocally topologically equivalent iff $\beta_2 = \tilde{\beta}_2$.

In this case we say that X has modulus of stability one, and that β_2 is a modulus for X. In some sense, we are parametrizing the equivalence classes in a neighborhood of X by the parameter β_2.

The next theorem is concerned with families of vector fields. Before stating it, we give a rough description of the vector fields that belong to the families with which we shall deal for heuristic purposes.

Due to the hyperbolicity of $\sigma_1(0)$ and non singularity of $dX(p(0))$, both σ_1 and p persist for nearby vector fields. We denote then by $\sigma_1(\mu_1,\mu_2)$ and $p(\mu_1,\mu_2)$ respectively.

When $\mu_1 > 0$ there shall appear new closed orbits $\sigma_2(\mu_1,\mu_2)$, unique for each value of (μ_1,μ_2), $\mu_1 > 0$, whose unstable manifold $W^u(\sigma_2)$ meets $W^s(\sigma_1)$ for some values of μ_2. These orbits σ_2 lie very near p, according to Hopf Theorem.

We prove that the intersection of those manifolds contains alternatively 0, 1 or 2 orbits. In case there is only 1 orbit, it shall be a quadratic tangency between $W^u(\sigma_2)$ and $W^s(\sigma_1)$ and this gives rise to a modulus of stability, as stated before.

For $\mu_1 < 0$ we also get tangencies when $W^u(p)$ meets $W^s(\sigma_1)$ again arising a modulus [14]. Then, every such family exhibits curves of vector fields with one modulus of stability each, and we have

<u>Theorem B</u>. Let $X_\mu \in E'$ be a generic 2-parameter family of vector fields. Then

a) the modulus of stability for mild topological equivalence is one, namely β_2,

b) the modulus of stability for strong topological equivalence is infinite, moreover, it is modelled over a space of germs of functions.

For the sake of completeness, let us now consider the "dual"

Hopf bifurcation, that is, a family X_μ where

a) p is a hyperbolic repellor for $\mu_1 < 0$

b) p is a "vague" repellor for $\mu_1 = 0$

c) p is a hyperbolic saddle point for $\mu_1 > 0$

then a hyperbolic periodic orbit σ_2 appears for $\mu_1 > 0$, and a curve of modulus one vector fields, namely $\{\mu_2 = 0, \mu_1 > 0\}$.

In this case we have, similarly:

<u>Theorem A'</u>. If X_μ and $\tilde{X}_\mu$ are C^r near enough, then:

a) $\tilde{X}_o$ is topologically equivalent to X_o

b) the families X_μ and $\tilde{X}_\mu$ are mildly equivalent

c) the modulus of stability for strong topological equivalence is infinite, and modelled over a space of germs of functions.

This paper is organized as follows:

Section 1 deals with the central bifurcation and there we prove the necessity of the modulus for those vector fields.

In Section 2 we give the bifurcation diagram and begin the proof of Theorem B. Section 3 completes the proof of Theorems A and B, by proving the sufficiency of modulus for the central bifurcation and for mild equivalence.

We are thankful to P.C. Carrião, M.J. Pacifico and J. Palis for stimulating conversations.

§1. <u>The vector fields in</u> E.

In this section we study the central bifurcation $X_o = X$, not considering it as a member of a family of vector fields.

By Takens [15] pp.144-145 we can take coordinates in $\mathbb{R}^3$ such that X is locally in standard form:

$$X = X_1(x_1,x_2)\frac{\partial}{\partial x_1} + X_2(x_1,x_2)\frac{\partial}{\partial x_2} + A(x_1,x_2)x_3\frac{\partial}{\partial x_3}$$

where

1) all eigenvalues of $(\partial X_i/\partial x_j)$ in $x_1 = x_2 = 0$ have real part zero.

2) $A(0,0) > 0$.

As X has only one hyperbolic eigenvalue, the coordinates can be taken C^r, for r arbitrarily high. In this way we get a splitting of our vector field, in a neighborhood L of the singularity.

Recalling a result of Takens [16] on normal forms we take the following as the expression of $X/W^s_\mu(0)$

$$\dot{\rho} = b\rho^3 + o(\rho^5), \quad b < 0 \quad \text{(vague attractor)}$$
$$\dot{\theta} = \alpha_2 + o(\rho^2).$$

Let us search a parametric expression for the orbits of this vector field.

<u>Proposition 1</u>. The orbits $(\rho(t),\theta(t))$ of X verify:

$$|\rho^2(t)(1-2bt)-1| \leq O(\sqrt{t})$$
$$|\theta(t)-\theta_0-\alpha_2 t| = O(\ln(t)).$$

<u>Proof</u>: It is clear that $\rho(t) \underset{t\to+\infty}{\to} 0$, because $b < 0$ and we can take $\rho(0)$ do as to get $b + \frac{o(\rho^5)}{\rho^3} < 0$. We suppose that this is possible with $\rho(0) = 1$, without loss of generality.

Then $\frac{\rho}{\rho^3} = b + o(\rho^2)$ and

$$\frac{1}{2} - \frac{1}{2\rho^2(t)} = bt + \int_0^t o(\rho^2(s))ds \qquad (1.1)$$

or $[1-2bt-2\int_0^t o(\rho^2(s))ds]^{-1} = \rho^2(t)$.

Clearly, $\rho^2(t) = o(t^{-1})$: in case $\int_0^t |o(\rho^2(1))|ds < \infty$ this is direct and if $\int_0^t |o(\rho^2(s))|ds \to +\infty$, an application of L'Hôpital rule to $\dfrac{\int_0^t |o(\rho^2(s))|ds}{t}$ proves this fact.

We claim that $\epsilon(\rho(t)) = 2\rho^2(t)\int_0^t o(\rho^2(s))ds = o(t^{-1/2})$.

Indeed, if $\int_0^t |o(\rho^2(s))|ds < \infty$, this follows from the fact that $\rho^2(t) = o(t^{-1})$. On the other hand, if $\int_0^t |o(\rho^2(s))|ds \to +\infty$, we change variables so that

$$ds = \frac{ds}{d\rho}\, d\rho,$$

and $\int_0^t |o(\rho^2(s))|ds = \int_1^{\rho(t)} |o(\rho^2)| \frac{1}{\dot\rho}\, d\rho = -\int_{\rho(t)}^1 \frac{|o(\rho 2)|}{\rho^3}\frac{1}{b+o(\rho^2)}\, d\rho =$

$= o(\ell n\ \rho(t))$.

Returning to the expression of ϵ,

$$\epsilon(\rho(t)) \le 2\rho^2(t)o(\ell n(\rho(t))) \le o(t^{-1/2})$$

because $\rho(t) \xrightarrow[t\to\infty]{} 0$, and $x\ \ell n\ x \xrightarrow[x\to 0]{} 0$.

So, replacing in (1.1),

$$\rho^2(t)(1-2bt)-1 = 2\rho^2(t)\int_0^t o(\rho^2(s))ds = \epsilon(\rho(t)) = O(\sqrt{t})$$

as we wished.

In the same way we prove the result for $\theta(t)$. □

<u>Notation</u>: With the definitions of Proposition 1, we take $\rho(t) = \sqrt{\dfrac{1+\epsilon}{1-2bt}}$ as the expression of the radial component of an orbit in $W^s(p(0))$.

<u>Remark</u>: The convergence is uniform, that means, it is independent of the spiral.

We shall now define some objects for each vector field, that will be used to prove our results. Take $\Lambda = \{(x_1,x_2,1)\} \subset L$, a section transversal to the flow at $(0,0,1)$.

Consider $\pi_1\colon L \to W^s_\mu(p)$ the trivial fibration obtained by projecting on the first coordinates. Each fiber is $\pi_1^{-1}(\mu_1,\pi_2)(x_1^o,x_2^o) = \{(x_1^o,x_2^o,x_3),\ x_3 \in \mathbb{R}\} = R$, a line. Let $F = \bigcup_{t\ge 0} X_t(R) \cap \Lambda$, this we call a linear spiral. A sector S_n of F shall be a connected component of $F-W^s(\sigma_1)$.

We reparametrize all fields near X_o, so that the periodic orbit $\sigma_1(\mu_1,\mu_2)$ has period 1. Also, we take a section Σ, transversal to all flows near X_o and near $\sigma_1(\mu_1,\mu_2)$ where the Poincaré transform is simultaneously C^2 linearized.

In $W^s(\sigma_1(\mu_1,\mu_2)) \cap \Sigma$ we fix fundamental domains $D(\mu_1,\mu_2)$ which vary continuously. We also fix an open set N, $D(\mu_1,\mu_2) \subset N \subset \Sigma$, which we cal a fundamental neighborhood. It has the following property: $X_{2,\mu}(N) \cap N = \emptyset$, where $X_{2,\mu}$ is the time two flow of X_μ, μ small.

Call $\psi\colon \Lambda \to N \subset \Sigma$ the Poincaré transform associated to the flow, $\psi = (\psi_1,\psi_2)$, that is, $\psi_2 = \pi_2(\psi)$ where $\pi_2(\mu_1,\mu_2)\colon\Sigma\to W^u(\sigma_1(\mu_1,\mu_2))$ is the trivial fibration obtained by projecting on the second linearizing coordinate, with fiber $\pi_2^{-1}(\mu_1,\mu_2)(c)$, $c \in W^u(\sigma_1)$. Identifying $W^u(\sigma_1(\mu_1,\mu_2))$ with R, we call S_n an "upper" sector if $\pi_2(\psi(S_n)) \subset \mathbb{R}^+$.

<u>Proposition 2</u>. Let (ρ_n,θ_n) denote the maximum of $\pi_2(\psi(S_n))$, in polar coordinates. Then

$$\theta_n \to \pi/2 \quad (\text{mod } 2\pi).$$

<u>Proof</u>: Let $\psi_2(x_1,x_2) = \psi_2(x_1(t),x_2(t))$ where $(x_1(t),x_2(t))$ is the parametrization of a sector.

The critical points shall be:

$$\frac{d\psi_2}{dt} = \langle \nabla\psi_2,(x_1',x_2')\rangle = \left\{\frac{(1+\epsilon)2b}{(1-2bt)^2} + \frac{\epsilon}{(1-bt)}\right\}\frac{1}{2}\sqrt{\frac{1-2bt}{1+\epsilon}}\,\langle\nabla\psi_2,(\cos,\text{sen})\rangle$$

$$+\ \theta'\langle\nabla\psi_2,(-\text{sen},\cos)\rangle\sqrt{\frac{1+\epsilon}{1-2bt}} =$$

$$= \sqrt{\frac{1+\epsilon}{1-2bt}}\left\{\left[\frac{2b}{(1-2bt)} + \frac{\epsilon'}{1+\epsilon}\right]\frac{\langle\nabla\psi_2(\cos,\text{sen})\rangle}{2} + \right.$$

$$\left. +\ [\alpha_2+O(\rho^2(t))]\ \langle\nabla\psi_2,(-\text{sen},\cos)\rangle\right\} = 0.$$

Remark that $\epsilon' \to 0$:

$$\epsilon = \rho^2(t)\int_0^t o(\rho^2(s))ds \quad \text{and}$$

$$\frac{d\epsilon}{dt} = 2\rho\dot{\rho}\int_0^t o(\rho^2(s))ds + \rho^2(t)o(\rho^2(t)).$$

As $\dot{\rho} = b\rho^3 + o(\rho^5)$, it follows

$$\frac{d\varepsilon}{dt} = 2\rho^4(b+o(\rho^2))\int_0^t o(\rho^2(s))ds + \rho^2(t)o(\rho^2(t)) \underset{t \to +\infty}{\to} 0 .$$

Returning to the expression of $\frac{d\psi_2}{dt}$, we observe that the first term tends to zero. The term given by

$$[\alpha_2+o(\rho^2(t))]\langle\nabla\psi_2,(-\text{sen},\cos)\rangle = [\alpha_2+o(\rho^2(t))]\langle(\frac{\partial\psi_2}{\partial x_1},\frac{\partial\psi_2}{\partial x_2}),(-\text{sen},\cos)\rangle .$$

As $\frac{\partial\psi_2}{\partial x_1}(0,0) = 0$, and $\frac{\partial\psi_2}{\partial x_2}(0,0) \neq 0$, the expression shall be zero if $\cos\theta \sim 0$, that is, $\theta \sim \frac{\pi}{2}$ (mod 2π). □

Our next claim has a neat geometrical sense. We prove that if some iterates $f^{m_k}(e_{n_k})$ of a sequence of maxima of sectors S_{n_k} accumulate on $z_0 \in W^u(\sigma_1)-\sigma_1$, that is, a point of the unstable manifold not in σ_1, then the maxima of $h(S_{n_k})$ shall accumulate on $h(z_0)$.

<u>Proposition 3</u>. Let (ρ_n,θ_n) be the maxima of the sectors S_n. Take $\pi_2(\psi(\rho_n,\theta_n)) = A_n\rho_n$. Suppose X is another vector field, semilocally equivalent to X by an homeomorphism h. Let us call ρ_n "the" maximum of the sector $\tilde{S}_n = h(S_n)$. If $A_{n_k}\rho_{n_k}\beta_2^{m_k} \to z_0$, then $\tilde{A}_{n_k}\rho_{n_k}\tilde{\beta}_2^{m_k} \to h(z_0)$.

<u>Proof</u>: $h(S_n)$ accumulates on an interval whose maximum is $h(z_0)$. □

Let $F = \bigcup_{t\geq 0} X_t(R) \cap \Lambda$ be parametrized by $(\rho(t),\theta(t),1)$ in polar coordinates. Consider the analogous objects for another vector field $\tilde{X}$ in E.

<u>Proposition 4</u>. Suppose the semilocal equivalence h is such that $h(\Lambda) \subset \tilde{\Lambda}$. Take $(\rho(\tilde{t}_n),\tilde{\theta}(\tilde{t}_n))$ "the" sequence of maxima of $h(F)$. Then $\lim_{n\to+\infty} \theta(t_n) = \pi/2$.

<u>Proof</u>: Near the focus, we can approximate h(F) by two linear spirals, thus getting a spiral neighborhood. Then h(F) is shrunk in an arbitrarily thin spiral neighborhood, which behaves as a thick linear spiral, and the result follows.

□

Nest, we prove that β_2 is a modulus of stability by topological equivalence, and the linearity of $h/W^u(\sigma_1(0))$ thus establishing the first part of Theorem A.

Proposition 5. Let $X,\tilde{X} \in E$ be semilocally equivalent by an homeomorphism h.

Then $\beta_2 = \tilde{\beta}_2$ and $h/W^u(\sigma_1)$ is linear.

Proof: Let F be a linear spiral for X, $c_1 \in W^u(\sigma_1) \cap \Sigma$, and $\{S_n\}_{n\geq 1}$ an ordered sequence of the "upper" sectors of F.

Define $N_m(F,\pi_2^{-1}(c_1),S_p) = \#\{S_j/S_j \cap X_{-m}(\pi_2^{-1}(c_1)) \neq \phi,\ j \geq p\}$.

Calling $e_m = \pi_2(\psi(\rho_m,\theta_m))$, by definition,

$$e_{1+N_m} \leq c_1\beta_2^{-m} \leq e_{N_m} .$$

Analogously $\tilde{e}_{1+N_m} \leq \tilde{c}_1\beta_2^{-m} \leq \tilde{e}_{N_m}$ where $\tilde{c}_1 = h(c_1)$. Then

$$\frac{e_{1+N_m}}{\tilde{e}_{N_m}} \leq \frac{c_1}{\tilde{c}_1}\left(\frac{\beta_2}{\tilde{\beta}_2}\right)^{-m} \leq \frac{e_{N_m}}{\tilde{e}_{1+N_m}} .$$

By Proposition 4, the angles are approximately $\frac{\pi}{2}$, for m big enough, implying $e_{N_m} \sim \rho_{N_m}A_{N_m}$ and $\tilde{e}_{N_m} \sim \tilde{\rho}_{N_m}\tilde{A}_{N_m}$, where A and $\tilde{A}$ stand for the normal derivatives $\frac{\partial\psi_2}{\partial x_2}$ and $\frac{\partial\tilde{\psi}_2}{\partial\tilde{x}_2}$.

So the last inequality becomes

$$\frac{A_{1+N_m}}{\tilde{A}_{N_m}}\sqrt{\frac{1+\epsilon_{1+N_m}}{1+\tilde{\epsilon}_{N_m}}\left[\frac{1-2\tilde{b}\tilde{t}_{N_m}}{1-2bt_{1+N_m}}\right]} \leq \frac{c_1}{\tilde{c}_1}\left[\frac{\beta_2}{\tilde{\beta}_2}\right]^{-m} \leq$$

$$\leq \frac{A_{N_m}}{\tilde{A}_{1+N_m}}\sqrt{\frac{1+\epsilon_{N_m}}{1+\tilde{\epsilon}_{1+N_m}}\,\frac{1-2\tilde{b}\tilde{t}_{1+N_m}}{1-2bt_{N_m}}} .$$

The expression under the radical tends to $\frac{\tilde{b}\alpha_2}{b\tilde{\alpha}_2}$ when m tends to infinity.

So there exists $\delta_1,\delta_2 > 0$ such that

$$0 < \delta_1 < \left[\frac{\beta_2}{\tilde{\beta}_2}\right]^{-m} < \delta_2 \quad \text{for all } m \in \mathbb{N}.$$

Hence $\beta_2 = \tilde{\beta}_2$ establishing the first statement.

In the limit, the inequality becomes

$$\frac{A}{\tilde{A}}\sqrt{\frac{\alpha_2}{\tilde{\alpha}_2}\,\frac{\tilde{b}}{b}} = \frac{c_1}{\tilde{c}_1} = \frac{A}{\tilde{A}}\sqrt{\frac{a_2\tilde{b}}{\tilde{a}_2 b}}$$

thus proving the linearity of $h/W^u(\sigma_1)$. □

Remarks: 1) Comparing with [13], [2] we see that we did not take into account any consideration about the irrationality of the invariant.

2) The same method of demonstration applied to the hyperbolic case does not lead to the same conclusion.

3) $h/W^u(\sigma_1)$ is extremely rigid, i.e., it is unique, given the normal forms of X and $\tilde{X}$.

Call now $D \subset \mathfrak{X}_1(M)$ the set of vector fiels analogous to those of E, except that the singularity p is hyperbolic.

Proposition 6. Take $X \in E$, $\tilde{X} \in D$. Then X is not semilocally equivalent to $\tilde{X}$.

Proof: With the same notation of the last proposition, we will see that $\frac{N_m}{\beta_2^{2m}} \to L \neq 0$ and $\frac{N_m}{m} \to \frac{2\pi\tilde{\alpha}_1}{\tilde{\alpha}_2 \ln \tilde{\beta}_2}$ and clearly both statements cannot be true at the same time.

1) By the definition of N_m, it follows

$$A_{1+N_m}\sqrt{\frac{1+\epsilon_{1+N_m}}{1-2bt_{1+N_m}}} \le C_1\,\beta_2^{-m} \le A_{N_m}\sqrt{\frac{1+\epsilon_{N_m}}{1-2bt_{N_m}}}\,.$$

Quadrating the inequality and multiplying by N_m, we get

$$A^2_{1+N_m}\,\frac{1+\epsilon_{1+N_m}}{\frac{1}{N_m}-2b\frac{t_{1+N_m}}{N_m}} \le C_1^2\,\frac{N_m}{\beta_2^{2m}} \le \frac{1+\epsilon_{N_m}}{\frac{1}{N_m}-2b\frac{t_{N_m}}{N_m}}\,A^2_{N_m}\,.$$

As $\frac{N_m}{t_{N_m}} \to \frac{\alpha_2}{2\pi}$, in the limit the inequality becomes

$$-\frac{\alpha_2}{2b2\pi} \le \frac{c_1^2}{A^2} \lim_{m\to+\infty} \frac{N_m}{\beta_2^{2m}} \le \frac{-\alpha_2}{4\pi b}$$

or $\frac{N_m}{\beta_2^{2m}} \to \frac{-A^2\alpha_2}{4\pi c_1^2 b} = L.$

2) A calculation of the same type, applied to $\tilde{X}$, leads to $\frac{N_m}{m} \to \frac{2\pi\tilde{\alpha}_1}{\tilde{\alpha}_2 \ln \tilde{\beta}_2}$.

Dividing both expressions, it follows that

$$\frac{m}{\beta_2^{2m}} \to \frac{L\tilde{\alpha}_2 \ln \tilde{\beta}_2}{2\pi\,\tilde{\alpha}_1} \neq 0 \quad \text{(absurd)}. \qquad \square$$

§2. The families in E′.

In order to give our bifurcation diagram, let us state Hopf Theorem, following [4].

Hopf Theorem in $\mathbb{R}^2$. Let X_λ be a C^k $(k\ge4)$ vector field on $\mathbb{R}^2$ such that $X_\lambda(0) = 0$ for all λ and $X = (X_\lambda, 0)$ is also C^k. Let $dX_\lambda(0,0)$ have two distinct, complex conjugate eigenvalues $\alpha_1(\lambda) \pm i\,\alpha_2(\lambda)$, $\alpha_1 > 0$ for $\lambda > 0$. Also let $\frac{d}{d\lambda}\alpha_1(\lambda)\Big|_{\lambda=0} > 0$. Then

A: there is a C^{k-2} function $\lambda: (-\varepsilon,\varepsilon) \to \mathbb{R}$ such that $(x_1, 0, \lambda(x_1))$ is on a closed orbit of period $\approx 2\pi/|\alpha_2(0)|$ and radius growing like $\sqrt{\alpha_1}$, of X for $x_1 \neq 0$ and such that $\lambda(0) = 0$.

B: there is a neighborhood U of $(0,0,0)$ in $\mathbb{R}^3$ such that any closed orbit in U is one of those above. Furthermore, if 0 is a "vague attractor" for X_0, then

C: $\lambda(x_1) > 0$ for all $x_1 \neq 0$ and the orbits are attracting.

In our case $X_\mu \in \mathfrak{X}_1(M^3)$, and it is hyperbolic (expansive) along the x_3-direction.

Corollary. For small (positive) values of α_1 the unique periodic orbit σ_2 which appears near p, is hyperbolic of saddle type.

We want to distinguish a subset of E'. Namely, those families that meet E only at one point, that is, only one member of the family is simultaneously non hyperbolic and non transversal.

To do so, consider a family $X_\mu \in E'$ and call $v(\mu_1,\mu_2) = \pi_2(W^{uu}(p(\mu_1,\mu_2)) \cap N$ where $W^{uu} = W^u$ when p is a saddle type singularity.

Identify $W^u(\sigma_1) \cap \Sigma$ with a neighborhood of $o \in R$, so that $v: U \to \mathbb{R}$, $v \in C^1$.

As before, $\alpha_1(\mu_1,\mu_2)$ is the real part of the complex eigenvalue of $dX(\mu_1,\mu_2)$ at $p(\mu_1,\mu_2)$.

Definition of E''. Let $X_\mu \in E'$. We say that $X_\mu \in E''$ if $J(\alpha_1(\mu_1,\mu_2),v(\mu_1,\mu_2))(0,0)$ is non singular, where J is the Jacobian matrix of (α_1,v). For families in E'' we shall consider, from now on, $\mu_1 = \alpha_1$ and $\mu_2 = v$, on account of the Inverse Function Theorem.

E'' is open and dense in E', so we shall call $X_\mu \in E''$ a generic family.

We want to establish the necessity of some moduli of stability for equivalence between members of E''.

In the first place, let us mention which kind of vector fields shall appear in a generic family (see Figure 1).

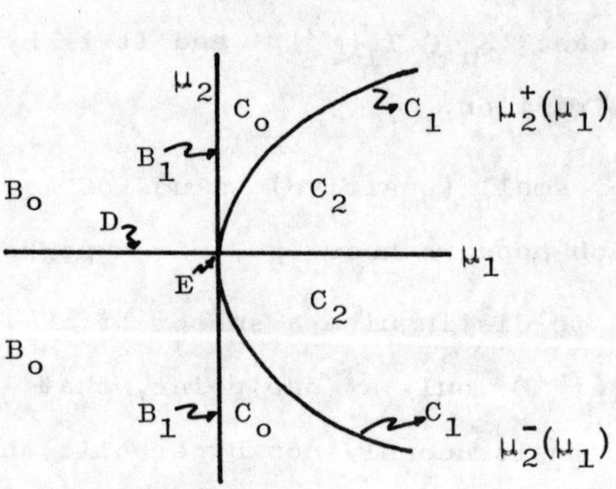

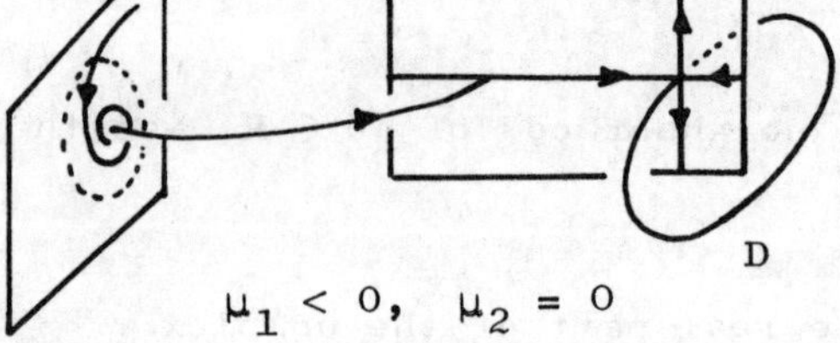

$\mu_1 < 0, \ \mu_2 = 0$

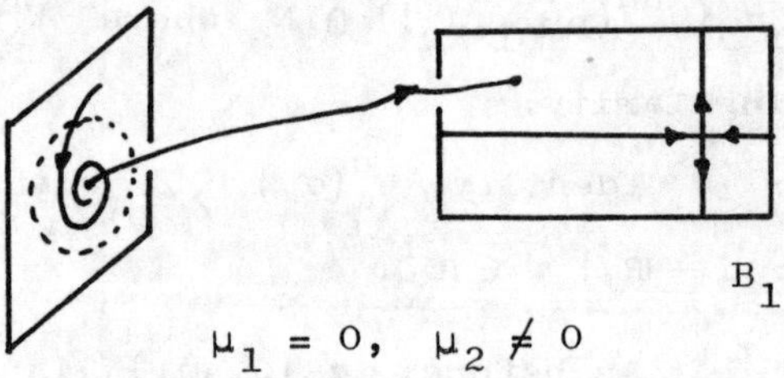

$\mu_1 = 0, \ \mu_2 \neq 0$

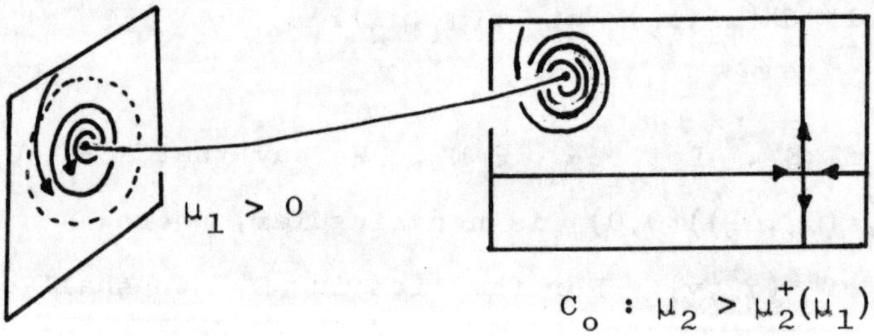

$C_o : \mu_2 > \mu_2^+(\mu_1)$

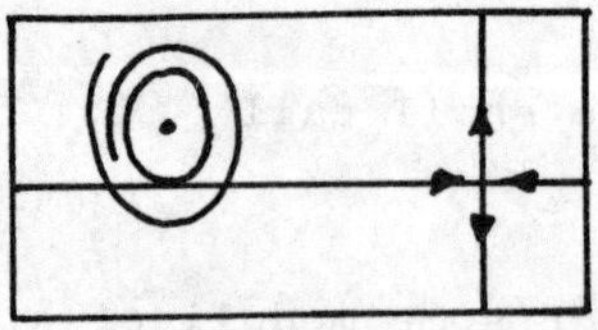

$C_1 : \mu_2 = \mu_2^+(\mu_1)$

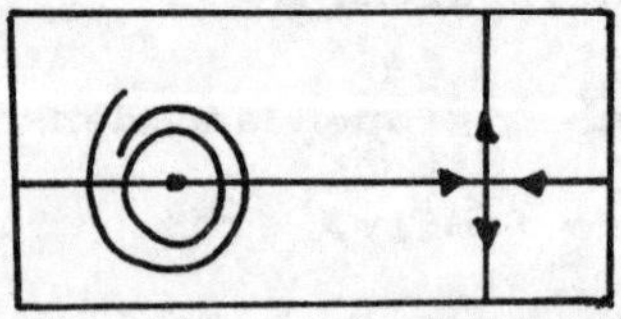

$C_2 : \mu_2^-(\mu_1) < \mu_2 < \mu_2^+(\mu_1)$

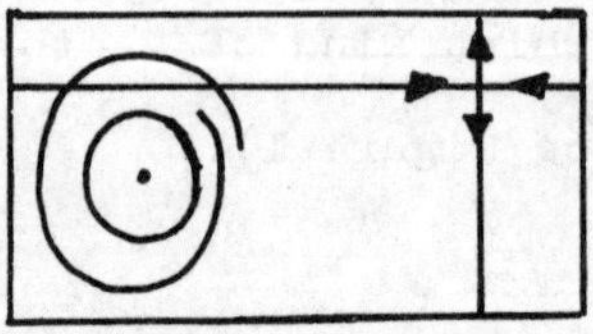

$C_o : \mu_2 < \mu_2^-(\mu_1)$

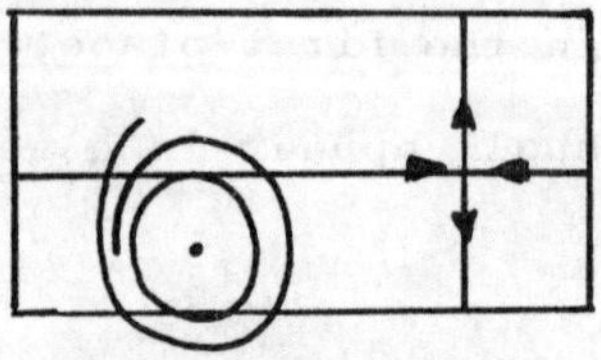

$\mu_2 = \mu_2^-(\mu_1) \subset C_1$

1) a unique vector field in E, the central bifurcation,

2) vector fields in D, which look like those in E, but are hyperbolic,

3) vector fields in C, which exhibit a singularity of the source type, hyperbolic and a periodic orbit σ_2 very near this singularity.

This set C we subdivide into:

a) C_1, the set of fields such that $W^u(\sigma_2) \cap W^s(\sigma_1)$ consists only of one orbit, giving rise to a quasi transversal intersection.

b) C_2, the set of vector fields such that $W^u(\sigma_2) \pitchfork W^s(\sigma_1)$ along 2 orbits.

c) C_o, the set of vector fields such that $W^u(\sigma_2) \cap W^s(\sigma_1) = \phi$.

4) Vector fields in B, with a singularity p of saddle type, such that $W^u(p) \cap W^s(\sigma_1) = \phi$, which we subdivide into: a) B_o the set of fields where p is hyperbolic; b) B_1 the set of fields where p is quasi-hyperbolic.

Observe that $W^u(\sigma_2) \cap W^s(\sigma_1)$ consists alternatively of none, one or two orbits, and that there might be tangencies only for those fields of C_1.

This description follows from the next

<u>Lemma 7</u> (unicity of the curve of tangencies). Let N be the fundamental neighborhood which we chose before, and consider it fibered by π_1 and π_2.

Then these fibers intersect each other along an unique orbit of tangencies $\Gamma(\mu_1,\mu_2)$ which is differentiable, and depends continuously on the parameters.

<u>Proof</u>: We are needing an expression of the vector field X_μ in a neighborhood of $(0,0,0)$.

Due to the hyperbolicity of $(0,0,0)$ along the x_3 direction (uniform in μ), the λ-Lemma, and Strong Stable Foliation we get a product structure for $X_\mu = X_{1,\mu}(x_1,x_2)\frac{\partial}{\partial x_1} + X_{2,\mu}(x_1,x_2)\frac{\partial}{\partial x_2} + A_\mu(x_1,x_2)x_3\frac{\partial}{\partial x_3}$ where $A(0,0) > 0$ and the eigenvalues of $(\partial X_{i,\mu}/\partial x_j)$ at $(0,0)$ are complex conjugate, whose real part is α_1. The differentiability of the manifold $\{x_3=0\}$ depends on x_1/α_3). If $\alpha_1 > 0$ is small enough, we get a high class of differentiability. For $\alpha_1 = 0$, we can take a C^∞ central manifold because p_o is a "vague attractor" in this direction.

Now, we change coordinates x_1, x_2 to polar ones. By [16] Theorem 2.1 applied to the angular part of X_μ and Proposition 2.3 of the same paper, we have $X_\mu = (\alpha_1\rho+b\rho^3)\frac{\partial}{\partial\rho} + (\alpha_2+o(\rho^2))\frac{\partial}{\partial\theta} + A(\rho,\theta)x_3\frac{\partial}{\partial x_3}$, where the coefficients and coordinates depend on (μ_1,μ_2).

Consider $\psi(\mu_1,\mu_2)\colon \Lambda \to \Sigma$ the Poincaré transform.

Take an arc $(x_1(t),x_2(t))$ corresponding to a sector, and $a(t) = \phi_2(\rho(t)\cos\theta(t),\rho(t)\operatorname{sen}\theta(t))$.

Examine the critical points:

$$\begin{aligned}\frac{d}{dt}a(t) &= \langle\nabla\psi_2(x_1(t),x_2(t)),(\dot\rho\cos\theta-\dot\theta\rho\operatorname{sen}\theta,\dot\rho\operatorname{sen}\theta+\dot\theta\rho\cos\theta)\rangle = \\ &= \langle\nabla\psi_2(x_1(t),y_1(t)),(\cos\theta,\operatorname{sen}\theta)\rangle\dot\rho + \langle\nabla\psi_2,\dot\theta(-\operatorname{sen}\theta,\cos\theta)\rangle\rho = \\ &= \{\langle\nabla\psi_2(\cos\theta,\operatorname{sen}\theta)\rangle[\mu_1+b\rho^2+O(\rho^4)]+\dot\theta\langle\nabla\psi_2(-\operatorname{sen}\theta,\cos\theta)\rangle\rho = 0.\end{aligned}$$

Recall that $\frac{\partial\psi_2}{\partial x_1} = 0$ when $x_1 = x_2 = 0$ and $\mu_2 = 0$, and $\frac{\partial\psi_2}{\partial x_1}(0,0)$ is bounded away from zero. So

$$\begin{aligned}&[(\mu_1+b\rho^2+o(\rho^3))\cos\theta-(\alpha_2+o(\rho^2))\operatorname{sen}\theta]\frac{\partial\psi_2}{\partial x_1} = \\ &= -[\alpha_2\cos\theta+(\mu_1+b\rho^2+o(\rho^3))\operatorname{sen}\theta]\frac{\partial\psi_2}{\partial x_2}\end{aligned}$$

for μ_1 and ρ small enough, we see that $\theta \sim \pi/2$, in order to satisfy the equation.

Let us examine the type of these critical points. Call H = Hessian ψ_2, and Φ the expression in brackets { }.

$$\frac{d^2}{dt^2} a(t) = \dot{\rho}\Phi + \rho \frac{d}{dt}\Phi .$$

As $\Phi = 0$ on critical points, we develop

$$\frac{d}{dt}\Phi = \dot{\theta}\langle\nabla\psi_2,(-\mathrm{sen}\theta,\cos\theta)\rangle(\mu_1+b\rho^2+o(\rho^3))+\dot{\theta}^2\langle\nabla\psi_2,(-\cos\theta,-\mathrm{sen}\theta)\rangle +$$
$$+ \langle\nabla\psi_2,(\cos\theta,\mathrm{sen}\theta)\rangle(2b\rho\dot{\rho}+3\rho\dot{O}(\rho^2)+\dot{\theta}o(\rho^3))+ \rho\{[(\mu_1+b\rho^2+o(\rho^3)]^2 \cdot$$
$$\cdot (\cos\theta,\mathrm{sen}\theta)H(\cos\theta,\mathrm{sen}\theta)^T + \dot{\theta}(\mu_1+b\rho^2+o(\rho^3))(-\mathrm{sen}\theta,\cos\theta)H(\cos\theta,\mathrm{sen}\theta)^T$$
$$+ (\mu_1+b\rho^2+o(\rho^3))(\cos\theta,\mathrm{sen}\theta)H(-\mathrm{sen}\theta,\cos\theta)^T+\dot{\theta}(-\mathrm{sen}\theta,\cos\theta)H(-\mathrm{sen}\theta,\cos\theta)^T\}$$
$$+ \ddot{\theta}\langle\nabla\psi_2,(-\mathrm{sen}\theta,\cos\theta)\rangle .$$

As $\ddot{\theta} = O(\dot{\rho}\rho)$, it follows that

$$sg\frac{d}{dt}\Phi = sg[\dot{\theta}^2\langle\nabla\psi_2,(-\cos\theta,-\mathrm{sen}\theta)\rangle+\rho\dot{\theta}(-\mathrm{sen}\theta,\cos\theta)H(-\mathrm{sen}\theta,\cos\theta)^T +$$
$$+ \dot{\theta}\langle\nabla\psi_2,(-\mathrm{sen}\theta,\cos\theta)\rangle(\mu_1+b\rho^2+o(\rho^3))], \quad \text{for } \rho \text{ small.}$$

We are only interested in the points for which $\frac{d}{dt}(t) = 0$, that is, $\dot{\theta}\langle\nabla\psi_2,(-\mathrm{sen}\theta,\cos\theta)\rangle = -\langle\nabla\psi_2,(\cos\theta,\mathrm{sen}\theta)\rangle[\mu_1+b\rho^2+o(\rho^3)]$. Replacing in the expression of $sg\frac{d}{dt}\Phi$, we obtain

$$sg\frac{d}{dt}\Phi = sg\{-[\dot{\theta}^2+(\mu_1+b\rho^2+o(\rho^3))^2]\langle\nabla\psi_2,(\cos\theta,\mathrm{sen}\theta)\rangle +$$
$$+ \rho\dot{\theta}(-\mathrm{sen}\theta,\cos\theta)H(-\mathrm{sen}\theta,\cos\theta)^T\}$$

and if ρ is smaller,

$$sg\frac{d}{dt}\Phi = -sg\langle\nabla\psi_2,(\cos\theta,\mathrm{sen}\theta)\rangle = -sg\frac{\partial\psi_2}{\partial x_2}\mathrm{sen}\theta, \quad \text{because } \cos\theta \sim 0.$$

So all the critical points for the "upper" sectors are extrema of the same type, for ρ and μ_1 small enough. In particular, if the periodic orbit $\sigma_2(\mu)$ is such that $\sigma_2(\mu) < \rho_0$, the tangency between $W^u(\sigma_2)$ and $W^s(\sigma_1)$ shall be unique or else will not exist.

From the last expression we get the continuity of $\Gamma(\mu_1,\mu_2)$.

□

We are now ready to prove the necessity of modulus for weak equivalence.

<u>Proposition 8</u>. Let $X_\mu, \tilde{X}_\mu \in E''$ be weakly equivalent. Then $\beta_2(0) = \tilde{\beta}_2(0)$.

<u>Proof</u>: For parameters in a small neighborhood of zero, each family exhibits unique vector fields X_o and $\tilde{X}_o$ that belong to E. They should be equivalent on account of the hypothesis and Proposition 6. In this case, we use Proposition 5 to prove our claim. □

Let us prove the necessity of infinite moduli for strong equivalence.

We recall that in [2] it is proven that the fields in D have modulus of stability 1, given by $\lambda_1 = \frac{\alpha_1}{\alpha_2} \ln \beta_2$. De Melo [5] and Palis [9] proved that a vector field exhibiting a quasi transversal connection between the invariant manifolds of periodic orbits has modulus 1, namely $\lambda_2 = \ln \beta_3 / \ln \beta_2$. Van Strien proves in [13] that a one parameter family in D′ has 2 moduli of stability for strong equivalence: $\lambda_{1,1} = \alpha_1/\alpha_2$: $\lambda_{1,2} = \ln \beta_2$, that is, the modulus splits. For a family in C_1 the 2 moduli are $\lambda_{1,2} = \ln \beta_2$, $\lambda_{1,3} = \ln \beta_3$. Obviously for weak equivalence we have modulus 1.

<u>Proposition 9</u>. Let $X_\mu \in E'$ be a generic family. Then there are infinitely many moduli of stability for continuous equivalence. Moreover, the moduli are modelled over a space of germs of functions.

<u>Proof</u>: Take X_μ as in the hypothesis. From the genericity of this family, $J(\alpha_1(\mu), v(\mu))(0,0)$ is non singular. So applying the Inverse Function Theorem, we can consider $\mu_1 = \alpha_1$ and $v(\mu) = \mu_2$.

Suppose then that $\tilde{X}_\mu$ is another 2-parameter family equivalent to X_μ by an equivalence $H(\mu)$, with reparametrization ρ.

According to [13] a 1-parameter family $X(\mu_1,0)$ has 2 moduli for strong equivalence:

$$\lambda_{1,1} = \mu_1/\alpha_2(\mu_1) \quad \text{and} \quad \lambda_{1,2} = \ell n\ \beta_2(\mu_1).$$

For $\mu_2 = 0$, μ_1^o small enough we can consider that $\lambda_{1,2}$ is a function of $\lambda_{1,1}$, that is, $\lambda_{1,2} = \lambda_{1,2}(\lambda_{1,1})$.

Given $\lambda_{1,1}$ there is only one value $\rho(\mu_1^o,0)$ such that

$$\tilde{X}(\rho(\mu_1^o,0)) \in D \quad \text{and} \quad \tilde{\lambda}_{1,1} = \lambda_{1,1}.$$

Then, as the families $X(\mu_1^o,\mu_2)$ and $\tilde{X}(\rho(\mu_1^o,\mu_2))$ are equivalent, we must have $\lambda_{1,2} = \tilde{\lambda}_{1,2}$. This means that if $\lambda_{1,1} = \tilde{\lambda}_{1,1}$ consequently will follow $\lambda_{1,2} = \tilde{\lambda}_{1,2}$. Then the functions $\lambda_{1,2}$, $\tilde{\lambda}_{1,2}$ must be identical on a left neighborhood of zero.

As it was demonstrated in the last Proposition, for $c > 0$, the one parameter family $X(c,\mu_2)$ must have two vector fields $X(c,\lambda_2^+(c))$ and $X(c,\lambda_2^-(c))$ which belong to C_1. Again, for a strong equivalence there shall be 2 moduli: $\ell n\ \beta_2$, $\ell n\ \beta_3$, where β_3 is the contracting eigenvalue associated to $\sigma_2(\mu)$. The same reasoning leads to an analogous conclusion: a) the functions $\beta_2^+(\ell n\ \beta_3)$ and $\tilde{\beta}_2^+(\ell n\ \tilde{\beta}_3)$ must be identical in a right neighborhood of zero and b) $\beta_2^-(\ell n\ \beta_3)$ and $\tilde{\beta}_2^-(\ell n\ \tilde{\beta}_3)$ must coincide in a right neighborhood of zero. □

So we have proven Theorem B, part b).

§3. Sufficiency of moduli for weak equivalence.

Now we have to prove that β_2 is the only modulus needed to construct an equivalence between two vector fields in E.

Proposition 10. Let $X,\tilde{X} \in E$ and $\beta_2 = \tilde{\beta}_2$. Then they are semi-locally equivalent.

Proof: Use Lemma 7 to apply essentially the same technique as in [2] which we outline.

Take a fundamental domain $D^s(p_1)$ and the fibrations π_1 and π_2. There is a unique curve of tangencies $C \subset N$. Consider the same objects for $\tilde{X}$.

Define h (the topological equivalence) on D^s, $h/D^s : D^s \to \tilde{D}^s$ by a rotation of small angle which induces a correspondence between the fibers of π_1 and $\tilde{\pi}_1$. Consider $h(C) \subset \tilde{C}$, and preserve $\tilde{\pi}_2$ in a neighborhood of $\tilde{C}$. Then the correspondence between π_1 and $\tilde{\pi}_1$ induces one between π_2 and $\tilde{\pi}_2$, through the definition of h on C.

By conjugation on Σ, we define $h(X_n(C)) = \tilde{X}_n(h(C))$, where X_n is time n flow, and extend it to $W^u(\sigma_1)$ as in Proposition

Now complete $\tilde{\pi}_2$ by linear segments "in the middle" of C and $\tilde{X}_1(\tilde{C})$, in order to make it compatible with h. □

This proposition and Proposition 8 prove Theorem A.

Let us end up the proof of Theorem B.

Proposition 11. Let $X_\mu, \tilde{X}_\mu \in E'$, generic, $\beta_2(0) = \tilde{\beta}_2(0)$. Then they are weakly equivalent.

Proof: On account of the genericity of X and $\tilde{X}$ we can suppose $\mu_1 = \alpha_1$, $\mu_2 = v$.

The axis $\mu_1 = 0$ corresponds to fields in B_1, if $\mu_2 \neq 0$.

The axis $\mu_2 = 0$ corresponds to fields in E (for $\mu_1 = 0$), fields in D (for $\mu_1 < 0$) and to fields in C_2 ($\mu_1 > 0$).

If $\mu_1 < 0$, $\mu_2 \neq 0$ we have fields of B_0, that is Morse-Smale ones. For $\mu_1 > 0$, there are two curves $\mu_1^+(\mu_2)$, $\mu_1^-(\mu_2)$ of vector fields such that $W^u(\sigma_2(\mu_1,\mu_2)) \cap W^s(\sigma_1(\mu_1,\mu_2))$ has only one orbit, and these manifolds meet quasi transversally along this orbit.

As $\tilde{X}_\mu$ is also generic, we get the same portrait. In order to prove our statement we shall define a reparametrization ρ.

For $\mu_1 < 0$, $\mu_2 = 0$, observe that there are fields in D or else Morse-Smale. In case $X(\mu_1, 0) \in D$, its modulus is

$$\frac{\mu_1}{\alpha_2(\mu_1,0)\ln \beta_2(\mu_1,0)},$$ which is a decreasing function of μ_1, for μ_1 small enough, due to the genericity of X_μ. So, for $(\mu_1, 0)$, define $\rho(\mu_1, 0)$ as the only value of $\tilde{\mu}_1$ such that

$$\frac{\mu_1}{\alpha_2(\mu_1,0)\ln \beta_2(\mu_1,0)} = \frac{\tilde{\mu}_1}{\tilde{\alpha}_2(\tilde{\mu}_1,0)\ln \tilde{\beta}_2(\tilde{\mu}_1,0)} .$$

For $\mu_1 > 0$, we have to take $\mu_1^+(\mu_2)$ onto $\tilde{\mu}_1^+(\tilde{\mu}_2)$ and $\mu_1^-(\mu_2)$ onto $\tilde{\mu}_1^-(\tilde{\mu}_2)$, respecting the modulus $\ln \beta_3/\ln \beta_2$, as proven in [10,11]. By a reparametrization ρ on $\tilde{\mu}_1 > 0$, in a way that

$$(\ln \beta_3/\ln \beta_2)(\mu_1^+(\mu_2),\mu_2) = (\ln \tilde{\beta}_3/\ln \tilde{\beta}_2)(P(\mu_1^+(\mu_2),\mu_2))$$

and

$$(\ln \beta_3/\ln \beta_2)(\mu_1^-(\mu_2),\mu_2) = (\ln \tilde{\beta}_3/\ln \tilde{\beta}_2)(P(\mu_1^-(\mu_2),\mu_2))$$

we accomplish our goal. Note that here we use the fact that

$$\beta_3^+ = \beta_3(\mu_1^+(\mu_2),\mu_2) \quad \text{and} \quad \beta_3^- = \beta_3(\mu_1^-(\mu_2),\mu_2)$$

are monotonous for families in E''. □

Let us now prove that β_3^+ and β_3^- are monotonous.

<u>Proposition 12</u>. The set of values (μ_1,μ_2) in U for which $W^u(\sigma_2(\mu_1,\mu_2)) \cap W^s(\sigma_1(\mu_1,\mu_2))$ is non transversal is formed by two curves $\mu_1^- = \mu_1(\mu_2)$ for $\mu_2 < 0$ and $\mu_1^+ = \mu_1(\mu_2)$ for $\mu_2 > 0$. Both are differentiable.

<u>Proof</u>: Let $(x_1(\mu_1,\mu_2), x_2(\mu_1,\mu_2)) \in \Lambda \cap W^u(\sigma_2)$ be such that $\psi_2(x_1(\mu_1,\mu_2), x_2(\mu_1,\mu_2))$ is a maximum (minimum). The dependence on (μ_1,μ_2) is differentiable.

For these points let us consider the equation $\lambda(\mu_1,\mu_2) = \psi_2(x_1(\mu_1,\mu_2),x_2(\mu_1,\mu_2),\mu_1,\mu_2) = 0$ where we are stressing the dependence of ψ_2 upon (μ_1,μ_2).

Differentiating λ with respect to μ_1, we obtain:

$$\frac{\partial\lambda}{\partial\mu_1} = \frac{\partial\psi_2}{\partial x_1}\frac{\partial x_1}{\partial\mu_1} + \frac{\partial\psi_2}{\partial x_2}\frac{\partial x_2}{\partial\mu_1} + \frac{\partial\psi_2}{\partial\mu_1}.$$

We know that $\frac{\partial\psi_2}{\partial x_1} \sim 0$, $\frac{\partial\psi_2}{\partial\mu_1}$ is bounded. Besides both $\frac{\partial x_1}{\partial\mu_1}$ and $\frac{\partial x_2}{\partial\mu_1}$ have the same order and tend to infinity when μ_1 approaches zero.

Hence we can express μ_1 as a function of μ_2, that is, $\mu_1^- = \mu_1^-(\mu_2)$ because $\frac{\partial\lambda}{\partial\mu_1}$ preserves its sign for μ_1 sufficiently small and for $\lambda(\mu_1,\mu_2) = 0$, the curve of maxima with value zero. By the Implicit Function Theorem, we get:

$$\frac{d\mu_1^-}{d\mu_2} = \frac{-\frac{\partial\psi_2}{\partial\mu_2} - \frac{\partial\psi_2}{\partial x_1}\frac{\partial x_1}{\partial\mu_1} - \frac{\partial\psi_2}{\partial x_2}\frac{\partial x_2}{\partial\mu_2}}{\frac{\partial\psi_2}{\partial x_1}\frac{\partial x_1}{\partial\mu_1} + \frac{\partial\psi_2}{\partial x_2}\frac{\partial x_2}{\partial\mu_1} + \frac{\partial\psi_2}{\partial\mu_1}}. \quad \square$$

<u>Proposition 13</u>. $\beta_3(\mu_1^-(\mu_2),\mu_2))$ is monotonous in a left neighborhood of zero.

<u>Proof</u>: We recall that for a periodic orbit σ_2 of X, with period T, and for the Poincaré transformation $\rho: \Sigma_o \to \Sigma_o$ on a transversal section Σ_o, we have

$$\rho'(q) = \exp\left[\int_0^T \operatorname{div} X(\sigma_2(t))dt\right] = \beta_3, \qquad q \in \sigma_2 \cap \Sigma_o.$$

Our goal shall be achieved if we see that $\frac{d}{d\mu_2}\beta_3^-$ preserves its sign.

This is the case:

$$\frac{d}{d\mu_2}\beta_3^- = \rho'(q)\frac{d}{d\mu_2}\int_0^T \operatorname{div} X(\sigma_2(t))dt =$$

$$= \rho'(q)\frac{dT}{d\mu_2}(\mu_1^-(\mu_2),\mu_2)\operatorname{div} X(\sigma_2(T)) + \int_0^T \frac{d}{d\mu_2}\operatorname{div}(X(\sigma_2(t)))dt .$$

Then the sign of $\int_0^T \frac{d}{d\mu_2}\operatorname{div}(X(\sigma_2(t))dt$ shall determinate the sign we are looking for

$$\int_0^T \frac{d}{d\mu_2}\operatorname{div}(X(\sigma_2(t)))dt =$$

$$= \int_0^T \left[\frac{d\mu_1^-}{d\mu_2}\frac{\partial}{\partial\mu_1}\operatorname{div}(X(\sigma_2(t))) + \frac{\partial}{\partial\mu_2}\operatorname{div}(X(\sigma_2(t)))\right]dt,$$

where $\int_0^T \frac{\partial}{\partial\mu_1}\operatorname{div}(X(\sigma_2(t)))\frac{\partial\mu_1}{\partial\mu_2}dt$ is dominating. As $\frac{\partial}{\partial\mu_1}\operatorname{div}(X(\sigma_2(t))) = 1 + o(r,\mu_1,\mu_2)$ and $T \sim 2\pi/\alpha_2$, we conclude that the positive sign is preserved. □

Remark. $\frac{\partial x_1}{\partial\mu_1}$ and $\frac{\partial x_2}{\partial\mu_1}$ have the same order because the periodic orbit σ_2 is almost a circle. They tend to infinity, because for the Hopf bifurcation the radius of σ_2 is $r \sim c\sqrt{|\mu_1|}$.

REFERENCES

[1] Arnold, V.I. - Lectures on bifurcations and versal families. In: Russian Math. Surveys 27; 57-123, (1972).

[2] Beloqui, J. - Modulus of stability for vector fields on 3-manifolds, J. Diff. Eq. (to appear).

[3] Brunovsky, P. - On one parameter families of diffeomorphisms I, II, Comment. mat. univ. Carolinae, 11, 559-582 (1970) and 12, 765-784 (1971).

[4] Marsden, J.E. and McCracken, M. - The Hopf bifurcation and its applications, Springer-Verlag 1976, Applied mathematical sciences, Vol. 19.

[5] de Melo, W. - Moduli of stability of two-dimensional diffeomorphisms, Topology, 19, 9-21 (1980).

[6] Newhouse, S. and Palis, J. - Bifurcations of Morse-Smale dynamical system, In: Dynamical Systems, ed. M.M. Peixoto, Acad. Press, N.Y. (1973).

[7] Newhouse, S. and Palis, J. - Cycles and bifurcation theory, Astérisque, 31, 43-140 (1976).

[8] Newhouse, S., Palis, J., and Takens, F. - Stable families of Diffeomorphisms, to appear in Publ. I.H.E.S. (1981).

[9] Palis, J. - A differentiable invariant of topological conjugacies and moduli of stability, Astérisque, 51 (1978), 335-346.

[10] Shoshitaishvili, A.M. - Bifurcations of topological type at singular points of parametrized vector fields, Funct. Anal. Appl. 6, 169-170 (1972).

[11] Sotomayor, J. - Generic one parameter families of vector fields on two dimensional manifolds, Publ. I.H.E.S., 43, 5-46 (1976).

[12] Sotomayor, J. - Generic bifurcations of dynamical systems. In: Dynamical Systems, ed. M.M. Peixoto, Acad. Press N.Y. (1973).

[13] Van Strien, S.J. - One parameter families of vector fields, Bifurcations near saddle connections (Thesis, 1982).

[14] Takens, F. - Moduli and bifurcations; non-transversal intersection of invariant manifolds of vector fields. "Functional differential equations and bifurcations". Ed. A.F. Izé, Springer-Verlag, 799 (1980), 368-384.

[15] Takens, F. - Partially hyperbolic fixed points, Topology, Vol. 10 (1971), p. 133-147.

[16] Takens, F. - Singularities of Vector Fields, Publications de l'IHES, v. 43 (1974), p. 47-100.

[17] Takens, F. - Global phenomena in bifurcations of dynamical systems with simple recurrence, Jber.d. Dt. Math. Verein 81 (1979), p. 87-96.

Instituto de Matemática e Estatística (IME)
Universidade de São Paulo
C.P. 20570
(01498) São Paulo, SP
Brasil

M I CAMACHO* & C F B PALMEIRA

Polynomial foliations of degree 3 in the plane

Introduction

A non singular differential equation in two real variables defines a foliation of the plane. It is well known that the topological classification of such foliations depends only on the number of inseparable leaves and the way they are distributed in the plane [K], [H-R]. Two leaves (or trajectories) L_1 and L_2 are said to be inseparable if for any arcs T_1 and T_2 respectively transversal to L_1 and L_2 there are leaves which intersect both T_1 and T_2 (figure 1).

For polynomial foliations of degree n, i.e. defined by equations $Pdx + Qdy = 0$ with P and Q of degree at most n, it is known that the number of inseparable leaves is at most 2n [M], [S-S] and there are examples of foliations with n inseparable leaves for all $n \geq 2$ in [M], and with $2n-4$ inseparable leaves for all $n \geq 4$, n even, in [S-S]. Actually, a construction leading to examples with $2n-4$ inseparable leaves for all $n \geq 4$ can be already found in [P]. It is our belief that $2n-4$ is the sharp bound for $n \geq 4$. In [C-P] we considered the case $n=2$ and now we look at $n=3$. Our goal was to get a method which could be applied to the general case. Although we did not obtain such a method, this

* partially supported by CNPq (Brazil).

paper fills a gap, since from [M] we know that there are foliations of degree 3 with 3 inseparable leaves, and from the general result there are at most 6 inseparable leaves. Here we show that for $n=3$, 3 is the sharp bound. The method we use is the same as in [C-P]. We look at $\mathbb{R}^2$ as the tangent plane to $S^2 \subset R^3$ at $(0,0,1)$ and we use central projection to get a foliation in the open north hemisphere, which extends to a foliation (with singularities on the boundary) of the closed hemisphere. We study these singularities (blowing up the degenerate ones) and from the topological behaviour of the foliation near the singularities we get the global behaviour.

In [C-P] we made a complete study (for $n=2$), listing all possible behaviours and giving examples which realized all cases. Here, due to the greater complexity, we have limited ourselves to show that the cases with more than 3 inseparable leaves can not occur. Since we have an example with 3 inseparable leaves, this shows that 3 is the sharp bound. In what follows, we do not present the computations involving degenerate singularities (blowing up) since they are long but straightforward. The interested reader may consult [C-P] where analogous computations are done for $n=2$.

1. Projecting on the north hemisphere.

As figure 2 indicates, we will use central projection to go from the tangent plane to S^2 at $(0,0,1)$ to its tangent plane at $(1,0,0)$. We will get a foliation in the upper half-plane which extends (with singularities) to the whole plane. If we call (x,y) the coordinates in $T_{(0,0,1)}S^2$ and (Y,Z) the coordinates in $T_{(1,0,0)}S^2$, we have the change of variables f defined by $(x,y) = f(X,Z) = (\frac{1}{Z},\frac{Y}{Z})$. If

$$w = P(x,y)dx + Q(x,y)dy,$$

then

$$f^*w = \frac{1}{Z^2}[ZQ(\tfrac{1}{Z},\tfrac{Y}{Z})dY - (P(\tfrac{1}{Z},\tfrac{Y}{Z}) + YQ(\tfrac{1}{Z},\tfrac{Y}{Z}))dZ].$$

Let

$$\tilde{P}(Y,Z) = P_3(1,Y) + ZP_2(1,Y) + Z^2P_1(1,Y) + Z^3P_0$$

and

$$\tilde{Q}(Y,Z) = Q_3(1,Y) + ZQ_2(1,Y) + Z^2Q_1(1,Y) + Z^3Q_0$$

where P_i and Q_i denote respectively the homogeneous part of degree i of P and Q. Then it is easy to see that $Z^5f^*w = Z\tilde{Q}dY - (\tilde{P}+Y\tilde{Q})dZ$. If $P_3(1,Y) + YQ_3(1,Y) \not\equiv 0$ then $Z = 0$ is an integral curve of Z^5f^*w, with singularities at the roots of $P_3(1,Y) + YQ_3(1,Y)$. So, we consider the vector field $V = (\tilde{P}+Y\tilde{Q}, Z\tilde{Q})$ defined in the whole plane and study its singularities. If $P_3(1,Y) + YQ_3(1,Y) \equiv 0$ we can consider $\bar{V} = \frac{1}{Z}V$ which extends naturally to $Z = 0$.

<u>Remark</u>: If $P_3 + YQ_3$ is of degree less than 4, this indicates the presence of singularities at infinity, i.e. at the points $(0,\pm1,0)$ in the sphere. A linear change of coordinates in the original (x,y)-plane will allow us to disregard this situation.

We know that to each singularity of V corresponds a pair of singularities at antipodal points of the half-sphere. The behaviour of the trajectories in the half-sphere near these singularities is the same as the behaviour of the trajectories of V near the corresponding singularity, taking into account that $Z \geq 0$ corresponds to one singularity and $Z \leq 0$ to the other. Inseparable leaves of the foliation in R^2 correspond to separatrices of hyperbolic sectors at singularities of the foliation in the half-sphere. So if we can bound the numbers of such separatrices, we will have bounded the number of inseparable leaves.

Let $(Y_0,0)$ be a singularity of V. We have:

$$DV(Y_0,0) = \begin{pmatrix} \frac{d}{dy}\left(P_3(1,Y)+YQ_3(1,Y)\right)\Big|_{Y=Y_0} & \frac{\partial}{\partial Z}(\tilde{P}+Y\tilde{Q}) \\ 0 & Q_3(1,Y_0) \end{pmatrix}$$

We see that difficulties arise only when Y_0 is both a root of $Q_3(1,Y)$ and a multiple root of $P_3 + YQ_3$ (remember that Y_0 is a root of $P_3 + YQ_3$). These are the cases where blowing up is needed.

<u>Terminology</u>: As usual, we will say that the singularity is hyperbolic if $\frac{d}{dY}(P_3+YQ_3) \neq 0 \neq Q_3(1,Y)$. If $\frac{d}{dY}(P_3+YQ_3) = 0$ and $Q_3(1,Y) \neq 0$ we will say that the singularity is semi-hyperbolic in the Z direction and if $\frac{d}{dY}(P_3+YQ_3) \neq 0 = Q_3(1,Y)$ we will say that the singularity is semi-hyperbolic in the Y direction. It is well known that hyperbolic singularities are either nodes or saddles and semi-hyperbolic singularities can be nodes, saddles or saddle-nodes.

<u>Remark</u>: Singularities which are semi-hyperbolic in the Z direction appear when $P_3 + YQ_3$ has a multiple root, so we know the behaviour of the trajectory along the Y axis, which is one of the invariant manifolds of the singularity, i.e. double and quadruple roots produce saddle-nodes, triple roots produce nodes or saddles. For singularities which are semi-hyperbolic in the Y direction, nothing can be said a priori.

If $P_3(1,Y) + YQ_3(1,Y) \equiv 0$, then $\bar{V} = \frac{1}{Z}V$ is transverse to the Y axis except at the roots of Q_3. When we have a common root with $P_2 + YQ_2$ it is a singularity, if not, it is a tangency point, which may also generate a pair of inseparable leaves, as we will see later on.

2. The case by case analysis.

There are five different situations according to the relation between P_3 and Q_3.

a. P_3 and Q_3 are relatively prime.

b. P_3 and Q_3 have one common factor (of degree 1), i.e.

$P_3(x,y) = (ax+by)A_2(x,y)$ and $Q_3(x,y) = (ax+by)B_2(x,y)$

where A_2 and B_2 are homogeneous polynomials of degree 2, relatively prime.

c. P_3 and Q_3 have two distinct common factors, i.e.

$P_3 = (ax^2+bxy+cy^2)(ex+fy)$ and $Q_3 = (ax^2+bxy+cy^2)(e'x+f'y)$

with $b^2-4ac \neq 0$ and $ef'-e'f \neq 0$.

d. P_3 and Q_3 have a common factor of multiplicity 2, i.e.

$P_3(x,y) = (ax+by)^2(cx+ey)$ and $Q_3(x,y) = (ax+by)^2(c'x+e'y)$

with $ce'-c'e \neq 0$.

e. P_3 and Q_3 have three common factors, i.e. $P_3 = mQ_3$.

Each of these cases divides into subcases according to the existence of multiple roots for $P_3(1,Y) + YQ_3(1,Y)$ and to whether multiple roots are common roots of $P_3(1,Y)$ and $Q_3(1,Y)$. We have then:

b1. $P_3 + YQ_3$ has four simple roots.

b2. $P_3 + YQ_3$ has one double root and two simple real roots, which subdivides in

b2.1 - The double root is the common root.

b2.2 - The double root is not the common root.

b3. $P_3 + YQ_3$ has one double root and two complex roots, in this case, the double root is the common root.

b4. $P_3 + YQ_3$ has two double roots.

b5. $P_3 + YQ_3$ has a triple root, which subdivides in

b5.1 - The triple root is not the common root.

b5.2 - The triple root is the common root.

b6. $P_3 + YQ_3$ has a quadruple root.

c1. $P_3 + YQ_3$ has four simple, real roots.

c2. $P_3 + YQ_3$ has a double root and two simple real roots, which subdivides in

c2.1 - The double root is a common root.

c2.2 - The double root is not a common root.

c3. $P_3 + YQ_3$ has two complex roots which are not the common roots.

c4. $P_3 + YQ_3$ has two complex root which are the common roots. This subdivides into two cases:

c4.1 - The two other roots are real and distinct.

c4.2 - There is a double root for $P_3 + YQ_3$.

c5. $P_3 + YQ_3 \equiv 0$, i.e., $P_3(x,y) = yA(x,y)$ and $Q_3(x,y) = -xA(x,y)$ for some polynomial A.

c6. $P_3 + YQ_3$ has a triple root.

c7. $P_3 + YQ_3$ has two double roots.

d1. The non common roots of $P_3 + YQ_3$ are real and distinct.

d2. The non common roots of $P_3 + YQ_3$ are real and equal.

d3. The non common roots of $P_3 + YQ_3$ are complex conjugate.

d4. $P_3 + YQ_3$ has a triple root.

d5. $P_3 + YQ_3$ has a quadruple root.

e1. $P_3 + YQ_3$ has four simple real roots.

e2. $P_3 + YQ_3$ has a double root and two simple real roots. Notice that the double root is a common root of P_3 and Q_3.

e3. $P_3 + YQ_3$ has two simple real roots and two complex roots.

e4. $P_3 + YQ_3$ has a triple root, which subdivides in e.4.1. the triple root is a triple root of P_3 and Q_3; e.4.2. the triple root is a double root of P_3 and Q_3.

e5. $P_3 + YQ_3$ has a quadruple root.

e6. $P_3 + YQ_3$ has two double roots.

<u>Remark</u>: One may notice that we did not consider the case "$P_3 + YQ_3$ has four complex roots". This case does not occur, since it would imply the existence of a non singular vector field in S^2.

We will say that cases a, b1, b2.2, b5.1, c1, c2.2, c3, c4.1, c4.2, e1, and e3 are non degenerate, all singularities are either hyperbolic or semi-hyperbolic. All other case are called degenerate except case c5 which is called dicritical case.

2.1 - <u>The non degenerate cases</u>:

a. P_3 and Q_3 are relatively prime.

All singularities of V are semi-hyperbolic in the Z-direction. So roots of $P_3 + YQ_3$ with odd multiplicity give rise to singularities of index ±1 (nodes or saddles); roots of even multiplicity give rise to singularities of index 0 (saddle-nodes). The sum of the indices of the singularities of V must be 1, in order to get a sum 2 in the half-sphere, since the foliation can easily be extended to the whole sphere. It is easy to see that with the above conditions it is impossible to get sum 1, so this case does not occur (remember that $P_3 + YQ_3$ has degree 4).

b. P_3 and Q_3 have one and only one common factor.

b.1 $P_3 + YQ_3$ has four simple roots. If they are all real, we have in the sphere three pairs of hyperbolic singularities and one pair semi-hyperbolic in the Y direction. In order to have 1 as the sum of indices for the singularities of V, we must then have two nodes, one saddle, and one saddle-node, which implies that there are at most three separatrices, so at most three inseparable leaves (each hyperbolic sector produces only one separatrix, since the other is along infinity (figure 3). If $P_3 + YQ_3$ has two complex roots, we have a pair semi-hyperbolic in the Y direction and a hyperbolic pair. As before, index considerations imply that we must have a saddle-node and a node, and it is clear that there are no inseparable leaves.

b.2.2 The double root is not the common root. V has one hyperbolic singularity, one singularity semi-hyperbolic in the Y direction and one semi-hyperbolic in the Z direction. Again, Hopf characteristic considerations imply that we must have one node and two saddle-nodes, so at most three inseparable leaves (figure 4).

b.5 $P_3 + YQ_3$ has a triple root.

b.5.1 The triple root is not the common root. We have, on the sphere, one pair of singularities semi-hyperbolic in the Z direction, with index 1, corresponding to the common root, so these must be saddle-nodes (index 0) and it is easy to see that there are no inseparable leaves. (Figure 5).

c.1 $P_3 + YQ_3$ has four simple roots, all real. We must have on the sphere two pairs of nodes, one pair of saddles, one pair of saddle nodes, (hyperbolic along infinity); so there are three separatrices, which implies that we have at most three inseparable leaves.

c.2 $P_3 + YQ_3$ has one double root and two simple, real roots.

c.2.2 The double root is not a common root. We have, on the sphere, one pair of hyperbolic singularities, one pair semi-hyperbolic in the Y-direction and one pair semi-hyperbolic in the Z-direction with index 0 (double root). So, the hyperbolic pair is a pair of nodes and the other pair is a pair of saddle-nodes. So there are only two separatrices, which implies that there are at most two inseparable leaves.

c.3 $P_3 + YQ_3$ has a pair of complex roots. We have two pairs of singularities hyperbolic in the Y-direction; so one pair has index 1 and the other has index 0, so there is only one separatrix, which implies that there are no inseparable leaves.

c.4.1 P_3 and Q_3 have two common roots which are complex and $P_3 + YQ_3$ has two distinct real roots.

We have

$$P_3(x,y) = (ax^2+bxy+cy^2)(ex+fy)$$

and

$$Q_3(x,y) = (ax^2+bxy+cy^2)(e'x+f'y) \quad \text{with} \quad ef'-e'f \neq 0$$

$b^2-4ac < 0$ and $e + (e'+f)Y + f'Y^2$ has two real distinct roots.

If $e \neq 0$, then no root of $e + (e'+f)Y + f'Y^2$ is also a root of $Q_3(1,Y)$ so we have, on the sphere, two pairs of hyperbolic singularities which are either pairs of nodes or pairs of saddles. In any case the sum of the Hopf characteristics is not two so this case does not occur. If $e = 0$ we have a hyperbolic pair and a pair semi-hyperbolic in the Y-direction. It is easy to see that we have then a pair of nodes and a pair of saddle-nodes, and the foliation is trivial since there is only one separatrix.

c.4.2 Same situation as above, except that now $e+(e'+f)Y+f'Y^2$ has a double root. We have, on the sphere, only one pair of singularities, semi-hyperbolic in the Z-direction, so it is a pair of saddle-nodes (index 0). So this case does not occur.

e. P_3 and Q_3 have three common factors.

e.1 $P_3 + YQ_3$ has four simple roots, all real. It is easy to see that the non common root corresponds to a pair of nodes and the other three pairs of singularities are semi-hyperbolic along infinity, so we must have a pair of nodes, a pair of saddles and a pair of saddle-nodes. We have, then, three separatrices, so at most three inseparable leaves.

e.3 $P_3 + YQ_3$ has four simple roots, two of which are complex. As before, we have one pair of nodes and a pair semi-hyperbolic along infinity, which must be a pair of saddle-nodes. Again we have figure 5.

2.2 - <u>The degenerate cases</u>

These are the cases where $P_3 + YQ_3$ has a multiple root which is a common root of P_3 and Q_3. From the local point of view there are essentially three different situations, according to whether the multiplicity is 2, 3 or 4. The existence of one or more pairs of singularities and of what kind, will allow us in each case to choose among the different possibilities, the ones that may occur.

2.2.1 Double root

By a simple change of coordinates we can suppose that the double root of $P_3 + YQ_3$ is $Y = 0$ i.e.

$$P = y(a_1x^2+b_1xy+c_1y^2) + e_1x^2 + f_1xy + g_1y^2 + h_1x + i_1y + j_1$$

$$Q = y(a_2x^2+b_2xy+c_2y^2) + e_2x^2 + f_2xy + g_2y^2 + h_2x + i_2y + j_2$$

Since 0 is a double root of $P_3 + YQ_3$, we have $a_1 = 0 \neq b_1 + a_2$. We have $V = (P+YQ, ZQ)$ with $V(0,0) = 0$. To study this singularity we will consider three different situations:

1. The one jet of V at $(0,0)$ which we denote by j^1V is not zero (this is the same as saying that $DV(0,0) \neq 0$) i.e. $e_1 \neq 0$.
2. $j^1V = 0 \neq j^2V$ and j^2V (jet of order two of V at $(0,0)$) is in general position.
3. $j^1V = 0 \neq j^2V$ is not in general position.

<u>Remark</u>: $j^2V = 0$ implies that $Y = 0$ is a triple (or quadruple) root.

<u>Case 1</u>. $j^1V \neq 0$. If $V = (V_1, V_2)$, we have:

$$V_1 = e_1Z + (b_1+a_2)Y^2 + (f_1+e_2)YZ + h_1Z^2 + (c_1+b_2)Y^3 +$$
$$+ (g_1+f_2)Y^2Z + (i_1+h_2)YZ^2 + j_1Z^3 + c_2Y^4 + g_2Y^3Z + i_2U^2Z^2 + j_2YZ^3$$

$$V_2 = a_2YZ + e_2Z^2 + b_2Y^2Z + f_2YZ^2 + h_2Z^3 + c_2Y^3Z + g_2Y^2Z^2 +$$
$$+ i_2YZ^3 + j_2Z^4.$$

Blowing up $(0,0)$ with $e_1 \neq 0 \neq (b_1+a_2)$ we get figures 6 to 10.

Figures 6, 7, 9 have index 1, figure 8 has index -1 and figure 10 has index 0.

Case 2. j^2V is in general position at (0,0). If j^kV is the first non-zero jet of V at (0,0), let

$$h(Y,Z) = \langle j^kV, Y\frac{\partial}{\partial Y} + Z\frac{\partial}{\partial Z}\rangle \quad \text{and} \quad g(Y,Z) = \langle j^kV, Y\frac{\partial}{\partial Z} - Z\frac{\partial}{\partial Y}\rangle$$

where $\langle\ ,\ \rangle$ is the scalar product.

Then, j^kV is said to be in general position if for all $(Y,Z) \neq$ $\neq (0,0)$ such that $h(Y,Z) = 0$, then $g(Y,Z) \neq 0 \neq dg(Y,Z)$. If this is the case, it is shown in [T] that the topological behaviour of V near (0,0) is the same as of j^kV. Supposing j^2V in general position, it is sufficient to study the vector field $((b_1+a_2)Y^2 +$ $+ (f_1+e_2)YZ + h_1Z^2; -a_2YZ + e_2Z^2)$ with the conditions $b_1 \neq 0$, $h_1^2 + f_1^2 \neq 0$ and $b_1e_2 - f_1a_2e_2 + h_1a_2^2 \neq 0$.

Blowing up we get figures 11 to 16. Figues 11 and 13 have index 2, 12 and 16 have index 0, 14 has index 4, 15 has index -2.

Case 3. j^2V is not in general position. We have three different situations:

3.1: $b_1 = 0$; 3.2: $b_1 \neq 0 = h_1 = f_1$; 3.3: $b_1 \neq 0 \neq h_1^2 + f_1^2$,
$b_1e_2^2 + f_1a_2e_2 + h_1a_2^2 = 0$.

Blowing up we get figures 11, 12, 13, 16, 17, 18, 19 in case 3.1; figues 7, 13, 16, 18, 19, 20, 21, 22, 23, 24 in case 3.2, and figures 21, 25, 26, 27 in case 3.3.

Now we can study the global situation in cases b.2.1, b.3, b.4, c.2.1, c.7, d.1, d.2, d.3, e.2 and e.6.

b.2.1 $P_3 + YQ_3$ has one double root which is a common root of P_3 and Q_3 and two simple roots. So we have, on the sphere, two pair of hyperbolic singularities and a pair of degenerate ones. The degenerate singularities will have index -1, 1 or 3 accoring to whether the hyperbolic pairs are both pairs of nodes, a pair of nodes and a pair of saddles or two pairs of saddles.

(i) If the index is -1 there are two possibilities: figues 8 and 21 and we see that there are at most three separatrices, so at most three inseparable leaves.

(ii) If the index is 1 there are two separatrices coming from the hyperbolic saddles and maybe two more if the local picture of the degenerate singularities is given by figure 7, or 18. Drawing the global picture it is easy to see that there are only two inseparable leaves.

Since there is no local picture with index 3, this concludes case b.2.1.

b.3 $P_3 + YQ_3$ has a double root which is a common root of P_3 and Q_3 and a pair of complex roots. We have only one pair of singularities, so they must have index 1. Looking at the local pictures we see that there are at most two separatrices, so at most two inseparable leaves.

b.4 $P_3 + YQ_3$ has two double roots, and one is a common root of P_3 and Q_3. We have a pair of singularities semi-hyperbolic in the Z-direction (index 0) and the degenerate pair. So, the degenerate singularities have index 1. Since the semi-hyperbolic pair is a pair of saddle-nodes, there is one separatrix, and since the degenerate pair produces at most two more separatrices, we have again at most three inseparable leaves.

c.2.1 P_3 and Q_3 have two common real roots. $P_3 + YQ_3$ has a double root which is one of the common roots, and two simple, real roots. So we have one pair of hyperbolic singularities, one pair semi-hyperbolic in the Y-direction and one pair of degenerate singularities. The hyperbolic singularities may have index 1 or -1. The semi-hyperbolic singularities may have index 1, 0 or -1, so the degenerate singularities will have index -1, 0, 1, 2, 3, and we know that there is no local picture with index 3. So we have:

(i) Two pairs of nodes, one pair of degenerate singularities with index -1. This is the same as b.2.1 (i).

(ii) One pair of nodes, one pair of saddles, one pair of degenerate singularities with index 1. This is the same as b.2.1 (ii).

(iii) One pair of nodes, one pair of saddle-nodes, one pair of degenerate singularities with index 0. We see that there is a local picture (figure 19) with four separatrices; this local picture corresponds to $j^1V = 0$ and j^2V not in general position. However it is easy to see that in order to have case c.2.1 one must have $b_1 \neq 0$ and then a simple change of variables will reduce us to the case $c_1 = c_2 = 0$. When one performs the blow up computations one sees that figure 19 occurs either if $b_1 = 0$ (which is not the case here) or if $P = b_1xy^2 + g_1y^2$ and $Q = a_2x^2y + b_2xy^2 + f_2xy + g_2y^2 + i_2y + j_2$ with $b_1 > 0$, $2a_2 > b_1$. It is clear that $y = 0$ is a solution of the equation $Pdx + Qdy = 0$ and the blow up computations show that two of the 4 separatrices of figure 19 are given exactly by the line $Y = 0$, so that the 4 separatrices are actually 3, and the global picture is figure 28, which has 3 inseparable leaves.

(iv) One pair of saddles, one pair of saddle-nodes and a degenerate pair of index 2 and we see that there are only 3 separatrices (no separatrices at the degenerate singularities).

c.7 P_3 and Q_3 have two distinct common roots, $P_3 + YQ_3$ has two double roots.

We have two pairs of degenerate singularities. Changing variable, we may consider $P = b_1xy^2 + e_1x^2 + f_1xy + g_1y^2 + h_1x + i_1y + j_1$ and $Q = a_2x^2y + e_2x^2 + f_2xy + g_2y^2 + h_2x + i_2y + j_2$ $b_1 \neq 0 \neq a_2 \neq -b_1$. Here the singularities of V are at $Y = 0$, $Z = 0$ and at $Y = \infty$, $Z = 0$. In order to study $Y = \infty$, we simply interchange x and y and study $Y = 0$. We have the following possibilities, described in the begining of section 2.2.1: $j^1V \neq 0$, $j^1V = 0$ and j^2V in general position, $j^1V = 0$ and j^2V not in general position. Each of these cases may occur at either singularity. The only case which does not occur is 3.1 $(b_1 = 0)$. Combining all the possible local pictures it is easy to see that, except for some combinations which we will study, there are at most 3 separatrices, so at most 3 inseparables leaves. The combinations to be analysed in detail are "figure a at $Y = 0$ and figure b at $Y = \infty$" where $(a,b) \in \{(7,16),(6,19),(7,19),(9,19),(7,26),(16,18),(17,19),(25,26)\}$.

If we have figure 7 and 16 it is easy to see that there are only 2 inseparable leaves in the global picture. In the same way, if we have figures 16 and 25 there are only 2 inseparable leaves.

Let us consider the 4 situations involving figure 19. Interchanging x and y if necessary, we can suppose that figure 19 occurs at $Y = 0$. The blow up computations show that two of the separatrices are exactly on the line $Y = 0$, and figure 19 occurs with $e_1=0$ and either $b_1=0$ (which is not the case here) or $b_1\neq 0$ and $h_1=f_1=j_1=0$. $j_1=0$ implies that $y=0$ is a solution of $Pdx+Qdy = 0$, so two of the four separatrices produced by figure 19 coincide and the global picture starts with figure 29, and there is no way to complete figure 29 with either figure 6, or 7, or 9, or 17 to get more than 3 inseparable leaves. In the other cases it is easy to see by drawing the global picture that there are only 2 inseparable leaves.

d. P_3 and Q_3 have a double root which is a common root, i.e.

$$P_3 = (ax+by)^2(\alpha x+\beta y) \qquad Q_3 = (ax+by)^2(\gamma x+\delta y)$$

$$P_3+YQ_3 = (a+bY)^2[\alpha+(\beta+\gamma)Y+\delta Y^2]$$

There are three cases according to whether $\alpha+(\beta+\gamma)Y+\delta Y^2$ has two distinct roots, a double root, two complex roots.

d.1 P_3+YQ_3 has one double and two simple real roots. This means that we have two pairs of hyperbolic singularities (index ±1), so the degenerate pair has index -1, or 3. This is the same as case b.2.1, so we have at most three inseparable leaves.

d.2 P_3+YQ_3 has one double root which is the common root of P_3 and Q_3 and another double root which is not a common root. This means we have a pair of singularities semi-hyperbolic in the Z-direction and a degenerate pair. The semi-hyperbolic singularities are saddle-nodes, so have index 0. Then the degenerate pair has index 1. Since there are local pictures with two separatrices we will have four separatrices, but drawing the global picture, we see that there are at most two inseparable leaves (figure 30).

d.3 P_3+YQ_3 has a double root which is a common root of P_3 and Q_3 and a pair of complex roots. So we have only the pair of degenerate singularities, which must have index 1 and there are at most two separatrices.

e.2 P_3 and Q_3 have three common factors, P_3+YQ_3 has two simple real roots and a double root, which is a common root of P_3 and Q_3.

We have two singularities semi-hyperbolic in the Y-direction and a degenerate singularity.

(i) If the semi-hyperbolic singularities are saddles or nodes, we are in the situation of b.2.1. So, there are at most three inseparable leaves.

(ii) If the semi-hyperbolic singularities are both saddle-nodes, the degenerate singularity must have index 1. In this case we might have four separatrices, but this means that we are in the case of figure 7, and we have

$$P = y(b_1xy+c_1y^2) + e_1x^2 + f_1xy + g_1y^2 + h_1x + i_1y + j_1$$

$$Q = y(a_2x^2+b_2xy+c_2y^2) + e_2x^2 + f_2xy + g_2y^2 + h_2x + i_2y + j_2$$

$b_1+a_2 \neq 0$, and since this is case e, we have also $a_2 = 0$ and $b_1c_2 = c_1b_2$. But the blow up computations show that figure 7 occur only if $a_2 \neq 0$.

(iii) If one of the semi-hyperbolic singularities is a node and the other is a saddle-node, the degenerate singularity has index 0. In order to have more than three separatrices, we must be in the case of figure 19. Drawing the global picture it is easy to see that there are only three inseparable leaves.

(iv) If one of the semi-hyperbolic singularities is a saddle and the other is a saddle-node, the degenerate singularity has index 2 and since the corresponding figure have no separatrices, we have only three separatrices in the global picture, so at most three inseparable leaves.

e.6 P_3 and Q_3 have 3 common factors and $P_3 + YQ_3$ has two double roots.

This is case c.7 with $a_2 = 0$, and this implies that if j^1V is zero at $Y = \infty$, then j^2V is not in general position and we are in case 3.1 of the begining of section 2.2.1. Now we have the following possibilities:

at $Y = 0$ we can have cases 1, 2, 3.2 and 3.3,

at $Y = \infty$ we can have cases 1 and 3.1.

Cases 1 and 1 ($j^1V \neq 0$ at both singularities), 2 and 1, 3.2 and 1, 3.3 and 1 were studied in c.7, so we may restrict ourselves to $j^1V = 0$, j^2V not in general position (case 3.1) at $Y = \infty$. As before we combine the local pictures and disregard those with total number of separatrices less then 4. The only combinations which will need a further analysis are then figure a at $Y = 0$ and figure b at $Y = \infty$ with $(a,b) \in \{(7,16), (18,16), (16,18), (22,18), (23,18), (26,18), (7,19), (20,19), (25,19), (6,19), (9,19), (20,19), (19,17), (19,20), (19,18)\}$

Let us start by looking at (a,b) with a = figure 19 at $Y = 0$; this means that we are in situation 3.2 i.e. $e_1 = 0 = h_1 = f_1$ and since blow up computations show that in order to get figure 19 we must have $j_1 = 0$, we have $P = b_1xy^2 + g_1y^2 + i_1y$ and $y = 0$ is a solution of $Pdx + Qdy = 0$; since two of the separatrices are on the line $Y = 0$, again we have figure 29 which can not be completed to give more than 3 inseparable leaves.

Now let us consider cases (a,b) with b = figure 19 at $Y = \infty$, so we are in case 3.1 and blow up computations show that figure 19 occurs only if $Q = e_2x^2$; so as before $x = 0$ is a solution of $Pdx + Qdy$ and again two of the separatrices are on the line $x = 0$, so we have figure 29 and no more than 3 inseparable leaves.

The combinations (7,16) and (18,16) have been studied in c.7. For case (22,18) it is easy to see that its global picture does not present more that two inseparable leaves. Let us analyse now cases (23,18) and (26,18).

In case (23,18) blow up computations show that we have $P = b_1xy^2 + i_1y + j_1$ and $Q = e_2x^2 + h_2x$. We see that $x = 0$ is a solution of $Pdx + Qdy = 0$ and as before the line $x = 0$ bounds both the hyperbolic and the eliptic sector at the singularity $X = 0$, i.e., $Y = \infty$ (figue 18), We have then figure 30 which can not be completed with figure 23 at the other pair of singularities.

In case (26,19) blow up computations show that in order to have figure 26 at $Y = 0$, we must have $b_1 \neq 0 \neq h_1^2 + f_1^2$ and $e_2 = 0$. In the other hand, in order to have figure 18 at $Y = \infty$ we must have $f_2 = g_2 = h_2 = i_2 = j_2 = 0$, so $Q = e_2x^2$ and since $e_2 = 0$, we have $Q = 0$ and the equation becomes $Pdx = 0$ which gives a trivial foliation.

2.2.2 <u>Triple root</u>.

These are case b.5.2, c.6, d.4 and e.4. It is easy to see that, except in case c.6, we have on the sphere a pair of hyperbolic singularities and the degenerate pair (corresponding to the triple root); so the degenerate singularities can only have index 0 or 2. In case c.6 the non degenerate singularities can also be a pair of saddle-nodes (index 0) and in this case the degenerate singularities may also have index 1.

As before, we blow up the degenerate singularity and disregard situations with index different from 0 or 2 in cases b.5.2, d.4, e.4, and with index different from 0, 1, 2 in case c.6. Again there are three cases to consider:

$$j^1V \neq 0, \qquad j^1V = 0 \neq j^2V, \qquad j^1V = j^2V = 0 \neq j^3V$$

If $j^1V \neq 0$ we get figures 32 in cases b.5.2 and c.6; 31 and 33 in case d.4; a node in case e.4.

If $j^1V = 0 \neq j^2V$, blowing up the singularity we obtain figures 32, 33, 34, 35 in case b.5.2; figures 31, 33 and 38 to 47 in case d.4; figures 31, 32, 33, 41 and 43 in case e.4; figures 32 to 37 or a node in case c.6.

If $j^1V = j^2V = 0 \neq j^3V$, we have:

$$P = c_1y^3 + g_1y^2 + i_1y + j_1$$

$$Q = b_2xy^2 + c_2y^3 + f_2xy + g_2y^2 + h_2x + i_2y + j_2 .$$

This implies that we are in case d.4 if $c_1 \neq 0$ and in case e.4 if $c_1 = 0$. In both cases we have a linear equation

$$\frac{dx}{dy} = - \frac{(b_2y^2+f_2y+h_2)x + c_2y^3 + g_2y^2 + i_2y + j_2}{c_1y^3 + g_1y^2 + i_1y + j_1}$$

and inseparable leaves can only occur at the roots of the denominator. So there are at most three inseparable leaves in case e.4 and at most two, in case d.4.

Let us now study the global situation

b.5.2 $P_3 + YQ_3$ has a triple root which is the only common root of P_3 and Q_3.

If the hyperbolic singularity is a node, the degenerate one is figure 32 which has only 2 separatrices.

If the hyperbolic singularity is a saddle, the degenerate one is either figure 33, 34 or 35 and the only case to analyse is figure 35 (total number of separatrices equal 4). It is easy to draw the global picture and see that the separatrices from the saddle do not generate inseparable leaves.

d.4 $P_3 + YQ_3$ has a triple root which is a common root of multiplicity 1 of P_3 and Q_3.

As before we count the total number of separatrices and look only at these cases when this number is at least 4. There are 3 such situations: a node and figure 38, a saddle and figure 42, a saddle and figure 46. Again it is easy to draw the global pictures and see that in case of figure 38 there are 3 inseparable leaves, in case of figure 46 there are 3 inseparable leaves, and in case of figure 42, although there is a global picture with 4 inseparable leaves, a tangency counting argument shows that such a picture can not occur with polynomials of degree 3.

e.4 $P_3 + YQ_3$ has a triple root which is a common root of multiplicity 3 of P_3 and Q_3.

In this case the total number of separatrices is always at most 2.

c.6 $P_3 + YQ_3$ has a triple root which is a common root of multiplicity 1 of P_3 and Q_3, and P_3 and Q_3 have another common root of multiplicity 1.

There is only one case of total number of separatrices biger than 3: figure 35 and a saddle, but this has already been analysed in b.5.2.

2.2.3 <u>Quadruple root</u>. These are case b.6, d.5 and e.5.

To avoid blowing up in the case $j^1V \neq 0$ we will use the following results from [S.S-2].

Let i be the index of the singularity, h the number of hyperbolic sectors, e the number of elliptic sectors, s the number of separatrices (trajectories bounding hyperbolic sectors), d the degree of the first non zero jet. We have

$$2i = 2 + e - h \qquad \text{(Bendixson formula)}$$

$$e \leq 2d - 1$$

$$h \leq 2d + 2$$

$$s \leq \begin{cases} 4 & \text{if } d = 1 \\ 6 & \text{if } d = 2 \\ 4d-4 & \text{if } d \geq 3 \end{cases}$$

Since the index of the singularity is 1, we have by Bendixson formula, $e-h = 0$.

If $j^1V \neq 0$ we have $e \leq 1$ and there are two possibilities: $e = h = 0$ or $e = h = 1$. In the first case there are no inseparable leaves and in the second case there are at most two inseparable leaves.

If $j^1V = 0 \neq j^2V$, blowing up the singularity, it is easy to see that either we get figures with $e \neq h$ or figures with $e = h = 1$, so there are at most two inseparable leaves.

If $J^1V = j^2V = 0 \neq j^3V$, we have:

$$P = c_1y^3 + g_1y^2 + i_1y + j_1$$

$$Q = -c_1xy^2 + c_2y^3 + g_2y^2 + h_2x + i_2y + j_2$$

Again we have a linear equation of the form

$$\frac{dx}{dy} = \frac{A(y)x + B(y)}{C(y)}$$

with $C(y)$ a polynomial of degree 3, so there are at most three inseparable leaves.

If $J^1V = J^2V = j^3V = 0 \neq j^4V$ we have $P = 0$ and

$$Q = c_2y^3 + g_2y^2 + i_2y + j_1 .$$

So if $P^2 + Q^2$ is always non zero, the foliation is the same as $dy = 0$.

2.3 <u>The dicritical case</u>: $P_3 = y(ax^2+bxy+cy^2)$ and $Q_3 = -x(ax^2+bxy+cy^2)$.

Since the vector field V is of the form $V = Z\bar{V}$ we consider the vector field $\bar{V}$ which in $Z > 0$ defines the same foliation as the vector field V. We have

$$\bar{V}_1 = P_2 + YQ_2 + Z(P_1+YQ_1) + Z^2(P_0+YQ_0)$$

$$\bar{V}_2 = Q_3 + ZQ_2 + Z^2Q_1 + Z^3Q_0 .$$

We see that $\bar{V}$ is transverse to $Z = 0$ except at the roots of $Q_3(1,Y)$ (at most three points, including infinity, since $Q_3(1,Y)$ has degree less than 3). To be able to obtain the global picture of the foliation we have to study not only the singularities of $\bar{V}$ but also its tangency points with the line $Z = 0$.

If $b^2-4ac < 0$, we see that $Q_3(1,Y)$ is always non zero, so the foliation in the semi-sphere is transverse to the equator except maybe at the points $(0,\pm 1,0)$ which correspond to infinity in the Y-direction in the Y-Z coordinates. To study these points we will use the change of variable $x = \frac{X}{Z}$, $y = \frac{1}{Z}$ getting a vector field $W = (W_1,W_2)$ with

$$W_1 = XP_3(X,1) + Q_3(X,1) + Z(XP_2(X,1)+Q_2(X,1)) + \\ + Z^2(XP_1(X,1)+Q_1(X,1)) + Z^3(XP_0+Q_0)$$

$$W_2 = ZP_3(X,1) + Z^2P_2(X,1) + Z^3P_1(X,1) + Z^4P_0$$

And since $XP_3(X,1) + Q_3(X,1) \equiv 0$, we divide by Z obtaining $\bar{W} = \frac{1}{Z}W$ which extends to the whole XZ plane.

We are interested in studying $\bar{W}$ at $(0,0)$ and it is easy to see that if $b^2-4ac < 0$, $\bar{W}(0,0) \neq 0$, so the foliation in the sphere is non singular, which implies that this case does not occur.

If $b^2-4ac = 0$, a simple change of variables reduces P_3 to y^3 and Q_3 to $-xy^2$. Let

$$P = y^3 + e_1x^2 + f_1xy + g_1y^2 + h_1x + i_1y + j_1$$

$$Q = -xy^2 + e_2x^2 + f_2xy + g_2y^2 + h_2x + i_2y + j_2 .$$

Since $Q_3(1,Y) = -Y^2$ we have transversality to $Z = 0$ at all points except zero and possibly infinity. We have

$$\bar{V}(0,0) = (e_1,0) \quad \text{and} \quad \bar{W}(0,0) = (g_2,1).$$

This implies $e_1 = 0$ since otherwise the foliation would be non-singular in the sphere. Let us study the singularity of $\bar{V}$:

$$D\bar{V}(0,0) = \begin{pmatrix} f_1+e_2 & * \\ 0 & e_2 \end{pmatrix}. \quad \text{If} \quad (f_1+e_2,e_2) \neq (0,0)$$

we have a semi-hyperbolic singularity, which is necessarily of index 1, so there are no inseparable leaves at the singularity and since there is no tangency, the foliation is trivial.

Let us suppose now that $f_1 = e_2 = 0$, then:

$$P = y^3 + g_1y^2 + h_1x + i_1y + j_1$$

$$Q = -xy^2 + f_2xy + g_2y^2 + h_2x + i_2y + j_2 .$$

If $h_1 \neq 0$ we can solve $P = 0$ in x, and replace x by its value in $Q = 0$, getting an equation of degree 5 in y, so $P = 0$ and $Q = 0$ intersect. If $h_1 = 0$ our equation becomes

$$(y^3+g_1y^2+i_1y+j)dx + (-xy^2+f_2xy+g_2y^2+h_2x+i_2y+j_2)dy = 0$$

which is of the form

$$\frac{dx}{dy} = \frac{A(y)x + B(y)}{C(y)} .$$

As before, it is a linear equation which has at most three inseparable leaves, corresponding to the roots of the denominator.

Let us suppose now $b^2-4ac > 0$, i.e.,

$$ax^2 + byx + cy^2 = (mx+ny)(px+qy).$$

Changing variables $(u = mn+ny,\ v = pn+qy)$ we may suppose $P_3 = xy^2$ and $Q_3 = -x^2y$. We have

$$\bar{V}_1(Y,0) = P_2(1,Y) + YQ_2(1,Y)$$

$$\bar{V}_2(Y,0) = -Y.$$

Again we can lose transversality only at $(0,0)$ and at infinity.

$$\bar{V}_1(0,0) = e_1 \quad \text{and} \quad \bar{W}(0,0) = (g_2,0).$$

If $g_2e_1 \neq 0$, the foliation is non-singular, so we must have $e_1g_2 = 0$, and there are three cases to consider: $e_1 = 0 \neq g_2$, $e_1 \neq 0 = g_2$, $e_1 = 0 = g_2$. If $e_1 = 0 \neq g_2$ we have a singularity at $Y = 0$, $Z = 0$ and a tangency at $X = 0$, $Z = 0$.

$$D\bar{V}(0,0) = \begin{pmatrix} f_1+e_2 & h_1 \\ -1 & e_2 \end{pmatrix}.$$

If the singularity is hyperbolic or semi-hyperbolic, it produces no inseparable leaves (its index is one), so there are only two inseparable leaves (generated by the tangency). Let us compute the eigenvalues of $D\bar{V}(0,0)$, we have:

$$\det\,(D\bar{V}(0,0)-xI) = x^2 - (f_1+2e_2)x + (f_1+e_2)e_2 + h_1 .$$

We see that if $f_1+2e_2 \neq 0$, at least one eigenvalue is real and non-zero, so the singularity is hyperbolic or semi-hyperbolic. If $f_1 = -2e_2$, there are two different cases:

(i) $e_2^2 < h_1$: eigenvalues $\pm \frac{i}{2}\sqrt{h_1-e_2^2}$.

In this case the trajectories either spiral around the singularity or are closed curves around the singularity, so we have figure 49.

(ii) $e_2^2 = h_1$.

We have

$$P = xy^2 - 2e_2xy + g_1y^2 + e_2^2x + i_1y + j_1$$

$$Q = -x^2y + e_2x^2 + f_2xy + g_2y^2 + h_2x + i_2y + j_2 .$$

Changing variables again $(\nu = y-e_2;\ \mu = x+g_1)$ or (which is the same thing) supposing $e_2 = 0 = g_1$, we have:

$$P = xy^2 + i_1y + j_1$$

$$Q = x^2y + f_2xy + g_2y^2 + h_2x + i_2y + j_2 .$$

If $j_1 \neq 0$, we can solve $P = 0$ in x, replace x by its value in $Q = 0$, obtaining an equation of degree 5 (with coefficient of y^5 equal to g_2). So, in order to avoid singularities for $Pdx+Qdy = = 0$, we must have $j_1 = 0$, which implies $j_2 \neq 0$. If $i_1 \neq 0$, again solving $P = 0$ in x and replacing x by its value in $Q = 0$, we get a 3rd degree equation in y, so $i_1 = 0$ and we have $P = xy^2$ and

$$Q = -x^2y + f_2xy + g_2y^2 + h_2x + i_2y + j_2$$

$$x = 0 \Rightarrow P = 0 \text{ and } Q = g_2y^2+i_2y+j_2, \quad \text{so} \quad i_2^2-4g_2j_2 < 0$$

$$y = 0 \Rightarrow P = 0 \text{ and } Q = h_2x+j_2, \quad \text{so} \quad h_2 = 0,$$

wo we have

$$P = xy^2$$

$$Q = -x^2y + f_2xy + g_2y^2 + i_2y + j_2$$

with

$$q_2 \neq 0 \neq j_1, \qquad i_2^2-4j_2g_2 < 0 .$$

Blowing up the origin for the vector field $\bar{V}$ we obtain figure 52 wich implies that there are no inseparable leaves generated by the singularity of $\bar{V}$, so there are only the two inseparable leaves generated by the tangency.

If $e_1 \neq 0 = g_2$, it is easy to see that interchanging x and y, we get the same situation as in the previous case $(e_1 = 0 \neq g_2)$.

If $e_1 = g_2 = 0$, as before all we have to consider is

$f_1 = 2e_2$ and $h_1 = e_2^2$, so:

$$P = xy^2 + i_1 y + j_1$$

$$Q = -x^2 y + f_2 xy + h_2 x + i_2 y + j_2 .$$

We have:

$$\bar{V} = (f_2 Y^2 + Z(i_1 Y + h_2 Y + i_2 Y^2) + Z^2(j_1 + j_2 Y); -Y + f_2 YZ + h_2 Z^2 + i_2 YZ^2 + j_2 Z^3).$$

$\bar{V}(Y,0) = (f_2 Y^2, -Y)$. So $\bar{V}$ has no tangencies and has singularities at $Y = 0$ and $Y = \infty$.

Blowing up $\bar{V}$ at $(0,0)$ we get figure 51 if $j_1 = 0 \neq i_1$, figure 52 if $j_1 = 0 = i_1$, and figure 48 if $j_1 \neq 0$. To study $Y = \infty$, we use the X-Z coordinates and blow up the origin for the vector field $\bar{W} = (f_2 X + Z(i_1 X + h_2 X + i_2) + Z^2(j_1 X + j_2); X + i_1 Z^2 + j_1 Z^3)$. Disregarding all cases with index different from 0 or 1 we get figures 48 and 52 to 60.

Combining the local pictures we see that in the half-sphere there are at most 2 separatrices, so no more then 2 inseparable leaves.

Bibliography

[K] W. Kaplan - Regular curve families filling the plane. Duke Math. Journal 7 (1970), 154-185.

[H-R] A. Haefliger and G. Reeb - Variétés (non separées) à une dimension et structures feuilletées du plan, Ens. Math. 3 (1957), 107-125.

[M] M.P. Müller - Quelques propriétés des feuilletages polynomiaux du plan, Bol. Soc. Mat. Mex., 21 #1 (1976).

[S-S] S. Schecter and M. Singer - Planar polynomial foliations, Proc. of A.M.S., 79, #4 (1980).

[P] F. Pluvinage - Colloq. Math. 18 (1967), 90-101.

[C-P] M.I.T. Camacho and C.F.B. Palmeira - Non singular quadratic differential equations in the plane, to appear in Trans. A.M.S.

[S-S-2] S. Schecter and M. Singer - Separatrices at singular points of planar vector fields, Acta Mathematica #145 (1980), 47-78.

M.I.T. Camacho

Instituto de Matemática
Universidade Federal do Rio
de Janeiro

Ilha do Fundão

Rio de Janeiro, RJ

Brasil

C.F.B. Palmeira

Departamento de Matemática
Pontifícia Universidade Católica
do Rio de Janeiro

Rua Marques de S. Vicente, 225

Rio de Janeiro, RJ

Brasil

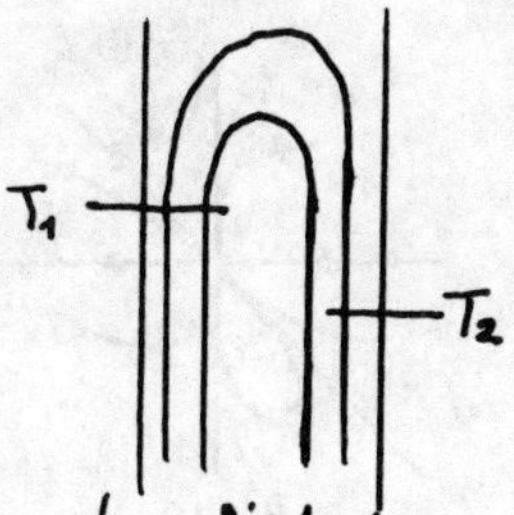

fig 1

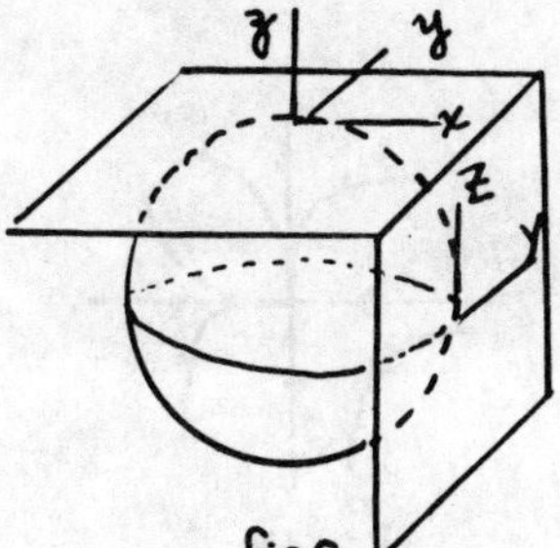

fig 2

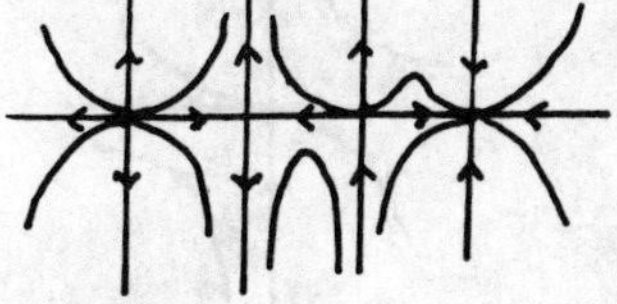
fig 3

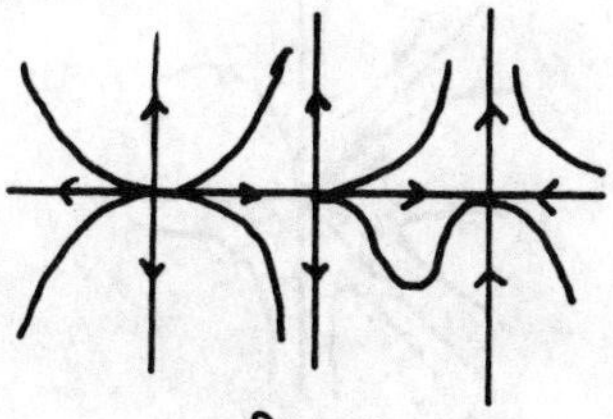
fig 4

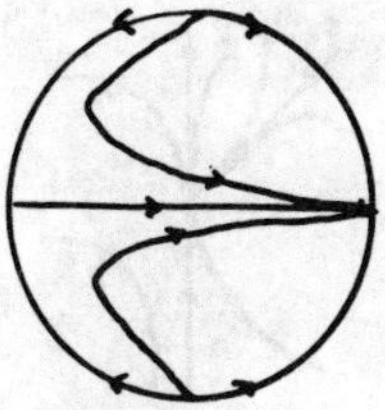
fig 5

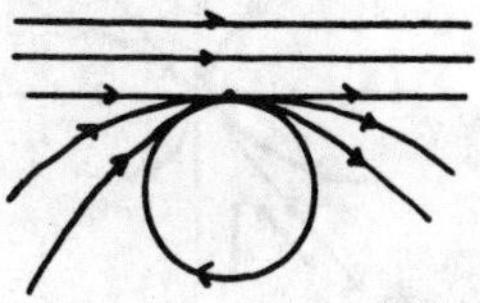
fig 6

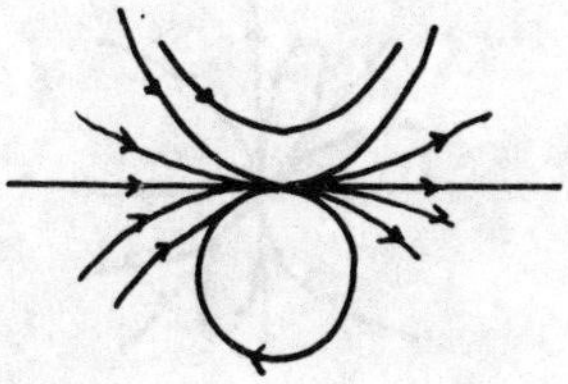
fig 7

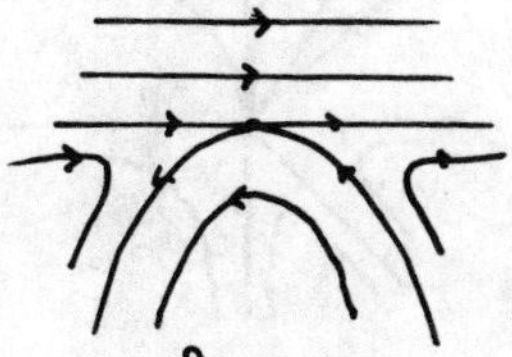
fig 8

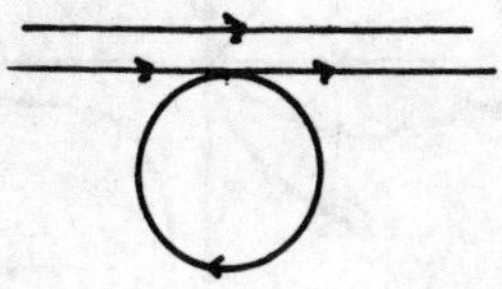
fig 9

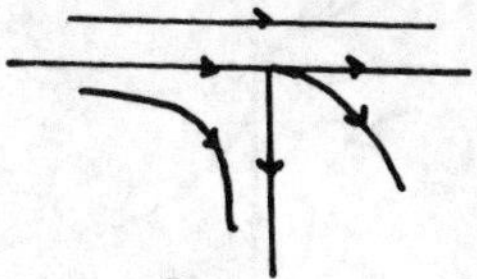
fig 10

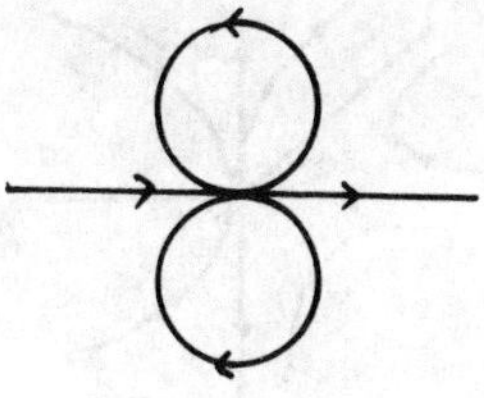
fig 11

fig 12

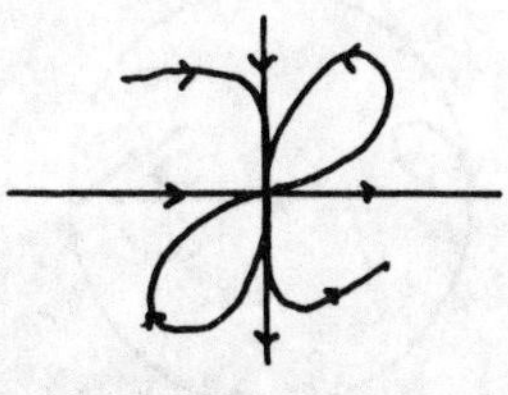
fig 13

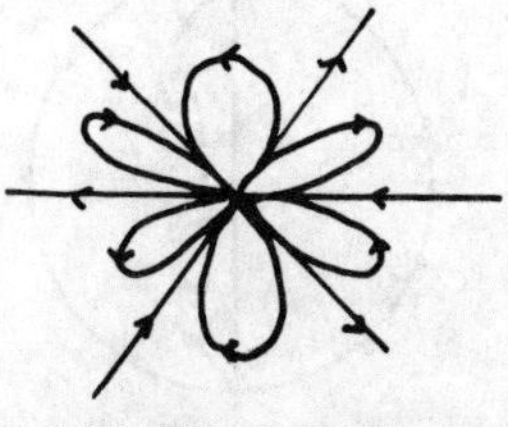
fig 14

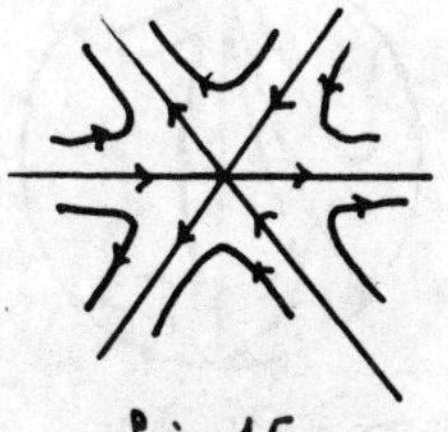
fig 15

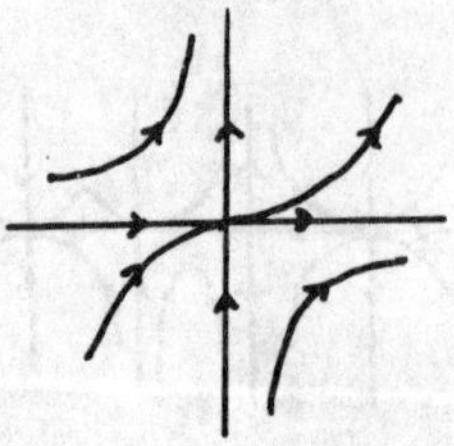
fig 16

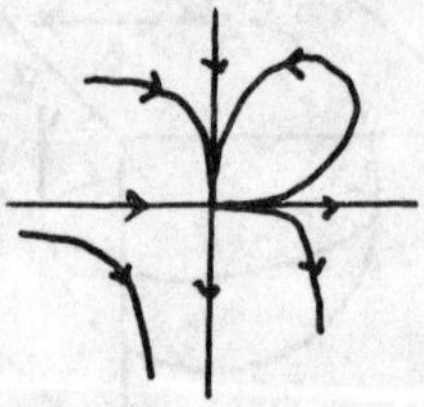
fig 17

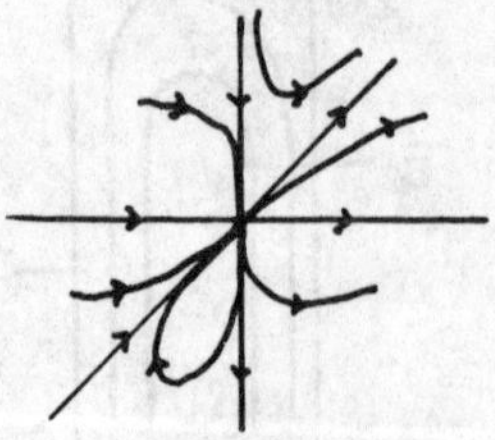
fig 18

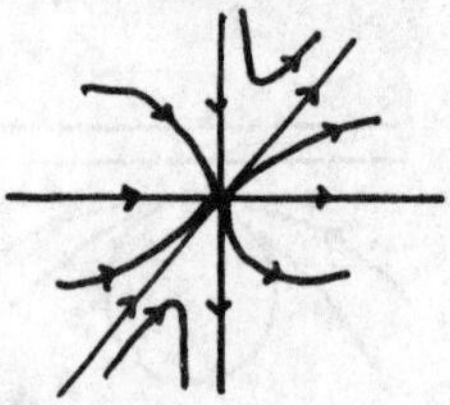
fig 19

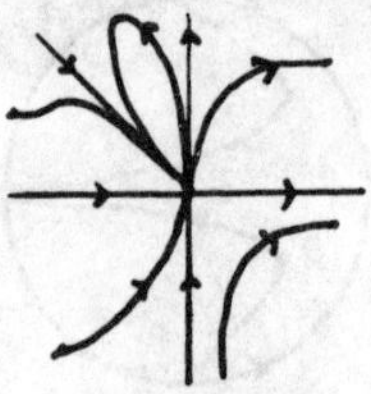
fig 20

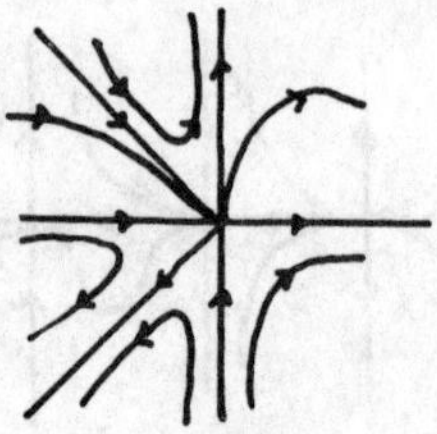
fig 21

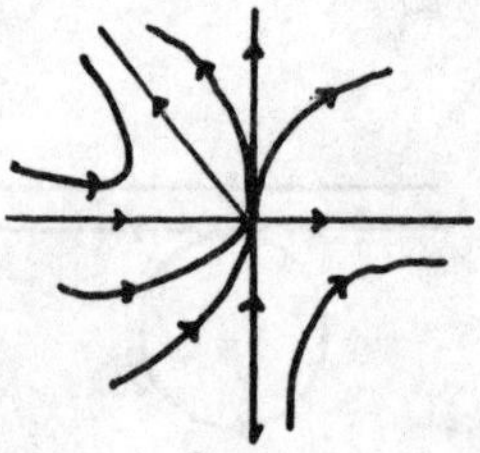
fig 22

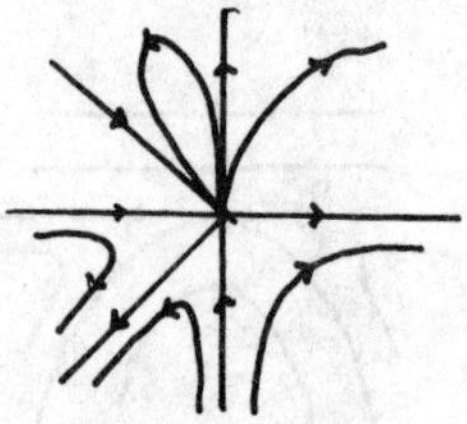
fig 23

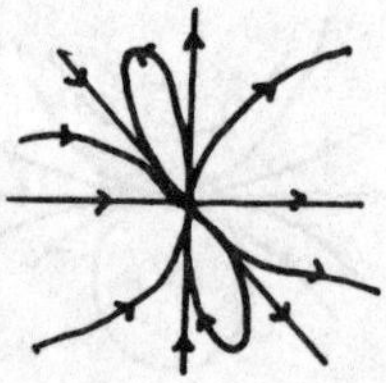
fig 24

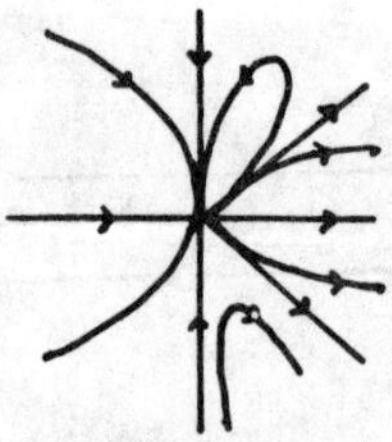
fig 25

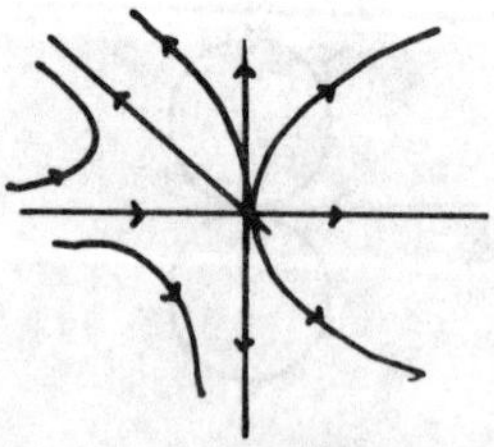
fig 26

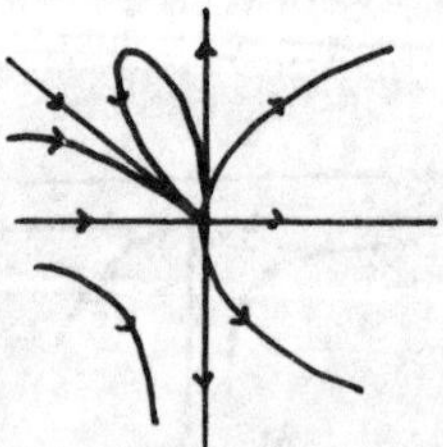
fig 27

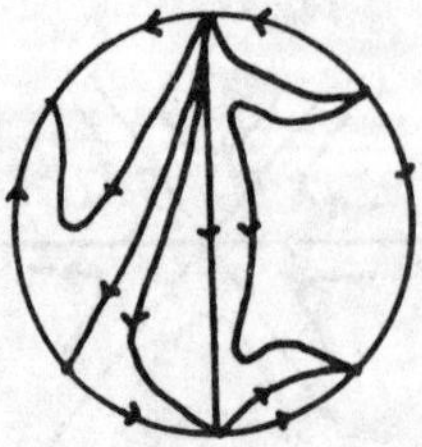
fig 28

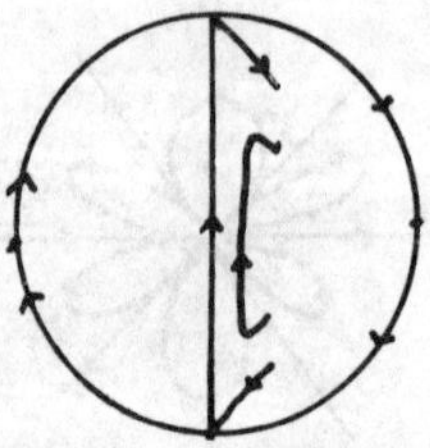
fig 29

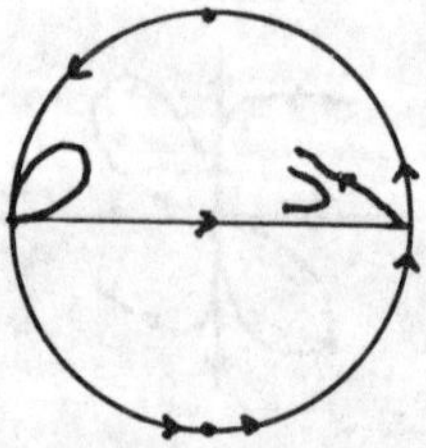
fig 30

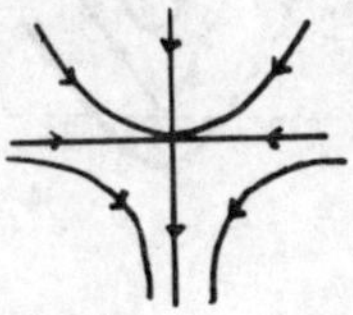
fig 31

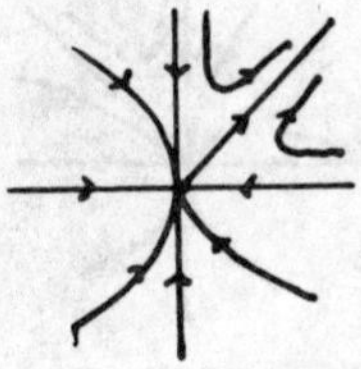
fig 32

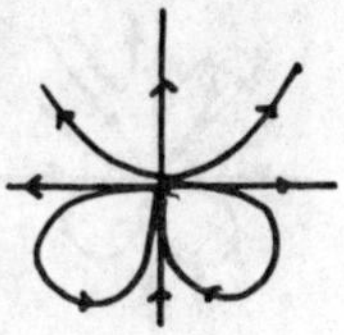
fig 33

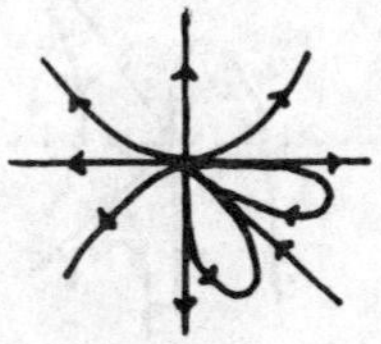
fig 34

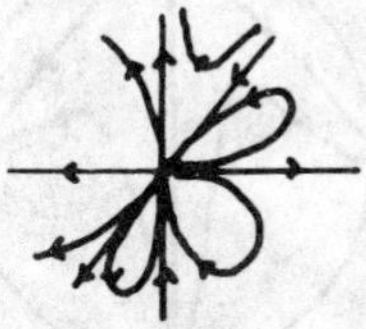
fig 35

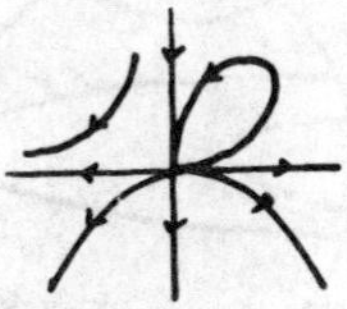
fig 36

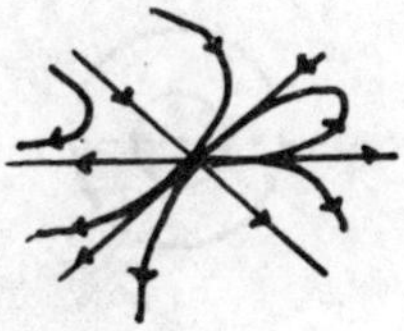
fig 37

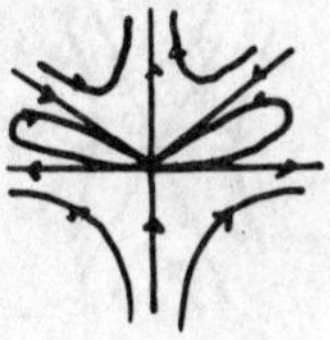
fig 38

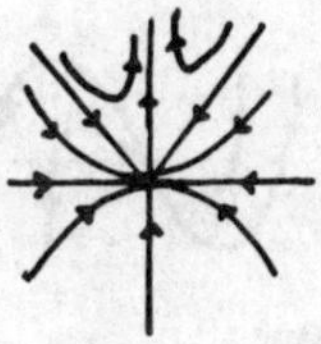
fig 39

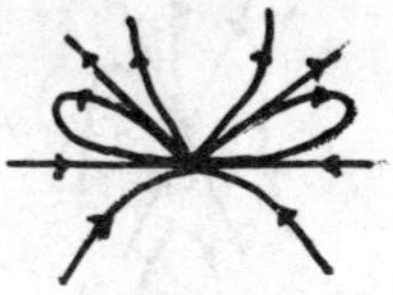
fig 40

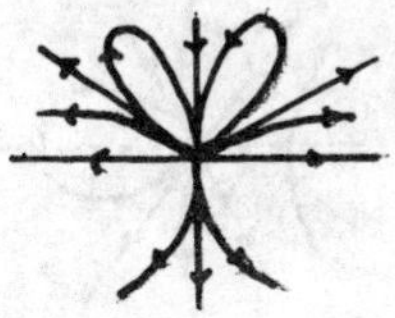
fig 41

fig 42

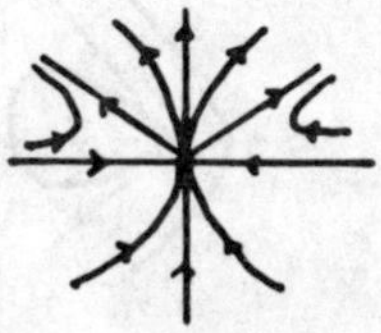
fig 43

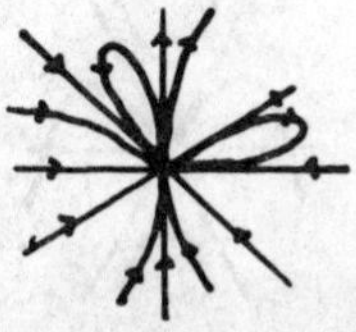
fig 44

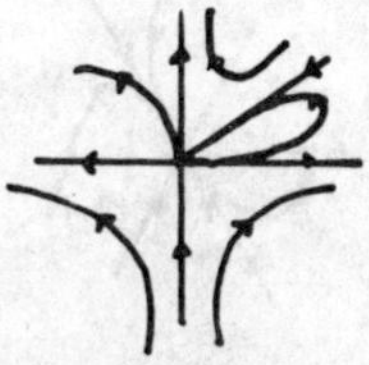
fig 45

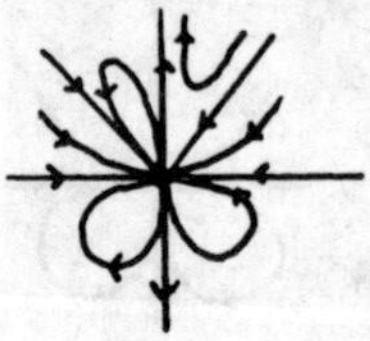

fig 46

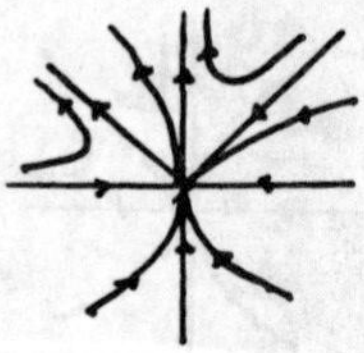

fig 47

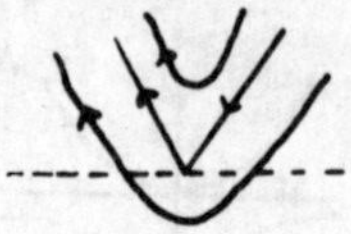

fig 48

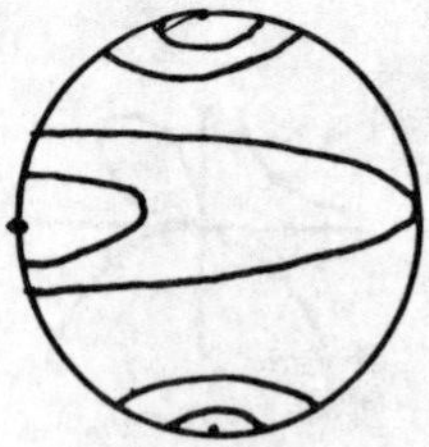

fig 49

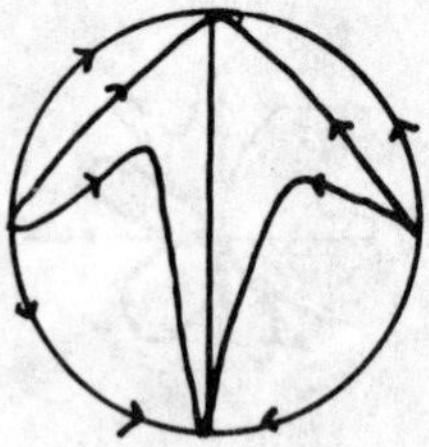

fig 50

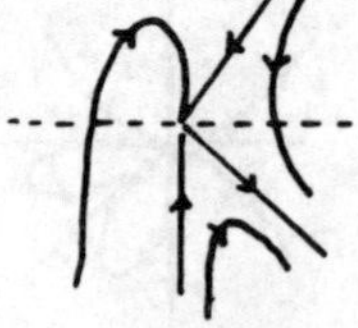

fig 51

fig 52

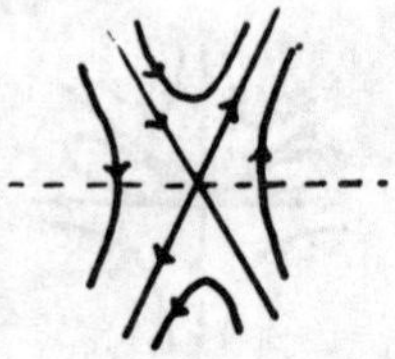

fig 53

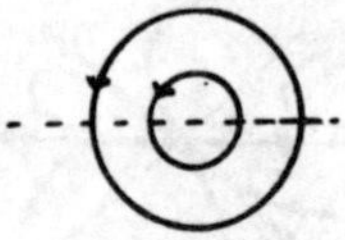

fig 54

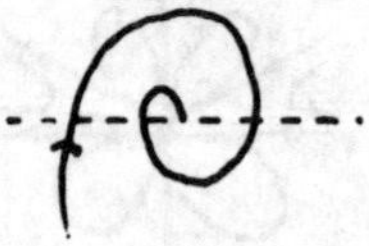

fig 55

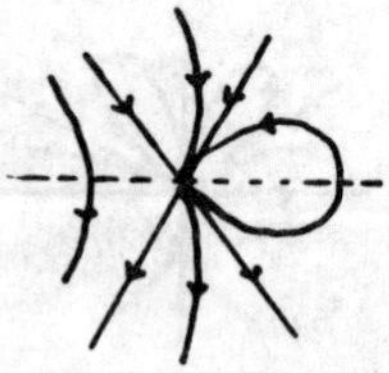

fig 56

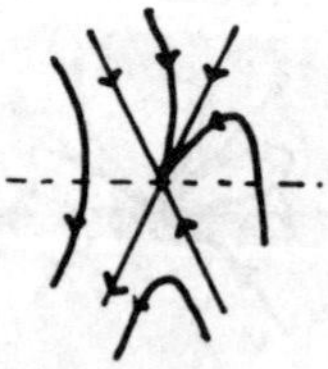

fig 57

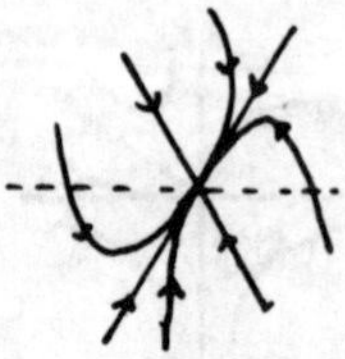

fig 58

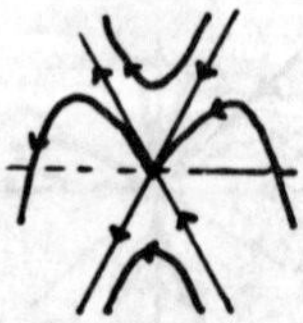

fig 59

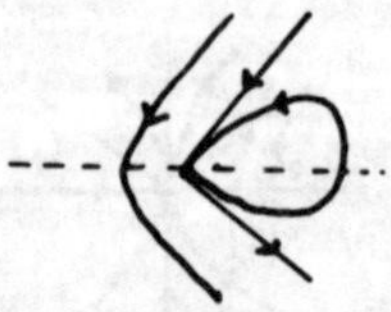

fig 60

C I DOERING

Persistently transitive vector fields on three–dimensional manifolds

Summary

Let M be a compact three-dimensional manifold. We prove that a C^1 vector field on M, without wandering points and with each singularity hyperbolic, is either an Anosov vector field or else can be C^1 approximated by vector fields exhibiting attracting or repelling periodic orbits. As a first corollary we prove that on M, contrary to what happens on higher dimensional manifolds, the persistently transitive vector fields are completely characterized by the transitive Anosov vector fields. As a second corollary we obtain a partial result concerning the Stability Conjecture for flows.

Introduction

Let M be a smooth, compact and boundaryless manifold. We will say that a C^1 (tangent) vector field X on M is transitive if there exists an orbit of X which is dense in M, and persistently transitive provided that every sufficiently C^1-close vector field is transitive.

Every (connected) manifold M with dim M ≥ 2 has transitive

We would like to thank IMPA for its hospitaliy during the period this paper was prepared.

vector fields [2]; on the other hand there are no persistently transitive vector fields on manifolds M with dim M = 2 [15]. For dim $M \geq 3$, however, there is an outstanding class of persistently transitive vector fields, namely the transitive Anosov vector fields [1].

The question naturally arises whether these transitive Anosov fields are the only persistently transitive ones.

For dim $M \geq 4$ the answer is negative: Mañé [13] has an example of an open set of diffeomorphisms of the torus T^3, each being (topologically mixing, hence) transitive but not Anosov; an example for T^4 had been previously constructed by Shub ([22],[7]), thus we have them on any T^n, $n \geq 3$. Suspending one of these diffeomorphisms we obtain persistently transitive vector fields which are not Anosov on n-dimensional manifolds, for all $n \geq 4$.

In this paper we prove that for dim M = 3, on the contrary, the answer is affirmative. In fact we shall prove a stronger statement; let $\Omega(X)$ denote the set of nonwandering points of X.

<u>Theorem</u>. Let dim M = 3 and X be a C^1 vector field on M such that all its singularities are hyperbolic. If $\Omega(X) = M$ then either X is Anosov or else X can be C^1-approximated by vector fields exhibiting attracting or repelling periodic orbits.

Let us initially derive the above characterization of persistently transitive vector fields on three-dimensional manifolds from the theorem.

Since persistently transitive vector fields have no wandering points and obviously cannot be approximated by vector fields exhibiting attracting or repelling periodic orbits, the whole point consists in observing that a persistently transitive vector field also cannot have any singularities. For suppose that such a vector field X does have singularities; approximating X by a (persistently transitive) vector field Y still having singularities but now with each one hyperbolic [15], the theorem leaves no options at all for Y, since Anosov fields do not have singularities. Thus X satisfies all the hypotheses of the theorem and we have:

<u>Corollary 1</u>. Let dim M = 3 and X be a C^1 vector field on M. X is a persistently transitive vector field if and only if X is a transitive Anosov vector field.

Since structurally stable vector fields without wandering points clearly satisfy the assumptions of the theorem, we also obtain the following:

<u>Corollary 2</u>. Let dim M = 3 and X be a C^1 vector field on M such that $\Omega(X) = M$. If X is structurally stable then $\Omega(X)$ has a hyperbolic structure for X.

This is a partial result concerning the Stability Conjecture for (continuous) dynamical systems. In 1968, Palis and Smale [16] conjectured that Smale's Axiom A plus certain global conditions is equivalent to structural or Ω-stability; by the mid-seventies, this conjecture had been reduced (see [15] and [23] for references) to the following, known as the Stability Conjecture: Ω-stable dynamical systems have hyperbolic nonwandering sets.

For discrete dynamical systems there has been so far a definite result, namely Mañé's [14] and Liao's [9] independent proofs for diffeomorphisms of two-dimensional manifolds, but for flows no such result is as yet available. The only other known partial result for flows is due to Liao [9], who proves the Stability Conjecture for vector fields without singularities on three-dimensional manifolds or, more precisely, with singularities isolated from the rest of the nonwandering set, a case which technically is very close to diffeomorphisms on surfaces. The main difference between our and Liao's result is that while we have to deal with singularities, his hypotheses avoid the complications arising from them. Now, the main feature that distinguishes the discrete from the continuous case is exactly the possibility of vector fields having singularities accumulated by nonwandering points. Let us make this more precise.

The proofs of Mañé, Liao and even the early preliminary results of Franks [4] and Pliss [19;20] concerning the Stability Conjecture, are all done for dynamical systems which have persistently hyperbolic compact orbits, i.e., are such that every sufficiently close system has each compact orbit hyperbolic; Ω-stable systems do have this property [4]. This approach, however, is doomed for vector fields, unless we a priori discard singularities: Guckenheimer's example of a Lorenz attractor [5] can be realized as a vector field on S^3 (adding a source singularity) which has persistently hyperbolic compact orbits but exhibits a saddle-type singularity accumulated by periodic orbits, thus preventing its nonwandering set from being hyperbolic.

Our approach to the theorem is similar to those above, except that we do have to take into account possible singularities, and the heart of our argument is to prove that the global nature of our hypothesis implies that the Lorenz phenomenon cannot occur: there cannot be any singularities at all. In order to prove this we construct a splitting of the tangent bundle in the complement of the singularities with certain properties that lead to a topological obstruction in a neighborhood of any singularity. Thus there aren't any singularities and the rest of the proof follows adapting the methods of Mañé [13;14] or Liao [9;10].

When $\dim M \geq 4$, the problem of characterizing the persistently transitive vector fields seems to be exceedingly difficult. There are no natural candidates and the property of persistent transitivity no longer implies the persistent hyperbolicity of periodic orbits, thus destroying the strategy followed in the three-dimensional case. Even for diffeomorphisms of M with $\dim M \geq 3$ there is a shortage of illuminating examples. Shub's and Mañé's examples are homotopic to Anosov diffeomorphisms: this seems to be what we find crossing the boundary of the class of Anosov diffeomorphisms in an adequate way. Does there exist a persistently transtive diffeomorphism not homotopic to an Anosov diffeomorphism? A puzzling and important question related to this is the following one [7]: if $f: T^4 \circlearrowleft$ is a linear diffeomorphism such that its linear lifting to $\mathbb{R}^4$ has two eigenvalues on S^1 which are not roots of unity and the other two eigenvalues are one on each side of S^1, is f persistently transitive?

The theorem of this paper corresponds to the author's doctoral thesis at IMPA, written under the direction of Prof. R. Mañé. It is a pleasure to acknowledge his encouragement, assistance and patience.

The structure of the paper is as follows. In the first section we introduce the concept of domination for vector fields and establish some general facts. Section 2 is the core of the paper, and in it we prove our key result concerning singularities, which enables us to prove the theorem in the last section.

§1. Dominating Splitting.

Throughout this paper, M is a smooth, compact and boundaryless three-dimensional Riemannian manifold and X is a C^1 tangent vector field on M. The flow generated by X will be denoted by $\varphi = (\varphi_t)_{t\in\mathbb{R}}$; each $\varphi_t : M \circlearrowleft$ is a C^1 diffeomorphism of M and sometimes we write $\varphi_t \cdot x$ or $\varphi(t,x)$ for $\varphi_t(x)$, $t \in \mathbb{R}$, $x \in M$. The tangent mappings $T\varphi_t : TM \circlearrowleft$ induced by φ define a tangent flow $T\varphi$ of vector bundle automorphisms of the tangent vector bundle TM of M.

Hyperbolic structures for a vector field are defined with the tangent flow. On regular points these structures are transversal to the vector field and we may as well define them with the quotient flow of X; this procedure will be essential for what follows. For general background, definitions and terminology associated to the qualitative theory of dynamical systems we refer the reader to [15] and [23].

The closed set of singularities of X will be denoted by Sing(X) and

$$R = R(X) = M\text{-Sing}(X)$$

denotes the open set of regular points of X. Given $x \in R$ let $[X(x)]$ denote the linear span of $X(x) \in T_xM$ and $N_x = [X(x)]^{\perp}$ its orthogonal complement in T_xM. We define $[X]_R$ and $N_R = N_R(X)$ to be the vector subbundles of the induced tangent bundle T_RM over R with fibers $[X(x)]$ and N_x, $x \in \mathbb{R}$, respectively. Let $\tilde{T}_RM = T_RM/[X]_R$ be the quotient vector bundle with its orthogonal norm $|v+[X(x)]| = \inf\{|v+\lambda X(x)|;\ \lambda \in \mathbb{R}\}$ on fibers, any $x \in R$, $v \in T_xM$. The restriction

$$\eta : N_R \to \tilde{T}_RM$$

of the quotient mapping is a vector bundle isomorphism, norm preserving on fibers.

Since R is φ-invariant and $[X]_R$ is $T\varphi$-invariant, the restriction $T\varphi|T_RM$ induces a flow $\tilde{T}\varphi$ of automorphisms $\tilde{T}\varphi_t|\tilde{T}_RM$ over

$\varphi_t|R$. For each $t \in \mathbb{R}$ let $P_t: N_R \circlearrowleft$ be the only automorphism of N_R over $\varphi_t|R$ such that the diagram below commutes.

$$\begin{array}{ccc} N_R & \xrightarrow{\;P_t\;} & N_R \\ \eta \downarrow & & \downarrow \eta \\ \tilde{T}_R M & \xrightarrow[\tilde{T}\varphi_t]{} & \tilde{T}_R M \end{array}$$

Clearly $P = (P_t)_{t \in \mathbb{R}}$ is a flow of automorphisms of the normal bundle N_R, which we shall call the linear Poincaré flow of X, or LPF for short.

The hyperbolic structures for X on compact sets, except on singularities, can be defined with the LPF of X.

<u>Proposition 1.1</u>. Let $\Lambda \subseteq R$ be a φ-invariant and compact set. Λ has a hyperbolic structure for X if and only if the LPF restricted over Λ is a hyperbolic vector bundle automorphism.

<u>Proof</u>. Since Λ is compact, it is easy to show that some continuous splitting of a vector bundle over Λ is a hyperbolic splitting for a flow $(F_t)_{t \in \mathbb{R}}$ of automorphisms if and only if it is a hyperbolic splitting for some (or any) F_t, with $t \neq 0$. Suppose that $T_\Lambda M = E^s \oplus [X]_\Lambda \oplus E^u$ is the hyperbolic splitting for $T\varphi_t|T_\Lambda M$. Then the projection $\tilde{T}_\Lambda M = \tilde{E}^s \oplus \tilde{E}^u$ is a hyperbolic splitting for $\tilde{T}\varphi_t|\tilde{T}_\Lambda M$ and $\eta|N_\Lambda$ induces a hyperbolic splitting $N_\Lambda = N^s \oplus N^u$ for $P_t|N_\Lambda$. Conversely, let $N_\Lambda = N^s \oplus N^u$ be the hyperbolic splitting for $P_t|N_\Lambda$. Then

$$0 \to [X]_\Lambda \to [X]_\Lambda \oplus N^u \xrightarrow{\;\pi\;} N^u \to 0$$

is a short exact sequence of vector bundles, and the automorphisms $T\varphi_t|[X]_\Lambda$, $T\varphi_t|[X]_\Lambda \oplus N^u$ and $P_t|N^u$ commute. According to [7; Lemma 2.18], there exists a unique $T\varphi_t$-invariant subbundle $\dot{E}^u$ of

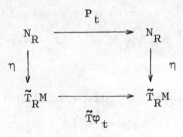

$[X]_\Lambda \oplus N^u$ such that $[X]_\Lambda \oplus E^u = [X]_\Lambda \oplus N^u$, provided that $t > 0$ is sufficiently large. Since $P_t|N^u$ and $T\varphi_t|E^u$ are conjugated by the projection, $T\varphi_t|E^u$ is expansive. Similarly we obtain E^s, implying that $T_\Lambda M = E^s \oplus [X]_\Lambda \oplus E^u$ is a hyperbolic splitting for $T\varphi_t|T_\Lambda M$. □

Let $\gamma = \gamma(x)$ be a periodic non singular orbit of X. If $\pi > 0$ is the period of γ then $P_\pi|N_x : N_x \circlearrowleft$ is precisely the tangent map of a Poincaré return map for γ at x: the eigenvalues of $P_\pi|N_x$ are the eigenvalues of γ. γ is a hyperbolic orbit of X if and only if its eigenvalues are off the unit circle, if and only if γ has a hyperbolic structure for X. In this case, we let $E^s(\gamma) \oplus E^u(\gamma) = N_\gamma$ denote the $P|N_\gamma$-invariant hyperbolic splitting for the LPF of X.

Now we define the main structure we will be using.

<u>Definition</u>. Let $\Lambda \subseteq R$ be a φ-invariant set. We say that Λ has a <u>dominating structure for</u> X if there exists a splitting of the normal bundle N_Λ over Λ into a continuous sum $N_\Lambda = N^s \oplus N^u$ of vector subbundles such that:

a) each fiber N_x^s and N_x^u, $x \in \Lambda$, is one-dimensional;

b) N^s and N^u are invariant under the LPF of X and

c) there exist constants $C > 0$ and $0 < \lambda < 1$ such that

$$\|P_t|N_x^s\| \cdot \|P_{-t}|N^u_{\varphi_t \cdot x}\| \le C\lambda^t$$

for each $x \in \Lambda$ and all $t \ge 0$.

If Λ has such a dominating structure for X we say that the ordered pair (N^s, N^u) is a <u>dominating splitting for</u> X <u>at</u> Λ.

Let us initially observe that a dominating structure for X is independent of the chosen Riemannian metric on M. Of course, the dominating splitting itself changes with the metric, as well as the LPF and even the normal bundle, but any dominating splitting determines a unique splitting of the quotient bundle, invariant under the quotient flow. Since M is compact, a C^0 change in the metric produces equivalent Finslers on M, hence equivalent norms on the quotient bundle, and we obtain for a new LPF of a new normal bundle another dominating splitting. The constant λ remains unaffected, whereas for C there is a factor depending only upon the relation between the equivalent Finslers on M.

Dominating splittings appear naturally in a context of normal hyperbolicity; for [7], P would be "eventually relatively ρ-pseudo hyperbolic", with continuous $\rho: \Lambda \to \mathbb{R}$. Our definition is an adaptation of Mañé's [11; Definition 1.2]. The main feature of a dominating splitting is the following: P_t may not contract N^s nor expand N^u, but P_t must contract N^s if P_{-t} does not expand N^u and P_t must expand N^u if P_{-t} does not contract N^s.

A hyperbolic saddle-type structure for X at $\Lambda \subseteq R$ clearly defines a dominating structure for X at Λ. Conversely, a dominating structure for X at Λ, although being a natural candidate for a hyperbolic saddle-type structure for X at Λ (see [9], [13] and [14]), does not necessarily define a hyperbolic structure, even if Λ is compact. As an example, it suffices to take a periodic orbit with eigenvalues $|\lambda_1| = 1 \neq |\lambda_2|$, which is not hyperbolic but does have a dominating structure. Actually, it is easily proved that a periodic orbit has a dominating structure if and only if its two eigenvalues have distinct modulus.

Existence of dominating structures will be considered in Section 3; in Section 2 and the rest of this section we are concerned with the consequences of an existing dominating structure. The basic properties of domination are of an assymptotic nature. In order to deal with convergence of subspaces of the tangent spaces, it is convenient to introduce the Grassmannians. Since we will also use domination with the tangent flow, the following basic facts will be presented in a more general setting.

Let $\Lambda \subseteq M$ be any subset of M, $E \to \Lambda$ a vector subbundle of $T_\Lambda M$ and $F = (F_t)_{t \in \mathbb{R}}$ a flow of automorphisms of E over $f = (f_t)_{t \in \mathbb{R}}$; no restriction on $\dim M$ is necessary.

Given $x \in \Lambda$, let $Gr(E_x)$ denote the compact and smooth Grassmannian manifold of vector subspaces of E_x and let $G(E_x)$ denote the connected component of one-dimensional subspaces of E_x. We shall denote by $Gr(E)$ the continuous fiber bundle over Λ with fibers $Gr(E_x)$ and by $G(E)$ its subbundle with fibers $G(E_x)$, $x \in \Lambda$. The flow F induces a continuous flow $F_t \cdot L = F_t(L)$ of homeomorphisms of $Gr(E)$ over f_t, leaving $G(E)$ invariant, and a continuous function $\|F_t|L\|$ of $\mathbb{R} \times Gr(E)$.

The local version of domination goes as follows.

<u>Definition</u>. Given $x \in \Lambda$ and $N^s, N^u \in G(E_x)$, we say that N^s <u>dominates</u> N^u if there exist constants $C > 0$ and $0 < \lambda < 1$ such that

$$\|F_t|N^s\| \cdot \|F_{-t}|F_t \cdot N^u\| \le C\lambda^t$$

for all $t \ge 0$. If this inequality is satisfied we may specify that N^s $(C;\lambda)$-dominates N^u; if we want to draw attention to the flow (or the base point) we may further specify that N^s $(F;C;\lambda)$-dominates N^u (at x).

<u>Remark</u>: In the next sections, the statement N^s dominates N^u will always be with respect to the LPF of X, with $x \in R$, $N^s, N^u \in$ $\in G(N_x)$; domination with respect to the tangent flow will be specified.

Lemma 1.2. Let $C > 0$, $0 < \lambda < 1$, $x \in \Lambda$ and $N^s, N^u \in G(E_x)$ be given.

a) If N^s $(C;\lambda)$-dominates N^u then $|F_t \cdot v| \cdot |w| \leq |v| \cdot |F_t \cdot w| C\lambda^t$ for each $v \in N^s$, $w \in N^u$, all $t \geq 0$.

b) $N^s \cap N^u = \{0\}$.

c) N^u does not dominate N^s.

d) Suppose that $N_n^s, N_n^u \in G(E)$ are such that N_n^s $(C;\lambda)$-dominates N_n^u, for each $n \in \mathbb{N}$. If $\lim_n N_n^\sigma = N^\sigma$, $\sigma = s,u$, then N^s $(C;\lambda)$-dominates N^u.

Proof: $|F_t \cdot v| \cdot |w| = |F_t \cdot v| \cdot |F_{-t} \cdot (F_t \cdot w)| \leq \|F_t|N^s\| \cdot |v| \cdot \|F_{-t}|F_t \cdot N^u\| \cdot$ $\cdot |F_t \cdot w| \leq |v| \cdot |F_t \cdot w| C\lambda^t$ for all $t \geq 0$. If $N^s \cap N^u \neq \{0\}$ we may choose $v = w \neq 0$ and obtain $1 \leq C\lambda^t$ for all $t \geq 0$. If N^u $(D;\mu)$-dominates N^s we obtain $|F_t \cdot v| \cdot |w| \leq |F_t \cdot v| \cdot |w| CD(\lambda\mu)^t$ from a) and choosing $v \neq 0 \neq w$ we may cancel and again $1 \leq CD(\lambda\mu)^t$ for all $t \geq 0$. Finally $N_n^u \to N^u$ implies $F_t \cdot N_n^u \to F_t \cdot N^n$, hence $\|F_t|N^s\| \cdot \|F_{-t}|F_t \cdot N^u\| = \lim_n \|F_t|N_n^s\| \cdot \|F_{-t}|F_t \cdot N_n^u\|$ for all $t \geq 0$. □

The essential picture for domination emerges from the next fact. Let $d: G(E) \times G(E) \to \mathbb{R}$ be a metric on $G(E)$ inducing its topology.

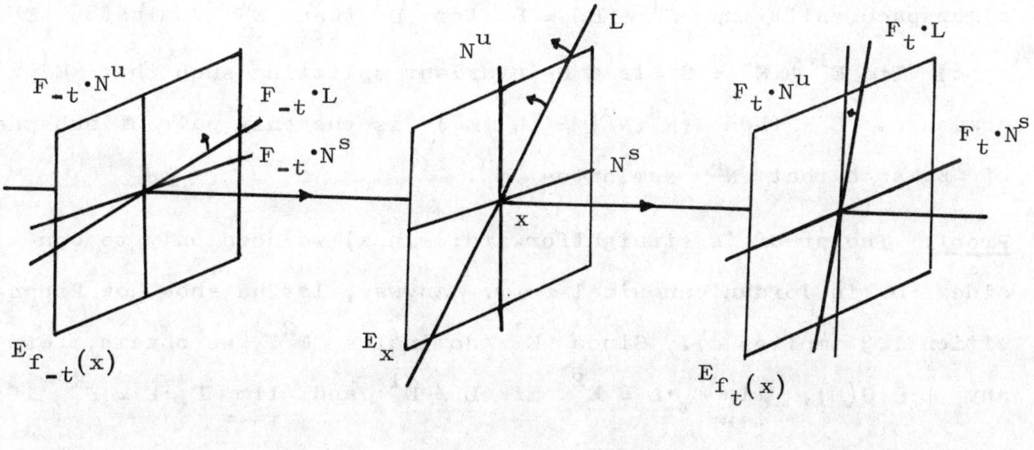

Proposition 1.3. Let $x \in \Lambda$ and $L, N^s, N^u \in G(E_x)$ be such that N^s dominates N^u and $L \subseteq N^s + N^u = N^s \oplus N^u$.

a) If $L \cap N^s = \{0\}$ then $\lim_{t \to +\infty} d(F_t \cdot L, F_t \cdot N^u) = 0$.

b) If $L \cap N^u = \{0\}$ then $\lim_{t \to -\infty} d(F_t \cdot L, F_t \cdot N^s) = 0$.

Proof: Given $t \geq 0$ let $C_t(\delta) = \{v+w \mid v \in F_t \cdot N^s, w \in F_t \cdot N^u, |v| \leq \delta |w|\}$ be the δ-cone neighborhood of $F_t \cdot N^u$ in $F_t \cdot (N^s \oplus N^u)$. If $L \subseteq N^s \oplus N^u$ and $L \cap N^s = \{0\}$ we obtain $\delta > 0$ such that $L \subseteq C_0(\delta)$. Lemma 1.2.a) clearly implies $F_t(C_0(\delta)) \subseteq C_t(\delta C \lambda^t)$ for all $t \geq 0$, hence $F_t \cdot L \subseteq C_t(\delta C \lambda^t)$ for every $t \geq 0$. Now if $t \to +\infty$ we have $\delta C \lambda^t \to 0$ and therefore $F_t \cdot L$ is assymptotic to $F_t \cdot N^u$, proving a). The proof of b) is analogous. □

Let E now be a single two-dimensional vector space ($\Lambda = \{0\}$) and $D: E \circlearrowleft$ a linear mapping. The flow associated to D is given by the isomorphisms $F_t = e^{tD}: E \circlearrowleft$. The proof of c) below illustrates in this simple setting the use of Proposition 1.3; further on such arguments will appear many times.

Lemma 1.4. a) If there exist $N^s, N^u \in G(E)$ such that N^s dominates N^u then the eigenvalues of D are real and distinct.

b) If $\lambda_1 < \lambda_2$ are the eigenvalues of D with associated eigenspace splitting $E^1 \oplus E^2 = E$ for D then E^1 dominates E^2.

c) If $E^1 \oplus E^2 = E$ is a D-invariant splitting such that E^1 dominates E^2 then $(N^s, N^u) = (E^1, E^2)$ is the only pair of subspaces of E such that N^s dominates N^u.

Proof: The proof is straightforward; in a) we need only to consider D in Jordan canonical form. Anyway, let us show how Proposition 1.3 implies c). Since E^1 dominates E^2, we obtain, for any $L \in G(E)$, $\lim_{t \to +\infty} F_t \cdot L = E^2$ if $L \neq E^1$ and $\lim_{t \to -\infty} F_t \cdot L = E^1$ if

$L \neq E^2$. Now let $N^s, N^u \in G(E)$ be such that N^s dominates N^u. If $N^s = E^2$ then $N^s \neq E^1 \neq N^u$ by Lemma 1.2.c) and therefore $E^1 = \lim_{t\to+\infty} F_t \cdot E^1 = \lim_{t\to+\infty} F_t \cdot N^u = E^2$ by Proposition 1.3.a), which is an impossibility. Also impossible is $E^1 \neq N^s \neq E^2$, for then $N^u = E^1$ implies $E^2 = \lim_{t\to+\infty} F_t \cdot E^2 = \lim_{t\to+\infty} F_t \cdot N^u = E^1$ and $N^u \neq E^1$ implies $E^1 = \lim_{t\to+\infty} F_t \cdot E^1 = \lim_{t\to+\infty} F_t \cdot N^u = E^2$, again by Proposition 1.3.a). The only choice left for N^s is $N^s = E^1$. Since $E^2 \neq N^u$ implies $E^2 = \lim_{t\to-\infty} F_t \cdot E^2 = \lim_{t\to-\infty} F_t \cdot N^s = E^1$ by Proposition 1.3.b), we also have $N^u = E^2$, thus $(N^s, N^u) = (E^1, E^2)$. □

§2. Singularities.

In this section we consider the interaction of singularities and dominating structures at regular points. The key result is:

Theorem 2.1. Let $\Lambda \subseteq R$ be a φ-invariant set of regular points of X and suppose that $x_o \in \mathrm{Sing}(X)$ is a hyperbolic saddle-type singularity of X. If Λ has a dominating structure for X then x_o is NOT an interior point of $\Lambda \cup \{x_o\}$.

Throughout this section let x_o be a hyperbolic saddle-type singularity of X, with (non-trivial) hyperbolic $T_{x_o}\varphi$-invariant splitting $E^s \oplus E^u = T_{x_o}M$. Let $W^s(x_o)$ and $W^u(x_o)$ be the stable and unstable manifolds of X at x_o; each $W^\sigma(x_o)$ is a φ-invariant injectively immersed Euclidean space, with tangent space $T_{x_o}W^\sigma(x_o) = E^\sigma$ at x_o. Set $W_o^\sigma = W^\sigma(x_o) - \{x_o\}$; clearly $W_o^\sigma \subseteq R$, the open set of regular points of X.

For the proof of Theorem 2.1 we need to know the location of a dominating splitting defined at a point of the two-dimensional invariant manifold.

Proposition 2.2. Let $\sigma = s,u$ be such that $\dim W^{\sigma}(x_o) = 2$. Suppose that $x \in W_o^{\sigma}$ and that $N^s, N^u \in G(N_x)$. If N^s dominates N^u then $N^{\sigma} = N_x \cap T_x W^{\sigma}(x_o)$.

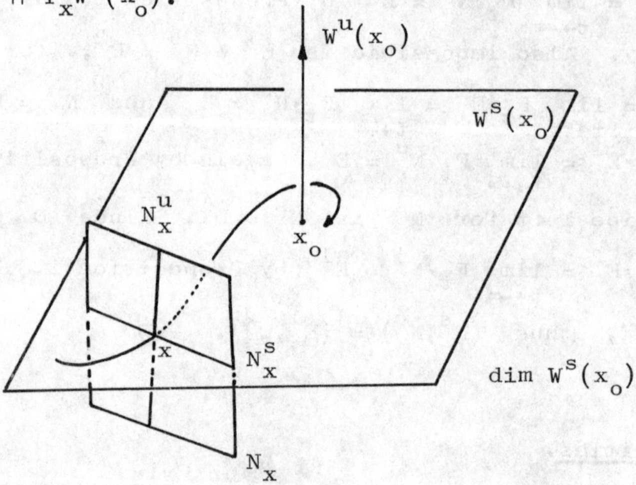

We postpone the proof of this proposition; first we prove the theorem.

Proof of Theorem 2.1: We will assume that $\dim W^s(x_o) = 2$; the proof for $\dim W^u(x_o) = 2$ is entirely analogous.

We start the proof constructing an embedded disc orthogonal to $W^u(x_o)$ and an embedded cylinder orthogonal to $W^s(x_o)$. More precisely, let $U \subseteq M$ be an open neighborhood of x_o contained in a domain of C^o linearization of X at x_o and let W^{σ} denote the connected component of $U \cap W^{\sigma}(x_o)$ that contains x_o, $\sigma = s,u$. We may assume that $U\text{-}W^s$ has two components. Now define C^1-embeddings $g: \mathbb{R}^2 \to U$ and $f: \mathbb{R}\times S^1 \to U$ such that, denoting $D = g(\mathbb{R}^2)$, $y_o = g(0)$, $C = f(\mathbb{R}\times S^1)$ and $S = f(\{0\}\times S^1)$, we have

a) $D \cap W^u = \{y_o\}$ and $C \cap W^s = S$;

b) $D \cap W^s = \emptyset = C \cap W^u$;

c') X is transversal to D and to C.

Clearly such embeddings exist; for f it suffices to let x_o

belong to the disc component of $W^s - S$. Since the theorem is independent of the particular Riemannian metric chosen on M (see Section 1), we may C^0-perturb the metric, in neighborhoods of C and D in U, in such a way that the transversality of c') is straightened into orthogonality, i.e., our embeddings g and f also satisfy:

c) X is orthogonal to D and to C.

For definiteness we also assume that y_o is in the same component of $U - W^s$ that contains $f(\{1\} \times S^1)$.

Given $t < 0$, let D_t denote the connected component of $U \cap \varphi_t(D)$ that contains $\varphi_t(y_o)$. According to Palis' Inclination Lemma [15; Lemma 7.2], the transversal disc D_t approaches W^s C^1-continuously under negative φ-iteration. Hence there exists $t_o < 0$ such that, for each $t \le t_o$, $D_t \cap C$ is homeomorphic to S^1; we choose an $r_o > 0$ such that, for every $0 < r < r_o$, $f(\{r\} \times S^1)$ is contained in the open cylinder in C bounded by S and $D_{t_o} \cap C$.

Denote $B = f((0, r_o) \times S^1)$ and define

$$\psi : B \to D$$

along trajectories: $y = \psi(x) = \varphi(t_x, x)$, where $t_x = \min\{t > 0 \mid \varphi_t(x) \in D\}$. By construction, ψ is a well defined mapping; in addition, B and D are C^1 submanifolds of M and the usual methods (flow boxes) imply that ψ is a C^1 embedding onto the open $\psi(B) \subseteq D - \{y_o\}$.

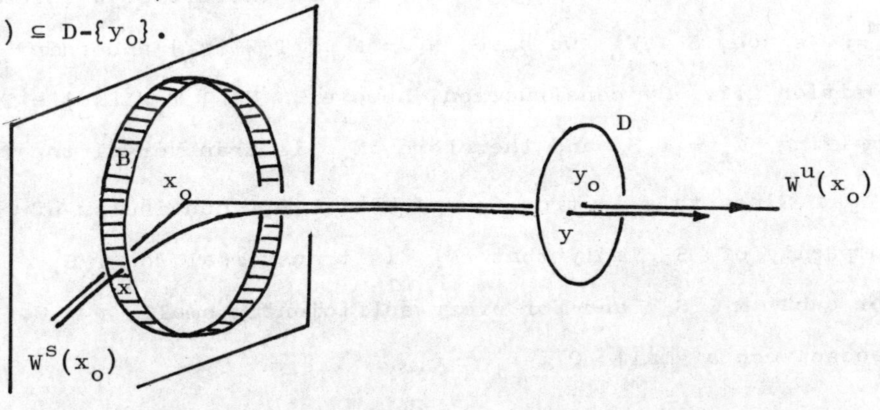

For $0 < r < r_o$ set $S_r = f(\{r\} \times S^1)$. We have

$$P_t \cdot T_x S_t = T_y \psi(S_r) \qquad (*)$$

for any $x \in S_r$ and $t > 0$ with $y = \psi(x) = \varphi_t(x)$. Indeed, let $W_r = \varphi(\mathbb{R} \times S_r)$ be the φ-saturation of S_r: W_r is a C^1 immersed and φ-invariant cylinder containing S_r and $\psi(S_r)$. It follows that $[X]_{W_r} \subseteq TW_r$ and therefore $N \cap TW_r$ is P-invariant, thus $P_t(N_x \cap T_x W_r) = N_y \cap T_y W_r$. But item c) of the construction of C and D implies that $T_x S_r \subseteq T_x C = N_x$ and $T_y \psi(S_r) \subseteq T_y D = N_y$ and accordingly we have $T_x S_r = N_x \cap T_x W_r$ and $T_y \psi(S_r) = N_y \cap T_y W_r$, proving (*).

Now let $\Lambda \subseteq R$ be a φ-invariant set and (N^s, N^u) a dominating splitting for X at Λ. Let us assume, contrary to our assertion, that x_o is an interior point of $\Lambda \cup \{x_o\}$ or, what amounts to the same, that

$$x_o \in U \subseteq \Lambda \cup \{x_o\}.$$

Clearly N^u_D is then a (one-dimensional) continuous tangent line bundle without singularities defined on the entire disc D, since $D \subseteq \Lambda$ and $T_y D = N_y$ by construction. We will presently show that this leads to a topological obstruction, thus proving the theorem.

Given $x \in S$ we have $x \in \Lambda$ and therefore N^s_x dominates N^u_x; since $S \subseteq W^s_o$ we have $N^s_x = N_x \cap T_x W^s(x_o)$ according to Proposition 2.2. By construction, however, $N_x \cap T_x W^s(x_o) = T_x S$, implying $N^s_x = T_x S$ and therefore N^u_x is transversal to $T_x S$ in N_x.

Since this is true for each $x \in S$, continuity of N^u and compacity of S imply that N^u_x is transversal to $T_x S_r$ in N_x for each $x \in S_r$ and for every sufficiently small $r > 0$. We choose such a small $0 < r_1 < r_o$.

Given any $x \in S_{r_1}$ we have $P_t \cdot N_x^u = N_y^u$ and, by (*) above, $P_t \cdot T_x S_{r_1} = T_y \psi(S_{r_1})$, where $y = \psi(x) = \varphi_t(x)$. But N_x^u is transversal to $T_x S_{r_1}$ in N_x and $P_t: N_x \to N_y$ is an isomorphism, hence N_y^u is transversal to $T_y \psi(S_{r_1})$ in N_y. Thus we have:

<u>Claim</u>: N_y^u is transversal to $T_y \psi(S_{r_1})$ in N_y for each $y \in \psi(S_{r_1})$.

By construction, $\psi(S_{r_1}) \subseteq D$ is homeomorphic to S^1 and the claim says that the continuous tangent line bundle N_D^u over the disc D is transversal to $\psi(S_{r_1})$: by the Hopf-Poincaré Index Theorem [6; Ch. 3, §5] there should be a singularity of N_D^u inside $\psi(S_{r_1})$, which is not the case, thus providing the topological impossibility we needed. The proof of the theorem is complete. □

Now we turn to the proof of Proposition 2.2, which is essentially a uniqueness result. At the end of this section we give the dual result for the one-dimensional invariant manifold. To prove these facts we push a dominating splitting from $x \in W_o^\sigma$ to x_o, where we have control through eigenspaces; since the LPF is not defined at x_o, we will use the tangent flow for iteration. First we estimate a dominating splitting in terms of the tangent flow; for convenience we state and prove the next result for the stable manifold, but an entirely similar result holds for the unstable manifold, with obvious modifications.

<u>Lemma 2.3</u>. Suppose that $x \in W_o^s$ and that $N^s, N^u \in G(N_x)$ are such that N^s dominates N^u. If N^s is not tangent to $W^s(x_o)$ at x then N^s $T\varphi$-dominates N^u.

<u>Proof</u>: Let $x \in R$ be given. For any $t \in \mathbb{R}$ and $0 \neq v \in T_x M$ let $c(t,v) = \langle T\varphi_t \cdot v, X(y)\rangle \cdot |T\varphi_t \cdot v|^{-1} \cdot |X(y)|^{-1}$ denote the cosine of the angle between $T\varphi_t \cdot v$ and $X(y)$ in $T_y M$, where $y = \varphi_t(x)$ and $\langle\ ,\ \rangle$ is the Riemannian metric on M. Clearly $|c(t,\lambda v)| = |c(t,v)|$

for any $\lambda \neq 0$ and we may define

$$C(L) = \sup\{|c(t,v)|;\ 0 \neq v \in L,\ t \geq 0\} \in [0,1]$$

for any $L \in G(T_xM)$.

Claim 1. Given $L_1, L_2 \in G(N_x)$ and $t \geq 0$ we have

$$(1-C(L_1))\|T\varphi_t|L_1\| \cdot \|T\varphi_{-t}|T\varphi_t \cdot L_2\| \leq$$
$$\leq \|P_t|L_1\| \cdot \|P_{-t}|P_t \cdot L_2\|.$$

Proof: The LPF is defined by orthogonal projection over the normal bundle of the tangent flow; given $x \in R$, $v \in N_x$ and $t \in \mathbb{R}$ we therefore have $P_t \cdot v = T\varphi_t \cdot v - \alpha(t,v)$, where $\alpha(t,v) = |X(y)|^{-2} \langle T\varphi_t \cdot v, X(y)\rangle X(y)$, with $y = \varphi_t(x)$. Since $\langle P_t \cdot v, \alpha(t,v)\rangle = 0$, we have $|P_t \cdot v| \leq |T\varphi_t \cdot v|$; if $t \geq 0$ we also have $|\alpha(t,v)| = |T\varphi_t \cdot v| \cdot |c(t,v)|$. Finally

$$\|P_{-t}|P_t \cdot L\|^{-1} = \|P_t|L\| = |P_t \cdot \frac{1}{|v|}v| = \frac{|P_t \cdot v|}{|v|}$$

for any $L \in G(N_x)$ and $0 \neq v \in L$, since L is one-dimensional; analogous relation holds for $T\varphi_t$.

Choose any $v \in L_1$ and $w \in L_2$ such that $v \neq 0 \neq w$. It follows

$$\|P_t|L_1\| \cdot \|P_{-t}|P_t \cdot L_2\| = \frac{|P_t \cdot v| \cdot |w|}{|v| \cdot |P_t \cdot w|} \geq \frac{|T\varphi_t \cdot v - \alpha(t,v)| \cdot |w|}{|v| \cdot |T\varphi_t \cdot w|} \geq$$

$$\geq \frac{|T\varphi_t \cdot v| \cdot |w|}{|v| \cdot |T\varphi_t \cdot w|} - \frac{|\alpha(t,v)| \cdot |w|}{|v| \cdot |T\varphi_t \cdot w|} = (1-c(t,v))\, \frac{|T\varphi_t \cdot v| \cdot |w|}{|v| \cdot |T\varphi_t \cdot w|} \geq$$

$$\geq (1-C(L_1))\|T\varphi_t|L_1\| \cdot \|T\varphi_{-t}|T\varphi_t \cdot L_2\|,$$

proving Claim 1.

$C(L)$ measures the sliding of $T\varphi_t \cdot L$ with respect to the $T\varphi$-invariant subbundle $[X]_R$ under forward iteration; $C(L) = 1$ implies that $T\varphi_t \cdot L$ and $[X(\varphi_t(x))]$ get arbitrarily close.

Claim 2. If $x \in W_o^s$ and $L \in G(T_xM)$ is not tangent to $W^s(x_o)$ at x then $0 \le C(L) < 1$.

Proof: Choose any $L_o \in Gr(T_xM)$ such that $L \subseteq L_o$ and L_o is transversal to $W^s(x_o)$ at x. Palis' Inclination Lemma [15] clearly implies that $\lim_{t\to+\infty} T\varphi_t \cdot L_o = E^u$. Since $T\varphi_t$ is an isomorphism, the angle between $T\varphi_t \cdot L$ and $[X(\varphi_t(x))]$ in $T\varphi_t \cdot xM$ is always non-zero; by the above, this angle approaches the non-zero angle between E^s and E^u, hence it is bounded away from zero, implying $0 \le C(L) < 1$ and proving Claim 2.

With these facts settled, we now prove the lemma. Let $x \in R$, $N^s, N^u \in G(N_x)$ be given. If N^s is not tangent to $W^s(x_o)$ at x then $0 \le C(N^s) < 1$ by Claim 2. If N^s $(C;\lambda)$-dominates N^u then N^s $(T\varphi;C_s;\lambda)$-dominates N^u by Claim 1, where $C_s = C(1-C(N^s))^{-1} > 0$. □

Proof of Proposition 2.2: We assume that $\dim W^s(x_o) = 2$; the proof for $\dim W^u(x_o) = 2$ is entirely analogous.

Let $x \in W_o^s$ be given and set $E = N_x \cap T_xW^s(x_o)$. Palis' Inclination Lemma [15] clearly implies that

$$\lim_{t\to+\infty} T\varphi_t \cdot L = E^u$$

for any $L \in G(N_x)-\{E\}$. By compactness of $G(TM)$ we may choose $t_n \to +\infty$ and $F \in G(T_{x_o}M)$ such that

$$\lim_n T\varphi_{t_n} \cdot E = F;$$

but $W^s(x_o)$ is φ-invariant, hence $F \subseteq T_{x_o}W^s(x_o) = E^s$.

Now let $N^s, N^u \in G(N_x)$ be such that N^s dominates N^u. First we prove that $N^u \ne E$. Assume, on the contrary, that $N^u = E$; then $N^s \ne E$ by Lemma 1.2.b) and N^s $T\varphi$-dominates N^u by Lemma 2.3. Choosing $L \in G(N_x) - \{E, N^s\}$, Proposition 1.3.a) and the two limits above imply

$$E^u = \lim_n T\varphi_{t_n} \cdot L = \lim_n T\varphi_{t_n} \cdot N^u = \lim_n T\varphi_{t_n} \cdot E = F \subseteq E^s,$$

which is impossible. Thus $N^u \neq E$. Now suppose that $N^s \neq E$. Again Lemma 2.3 applies and N^s $T\varphi$-dominates N^u; by Proposition 1.3.a) and the two limits above

$$E^s \supseteq F = \lim_n T\varphi_{t_n} \cdot E = \lim_n T\varphi_{t_n} \cdot N^u = E^u,$$

which is the same impossibility. The only choice left is $N^s = E$, proving Proposition 2.2. □

We close this section with a result that gives information on domination at the one-dimensional invariant manifold. For convenience we state and prove it for the stable manifold; the same result is true for the unstable manifold with obvious modifications.

Let $D = DX(x_o): T_{x_o}M \circlearrowleft$ be the derivative of X at x_o; $E^s \oplus E^u = T_{x_o}M$ is the hyperbolic splitting for D and $e^{tD} = T_{x_o}\varphi_t$ for all $t \in \mathbb{R}$.

<u>Proposition 2.4</u>. Suppose that $\dim W^s(x_o) = 1$ and let $x \in W^s_o$, $C > 0$, $0 < \lambda < 1$ and $N^s, N^u \in G(N_x)$ be given. If $P_r \cdot N^s$ $(C;\lambda)$-dominates $P_r \cdot N^u$, for all $r \geq 0$, then there exists $C_o > 0$ such that $T\varphi_r \cdot N^s$ $(T\varphi;C_o;\lambda)$-dominates $T\varphi_r \cdot N^u$ for all $r \geq 0$. In particular, the two eigenvalues of $D|E^u$ are real, positive and distinct: $0 < \lambda_1 < \lambda_2$; moreover, if $E^1 \oplus E^2 = E^u$ is the associated eigenspace splitting then

$$\lim_{t\to+\infty} T\varphi_t \cdot N^s = E^1 \quad \text{and} \quad \lim_{t\to+\infty} T\varphi_t \cdot N^u = E^2.$$

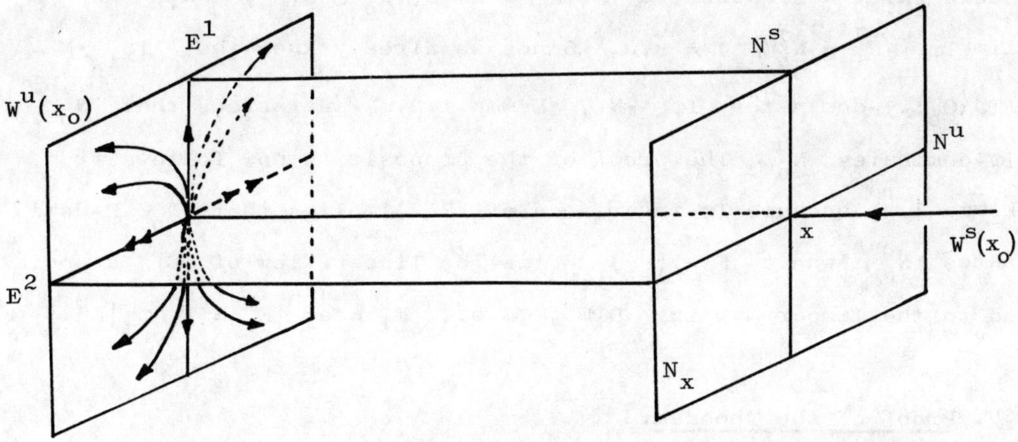

Proof: Given $x \in R$, $L \in G(N_x)$ and $t, r \in \mathbb{R}$, we have $P_t|P_r\cdot L = (P_{t+r}|L)\circ(P_r|L)^{-1} = (P_{t+r}|L)\circ(P_{-r}|P_r\cdot L)$ and therefore, since L is one-dimensional, $\|P_t|P_r\cdot L\| = \|P_{t+r}|L\|\cdot\|P_{-r}|P_r\cdot L\|$. It follows that, for any $L_1, L_2 \in G(N_x)$ and $t, r \in \mathbb{R}$ we have

$$\|P_t|P_r\cdot L_1\|\cdot\|P_{-t}|P_{t+r}\cdot L_2\| =$$

$$= [\|P_{t+r}|L_1\|\cdot\|P_{-(t+r)}|P_{t+r}\cdot L_2\|]\cdot[\|P_r|L_2\|\cdot\|P_{-r}|P_r\cdot L_1\|];$$

analogous relation holds for $T\varphi$.

Now let $x \in W^s_o$ and $N^s, N^u \in G(N_x)$ be given and suppose that $P_r\cdot N^s$ $(C;\lambda)$-dominates $P_r\cdot N^u$ for some $C > 0$, $0 < \lambda < 1$ and all $r \geq 0$. We will use the notation and the claims of the proof of Lemma 2.3. Since N^s and N^u are obviously not tangent to $W^s(x_o)$ at x, Claim 2 of 2.3 implies $0 \leq C(N^s) < 1$ and $0 \leq C(N^u) < 1$, thus $C_o = C[(1-C(N^s))(1-C(N^u))]^{-1} > 0$. Given $r \geq 0$, it suffices to apply Claim 1 of 2.3 to each factor of the above relation to conclude that $T\varphi_r\cdot N^s$ $(T\varphi;C_o;\lambda)$-dominates $T\varphi_r\cdot N^u$.

Finally, Palis' Inclination Lemma again implies that

$$\lim_{t\to+\infty} T\varphi_t\cdot N_x = E^u, \qquad \text{hence}$$

there exist a sequence $t_n \to +\infty$ and $N_o^s, N_o^u \in G(E^u)$ such that $\lim_n T\varphi_{t_n} \cdot N^\sigma = N_o^\sigma$, $\sigma = s,u$. Since we already know that $T\varphi_{t_n} \cdot N^s$ $(T\varphi; C_o; \lambda)$-dominates $T\varphi_{t_n} \cdot N^u$, Lemma 1.2.d) guarantees that N_o^s $T\varphi$-dominates N_o^u. The proof of the proposition now follows from Lemma 1.4, because $T\varphi$-invariance of E^u implies that N_o^s F-dominates N_o^u, where $F = (F_t)_{t \in \mathbb{R}}$ is the linear flow of E^u associated to the linear mapping $D|E^u$, i.e., $F_t = e^{tD}|E^u = T_{x_o}\varphi_t|E^u$. □

§3. Proof of the Theorem.

In this final section we prove the theorem stated in the Introduction.

Let $\mathfrak{X}(M)$ denote the linear space of the C^1 tangent vector fields on M, endowed with the C^1-norm topology. Let $\mathcal{S}(M)$ denote the interior of the class of vector fields $Y \in \mathfrak{X}(M)$ satisfying: a) each singularity of Y is hyperbolic and b) each non-singular periodic orbit of Y is a hyperbolic saddle. Since $\dim M = 3$, it is obvious that $X \in \mathcal{S}(M)$ if and only if each singularity of X is hyperbolic and X cannot be C^1-approximated by vector fields exhibiting attracting or repelling non-singular periodic orbits.

The theorem stated in the Introduction is therefore equivalent to the following:

<u>Theorem 3.1</u>. If $X \in \mathcal{S}(M)$ and $\Omega(X) = M$ then M has a hyperbolic structure for X.

<u>Corollary</u>. The following are equivalent statements for any $X \in \mathfrak{X}(M)$ such that $\Omega(X) = M$:

a) X is Anosov.

b) X is structurally stable.

c) X is topologically Ω-stable.

d) X is persistently transitive.

Proof: a) ⇒ b) is Anosov's Theorem [1], from which also follows a) ⇒ d). Both b) ⇒ c) and d) ⇒ c) are obvious; topological Ω-stability for such X of course means that $\Omega(Y) = M$ for every sufficiently C^1-close Y. Finally c) ⇒ a) follows from Theorem 3.1 by the same argument used in the Introduction to prove d) ⇒ a). □

To prove Theorem 3.1 we need the following results.

Theorem 3.2. If $X \in \mathcal{S}(M)$ then $\Omega(X)$-Sing(X) has a dominating structure for X.

Corollary 3.3. If $X \in \mathcal{S}(M)$ satisfies $\Omega(X) = M$ then Sing$(X) = \emptyset$.

Proof: We have $R = M$-Sing$(X) = \Omega(X)$-Sing(X). Assume, contrary to our assertion, that $x_o \in$ Sing(X); then necessarily x_o is a hyperbolic saddle and x_o is an interior point of $R \cup \{x_o\}$. But the theorem guarantees that R has a dominating structure for X. According to Theorem 2.1 this is not possible. Thus Sing$(X) = \emptyset$. □

Theorem 3.4. Suppose that $X \in \mathcal{S}(M)$ satisfies $\Omega(X) = M$ and Sing$(X) = \emptyset$. Then $\Omega(X)$ has a hyperbolic structure for X.

This result is contained in Liao's Theorem A in [10]. Actually he proves it for any $\Omega(X)$, as long as $\Omega(X)$-Sing(X) is closed, and for a class of vector fields larger than $\mathcal{S}(M)$. In [9; Theorem 4.1] he already proved this with the additional assumption that the periodic orbits of X are dense in $\Omega(X)$; it is not known whether this assumption follows from $X \in \mathcal{S}(M)$, but it does for discrete dynamical systems [13; Lemma 3.1]. For another proof of Theorem 3.4, see [3], where we use the theory of pseudo hyperbolicity and prelaminations of Hirsch, Pugh and Shub [7;§5] and techniques from Mañé [13].

The proof of Theorem 3.1 is clearly a straightforward consequence of Corollary 3.3 and Theorem 3.4; the rest of this section is therefore devoted to the proof of Theorem 3.2.

We need some more notation. Given $Y \in \mathfrak{X}(M)$ let $\varphi^Y = (\varphi_t^Y)_{t \in \mathbb{R}}$ denote the flow of Y, $\mathrm{Per}(Y)$ the set of non-singular periodic orbits of Y, $R(Y)$ the open set of regular points of Y and $N(Y)$ the normal bundle of Y over $R(Y)$ with linear Poincaré flow $P^Y = (P_t^Y)_{t \in \mathbb{R}}$. If $Y \in \mathcal{S}(M)$ then each $q \in \mathrm{Per}(Y)$ defines a hyperbolic saddle-type periodic orbit $\gamma = \gamma(q;Y)$; let $E^s(\gamma;Y) \oplus E^u(\gamma;Y) = N_\gamma(Y)$ denote the P^Y-invariant hyperbolic splitting of the restricted normal bundle, with fibers $E^\sigma(y;Y)$, $\sigma = s,u$, $y \in \gamma$. These symbols define for a variable $Y \in \mathfrak{X}(M)$ what was done for X in Section 1; for X we maintain the same unspecified notation of that section.

Theorem 3.2 is an existence theorem and the basic step in its proof is due to Mañé ([12; Lemma 6] and [13; Proposition 2.1]) and Pliss [18; 1.13]. They proved that although there may not (yet) be uniform hyperbolicity constants for all periodic orbits of $X \in \mathcal{S}(M)$, there necessarily exist uniform domination constants for all periodic orbits of every vector field sufficiently C^1-close to X, as follows.

<u>Proposition 3.5</u>. Suppose that $X \in \mathcal{S}(M)$. There exist constants $C > 0$, $0 < \lambda < 1$ and an open neighborhood $\mathcal{U}$ of X in $\mathcal{S}(M)$ such that $E^s(\gamma;Y)$ $(P^Y;C;\lambda)$-dominates $E^u(\gamma;Y)$ for any $Y \in \mathcal{U}$ and each $q \in \mathrm{Per}(Y)$, with $\gamma = \gamma(q;Y)$.

This is essentially a statement on uniformly hyperbolic families of periodic sequences of linear isomorphisms ([17; Theorem 3] and [14; Lemma II.3.b]). Our family is given by all the hyperbolic isomorphisms $P_{\pi_q}^Y | N_q(Y)$, $Y \in \mathcal{S}(M)$ and $q \in \mathrm{Per}(Y)$, where $\pi_q > 0$ is the Y-period of q; that this family is uniformly hyperbolic follows adapting a lemma of Franks [4; Lemma 1.1]. For detailed proofs of Proposition 3.5, see [8] or [3].

Proof of Theorem 3.2: Suppose that $X \in \mathcal{S}(M)$ and let $C > 0$, $0 < \lambda < 1$ and $X \in \mathcal{U} \subseteq \mathcal{S}(M)$ be given by Proposition 3.5.

Let $x \in \Omega(X) \cap R$ be a given regular nonwandering point of X. Of course we may assume that $\mathcal{U}$ is small enough to ensure $Y(x) \neq 0$ for each $Y \in \mathcal{U}$. Let $(\mathcal{U}_n)_n$ be a countable local basis of $\mathfrak{X}(M)$ at X, and again we assume that $\mathcal{U}_n \subseteq \mathcal{U}$, $n \in \mathbb{N}$.

For each $n \in \mathbb{N}$, Pugh's Closing Lemma [21] assures the existence of $Y_n \in \mathcal{U}_n$ such that $x \in Per(Y_n)$; since $Y_n \in \mathcal{U} \subseteq \mathcal{S}(M)$ we know that x is a hyperbolic saddle. Let $E^s(\gamma_n;Y_n) \oplus E^u(\gamma_n;Y_n) = N_{\gamma_n}(Y_n)$ be the P^{Y_n}-invariant hyperbolic splitting over the Y_n-orbit γ_n of x.

By compacity of $G(T_xM)$ we may choose $N_x^s, N_x^u \in G(T_xM)$ such that $\lim_n E^\sigma(x;Y_n) = N_x^\sigma$, $\sigma = s,u$. Since obviously $Y_n \to X$, we have $Y_n(x) \to X(x)$ and $N_x(Y_n) \to N_x$, hence $N_x^s, N_x^u \in G(N_x)$.

Now define N^σ, $\sigma = s,u$, to be the P-saturation of N_x^σ: $N_Y^\sigma = P_r \cdot N_x^\sigma$ for any $Y = \varphi_r(x)$, $r \in \mathbb{R}$. Clearly N^s and N^u are one-dimensional, P-invariant and continuous subbundles of the normal bundle $N_{\gamma(x)}$ over the X-orbit $\gamma(x)$ of x. We claim that (N^s,N^u) is a (C,λ)-dominating splitting for X at $\gamma(x)$.

Let $t,r \in \mathbb{R}$ be given, with $t \geq 0$ and set $Y = \varphi_r(x)$, $z = \varphi_t(y)$, $q_n = \varphi_r^{Y_n}(x)$ and $p_n = \varphi_t^{Y_n}(q_n)$. But $q_n \in Per(Y_n)$ and $Y_n \in \mathcal{U}$, and therefore Proposition 3.5 implies that

$$\|P_t^{Y_n}|E^s(q_n;Y_n)\| \cdot \|P_{-t}^{Y_n}|E^u(p_n;Y_n)\| \leq C\lambda^t ,$$

for all $n \in \mathbb{N}$. Since $Y_n \to X$ in $\mathfrak{X}(M)$, we have $q_n \to y$ and $p_n \to z$; since $E^\sigma(x;Y_n) \to N_x^\sigma$, $\sigma = s,u$, it is easy to show that we may take the limit as $n \to +\infty$, obtaining $\|P_t|N_y^s\| \cdot \|P_{-t}|N_z^u\| \leq C\lambda^t$

and proving the claim.

Thus for each $x \in \Omega(X) \cap R = \Omega(X)\text{-Sing}(X)$ we have a $(C;\lambda)$-dominating splitting for X at the orbit of x.

Lemma 3.6. Let $\Lambda \subseteq R$ be a φ-invariant set of regular points of X and let $C > 0$ and $0 < \lambda < 1$ be given constants. Suppose that for each $x \in \Lambda$ we have a $(C;\lambda)$-dominating splitting for X at the orbit of x. If every singularity of X in the closure of Λ is a hyperbolic saddle then Λ has a dominating structure for X.

Since $X \in \mathcal{S}(M)$, it is evident that any singularity of X in the closure of $\Omega(X) \cap R$ is necessarily a hyperbolic saddle, completing the proof of Theorem 3.2, except for the proof of Lemma 3.6, which we now consider. □

Proof of Lemma 3.6: This lemma is an adaptation of Mañé's Proposition 1.3 in [11], the main difference being that our base space Λ is not necessarily compact and that we have to consider the singularities of X. These are taken into account by the following.

Lemma 3.7. Let $x \in R$ be a regular point of X and suppose that the only singularities in the closure of the orbit of x are hyperbolic saddles. Then there exists at most one dominating splitting for X at the orbit of x.

We postpone the proof of Lemma 3.7 but use it to prove Lemma 3.6. Let $(N^s(x), N^u(x))$ denote the $(C;\lambda)$-dominating splitting that exists for X at the orbit of x, with fibers $N_t^\sigma(x)$ at $\varphi_t(x)$, $\sigma = s,u$, $t \in \mathbb{R}$. Set $N_x^\sigma = N_0^\sigma(x)$ for $x \in \Lambda$ and define

$$N^\sigma = \bigcup_{x\in\Lambda} N_x^\sigma ,$$

for $\sigma = s,u$. All we need to prove is that (N^s, N^u) is a $(C;\lambda)$-dominating splitting for X at Λ.

Let $x \in \Lambda$ and $t \in \mathbb{R}$ be given and denote $y = \varphi_t(x)$. Since both $(N^s(x), N^u(x))$ and $(N^s(y), N^u(y))$ are dominating splittings for X at the orbit of x, Lemma 3.7 implies that $N^\sigma(x) = N^\sigma(y)$ and therefore its fibers at y coincide:

$$N_y^\sigma = N_0^\sigma(y) = N_t^\sigma(x) = P_t \cdot N_0^\sigma(x) = P_t \cdot N_x^\sigma ,$$

for $\sigma = s,u$. Thus N^s and N^u are P-invariant; moreover for each $x \in \Lambda$, N_x^s and N_x^u are one-dimensional and N_x^s $(C;\lambda)$-dominates N_x^u. All there is left to prove is that N^s and N^u are subbundles of N_Λ or, what amounts to the same, that N^s and N^u are closed in N_Λ.

Let $(x_n)_n$ be a sequence in Λ such that $x_n \to x$ and suppose that $(N^s_{x_n})_n$ converges in $G(TM)$ to $L_0^s \in G(TM)$; we have to show that $L_0^s = N_x^s$. Clearly $L_0^s \in G(N_x)$ and we may also assume that $(N^u_{x_n})_n$ converges in $G(TM)$ to $L_0^u \in G(N_x)$. Now define L^σ to be the P-saturation of L_0^σ with fibers $L_r^\sigma = P_r \cdot L_0^\sigma$, $r \in \mathbb{R}$, $\sigma = s,u$. Given $r \in \mathbb{R}$, we already know that $N^s_{\varphi_r(x_n)}$ $(C;\lambda)$-dominates $N^u_{\varphi_r(x_n)}$, all $n \in \mathbb{N}$; but $N^\sigma_{\varphi_r(x_n)} = P_r \cdot N^\sigma_{x_n} \to P_r \cdot L_0^\sigma = L_r^\sigma$, $\sigma = s,u$, hence L_r^s $(C;\lambda)$-dominates L_r^u by Lemma 1.2.d). Thus (L^s, L^u) is a dominating splitting for X at the orbit of x and Lemma 3.7 guarantees that $L^\sigma = N^\sigma(x)$, $\sigma = s,u$; in particular, $L_0^s = N_0^s(x) = N_x^s$.

It follows that N^s is closed in N_Λ, and similarly we prove that N^u is closed in N_Λ, completing the proof of Lemma 3.6 except for the proof of Lemma 3.7. □

<u>Proof of Lemma 3.7</u>: Suppose that (N^s, N^u) and (L^s, L^u) are both dominating splittings for X at the orbit of x. We will show that $N_x^s = L_x^s$, which implies $N^s = L^s$ by P-invariance; the proof of $N^u = L^u$ is entirely analogous.

Let $w(x)$ be the ω-limit set of x in M. First suppose that $w(x) = \{x_o\}$, i.e., $x \in W^s(x_o) - \{x_o\}$ with x_o a hyperbolic saddle singularity of X. If $\dim W^s(x_o) = 2$, Proposition 2.2 implies $N_x^s = N_x \cap T_x W^s(x_o) = L_x^s$. If $\dim W^s(x_o) = 1$, Proposition 2.4 assures the existence of a splitting $E^1 \oplus E^2 = T_{x_o} W^u(x_o)$ such that $\lim_{t\to+\infty} T\varphi_t \cdot N_x^s = E^1$ and $\lim_{t\to+\infty} T\varphi_t \cdot L_x^u = E^2$; moreover, L_x^s $T\varphi$-dominates L_x^u. Now if $N_x^s \neq L_x^s$ it follows from Proposition 1.3.a) that

$$E^1 = \lim_{t\to+\infty} T\varphi_t \cdot N_x^s = \lim_{t\to+\infty} T\varphi_t \cdot L_x^u = E^2,$$

which is impossible. Thus $N_x^s = L_x^s$.

Now assume that $\# w(x) > 1$ and choose $(t_n)_n$ and $y \in w(x)$ such that $y \in R$, $t_n \to +\infty$ and $y_n = \varphi_{t_n}(x) \to y$. By compactness of $G(TM)$ we may choose $N_y^\sigma, L_y^\sigma \in G(N_y)$ such that $\lim_n N_{y_n}^\sigma = N_y^\sigma$ and $\lim_n L_{y_n}^\sigma = L_y^\sigma$, $\sigma = s,u$. From our hypothesis we know that $N_{y_n}^s$ and $L_{y_n}^s$ dominate $N_{y_n}^u$ and $L_{y_n}^u$, respectively, for all $n \in \mathbb{N}$, with constants independent of n, thus Lemma 1.2.d) implies that N_y^s dominates N_y^u and L_y^s dominates L_y^u.

Let us assume, contrary to our assertion, that $N_x^s \neq L_x^s$. Then Proposition 1.3.a) implies that

$$L_y^s = \lim_n L_{y_n}^s = \lim_n P_{t_n} \cdot L_x^s = \lim_n P_{t_n} \cdot N_x^u = \lim_n N_{y_n}^u = N_y^u .$$

If $L_x^u \neq N_x^s$, the same argument analogously implies $L_y^u = N_y^u$, hence $L_y^s \cap L_y^u = N_y^u \neq \{0\}$, which is impossible by Lemma 1.2.b). Therefore $L_x^u = N_x^s$, implying

$$L_y^u = \lim_n L_{y_n}^u = \lim_n P_{t_n} \cdot L_x^u = \lim_n P_{t_n} \cdot N_x^s = \lim_n N_{y_n}^s = N_y^s ,$$

and we arrive at $(N_y^s, N_y^u) = (L_y^u, L_y^s)$. But this is impossible by Lemma 1.2.c), since it implies that both L_y^s dominates L_y^u and L_y^u dominates L_y^s. We have run out of possibilities, thus $N_x^s = L_x^s$, proving Lemma 3.7. □

References

[1] D.V. Anosov, Geodesic flows on closed Riemannian manifolds with negative curvature, Proc. Steklov Inst. Math., 90 (1967); Amer. Math. Soc. Transl. (1969), MR 39#3527.

[2] D.V. Anosov, Existence of smooth ergodic flows on smooth manifolds, Math. USSR Izv., 8 (1974), 525-552, MR 50#11322.

[3] C.I. Doering, Vector fields without wandering points on three-dimensional manifolds (Portuguese), Thesis, IMPA (1979).

[4] J.M. Franks, Necessary conditions for stability of diffeomorphisms, Trans. Amer. Math. Soc., 158 (1971), 301-308, MR 44#1042.

[5] J. Guckenheimer, A strange, strange attractor. The Hopf bifurcation and its applications, Applied Math. Series 19, Springer Verlag, 1976, 368-381, MR 58#13209.

[6] V. Guillemin, A. Pollack, Differential Topology, Prentice-Hall, 1974, MR 50#1276.

[7] M.W. Hirsch, C.C. Pugh, M. Shub, Invariant Manifolds, Lecture Notes in Mathematics 583, Springer Verlag, 1977, MR 58#18595.

[8] S.-T. Liao, A basic property of a certain class of differential systems (Chinese), Acta Math. Sinica, 22 (1979), 316-343, MR 81c:58045.

[9] S.-T. Liao, On the Stability Conjecture, Chinese Ann. Math., 1 (1980), 9-30, MR 82c:58031.

[10] S.-T. Liao, On hyperbolicity properties of nonwandering sets of certain 3-dimensional differential systems, Acta Math. Scientia, 3 (1983), 361-368.

[11] R. Mañé, Persistent manifolds are normally hyperbolic, Trans. Amer. Math. Soc., 246 (1978), 261-283, MR 80c:58019.

[12] R. Mañé, Expansive diffeomorphisms. Dynamical Systems, Warwick, 1974, Lecture Notes in Mathematics 468, Springer Verlag 1975, 162-174, MR 58#31263.

[13] R. Mañé, Contributions to the Stability Conjecture, Topology, 17 (1978), 383-396, MR 84b:58061.

[14] R. Mañé, An ergodic closing lemma, Ann. of Math., 116 (1982), 503-540, MR84f:58070.

[15] J. Palis, W. de Melo, Geometric theory of dynamical systems: an introduction, Springer Verlag, 1982, MR 84a:58004.

[16] J. Palis, S. Smale, Structural stability theorems, Global Analysis, Proc. Symp. Pure Math. 14, Amer. Math. Soc., 1970, 223-231, MR 42#2505.

[17] V.A. Pliss, The coarseness of a sequence of linear systems of second-order differential equations with periodic coefficients, Diff. Eq., 7 (1971), 205-212, MR 44#5572.

[18] V.A. Pliss, The location of separatrices of periodic saddle-point motion of systems of second-order differential equations, Diff. Eq., 7 (1971), 906-927, MR 44#2987.

[19] V.A. Pliss, A hypothesis due to Smale, Diff. Eq., 8 (1972), 203-214, MR 45#8957.

[20] V.A. Pliss, Properties of solutions of a periodic system of two differential equations having an integral set of zero measure, Diff. Eq., 8 (1972), 421-426, MR 45#7167.

[21] C.C. Pugh, An improved closing lemma and a general density theorem, Amer. Journal Math., 89 (1967), 1010-1022, MR 37#2257.

[22] M. Shub, Topologically transitive diffeomorphisms of T^4. Proceed. Symp. Eq. Dyn. Syst. Warwick 1968/69, Lecture Notes in Mathematics, 206, Springer Verlag, 1971, 28-29, 39-40, MR 51#6004.

[23] S. Smale, Differentiable dynamical systems, Bull. Amer. Math. Soc., 73 (1967), 747-817, MR 37#3598, erratum MR 39, p.1593. With notes in: The Mathematics of time, Springer Verlag, 1980, MR 83a:01068.

Instituto de Matemática U.F.R.G.S.
Av. Bento Gonçalves, 9500
91.500 - Porto Alegre, R.S. - Brasil

I GUADALUPE[1], C GUTIERREZ[2], J SOTOMAYOR[3] & R TRIBUZY[4]

Principal lines on surfaces minimally immersed in constantly curved 4–spaces

1. Introduction

This paper is devoted to the study of the possible configurations of lines of principal curvature and umbilical points on surfaces which are minimally immersed in Riemannian four-dimensional manifolds with constant sectional curvature.

The notion of principal direction for an immersion $\alpha: M \to Q^4$, of a surface M into a four-dimensional Riemannian manifold Q^4, is a natural extension of the classical three dimensional idea. [G-S, 1,2,3;Sp]. Following Little [Li], we will say that a principal direction at $p \in M$ is a line in T_pM, generated by a unitary vector which makes extremal the length of $B(X,X)$, where B is the second fundamental form of the immersion α at p, and X varies on the unitary circle in T_pM. The set $\mathcal{E}$ of values of $B(X,X)$ is an ellipse called ellipse of curvature, which can degenerate into a line segment, a circle or a point. Also, we may easily see that as X goes once around the circle $B(X,X)$ goes twice around $\mathcal{E}$. Therefore, when $\mathcal{E}$ is either an ellipse or a line segment, there are four principal directions at p. When $\mathcal{S}$ is either a circle or a point. p will be said to be a semi-umbilical point of α. The principal lines of curvature of α are those curves in M, disjoint from semi-umbilical points, which are tangent to principal lines. A point p at which $\mathcal{S}$ degenerates to a point will be called an umbilical point of α.

Assume now that α is a minimal immersion and that $Q^4 = Q^4_c$ has constant sectional curvature, equal to $c \in R$. The semi-umbilical points in this case are either isolated or form a connected component of M. [Hp, 1,2; Chr; Hff]. In this work we will give a precise analytical model for the family of principal lines near an isolated semi-umbilical point. This model is related to a result in [Hp, 1; Chr; Li] that permits the computation of the index of the semi-umbilical point. We will prove that the index determines the local configuration of principal lines around semi-umbilical points.

The behavior of principal lines at infinity, near particular types of ends called here elementary ends of the immersion α, will also be described through precise analytical models.

Locally, away from semi-umbilical points, the principal lines determine four foliations.

Around semi-umbilical points the principal lines may not determine foliations but nets. Nevertheless, to study the global behavior of the principal lines, we consider two natural double coverings of the complement of the semi-umbilical points in a connected component of the surface. Each covering surface is endowed with a foliation, with a natural smooth transversal invariant measure which is carried by the covering map onto the family of pairs of principal lines which pass through every point in M.

As in the codimension one case [G-S,3] the presence of the invariant measure implies that the closures of recurrent principal lines (that is lines contained in their own limit sets) have non-empty interior. It also implies that the compact principal lines (cycles) appear forming cylindrical open sets. Examples illustrating all possible situations studied in this work will be given.

This paper is organized as follows. Sections 2, 3 and 4 are devoted to the study of quartic differentials on arbitrary Riemann surfaces. In Section 2 we study the local structure of certain important curves associated with them which we call maximal and minimal principal lines. In Section 3 we consider the principal lines around the zeroes and poles of a quartic differential. In Section 4 we establish the possible global configurations of the families of principal lines. In Section 5 we show that the principal lines of curvature of surfaces minimally immersed into Riemannian 4-manifolds with constant sectional curvature are precisely the principal lines of a well determined quartic differential on M, globally associated to the minimal immersion. This differential was defined in [Chr; Hff]. Sections 6 and 7 are devoted to the construction of examples of minimal surfaces in all the situations studied in this work.

2. Quartic differentials.

We shall study quartic differentials and the local structure of certain important curves associated with them. In this section, we shall follow the presentation of Jenkins [Je, 3] on quadratic differentials.

(2.1) Definition. Let $\mathcal{R}$ be an oriented Riemann surface. By a quartic differential defined on $\mathcal{R}$ we mean an entity Ω which assigns to every local uniformizing parameter z of $\mathcal{R}$ an expression $\Omega(z) = Q(z)dz^4$ where $Q(z)$ is a meromorphic function which satisfies the following condition: If z^* is a second local uniformizing parameter of $\mathcal{R}$ and $Q^*(z^*)$ is the corresponding meromorphic function associated with z^* and if the neighborhoods on $\mathcal{R}$ for z^* and for z overlap, then at common points of these neighborhoods, we have

$$Q^*(z^*) = Q(z)\left(\frac{dz}{dz^*}\right)^4.$$

Sometimes, quartic differentials will be denoted generically by symbols such as $Q(z)dz^4$. A point $p \in \mathcal{R}$ is called a zero or pole of order k of $Q(z)dz^4$ if for some (and therefore for every) local uniformizing parameter z it is represented by a point having this property relative to $Q(z)$.

(2.2) Definition. Let $\mathcal{R}$ be a Riemann surface and $Q(z)dz^4$ be a quartic differential given on $\mathcal{R}$. A maximal regular curve on which $Q(z)dz^4 > 0$ is called a maximal principal line of $Q(z)dz^4$. A maximal regular curve on which $Q(z)dz^4 < 0$ is called a minimal principal line of $Q(z)dz^4$.

It can be seen at once that maximal and minimal principal lines are intrinsically associated with the quartic differential, i.e., independently of the choice of local uniformizing parameters.

(2.3) Nets and foliations induced by a quartic differential:

Let $\mathcal{R}$ be a Riemann surface, Ω be a quartic differential on $\mathcal{R}$ and C be the set of zeros and poles of Ω. The maximal (resp. minimal) principal lines of Ω determine the maximal (resp. minimal) net $\mathfrak{F}$ (resp. f) on $\mathcal{R}$. By this we mean that any $p \in \mathcal{R}-C$ has a neighborhood V such that $\mathfrak{F}$ (resp. f) restricted to V is made up of two pairwise transversal foliations. In general, $\mathfrak{F}$ (resp. f) is not globally the union of two transversal foliations because global maximal (resp. minimal) principal lines may self intersect transversally, as will be seen to happen in the neighborhood of the points in C.

Let $\mathcal{R}_M$ be the set of pairs (p,L) such that $p \in \mathcal{R}$ and L is a one-dimensional subspace of $T_p\mathcal{R}$ tangent to a maximal principal line of Ω, where $T_p\mathcal{R}$ is the tangent space of $\mathcal{R}$ at p. Under these conditions, there is a unique Riemann surface structure on $\mathcal{R}_M$ such that:

a) $\pi_1: \mathcal{R}_M \to \mathcal{R} - C$, taking (p,L) to p, is a holomorphic double covering of $\mathcal{R} - C$.

b) $\pi_2: \mathcal{R}_M \to G^1(\mathcal{R}-C)$, taking (p,L) to L, is continuous, where $G^1(\mathcal{R}-C)$ is the one-dimensional Grasmanian bundle over $\mathcal{R} - C$.

The maximal net $\mathfrak{F}$ induces via $\pi_1: \mathcal{R}_M \to \mathcal{R} - C$, a foliation $\mathfrak{F}_M$ on $\mathcal{R}_M$. Similarly the minimal net f can be used to construct a double covering $\mathcal{R}_M$ of $\mathcal{R} - C$ and a foliation f_m on $\mathcal{R}_m$.

(2.4) <u>Transversal measures on</u> $\mathcal{R}_M$ <u>and</u> $\mathcal{R}_m$: A precise definition of transversal measures and measured foliations can be found in [H-M, Chap. I].

Let $\mathcal{R}$ be a Riemann surface, $\Omega = \Omega(z)dz^4$ be a quartic differential on $\mathcal{R}$ and C be the set of zeros and poles of $Q(z)dz^4$. Then, on $\mathcal{R} - C$,

$$\zeta = \int Q(z)^{1/4}\, dz$$

will be defined locally as a regular holomorphic function independently of the choice of local uniformizing parameters. In general, this function will not be single-valued in the large.

Any two determinations ζ_1 and ζ_2 of ζ in the neighborhood of a point of $\mathcal{R} - C$ are related by an equation of the form

(*) $$\zeta_1 = \pm i\zeta_2 + \delta,$$

where δ is a constant depending on the particular determinations.

If $\zeta(z) = \int_{z_0}^{z} Q(z)^{1/4}\,dz$, then in a sufficiently small simply connected neighborhood U of z_0, ζ is also a local uniformizing parameter of $\mathcal{R}$ and $\Omega(\zeta) = d\zeta^4$. In particular, if $\zeta(z) = u+iv$, where u and v are real-valued functions, then in the ζ-parameter any maximal principal line is tangent either to the kernel of (the 1-form) du or to the kernel of dv. Therefore du and dv induce a transversal measure for $\mathcal{F}_M$ restricted to $\pi_1^{-1}(U) \subset \mathcal{R}_M$, where π_1 is as in (2.3). It follows from relation (*) that this measure transversal to $\mathcal{F}_M$ does not depend on the local parameter and so it is globally defined on $\mathcal{R}_M$. Summarizing $\mathcal{F}_M$ is a measured foliation.

3. Zeroes and poles of a quartic differential

A minor modification in the proof of [G-S, 3; Prop. 4.2 and 4.3] gives the following two results.

(3.1) Proposition. Let $\mathcal{R}$ be an oriented Riemann surface. Let $p \in \mathcal{R}$ be either a zero of arbitrary order n or a pole or order $n \le 3$ of a quartic differential ω on $\mathcal{R}$. There exists a local uniformizing parameter z of $\mathcal{R}$, around p, such that

a) z takes p to 0

b) $\omega(z) = z^n dz^4$.

(3.2) Proposition. Let $\mathcal{R}$ be an oriented Riemann surface. Let $p \in \mathcal{R}$ be a pole of order $n \ge 4$ of a quartic differential $\omega = \phi(z)dz^4$ on $\mathcal{R}$. If $n-4$ is not a positive multiple of 4, there exists a local uniformizing parameter z of $\mathcal{R}$ around p such that

a) z takes p to 0

b) $\omega(z) = az^{-n}dz^4$, where $a = \lim z^4\varphi(z)$, $z \to 0$, if $n = 4$ and $a = 1$ otherwise.

Proposition (3.2) canot be extended to the case in which p is a pole of arbitrary order n. This is due to the appearence of resonances in the application of Poincaré-Dulac Theorem [Ar.,Ch.5]. This situation was also encountered in [G-S, 3, Prop. 4.5]. Nevertheless, the same argument of [Je, 3, Th. 3.3] proves the following.

(3.3) <u>Proposition</u>. Let $\mathcal{R}$ be an oriented Riemann surface. Let $p \in \mathcal{R}$ be a pole of order $n > 5$ of a quartic differential ω on $\mathcal{R}$. There exist neighborhoods $V \subset \mathcal{R}$, of p, and $U \subset \mathbb{R}^2$, of 0, and a homeomorphism $h: (V,p) \to (U,0)$ taking the maximal principal lines of $\omega|_V$ onto the maximal principal lines of $z^{-n}dz^4$, restricted to U.

(3.4) <u>Corollary</u>. Under the hypothesis of (3.1), there are exactly $n+4$ maximal principal lines $S_0, S_1, \ldots, S_{n+3}$ approaching p.

When $n \neq -3$, two lines S_i, S_{i+2}, $i = 0,1,2,\ldots,n+3$ ($S_{n+4} = S_0$ and $S_{n+5} = S_1$) bound an open sector $\Delta(i)$ on which the maximal principal lines determine two foliations. One of these, say $\mathfrak{F}_i$ is a hyperbolic sector having S_i and S_{i+2} as separatrices. The separatrix S_{i+1} is contained in $\Delta(i)$ and is transversal to $\mathfrak{F}_i$. See fig. 3.1. The union $\bigcup_{i=1}^{n+3} \Delta(i)$ and the foliations $\mathfrak{F}_1, \mathfrak{F}_2, \ldots, \mathfrak{F}_{n+3}$ determine the maximal net around p. See fig. 3.2 for an illustration.

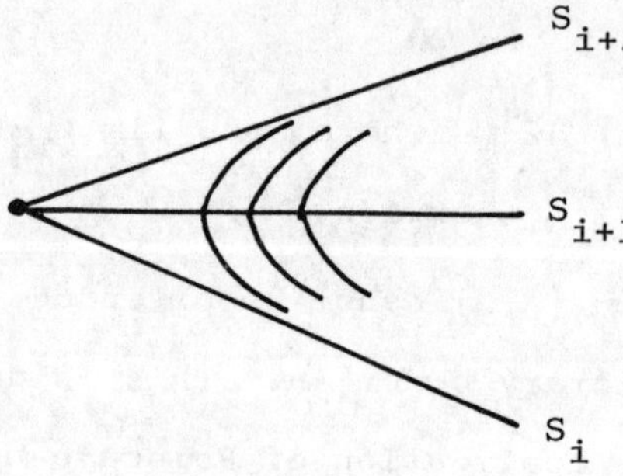

Figure 3.1

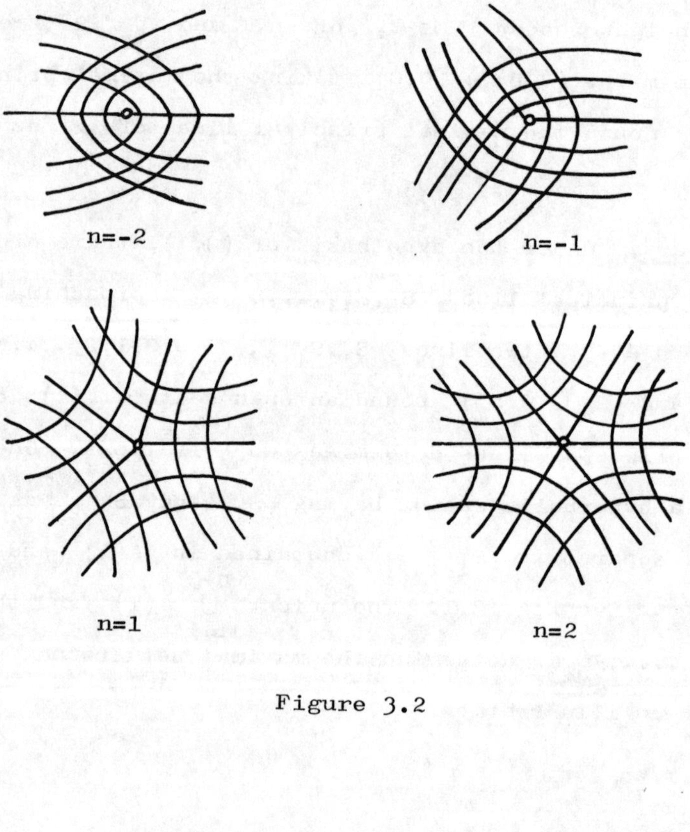

Figure 3.2

When $\mathcal{R}$ has a metric and the local uniformizing parameter z is conformal, there are $n+4$ rays $L_0, L_1, \ldots, L_{n+3}$, through $0 \in T_p\mathcal{R}$ two consecutive of which make an angle of $\frac{2\pi}{n+4}$. The separatrix S_i is tangent at p to each ray L_i.

When $n = -3$ and we are considering z-coordinates, with $z = x+iy$, the line S_0 is precisely the positive x half-axis. Also, any maximal principal line different from S_0 makes a loop around $z(p) = 0$ intersecting S_0 exactly once and intersecting itself exactly once. See figs. 3.3 and 3.4. The maximal net (around $z(p) = 0$) is invariant under a reflexion relative to the x-axis. It determines two pairwise orthogonal foliations on the complement of S_0.

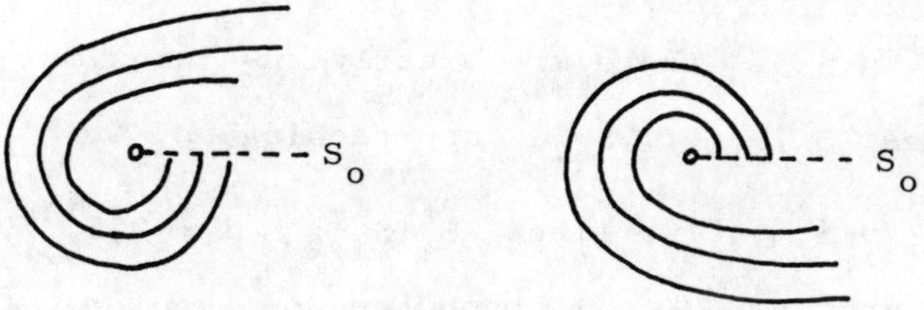

Figure 3.3

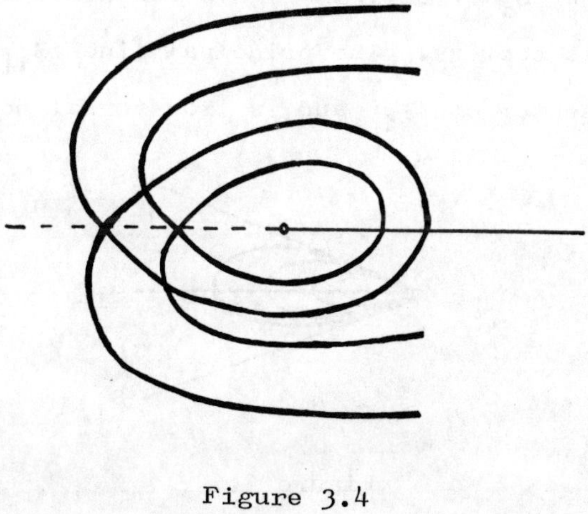

Figure 3.4

(3.5) Corollary. Under the hypotheses of (3.2) and (3.3), the net determined by the maximal principal lines is described as follows:

i) Let $n=4$. In case $a > 0$, the net $\mathfrak{F}$ is made up of circles centered at 0 and rays tending to p. In case $a \in \mathbb{C}-[0,\infty)$, the net $\mathfrak{F}$ is made up of two foliations. On each of these foliations the lines approach the origin. See fig. 3.5.

Figure 3.5

ii) For $n \geq 5$, there are exactly $n-4$ maximal principal lines $S_0, S_1, \ldots, S_{n-5}$ approaching p.

When $n > 5$, two lines S_i, S_{i+2}, $i = 0,1,\ldots,n-5$ ($S_{n-4} = S_0$ and $S_{n-3} = S_1$) bound an open sector $\Delta(i)$ on which the maximal principal lines determine two foliations. One of them, say $\mathfrak{F}_i$, is the union of an elliptic sector with two parabolic sectors. The line S_i (resp. S_{i+2}) is contained in exactly one of these parabolic sectors. The principal line S_{i+1} is contained in the elliptic sector of $\mathfrak{F}_i$ and is transversal to it. See fig. 3.6.

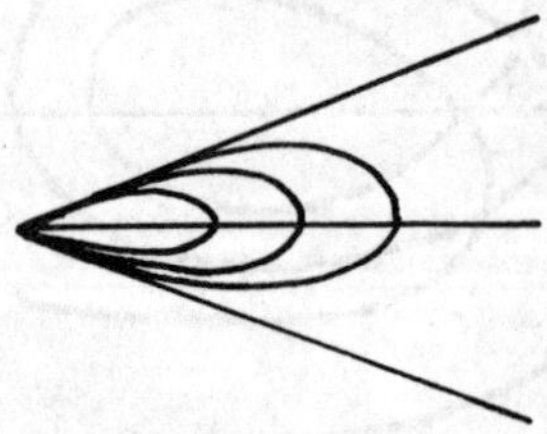

Figure 3.6

The union $\bigcup_{i=1}^{n-5} \Delta(i)$ and the foliations determine the maximal net around p. See fig. 3.7 for an illustration.

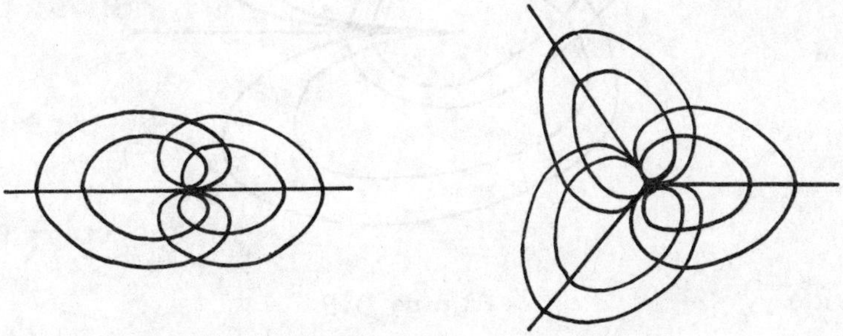

n=6 n=7

Figure 3.7

When n=5 and we are considering z-coordinates, with z = x+iy, the line S_0 is precisely the positive x half-axis. Also, any maximal principal line different from S_0 starts at p, tangentially to S_0, it makes a loop around z(p) = 0 intersecting once S_0 and intersecting itself three times before ending at p again tangentially to S_0. See fig. 3.9. The maximal net is invariant under reflection relative to the x-axis and determines two pairwise orthogonal foliations in the complement of S_0. See fig. 3.8.

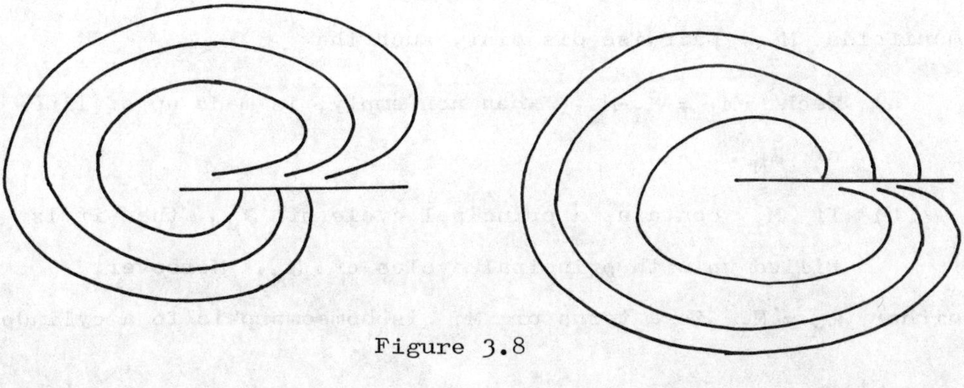

Figure 3.8

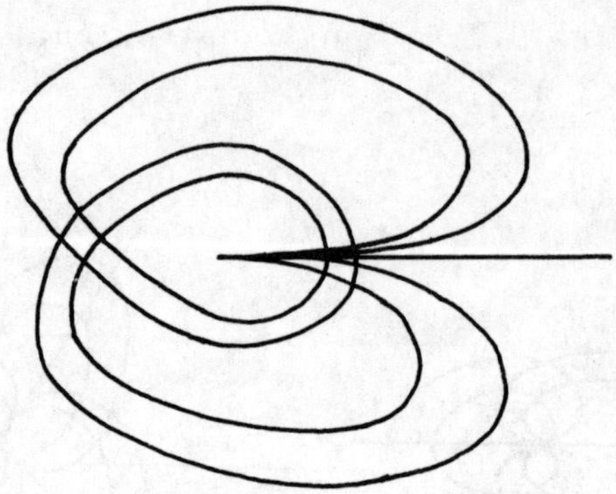

Figure 3.9

4. A Decomposition Theorem

From the local configuration of principal lines of quartic differentials around zeroes and poles and from the existence of a transversal measure to the foliation induced by the maximal net (resp. minimal net), using the same argument as in [G-S, 3, Theorem 7.1], follows:

(4.1) Theorem. Let S be a compact connected oriented two-dimensional manifold and $\{p_1,p_2,\ldots,p_n\} \subset S$. If $\mathcal{R} = S-\{p_1,p_2,\ldots,p_n\}$ is a Riemann surface and Ω is a quartic differential on $\mathcal{R}$ then, using the notation in (2.3), the Riemann surface $\mathcal{R}_M$ can be decomposed into the finite or countable union of the closure of open submanifolds M_i, pairwise disjoint, such that

a) Each $\partial M_i = \bar{M}_i - M_i$, when non empty, is made up of lines of $\mathfrak{F}_M$.

b) If M_i contains a principal cycle of $\mathfrak{F}_M$, then it is filled up with principal cycles of $\mathfrak{F}_M$. Moreover, either $\mathcal{R}_M = M_i$ is a torus or M_i is homeomorphic to a cylinder.

c) If M_i contains a non-trivial recurrent line of $\mathcal{F}_M$, then M_i is the interior of the closure of this line.

d) If M_i contains no recurrence of $\mathcal{F}_M$ then either $\mathcal{R}_M = M_i$ and the foliation $\mathcal{F}_M$ on $\mathcal{R}_M$ is topologically equivalent to the foliation on $\mathbb{R}^2-\{(0,0)\}$ determined by the integral curves of $x\frac{\partial}{\partial x} + y\frac{\partial}{\partial y}$, or $\mathcal{F}_M|M_i$ is topologically equivalent to the foliation on $\mathbb{R}^2$ determined by the integral curves of $\frac{\partial}{\partial x}$.

The surface $\mathcal{R}_M$, endowed with the foliation f_m, admits an analogous decomposition.

5. Minimal Immersions

In this section $\alpha: M \to Q_c^4$ will denote a minimal isometric immersion of the surface M into the four dimensional Riemannian manifold Q_c^4 of constant sectional curvature c. In [Ch r], S.S. Chern proved the existence of a quartic differential Ω defined globally in M and which is characterized by the following properties: If $\{x_1,x_2\}$ are local isothermal coordinates in M and $\Omega(z) = \varphi(z)dz^4$, with $z = x_1+ix_2$, then $\varphi(z)$ is holomorphic and given by

$$\varphi(z) = \langle\frac{B_{11}-B_{22}}{2}, \frac{B_{11}-B_{22}}{2}\rangle - \langle B_{12},B_{12}\rangle - 2i\langle\frac{B_{11}-B_{22}}{2}, B_{12}\rangle$$

where $\langle\ ,\ \rangle$ denotes the scalar product in the tangent bundle of Q_c^4, B is the second fundamental form of the immersion α, and $B_{ij} = B(\frac{\partial}{\partial x_i}, \frac{\partial}{\partial x_j})$, $i,j = 1,2$. We shall refer to Ω as the quartic differential associated to α. In connection with this differential, see also, [Hff]. The main result of this section is the following:

(5.1) Proposition. Let Ω be the quartic differential associated to the minimal immersion α. The zeroes of Ω are precisely the semi-umbilical points of α. Moreover, the principal lines of Ω, in the sense of Section 2, are precisely the principal lines of curvature of α.

Before proving this proposition, we shall review the concepts involved in its statement.

(5.2) <u>Some definitions</u>: For simplicity in the notation, we will identify locally M with $\alpha(M) \subset Q_c^4$.

Let T_pM, TM, N_pM and NM denote the tangent space of M at p, the tangent bundle of M, the normal space of M at p and the normal bundle of M, respectively. Let ∇ and $\bar{\nabla}$ be the covariant differentiations of M and Q_c^4, respectively. Let X and Y be on TM, then the second fundamental form B is given by

(a) $$\bar{\nabla}_X Y = \nabla_X Y + B(X,Y)$$

It is well known that $B(X,Y)$ is a symmetric bilinear form. For ξ in NM, we write

(b) $$\bar{\nabla}_X \xi = -A_\xi(X) + \nabla_X^\perp \xi$$

where $-A_\xi(X)$ and $\nabla_X^\perp \xi$ denote the tangential and normal components of $\bar{\nabla}_X \xi$, respectively. Then we have

(c) $$\langle A_\xi(X), Y\rangle = \langle B(X,Y), \xi\rangle$$

where $\langle\ ,\ \rangle$ denotes the scalar product in TM and NM.

The mean curvature vector H is defined by

(d) $$H = \frac{1}{2}\operatorname{trace} B.$$

The immersion $\alpha: M \to Q_c^4$ is said to be a minimal immersion if $H \equiv 0$.

Let R be the Riemann curvature tensor associated with ∇, defined by

(e) $$R(X,Y)Z = \nabla_X \nabla_Y Z - \nabla_Y \nabla_X - \nabla_{[X,Y]} Z$$

where X, Y, Z are on TM. We note that $\langle R(X,Y)Y,X\rangle = K|X\wedge Y|^2$ where K is the Gaussian curvature and $|X\wedge Y|^2 = \langle X,X\rangle\langle X,Y\rangle - \langle X,Y\rangle^2$. We define $R^\perp$, the curvature of NM relative to $\nabla^\perp$ by the equation

(f) $$R^\perp(X,Y)\xi = \nabla^\perp_X \nabla^\perp_Y \xi - \nabla^\perp_Y \nabla^\perp_X \xi - \nabla^\perp_{[X,Y]} \xi$$

where X, Y are on TM and ξ is on NM.

Suppose now the Q^4_c has a given orientation. Then we can define the <u>Normal Curvature</u> K_N of M by

(g) $$K_N = \langle R^\perp(X,Y)e_4, e_3\rangle$$

where $\{X,Y\}$ and $\{e_3, e_4\}$ are orthonormal oriented bases of T_pM and N_pM, respectively. Therefore $K_N > 0$ or $K_N < 0$ according to the orientation of the normal bundle NM.

Given $p \in M$, the <u>ellipse of curvature</u> $\mathcal{E}_p$ at p is the set $\{B(X,X) \in N_pM / \langle X,X\rangle = 1\}$. To see that it is an ellipse, we just have to look at the following formula, for $X = \cos\theta\, e_1 + \sin\theta\, e_2$

(h) $$B(X,X) = H + \cos 2\theta u + \sin 2\theta v$$

where $u = (B(e_1,e_1) - B(e_2,e_2))/2$, $v = B(e_1,e_2)$ and $\{e_1,e_2\}$ is tangent frame. So we see that, as X goes once around the unit tangent circle, B(X,X) goes twice around the ellipse. This ellipse can degenerate into a line segment, a circle or a point. The points where one of the last two possibilities hold are called <u>semi-umbilical points</u> of the immersion. Wherever the ellipse is not a circle or a point, we can choose the frame $\{e_1,e_2\}$, orthonormal, and so that u and v are orthogonal, and span the semi-axes of the ellipse.

(5.3) Proof of Proposition 5.1: Observing the properties characterizing Ω, we may see that Ω vanishes at p if and only if the ellipse of curvature at p is either a circle or a point. That is, the zeroes of Ω are precisely the semi-umbilical points of M.

Let $z = x_1+ix_2$ and $\Omega(z) = \varphi(z)dz^4$, where $\{x_1,x_2\}$ are local isothermal coordinates in M. Assume that $\varphi(z)dz^4$ does not vanish identically. To prove this proposition we only need to show that the differential equation of the principal lines of curvature of M in $\{x_1,x_2\}$ coordinates is given by:

(a) $$\mathrm{Im}(\varphi(z)dz^4) = 0.$$

First, we shall prove that if $w(\theta) = \cos\theta\, e_1 + \sin\theta\, e_2$ is a tangent vector, where e_1, e_2 are unitary vectors along $\frac{\partial}{\partial x_1}$, $\frac{\partial}{\partial x_2}$ then (a) is given by

(b) $$\sin(4\theta)\cdot \mathrm{Re}\,\varphi(z) + \cos(4\theta)\cdot \mathrm{Im}\,\varphi(z) = 0.$$

In fact, the vector $w(\theta)$ is tangent to a principal line of curvature of M if, and only if, θ is an isolated critical point of the function $f(\theta) = \|B(w(\theta),w(\theta))\|^2$. Since the immersion α is minimal, $B(\frac{\partial}{\partial x_1},\frac{\partial}{\partial x_1}) = -B(\frac{\partial}{\partial x_2},\frac{\partial}{\partial x_2})$ and therefore

$$\begin{aligned} B(w(\theta),w(\theta)) &= B_{11}\cos^2\theta + B_{22}\sin^2\theta + 2B_{12}\sin\theta\cos\theta \\ &= B_{11}\cos(2\theta) + B_{12}\sin(2\theta), \end{aligned}$$

where $B_{ij} = B(\frac{\partial}{\partial x_i},\frac{\partial}{\partial x_j})$, with $i,j \in \{1,2\}$.

Differentiating f we have

$$\begin{aligned} f'(\theta) &= 4\{-\cos 2\theta \sin 2\theta\|B_{11}\|^2 + (\cos^2 2\theta - \sin^2 2\theta)\langle B_{11},B_{12}\rangle \\ &\quad +\cos 2\theta \sin 2\theta\|B_{12}\|^2\} \\ &= 4\{\cos 4\theta\,\langle B_{11},B_{12}\rangle + \tfrac{1}{2}\sin 4\theta(\|B_{12}\|^2 - \|B_{11}\|^2)\}. \end{aligned}$$

Therefore $f'(\theta) = 0$ implies that

(c) $$\frac{1}{2} \sin 4\theta (\|B_{11}\|^2 - \|B_{12}\|^2) - \cos 4\theta \langle B_{11}, B_{12}\rangle = 0,$$

which is equivalent to (b).

Now, let us prove that the differential equation of principal lines is given by (a). Observe that

$$dz(w(\theta)) = \cos \theta + i \sin \theta$$

and therefore

$$\begin{aligned}(\varphi(z)(dz)^4)(w(\theta)) &= \varphi(z)(\cos \theta + i \sin \theta)^4 \\ &= \cos 4\theta \operatorname{Re} \varphi(z) - \sin 4\theta \operatorname{Im} \varphi(z) \\ &\quad + i\{\sin 4\theta \operatorname{Re} \varphi(z) + \cos 4\theta \operatorname{Im} \varphi(z)\}.\end{aligned}$$

This and (b) or (c) imply that (a) is the differential equation of principal lines. □

6. A Local Existence Theorem

In view of Proposition 5.1, concerning configurations of principal lines of curvature, the following result gives a complete answer to local questions about the existence of minimal surfaces studied in this work.

(6.1) Theorem. For all $c \in \mathbb{R}$, and for any holomorphic function $\varphi: V \to \mathbb{C}$, where V is an open neighborhood of $0 \in \mathbb{C}$, there exists a minimal isometric immersion of an open neighborhood $U \subset V$ of $0 \in \mathbb{R}^2 \cong \mathbb{C}$ into a Riemannian 4-manifold of sectional curvature c such that $\{\operatorname{Re} z, \operatorname{Im} z\}$ are isothermal coordinates in U and the quartic differential associated to the immersion is given by $\varphi(z)dz^4$.

Proof: Let $x_1 = \operatorname{Re} z$ and $x_2 = \operatorname{Im} z$. It will be proved that there is an analytic function $E = E(x_1, x_2) > 0$ defined in a disk with center $0 \in R^2$ and functions K and K_N, such that

(a) $$K = -\frac{1}{2E^3}\{E(E_{x_1x_1} + E_{x_2x_2}) - (E_{x_1}^2 + E_{x_2}^2)\}$$

(b) $$K_N = -\frac{1}{4E}\,\Delta\,\log\frac{c-K+K_N}{c-K-K_N}$$

(c) $$\frac{\|\varphi\|^2}{E^4} = (c-K+K_N)(c-K-K_N)$$

(d) $$c-K-|K_N| \geq 0,$$

where Δ denotes the Laplacian of the "flat metric". By Cauchy-Kowalewsky Theorem [Sp. Vol. 5] it follows that there are analytic functions $U_1, U_2, U_3, U_4, U_5, U_6$ which satisfy the following system of partial differential equations

(e) $$\begin{aligned} \frac{\partial U_1}{\partial x_2} &= U_2 \\ \frac{\partial U_2}{\partial x_2} &= \frac{U_3^2}{U_1} + \frac{U_2^2}{U_1} - \frac{\partial U_3}{\partial x_1} + \frac{|\varphi|(U_4+1)}{U_4^{1/2}} - 2cU_1^2 \\ \frac{\partial U_3}{\partial x_2} &= \frac{\partial U_2}{\partial x_1} \\ \frac{\partial U_4}{\partial x_2} &= U_5 \\ \frac{\partial U_5}{\partial x_2} &= \frac{U_6^2}{U_4} + \frac{U_5^2}{U_4} - \frac{\partial U_6}{\partial x_1} + \frac{2|\varphi|U_4^{1/2}(1-U_4)}{U_1} \\ \frac{\partial U_6}{\partial x_2} &= \frac{\partial U_5}{\partial x_1} \end{aligned}$$

with the initial conditions

$$U_1(x_1,0) = U_2(x_1,0) = 1, \quad U_3(x_1,0) = 0$$

$$U_4(x_1,0) = U_5(x_1,0) = 1, \quad U_6(x_1,0) = 0.$$

Take $E = U_1$. It follows that $U_2 = E_{x_2}$. Also a little computation shows that $U_3 = E_{x_1}$ (see [G-S, 3, Prop. 6.1,c]). Similarly by taking $F = U_4$, we have that $U_5 = F_{x_2}$ and $U_6 = F_{x_1}$. Under these conditions we define the functions K and K_N by the equations

$$K = c - \frac{|\varphi|(F+1)}{2E^2F^{1/2}} \tag{f}$$

$$K_N = \frac{|\varphi|(F-1)}{2E^2F^{1/2}}. \tag{g}$$

We can see that the second equation of (e) is equivalent to equation (a). The fifth equation of (e) is equivalent to equation (6). From (f) and (g) follows (c) and (d).

Observe that taking

$$X = c-K+K_N \quad \text{and} \quad Y = c-K-K_N \tag{h}$$

the equations (a) and (b) are equivalent to

$$\text{(i)} \qquad \begin{aligned} \Delta \log \sqrt{X} &= 2K - K_N \\ \Delta \log \sqrt{Y} &= 2K + K_N, \end{aligned}$$

whenever X and Y do not vanish. Moreover $X = \frac{|\varphi|F^{1/2}}{E^2}$ and $Y = \frac{|\varphi|}{E^2F^{1/2}}$. So the absolute value of X and Y are equal to the absolute value of a holomorphic function times a never vanishing function. Such a function we call absolute value type function. In [E-T], Eschenburg and Tribuzy proved that if a function K_N is defined over a Riemann surface $\mathfrak{R}$ such that X and Y, defined by (h), are nonnegative of absolute value type and satisfy (i) too, there exists a one parameter family of local isometric minimal immersions of $\mathfrak{R}$ in Q_c^4 such that the normal curvature is K_N. For each immersion of this family the associated quartic differential is given by a φ which is a holomorphic function whose absolute value is

$$\|\varphi\| = \sqrt{XY}\, E^2.$$

Moreover it is also proved in [E-T] that for one of these immersions, the holomorphic function is the one given above. □

7. Elementary Ends and Non-Trivial Recurrence

In this section we shall construct minimal isometric immersions of surfaces into $\mathbb{R}^4$ which have umbilical points and elementary ends with arbitrary index. Also, we shall show the existence of non-trivial recurrent lines of curvature for this sort of immersions. Recall that a recurrent line is one which is contained in its limit set. When this line is not a cycle, it is called a non-trivial recurrent line. The most common examples of these type of lines arise in irrational flows on the torus [P-M]. Examples of non-trivial recurrent principal lines have been given in [G-S, 2,3].

An end of the manifold M, defined by the system of open sets

$$U_j = \{0 < u^2+v^2 < \frac{1}{j}, \quad j \in \mathbb{N}\},$$

where (u,v) are isothermic coordinates for the immersion $\alpha: M \to Q_c^4$, is said to be an elementary end of α if the associated quartic differential extends (meromorphically) to $(u,v) = (0,0)$, but M cannot be extended to $(u,v) = (0,0)$. This elementary end is said to be complete if the distance to $(0,0)$ is infinite from any point in the punctured disk U_1.

To construct elementary ends, whether complete or not, we shall use the following partial version of a more general result due to D.A. Hoffman and R. Osserman [H-O, 2].

(7.1) Proposition. Let D be an open connected set of $\mathbb{R}^2$. Let ψ, f and g be holomorphic functions on D. Assume that for each loop $\gamma \subset D$, $\mathrm{Re}[\int_\gamma \varphi_k(z)dz] = 0$, $k = 1,2,3,4$, where

$$(\varphi_1,\varphi_2,\varphi_3,\varphi_4) = \frac{\psi}{2}\,(1+fg,\ i(1-fg),\ f-g,\ -i(f+g)).$$

For a point $z_0 \in D$, define $\alpha_k(z) = \mathrm{Re}[\int_{z_0}^{z} \varphi_k(w)dw]$, $k = 1,2,3,4$. Then $\alpha = (\alpha_1,\alpha_2,\alpha_3,\alpha_4)$ is a minimal isometric immersion into $\mathbb{R}^4$ and $\{\mathrm{Re}\ z, \mathrm{Im}\ z\}$ are isothermic coordinates for α. Moreover,

$$E_\alpha = \langle \frac{\partial\alpha}{\partial x_1}, \frac{\partial\alpha}{\partial x_1}\rangle = \langle \frac{\partial\alpha}{\partial x_2}, \frac{\partial\alpha}{\partial x_2}\rangle$$

is given by

$$\frac{|\psi|^2}{4}\,(1+|f|^2)(1+|g|^2).$$

(7.2) <u>Corollary</u>. Let α be given by (7.1). Then the associated quartic differential in z-coordinates is given by

$$\varphi(z)dz^4 = \frac{c}{4}\,\psi^2\ f'(z)g'(z)dz^4,$$

where $c \in \{-1,1\}$ is determined by the orientation of the immersed surface $\alpha(D)$.

<u>Proof</u>: If $B_{ij} = B(\frac{\partial}{\partial x_i}, \frac{\partial}{\partial x_j})$, with $i,j \in \{1,2\}$, then by the definition of B,

$$cE_\alpha B_{ij} = E_\alpha \frac{\partial^2\alpha}{\partial x_i \partial x_j} - \langle \frac{\partial^2\alpha}{\partial x_i\partial x_j}, \frac{\partial\alpha}{\partial x_1}\rangle \frac{\partial\alpha}{\partial x_1} - \langle \frac{\partial^2\alpha}{\partial x_i\partial x_j}, \frac{\partial\alpha}{\partial x_2}\rangle \frac{\partial\alpha}{\partial x_2}$$

where $c \in \{-1,1\}$. Since,

$$\frac{\partial\alpha}{\partial x_1} = (\mathrm{Re}\ \varphi_1,\ \mathrm{Re}\ \varphi_2,\ \mathrm{Re}\ \varphi_3,\ \mathrm{Re}\ \varphi_4),$$

$$\frac{\partial\alpha}{\partial x_2} = (-\mathrm{Im}\ \varphi_1,\ -\mathrm{Im}\ \varphi_2,\ -\mathrm{Im}\ \varphi_3,\ -\mathrm{Im}\ \varphi_4),$$

$$\frac{\partial^2\alpha}{\partial x_1\partial x_1} = (\mathrm{Re}(\varphi_1'),\ \mathrm{Re}(\varphi_2'),\ \mathrm{Re}(\varphi_3'),\ \mathrm{Re}(\varphi_4')),$$

$$\frac{\partial^2\alpha}{\partial x_1\partial x_2} = (-\mathrm{Im}(\varphi_1'),\ -\mathrm{Im}(\varphi_2'),\ -\mathrm{Im}(\varphi_3'),\ -\mathrm{Im}(\varphi_4')),$$

$$\frac{\partial^2 \alpha}{\partial x_2 \partial x_2} = (-\mathrm{Re}(\varphi_1'), -\mathrm{Re}(\varphi_2'), -\mathrm{Re}(\varphi_3'), -\mathrm{Re}(\varphi_4'))$$

and $\sum_{i=1}^{4} (\varphi_i)^2 = 0.$

We have that, by definition of φ,

$$E_\alpha^2 \varphi = cE_\alpha^2 \left(\sum_{i=1}^{4} (\varphi_i')^2 \right).$$

Therefore

$$\varphi = \frac{c}{4} \psi^2 f' g'. \qquad \square$$

(7.3) <u>Example</u>: For every integer n there is a minimal immersion with a complete elementary end of index $-\frac{n}{4}$.

In fact, take $\psi \equiv 1$, $f(z) = -4 \frac{z^{-m}}{m}$, $g(z) = \frac{e^{\lambda z^k}}{k}$ and $z_0 = 1$, where $\lambda \in \mathbb{C}-\{0\}$ is a constant and $k \geq 2$, $m \geq 2$, $m \geq 2$ are integers such that $n = k-m-2$. By 7.2, $\varphi(z) = \lambda z^n e^{\lambda z^k}$. By Section 3, the index and the local configuration is determined by λz^n. See in Fig. 7.1 and end of index 0.

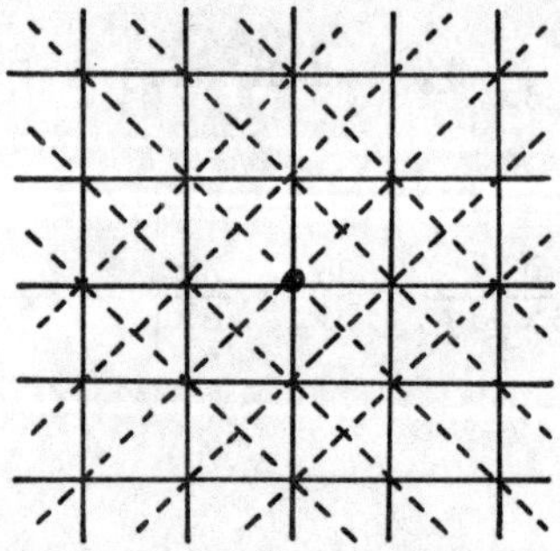

Figure 7.1

(7.4) <u>Example</u>: For every integer n there is a minimal immersion with a non-complete elementary end of index $-\frac{n}{4}$.

In fact, take $\psi \equiv 1$, $f(z) = \frac{z^k}{m} e^{\lambda z^{-m}}$, $g(z) = -4 \frac{e^{-\lambda z^{-m}}}{m}$ and $z_0 = 1$ where $\lambda \in \mathbb{C}-\{0\}$ is a constant and k, m are positive integers such that $n = k-2m-2$. We may apply Proposition 7.1, because

$$\int_{|z|=1} z^k e^{\lambda z^{-k}} dz = i\int_0^{2\pi} e^{\lambda e^{-im\theta}} e^{i(k+1)\theta}\, d\theta$$

$$= -i\int_0^{-2\pi} e^{\lambda e^{-m\theta}} e^{-i(k+1)\theta}\, d\theta$$

$$= \int_{|z|=1} e^{\lambda z} z^{-k-2}\, dz = 0.$$

By 7.2, $\varphi(z) = \lambda^2 z^n \left(1 - \frac{k}{\lambda^m} z^m\right)$. Using Proposition 7.1, we may check that these ends fail to be complete along the curves $\gamma(t) = t(\lambda/i)^{1/m}$.

(7.5) Example: For every positive integer n there is a minimal immersion with an umbilical point of index $-\frac{n}{4}$.

In fact, take $\psi \equiv 1$, $f(z) = 4z$, $g(z) = \frac{z^{n+1}}{n+1}$, $z_0 = 0$. Then by 7.2, $\varphi(z) = cz^n$.

The results about rotation number used in the next proposition can be found in [P-M].

(7.6) Proposition. There exists a minimal isometric immersion $\alpha: M \to \mathbb{R}^4$ of a complete surface M homeomorphic to a four times punctured sphere and such that

1) there are principal lines of curvature of α which are dense in M.

2) The immersion α has no semi-umbilical points.

Proof: Let $b \in (-1/4, 1/4)$ and $D_b = C - \{1, -1, -2, 2-ib\}$. For $z \in D_b$ we define $g_b(z) = f_b(z) = \int_0^z P_b(w)dw$ and $\psi_b(z) = 2(P_b(z))^{-2}$ where $P_b(z) = (z-1)(z+1)(z+2)(z-2+ib)$. We may check that, if we take $(\psi,f,g,D) = (\psi_b,f_b,g_b,D_b)$, Proposition 7.1 can be used to construct a minimal isometric immersion $\alpha = \alpha_b$. By Corollary 7.2 the associated quartic differential of α_b, in z-coordinates, is given by

$$\varphi_b(z)dz^4 = (P_b(z))^{-2}\, dz^4.$$

Therefore, the integral curves of the differential equation

$$\mathrm{Im}((P_b(z))^{-1}dz^2) = 0$$

determine two foliations $\mathfrak{F}_b$ and f_b which are part of the net determined by the principal lines of $\varphi_b(z)dz^4 = (P_b(z))^{-2}dz^4$.

We shall prove that, for $b \neq 0$ small enough

(1) Either $\mathfrak{F}_b$ or f_b has non-trivial recurrent principal lines.

In order to prove (1) we shall first notice the following facts:

(2.1) $\alpha_b\colon D_b \to \mathbb{R}^4$ extends to an isometric minimal immersion $\tilde{\alpha}_b\colon D_b \cup \{\infty\} \to \mathbb{R}^4$ of the four times punctured sphere $D_b \cup \{\infty\}$ (which is part of the Riemann sphere $C \cup \{\infty\}$).

(2.2) The immersion $\tilde{\alpha}_b$ has four complete elementary ends of index $\frac{1}{2}$.

(2.3) The foliations $\mathfrak{F}_b$ and f_b extend to the foliations $\tilde{\mathfrak{F}}_b$ and $\tilde{f}_b$ on $C \cup \{\infty\}$ whose set of singularities is $\{1, -1, -2, 2-ib\}$.

The proof of (2.1) follows at once by taking coordinates $w = \frac{1}{z}$ around $\{\infty\}$. The proof of (2.2) and (2.3) are immediate. The phase portrait of the lines of principal curvature of $\tilde{\alpha}_b$ around $\{\infty\}$ is that of fig. 7.1.

Observe that the interval $(-1,1)$ of the real axis is an integral curve of the differential equation $\mathrm{Im}(P_0(z)^{-1}dz^2) = 0$.

Let us assume that

(3) $\mathcal{F}_0$ denotes the foliation for which the inverval (-1,1) is a principal line.

We may easily check that

(4) If $b \neq 0$, away from the set $\{1, -1, -2, 2-ib\}$, the real axis is transversal to both $\mathcal{F}_b$ and f_b.

It follows from Theorem 4.1 that the phase portraits of $\mathcal{F}_0$ and f_0, put together, are as in fig. 7.2. The principal lines of f_0 are drawn as continuous curves. Most of the principal lines of $\mathcal{F}_0$ and f_0 are cycles.

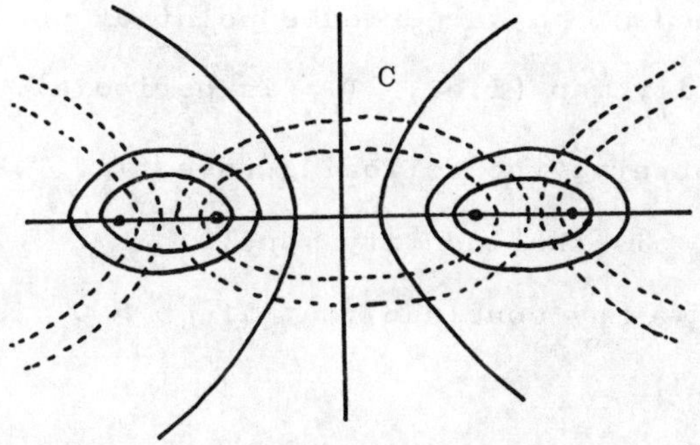

Figure 7.2

Let C be a circle which is a trajectory of $\mathcal{F}_0$ and which encloses an open disk $U \cap \{-1,1,-2,2\} = \{-1,1\}$. As C is transversal to f_0, it is also transversal to f_b for b small enough.

Given $p \in C$ and b small, let $\gamma^+_{p,b}$ be oriented half-principal line of f_b which starts at p and then enters U. Denote by $T_b(p)$ the point of C which is the second time, after p, that $\gamma^+_{p,b}$ meets C. See fig. 7.3. In this way we have established a map $T_b: C \to C$ defined everywhere except at finitely many points (T_b is not defined at the point $\tilde{p}$ of fig. 7.3). It is easy to see that T_b extends continuously to the whole C as an orientation preserving homeomorphism.

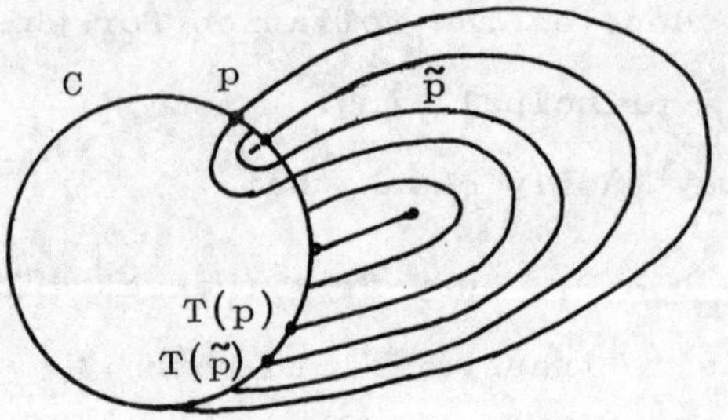

Figure 7.3

It follows from Theorem 4.1 that

(5) If $T_b: C \to C$ has a periodic point of period n, T_b^n is the identity map (i.e., T_b is periodic).

Also, we observe the following (see fig. 7.2).

(6) $T_0: C \to C$ is the identity map.

Using (4) we may conclude that if $\tilde{b} > 0$ is small enough, then (see fig. 7.4)

(7) $T_{\tilde{b}}: C \to C$ is not the identity map.

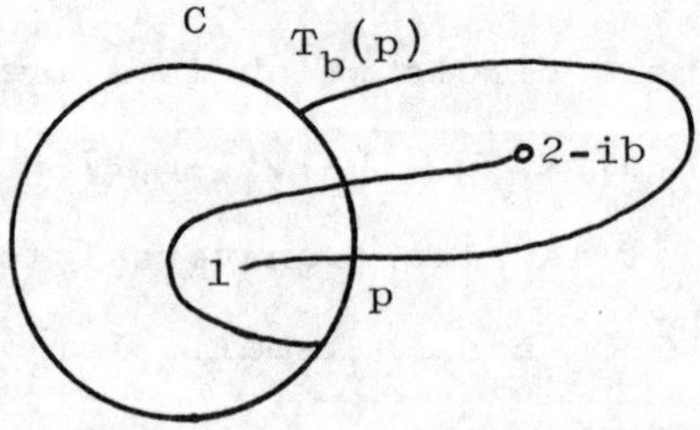

Figure 7.4

As the rotation number of T_b depends continuously on b, it follows from (6) and (7) that there exists $b' \in [0,b]$ such that the rotation number of $T_{b'}$ is irrational. This implies that $T_{b'}$ has no periodic points, that is, $T_{b'}$ has non-trivial recurrent points and therefore the foliation $_{b'}$, has non-trivial recurrent principal lines. This proves (1). It follows from Theorem 4.1 that all principal lines of $_{b'}$ must be dense. We observe that, as $T_{b'}$ is not periodic, $_{b'}$ cannot have a principal line connecting two of its singularities. See fig. 7.5. The immersion $\alpha = \tilde{\alpha}_{b'}$ satisfies the conditions required in this proposition. □

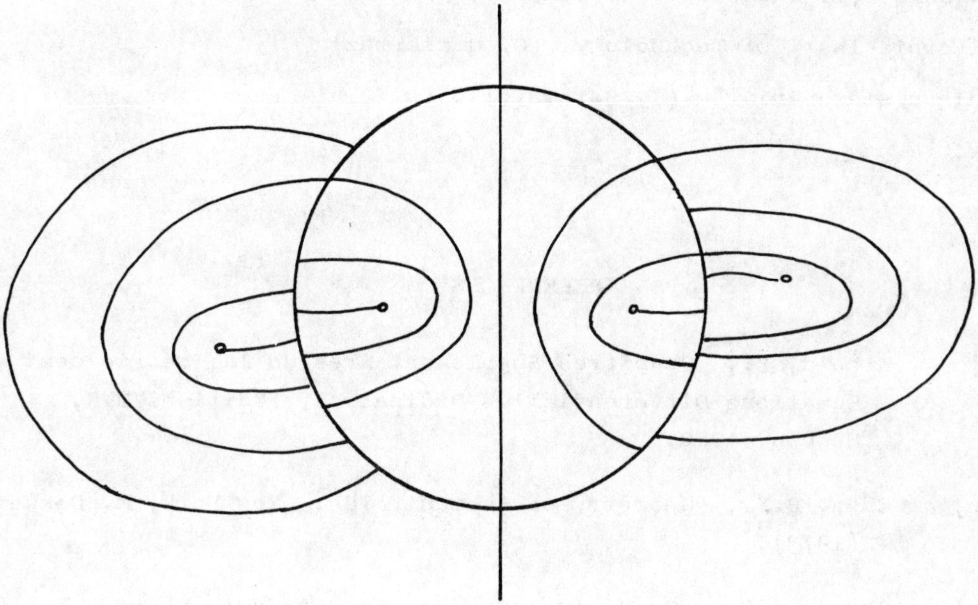

Figure 7.5

(7.7) Remark. The immersions α_b of (7.6), whose images are actually contained in $\mathbb{R}^3 = \{x_3=0\}$, provide examples of minimal immersions with non-trivial recurrent principal lines. The existence of such immersions was left open in [G-S 3; Remark 8.1].

8. Acknowledgement

The authors are grateful to the following institutions for the partial support received during the preparation of this work:

IMPA-CNPq (I. Guadalupe, R. Tribuzy)

California Inst. of Technology (C. Gutierrez)

Michigan State Univ. (J. Sotomayor).

REFERENCES

[Ar] Arnold, V., Chapitres Supplementaires de la Théorie des Équations Différentielles Ordinaires, Éditions MIR, Moscou, 1980.

[Chn] Chen, B.Y., Geometry of Submanifolds, New York, M. Dekker (1973).

[Chr] Chern, S.S., On Minimal Spheres in the four-sphere, In Studies and Essays. Presented to Y.W. Chen, Taiwan (1970), 137-150.

[E-T] Eschenburg, J. and Tribuzy, R., A Note on Surfaces with Parallel Mean Curvature Vector in Constantly Curved 4-Space. Preprint, Univ. of Munster.

[G-S.1] Gutierrez, C., Sotomayor, J., Structurally Stable Configurations of Lines of Principal Curvature, Astérisque 98-99, pp. 195-215, 1982.

[G-S.2] Gutierrez, C., Sotomayor, J., An Approximation Theorem for Immersions with Stable Configurations of Lines of Principal Curvature. Lecture Notes in Mathematics 1007, Geometric Dynamics. Proc. Rio de Janeiro 1981, pp. 332-368, Springer-Verlag.

[G-S.3] Gutierrez, C., Sotomayor, J., Principal Lines on Surfaces Immersed with Constant Mean Curvature. Trans. Amer. Math. Soc. 239, pp. 751-766, 1986.

[Hff] Hoffman, D.A., Surfaces of Constant Mean Curvature in Manifolds of Constant Curvature. Jour. Diff. Geom. 8 (1973), pp. 161-176.

[H-O,1] Hoffman, D.A., Osserman, R., The Gauss Map of Surfaces in $\mathbb{R}^n$. Jour. Diff. Geom. 18 (1983), pp. 733-754.

[H-O,2] Hoffman, D.A., Osserman, R., The Geometry of the Generalized Gauss Map. Memoirs of the Amer. Math. Soc. November 1980, Vol. 28 Number 236 (first of 2 numbers).

[Hp.1] Hopf, H., Lectures on Differential Geometry in the Large, Notes by J.M. Gray, Standord University, 1954. Reprinted in Springer-Verlag Lecture Notes in Math. 1000.

[Hp.2] Hopf, H., Über Flächen mit einer Relation zwischen den Hauptkrümungen, Math. Nachrichten, 4B, 1950/51.

[H-M] Hubbard, J., Masur, H., Quadratic Differential and Foliations, Acta Mathematica 142, pp. 221-274, 1979.

[Je 1] Jenkins, J.A., On the Local Structure of the Trajectories of a Quadratic Differential, Proc. Amer. Math. Soc. 5, 357-362 (1954).

[Je 2] Jenkins, J.A., A General Coefficient Theorem, Trans. Amer. Math. Soc. 77, 262-280 (1954).

[Je 3] Jenkins, J.A., Univalent Functions and Conformal Mappings, Springer-Verlag, Berlin, 1958.

[Li] Little, J.A., On Singularities of Submanifolds of a Higher Dimensional Euclidean Space, Ann. Mat. Pura Appl. 83 (1969), pp. 261-335.

[Oss] Osserman, R., A Survey of Minimal Surfaces, New York, Van Nostrand Reinhold Co., 1969.

[P-M] Palis, J., de Melo, W., Geometric Theory of Dynamical Systems, Springer-Verlag, New York, 1982.

[Sp] Spivak, M., A Comprehensive Introduction to Differential Geometry, Publish or Perish, Inc., Berkeley, 1979.

[St] Strebel, K., On Quadratic Differentials and Extremal Quasiconformal Mappings. Lecture Notes, University of Minnesota, 1967.

[T-G] Tribuzy, R., Guadalupe, J.V., Minimal Immersions of Surfaces into 4-dimensional Space Forms, Rend. Sem. Mat. Univ. Padova, Vol. 73 (1985), pp. 1-13.

(1) Universidade Estadual de Campinas, IMECC, Campinas 13100, SP, Brazil.

(2)-(3) IMPA; Estrada Dona Castorina 110, Rio de Janeiro, 22460, RJ, Brazil.

(4) Universidade Federal do Amazonas, Departamento de Matemática, I.C.E., Manaus, 69000, AM, Brazil.

R LARBARCA

Stability of parametrized families of vector fields

Abstract

We describe an open set of structurally stable families of vector fields in the space of one parameter families of vector fields without recurrence and non-cycles. With these results we go a long way to classify the stable one parameter families of vector fields without recurrence and non-cycles, at least in dimension less than five.

Introduction

In this work we present some contributions to the study of Global Structural Stability of one parameter families of vector fields. These results are an extension of those about one parameter families of gradient vector fields by J. Palis and F. Takens in [P-T]. To some extent, they are also a version for vector fields of the results for one parameter families of diffeomorphisms obtained by S. Newhouse, J. Palis and F. Takens in [N-P-T]. We point out that our results go a long way to classify the stable one parameter families of vector fields with simple recurrence and no-cycles, at least in dimension less than five. Some of them require new techniques different from those developed in [P-T] or [N-P-T].

In order to describe the results in a more precise way, we

recall some well known results and definitions. Let M be a C^∞, closed boundaryless, n-dimensional manifold. $\mathfrak{X}^\infty(M)$ denotes the space of C^∞ vector fields on M. $\mathfrak{X}_1^\infty(M)$ denotes the set of C^∞ arcs, $\xi: [0,1] = I \to \mathfrak{X}^\infty(M)$. We endow these sets with the usual C^∞ Whitney topology.

It is well-known (see [SO1], [SO2], [Br.]) that there exists a residual set $\Gamma_1 \subset \mathfrak{X}_1^\infty(M)$ such that $\xi \in \Gamma_1$ if and only if: (a) the set $B(\xi)$, of parameter values μ for wich $\xi(\mu)$ is not Kupka-Smale, is at most countable; (b) for each $\bar{\mu} \in B(\xi)$ we have that the vector field $\xi(\bar{\mu})$ has exactly one orbit $\theta \subset M$ along wich $\xi(\bar{\mu})$ is not Kupka-Smale in one of the following ways: (1) θ is a non-hyperbolic singularity (saddle-node or Hopf bifurcation) which unfolds generically; (2) θ is a non-hyperbolic periodic orbit (saddle-node or flip or Hopf bifurcation) which unfolds generically; (3) θ is an orbit along which there is a non-transversal intersection (which is quasi-transversal) of stable and unstable manifolds of critical elements which unfolds generically.

Let $\xi, \bar{\xi} \in \mathfrak{X}_1^\infty(M)$, $\mu_o, \bar{\mu}_o \in I$. We say that the family ξ at μ_o is equivalent to the family $\bar{\xi}$ at $\bar{\mu}_o$ if: There exists a reparametrization $\rho: (I,\mu_o) \to (I,\bar{\mu}_o)$ and an one parameter family of homeomorphisms $h_\mu: M \to M$ such that: (a) h_μ sends orbits of the field $\xi(\mu)$ into orbits of the field $\bar{\xi}(\rho(\mu))$, preserving their orientation for μ near μ_o and (b) the map $I \to \text{Homeo}(M,M)$, $\mu \to h_\mu$ is continuous.

We say that the family ξ is (structurally) stable at μ_o iff there exists a neighborhood $V(\xi) \subset \mathfrak{X}_1^\infty(M)$ of ξ such that for each $\bar{\xi} \in V(\xi)$ there is a parameter value $\bar{\mu}_o$, near μ_o, such that ξ at μ_o is equivalent to $\bar{\xi}$ at $\bar{\mu}_o$.

Let $\Gamma_2 \subset \mathfrak{X}_1^\infty(M)$ be the following subset: $\xi \in \Gamma_2$ if and only if $\forall \mu \in I$, the nonwandering set of $\xi(\mu)$ consists of a finite number of critical elements (singularities or periodic orbits) and there are not cycles between the critical elements of $\xi(\mu)$.

Let $\xi \in \Gamma_2$ and $\bar{\mu} \in B(\xi)$ be such that $\xi(\bar{\mu})$ has a quasi-transversal intersection between W^u_α and W^s_β, $\alpha \neq \beta$, where α and β are critical elements of the vector field $\xi(\bar{\mu})$. Then the stability of ξ at $\bar{\mu}$ depends on (i) if α and β are both singularities or one is singularity and the other a periodic orbit or both periodic orbit; (ii) the weakest contraction at α and the weakest expansion at β are real or complex numbers. All the different possibilities for α and β together yield 16 possibilities. We are going to work only with those cases that have a chance of being stable (see [V-S]). They are the following:

(I): α is a singularity with real weakest contraction and β is a periodic orbit with real weakest expansion.

(II): α is a singularity with complex weakest contraction and β is a singularity with real weakest expansion.

(III): α is a singularity with real weakest contraction and β is a singularity with real weakest expansion.

For the first case there exists a unique codimension one foliation F^{uu}_β of W^u_β with smooth leaves and invariant by the flow of the vector field $\xi(\bar{\mu})$. We call β u-critical if there is some critical element η of $\xi(\bar{\mu})$ such that W^s_η intersects some leaf of F^{uu}_β non-transversally. We call this criticallity simple if $W^s_\eta \cap W^u_\beta$ is a unique orbit of the vector field $\xi(\bar{\mu})$. We call this criticallity of codimension one if $W^s_\eta \cap W^u_\beta$ is a codimension one submanifold of W^u_β.

The following theorems are the purpose of this paper. With some additional conditions (which will be given at Chapter I) we have: Let $\xi \in \Gamma_2$ and $\bar{\mu} \in B(\xi)$ such that $\xi(\bar{\mu})$ has an orbit of quasi-transversal intersection between W^u_α and W^s_β $(\alpha \neq \beta)$.

__Theorem A__. Suppose the quasi-transversal intersection is as in Case (I):

(i) if β is not u-critical then ξ is stable at $\bar{\mu}$;

(ii) if β has only a simple criticallity with η and η is either a periodic orbit or a singularity with real weakest expansion or a singularity with complex weakest expansion and $\dim(W^u_\beta) = 3$ then ξ is stable at $\bar{\mu}$.

__Theorem B__. Suppose the quasi-transversal intersection is as in Case (I). If β has only a unique codimension one u-criticallity then ξ is stable at $\bar{\mu}$.

__Theorem C__. Suppose the quasi-transversal intersection is as in Case (II). If either $\dim(W^s_\alpha) = 2$ or $\dim(M) = 4$, then ξ is stable at $\bar{\mu}$.

In particular we have the corollary:

__Corollary__: For $\xi \in \Gamma_2(M^n)$, $n \leq 4$, let $\bar{\mu} \in B(\xi)$ be such that $\xi(\bar{\mu})$ has an orbit of quasi-transversal intersection between W^u_α and W^s_β as in Case (I) or (II). Then ξ is stable at $\bar{\mu}$.

__Remarks__

(1) The stability of ξ at $\bar{\mu} \in B(\xi)$ as in Case (III) (with some additional conditions) or when $\xi(\bar{\mu})$ has a saddle-node singularity (with some additional conditions) is possible to prove with similar techniques as in [P-T] ($\dim(M) = n$).

(2) The stability of ξ at $\bar{\mu} \in B(\xi)$ when $\xi(\bar{\mu})$ has a Hopf singularity or a flip-bifurcation (with some additional conditions) is possible to prove with similar techniques as in [N-P-T] ($\dim(M) = n$).

(3) The non-stability of ξ at $\bar{\mu} \in B(\xi)$ when $\xi(\bar{\mu})$ has a periodic orbit of Hopf type can be deduced from [N-P-T].

(4) The stability or non-stability of ξ at $\bar{\mu} \in B(\xi)$ when $\xi(\bar{\mu})$ has a periodic orbit of saddle-node type is, in the general case, an open problem (some partial results are in [N-P-T]).

This work is part of my doctoral Thesis at IMPA under the guidance of Jacob Palis. I whish to thank him for many helpful conversations. Also, I whish to thank to J. Gheiner, W. de Melo and F. Takens for many helpful conversations.

CHAPTER I

Results about local and global stability of one parameter families of vector fields

(I.1) Basic results and definitions (see also [dM-P]-1).

Let M be a C^∞, closed, boundaryless, n-dimensional manifold. $\mathfrak{X}^\infty(M)$ denotes the space of C^∞ vector fields on M endowed with the usual C^∞ topology. If $X \in \mathfrak{X}^\infty(M)$, X_t will denote its flow.

Let $p \in M$ be a singularity of X. The stable (unstable) set of p; W_p^s (resp. W_p^u) is the set: $W_p^s = \{x \in M / X_t(x) \to p,\ t \to +\infty\}$ (resp. $W_p^u = \{x \in M / X_t(x) \to p,\ t \to -\infty\}$). A singularity is called hyperbolic if $DX(p)$ has no purely imaginary eigenvalue. Let $\sigma \subset M$ be a periodic orbit of X. The stable (unstable) set of σ, W_σ^s (resp. W_σ^u) is the set: $W_\sigma^s = \{x \in M / X_t(x) \to \sigma,\ t \to +\infty\}$ (resp. $W_\sigma^u = \{x \in M / X_t(x) \to \sigma,\ t \to -\infty\}$). Let $q \in \sigma$ and Σ be a small cross-section at q. Let P be the associated Poincaré map. σ is called hyperbolic if $DP(q)$ has no eigenvalues on the unit circle $S^1 \subset \mathbb{R}^2$. A singularity or a periodic orbit of X is called a critical element. Ω_X will denote the set of non-wandering points of the vector field X i.e. $x \in \Omega_X$ iff for every neighborhood U of x in M and $t_o > 0$ there exists $t_1 > t_o$ such that: $X_{t_1}(U) \cap U \neq \emptyset$. A non trivial recurrence for the vector field X will be an invariant subset of Ω_X wich does not contain critical elements.

Let α and β be critical elements of the vector field X. We say that W^u_α and W^s_β are transversal at $x \in W^u_\alpha \cap W^s_\beta$ if $T_x(W^u_\alpha) + T_x(W^s_\beta) = T_x(M)$. We say that W^u_α is transversal to W^s_β if they are transversal at every point $y \in W^u_\alpha \cap W^s_\beta$. A cycle for the vector field X is a sequence $\alpha_1, \alpha_2, \ldots, \alpha_k = \alpha_1$ $(k \geq 2)$ of critical elements such that $W^s_{\alpha_i} \cap W^s_{\alpha_{i+1}} \neq \emptyset$.

Let us denote by $K\text{-}S^\infty(M) \subset \mathfrak{X}^\infty(M)$ the set of vector fields $X \in \mathfrak{X}^\infty(M)$ such that: (i) all the critical elements of the vector field X are hyperbolic and (ii) if σ_1 and σ_2 are critical elements then $W^u_{\sigma_1}$ is transversal to $W^s_{\sigma_2}$.

The following important theorem was obtained independently by I. Kupka [Ku] and S. Smale [Smℓ].

<u>Theorem</u>: The set $K\text{-}S^\infty(M)$ is residual in $\mathfrak{X}^\infty(M)$ [i.e. it contains a countable intersection of open and dense sets of $\mathfrak{X}^\infty(M)$].

We recommend Chapters 3 and 4 of [dM-P]-1 for a proof, much simpler than the originals, of this theorem.

(I.2) Genericity and local stability of one parameter families of vector fields.

Let $I = [0,1]$ be the unit interval. $\mathfrak{X}^\infty_1(M)$ denotes the set of C^∞ arcs $\xi: I \to \mathfrak{X}^\infty(M)$, endowed with the usual C^∞ topology.

<u>Singularities</u>

We say that a vector field $X \in \mathfrak{X}^\infty(M)$ has a saddle-node (resp. Hopf bifurcation) in $p \in M$ if $DX(p)$ has one eigenvalue zero (resp. two eigenvalues on the imaginary axes) and no other eigenvalues on the imaginay axes and if, moreover, X, restricted to a center manifold through p, has the form $X = ax^2 \frac{\partial}{\partial x} + \theta(|x|^3)$ with $a \neq 0$; (resp. $X = iwz + \gamma|z|^2 z + \theta(|z|^3)$ with $w \in (\mathbb{R}-\{0\})$, $Re(\gamma) \neq 0$, $z \in \mathbb{C}$) for the definition and existence of center manifold see [H-P-T], [L]. If $X_{\bar{\mu}}$, belonging to a one-parameter family $\{X_\mu\}$ of vector fields, has a saddle-node (resp. Hopf bifurcation) in p, we say that it unfolds generically if there is a μ-dependent center manifold passing through p (for $\mu = \bar{\mu}$) so that X_μ, restricted to this center manifold, has the form:

$$X_\mu = (ax^2+b(\mu-\bar{\mu}))\frac{\partial}{\partial x} + \theta(|x|^3+|x\cdot(\mu-\bar{\mu})|+|(\mu-\bar{\mu})|^2)$$

with $a,b \neq 0$

[resp. $X_\mu = (\alpha(\mu)+iw(\mu))z + \gamma(\mu)|z|^2 z + \theta(|z|^3)$ with $\alpha(\bar{\mu}) = 0$, $\alpha(\mu) \in R$, $\frac{d\alpha}{d\mu}(\bar{\mu}) \neq 0$, $w(\mu) \in (\mathbb{R}-\{0\})$, $Re(\gamma(\mu)) \neq 0$, $z \in \mathbb{C}$)].

Periodic orbits

We say that a periodic orbit $\sigma \subset M$ of the vector field $X \in \mathfrak{X}^\infty(M)$ is of the saddle-node type (resp. flip, Hopf) if there exists a codimension one section, Σ_q, $q \in \sigma$, transversal to the vector field such that for the associated Poincaré map $P: (V_q, q) \to (\Sigma_q, q)$ we have that $DP(q)$ has one eigenvalue 1 (resp. -1, two complex eigenvalues $\lambda, \bar{\lambda}$ on $S^1-\{-1,1\}$ with $\lambda, \lambda^2, \lambda^3, \lambda^4 \neq 1$) and no other eigenvalues on the unit circle and if, moreover, P, restricted to a center manifold through q, has the form $P = x+ax^2+\theta(|x|^2)$ with $a \neq 0$ (resp. $P = -x+bx^3+\theta(x^3)$, $b\neq 0$; $P = \lambda z+\alpha|z|^2 z+\theta(|z|^3)$, $Re(\lambda/\alpha) \neq 0$, $z \in \mathbb{C}$). If $X_{\bar{\mu}}$, belonging to a one-parameter family $\{X_\mu\}$ of vector fields, has a periodic orbit of saddle-node type (resp. flip, Hopf) in σ, we say that it unfolds generically if there is a μ-dependent center manifold passing through $q \in \sigma$ (for $\mu = \bar{\mu}$) so that P_μ, restricted to this center manifold, has the form

$$P_\mu = b(\mu-\bar{\mu}) + v_1(\mu)x + v_2(\mu)x^2 + \theta(x^2)$$

where $v_1(\bar{\mu}) = 1$, $\frac{dv_1}{d\mu}(\bar{\mu}) \neq 0$, $b \neq 0$ and $v_2(\mu) \in (\mathbb{R}-\{0\})$, [resp. $P_\mu = (\lambda(\mu)-1)x + bx^3 + \theta(x^4) + \theta(x^2|\mu-\bar{\mu}|)$, where $\lambda(\bar{\mu}) = 0$, $\frac{d\lambda}{d\mu}(\bar{\mu}) \neq 0$, $b \neq 0$; $P_\mu = \lambda(\mu)z + \alpha(\mu)|z|^2 z + \beta(\mu)\bar{z}^4 + \theta(|z|^4)$ where $\frac{d}{d\lambda}(|\lambda(\mu)|) \neq 0$, $\mathrm{Re}(\frac{\lambda(\mu)}{\alpha(\mu)}) \neq 0$.]

Quasi-transversalities

Let $X \in \mathfrak{X}^\infty(M)$ and α, β be critical elements of the vector field X. We say that W^u_α and W^s_β are quasi-transversal at $z_0 \in W^u_\alpha \cap W^s_\beta$ if there is a C^∞ coordinate chart $\xi: (U_{z_0}, z_0) \to$ $\to (\mathbb{R}^n, 0)$ (U_{z_0} = small neighborhood of z_0 in M) such that

(1) $\xi_*(X) = \frac{\partial}{\partial y_n}$

(2) $\xi(W^u_\alpha \cap U_{z_0}) = \{y_1 = \dots = y_{n-u} = 0\}$

(3) if $n-u = s$: $\xi(W^s_\beta \cap U_{z_0}) = \{y_{s+1} = \dots = y_{n-1} = 0\}$

if $n-u \leq s-1$: $\xi(W^s_\beta \cap U_{z_0}) = \{y_{s+1} = \dots = y_{n-1} = 0,$

$y_1 = Q(y_{n-u+1}, \dots, y_s)\}$.

Here Q is a non-degenerated homogeneous function. For an intrinsic definition see [N-P-T] or [P-T]. We say that W^u_α and W^s_β are quasi-transversal if they are not transversal and they are quasi-transversal at every point $z_0 \in W^u_\alpha \cap W^s_\beta$. If $X_{\bar{\mu}}$, belonging to a one-parameter family $\{X_\mu\}$ of vector fields, has a quasi-transversal intersection between W^u_α and W^s_β, we say that it unfolds generically at $\bar{\mu}$ if for each $z_0 \in W^u_\alpha \cap W^s_\beta$ there exists a μ-dependent coordinate chart $\xi_\mu: (U_{z_o}, z_o) \to (\mathbb{R}^n, 0)$ such that

(1) $(\xi_\mu)_*(X_\mu) = \frac{\partial}{\partial y_n}$

(2) $\xi_\mu(W^u_{\alpha_\mu} \cap U_{z_o}) = \{y_1 = \ldots = y_{n-u} = 0\}$

(3) if $n-u=s$: $\xi_\mu(W^s_{\beta_\mu} \cap U_{z_o}) = \{y_{s+1} = \ldots = y_{n-1} = 0,\ y_1 = \varepsilon(\mu)\}$

if $n-u \leq s-1$: $\xi_u(W^s_{\beta_\mu} \cap U_{z_o}) = \{y_{s+1} = \ldots = y_{n-1} = 0,$

$y_1 = Q(y_{n-u+1}, \ldots, y_s) + \varepsilon(\mu)\}$.

Here α_μ and β_μ are the corresponding critical elements of X_μ (for μ near $\bar{\mu}$), Q is a non-degenerate homogeneous function and ε is a smooth map such that $\varepsilon(\bar{\mu}) = 0$, $\frac{d\varepsilon}{d\mu}(\bar{\mu}) \neq 0$. For an intrinsic definition see [N-P-T] or [P-T].

Let $\{X_\mu\} \in \mathfrak{X}^\infty_1(M)$. We say that $\mu \in I$ is a regular value of the family if $X_\mu \in K\text{-}S^\infty(M)$. Otherwise we call it non-regular. It is clear that if $\mu = \bar{\mu}$ is a non-regular value of the family $\{X_\mu\}$ then there exists an orbit θ of the vector field $X_{\bar{\mu}}$ along wich $X_{\bar{\mu}}$ is not Kupka-Smale in at least one of the following ways:

(a) θ is a non-hyperbolic singularity

(b) θ is a non-hyperbolic periodic orbit

(c) θ is an orbit of quasi-transversal intersection between stable and unstable manifolds of hyperbolic critical elements of the vector field $X_{\bar{\mu}}$.

We denote by $\Gamma_1 \subset \mathfrak{X}^\infty_1(M)$ the set of one-parameter families $\xi \in \mathfrak{X}^\infty_1(M)$ such that

(1) the set $B(\xi) = \{\mu \in I \,/\, \mu$ is a non-regular value of the family$\}$, is at most countable.

(2) For each $\bar{\mu} \in B(\xi)$ we have that the vector field $\xi(\bar{\mu})$ has exactly one orbit $\theta \subset M$ along wich $\xi(\bar{\mu})$ is not Kupka-Smale in

one of the following ways (i) θ is a non-hyperbolic singularity (saddle-node or Hopf bifurcation) which unfolds generically; (ii) θ is a non-hyperbolic periodic orbit (saddle-node or flip or Hopf bifurcation) which unfolds generically; (iii) θ is an orbit along wich there is a non-transversal intersection (which is quasi-transversal) of stable and unstable manifolds of hyperbolic critical elements which unfolds generically.

It is well-known (see [SO1], [SO2], [Br.]) that Γ_1 is a residual set in $\mathfrak{X}_1^\infty(M)$. Provided with this residual set we consider the problem of structural stability of one-parameter families.

<u>Definition</u>. Let $\xi, \bar{\xi} \in \mathfrak{X}_1^\infty(M)$, $\mu_o, \bar{\mu}_o \in I$, θ an orbit of the vector field $\xi(\mu_o)$, $\bar{\theta}$ an orbit of the vector field $\bar{\xi}(\bar{\mu}_o)$, V a neighborhood of the closure of θ, $\bar{V}$ a neighborhood of the closure of $\bar{\theta}$. We say that the family ξ at (θ,μ_o) is equivalent to the family $\bar{\xi}$ at $(\bar{\theta},\bar{\mu}_o)$ if there exists a reparametrization $\rho: (I,\mu_o) \to (I,\bar{\mu}_o)$ and a family of homeomorphisms $h_\mu: M \to M$ such that

(1) $h_{\mu_o}(\theta) = \bar{\theta}$;

(2) $h_\mu|_V$ send orbits of $\xi(\mu)|_V$ into orbits of $\xi(\rho(\mu))|_{\bar{V}}$, for μ near μ_o;

(3) the map $I \to \mathrm{Homeo}(M,M)$, $\mu \longmapsto h_\mu$ is continuous.

We say that the family ξ at μ_o is equivalent to the family $\bar{\xi}$ at $\bar{\mu}_o$ if there exists a reparametrization $\rho: (I,\mu_o) \to (I,\bar{\mu}_o)$ and a family of homeomorphisms $h_\mu: M \to M$ such that

(1) h_μ send orbits of the vector field $\xi(\mu)$ into orbits of the vector field $\bar{\xi}(\rho(\mu))$ preserving their orientation for μ near μ_o;

(2) the map $I \to \mathrm{Homeo}(M,M)$, $\mu \longmapsto h_\mu$ is continuous.

We say that the family ξ is (structurally) stable at (θ,μ_o) (θ an orbit of the vector field $\xi(\mu_o)$) if there exists a neighborhood $V(\xi) \subset \mathfrak{X}_1^\infty(M)$ of ξ such that: for any $\bar{\xi} \in V(\xi)$ there are (a) a parameter value $\bar{\mu}_o$ near μ_o and (b) an orbit $\bar{\theta}$ of the vector field $\bar{\xi}(\bar{\mu}_o)$, near μ_o such that: ξ at (θ,μ_o) is equivalent to $\bar{\xi}$ at $(\bar{\theta},\bar{\mu}_o)$.

We say that the family ξ is (structurally) stable at μ_o iff there exists a neighborhood $V(\xi) \subset \mathfrak{X}_1^\infty(M)$ of ξ such that for each $\bar{\xi} \in V(\xi)$ there is a parameter value $\bar{\mu}_o$, near μ_o, such that ξ at μ_o is equivalent to $\bar{\xi}$ at $\bar{\mu}_o$.

The local stability of one-parameter families $\xi \in \Gamma_1$ was extensively studied, see for instance [SO1], [N-P-T], [P-T], [V-S], [Ma-P]. We present now a small review of them.

Let $\bar{\mu} \in B(\xi)$ and $\theta \subset M$ be the unique orbit along wich the vector field $\xi(\bar{\mu})$ is not Kupka-Smale; we have:

(1) if θ is a non-hyperbolic singularity of the saddle-node type or a Hopf bifurcation then ξ is stable at $(\theta,\bar{\mu})$ ([SO1],[P-T]);

(2) if θ is a non-hyperbolic periodic orbit of the saddle-node type or a flip then ξ is stable at $(\theta,\bar{\mu})$ ([N-P-T], [Ma-P]);

(3) if θ is a non-hyperbolic periodic orbit of the Hopf type then each neighborhood $V(\xi) \subset \mathfrak{X}_1^\infty(M)$ of ξ contains an element $\hat{\xi} \in \Gamma_1 \cap V(\xi)$ and we can find a parameter value $\hat{\mu} = \hat{\mu}(\hat{\xi})$, near $\bar{\mu}$, such that for the vector field $\hat{\xi}(\hat{\mu})$ there exists a unique non-hyperbolic periodic orbit of the Hopf type, $\hat{\theta}$, near θ, such that $\hat{\xi}$ at $(\hat{\theta},\hat{\mu})$ is not equivalent to ξ at (θ,μ) ([N-P-T]).

To describe the results about quasi-transversal intersections we give another definitions and conditions over the elements of Γ_1.

<u>Definition</u>. We say that $X \in \mathfrak{X}^\infty(M)$ has a weakest contracting

(expanding) eigenvalue, say A, at a singularity p if: (i) $Re(A) < 0$ (resp. $Re(A) > 0$), (ii) A has multiplicity one and (iii) for all eigenvalue B of $DX(p)$ with $Re(B) < 0$ (resp. $Re(B) > 0$), $B \neq A, \bar{A}$, one has $Re(B) < Re(A)$ (resp. $Re(B) > Re(A)$). Similarly, let us suppose that X has a periodic orbit σ. Let Σ be a transversal section at $q \in \sigma$ and P the corresponding Poincaré map. We say that X has a weakest contracting (expanding) eigenvalue at σ, say A if (i) $|A| < 1$ (resp. $|A| > 1$), (ii) A has multiplicity one and (iii) for all eigenvalue B of $DP(q)$ with $|B| < 1$ (resp. $|B| > 1$), $B \neq A, \bar{A}$, one has $|B| < |A|$ (resp. $|B| > |A|$).

Let $\alpha \in M$ be a hyperbolic singularity of the vector field X. Suppose that it is defined the weakest contraction (expansion) at α. In this situation it is possible to define a C^1 center unstable manifold (resp. C^1 center stable manifold), not unique, invariant by the flow of the vector field X and tangent, at α, to the direct sum of the expansive subspace (resp. contractive subspace) and the weakest contractive subspace (resp. weakest expansive subspace). We denote this submanifold by W_α^{cu} (resp. W_α^{cs}). Also if $\alpha \subset M$ is a hyperbolic periodic orbit of the vector field X such that it is defined the weakest contraction (expansion) at α then it is possible to define a C^1 center unstable manifold (resp. C^1 center stable manifold), not unique, invariant by the flow of the vector field X and tangent, at $q \in \alpha$, to the direct sum of the expansive subspace of the Poincaré map, (resp. contractive subspace of the Poincaré map) the weakest contractive subspace of the Poincaré map (resp. the weakest expansive subspace of the Poincaré map) and the subspace which corresponds to the flow-direction. We denote this submanifold by W_α^{cu} (resp. W_α^{cs}) and their restriction to the cross section Σ at $q \in \alpha$ by W_q^{cu} (resp. W_q^{cs}).

Let $\Gamma_2 \subset \Gamma_1$ be the set defined by: $\xi \in \Gamma_2$ iff for each $\bar{\mu} \in B(\xi)$ such that $\xi(\bar{\mu})$ has an orbit of quasi-transversal intersection between W_α^u and W_β^s then it is defined the weakest contraction at α and the weakest expansion at β. Also we have: W_α^{cu} is transversal to W_β^s and W_α^{cs} is transversal to W_α^u. It is clear that Γ_2 is dense in Γ_1.

(4) Let $\xi \in \Gamma_2$ and $\bar{\mu} \in B(\xi)$ be such that $\xi(\bar{\mu})$ has an orbit, θ, of quasi-transversal intersection between $W^u_{\alpha_{\bar{\mu}}}$ and $W^s_{\beta_{\bar{\mu}}}$, $\alpha_{\bar{\mu}} \neq \beta_{\bar{\mu}}$. In this case the stability (or not) of ξ at $(\theta,\bar{\mu})$ depends on (i) if $\alpha_{\bar{\mu}}$ and $\beta_{\bar{\mu}}$ are both singularities or one is a singularity and the other a periodic orbit; (ii) the weakest contraction at α and the weakest expansion at β are real or complex. All the different possibilities for α and β together yield 16 possibilities. In the following table we indicate the results in the different cases and the respective references:

$\alpha_{\bar{\mu}}$ \ $B_{\bar{\mu}}$		SINGULARITY		PERIODIC ORBIT	
		Real weakest expansion	Complex weakest expansion	Real weakest expansion	Complex weakest expansion
SINGULARITY	real weakest contraction	ξ is stable at $(\theta,\bar{\mu})$ [P-T].	ξ is stable at $(\theta,\bar{\mu})$ [VS] and this paper.	ξ is stable at $(\theta,\bar{\mu})$ [Be], [VS] and this paper.	ξ is not stable at $(\theta,\bar{\mu})$ [VS]. It is not hand to prove this.
	complex weakest contraction	ξ is stable at $(\theta,\bar{\mu})$ [VS] and this paper.	ξ is not stable at $(\theta,\bar{\mu})$, [TA1], [VS].	ξ is not stable at $(\theta,\bar{\mu})$ [Be], [VS].	ξ is not stable at $(\theta,\bar{\mu})$, [VS].
PERIODIC ORBIT	real weakest contraction	ξ is stable at $(\theta,\bar{\mu})$ [Be], [VS] and this paper.	ξ is not stable at $(\theta,\bar{\mu})$, [Be], [V-S].	ξ is not stable at $(\theta,\bar{\mu})$, [P2], [M2], [Dm-P-VS], [N-P-T].	ξ is not stable at $(\theta,\bar{\mu})$, [dM-P-VS], [N-P-T].
	complex weakest contraction	ξ is not stable at $(\theta,\bar{\mu})$, [VS]. It is	ξ is not stable at $(\theta,\bar{\mu})$, [VS]	ξ is not stable at $(\theta,\bar{\mu})$ [dM-P-VS], [N-P-T].	ξ is not stable at $(\theta,\bar{\mu})$, [dM-P-Vs], [N-P-T].

(I.3) Global Stability

With respect to Global Stability there are very goods results in relation to one parameter families of Gradient vector fields (see [P-T]) and one parameter families of diffeomorphisms (see [N-P-T]). Now we are going to work with those cases for which the local stability was guarantized in the last section.

We recall some facts about differentiable linearizations of vector fields at either singularities or periodic orbits. Let $X \in \mathfrak{X}^{\infty}(M)$ and $p \in M$ a singularity of X. The linear part of X in p, $DX(p)$, is an endomorphism of $T_p(M)$. Suppose that p is a hyperbolic singularity of X. In this situation there is a result of Sternberg ([ST1], [ST2]) accoring to which there is, for each positive integer $s \in \mathbb{N}$, an open and dense subset O_s in the space of linear endomorphisms of $T_p(M)$ such that if $DX(p)$ belongs to O_s, there is a C^s-coordinate system near p, with respect to which X is linear. Here we want somewhat more: we would like the linearization to depend smoothly on μ, when X depends smoothly on μ. Also, in the case of saddle-node or Hopf bifurcation, where we have no hyperbolicity and so we cannot apply Sternberg's result, we still want some partial linearization. For this we need a linearization theorem for partially hyperbolic singularities ([TA2]).

For positive integers s and ℓ, with $\ell \leq m = \dim(M)$, there is an open and dense subset $O_{s,\ell}$ in the set of ℓ-tuples of real numbers such that, if a C^{∞} vector field on M has a singularity p such that $DX(p)$ has $(m-\ell)$ purely imaginary (or zero) eigenvalues, while for the non-purely imaginary eigenvalues $\lambda_1,\ldots,\lambda_\ell$, $(\mathrm{Re}(\lambda_1),\ldots,\mathrm{Re}(\lambda_\ell)) \in O_{s,\ell}$, then there are C^s-coordinates $x_1,\ldots,x_m$ on a neighborhood of p, with respect to which X has the form:

$$X = \sum_{i,j=1}^{\ell} A_{ij}(x_{\ell+1},\ldots,x_m)x_i \frac{\partial}{\partial x_j} + \sum_{n=\ell+1}^{m} X_n(x_{\ell+1},\ldots,x_m)\frac{\partial}{\partial x_n}$$

with all eigenvalues of $\left(\frac{\partial X_i}{\partial x_j}(0)\right)_{i,j\geq \ell+1}$ purely imaginary. Let now $\{X_\mu\} \in \mathfrak{X}_1^\infty(M)$ and suppose that $p \in M$ is a hyperbolic singularity of the vector field $X_{\bar{\mu}}$. We have then that $(p,\bar{\mu})$ is a partially hyperbolic singularity of the vector field $X(x,\mu) = (X_\mu(x),0)$ and we can apply the above result. Then we know that there exist C^s-coordinates $x_1,\ldots,x_m$ on a neighborhood of p, with respect to which the vector field X_μ, for μ near $\bar{\mu}$, has the form:

$X_\mu = \sum_{i,j=1}^{m} x_i A_{ij}(\mu)\frac{\partial}{\partial x_j}$; i.e. linearization of the family wich depends C^s on the parameter μ (obviously we suppose $(\mathrm{Re}(\lambda_1),\ldots,\mathrm{Re}(\lambda_m)) \in O_{s,m}$).

Let now $\{X_\mu\} \subset \mathfrak{X}_1^\infty(M)$ and suppose that $p \in M$ is a non-hyperbolic singularity of the vector field $X_{\bar{\mu}}$. If $DX_{\bar{\mu}}(p)$ has $(m-\ell)$ purely imaginary eigenvalues (or zeros) and the other eigenvalues form a ℓ-tuple in $O_{s,\ell}$ then there are μ-dependent C^s-coordinates $x_1,\ldots,x_m$ on a neighborhood of p in which X_μ, for μ near $\bar{\mu}$, has the form

$$X_\mu = \sum_{i,j=1}^{\ell} x_i A_{ij}(x_{\ell+1},\ldots,x_m)\frac{\partial}{\partial x_j} + \sum_{n=\ell+1}^{m} X_n(x_{\ell+1},\ldots,x_m)\frac{\partial}{\partial x_n}$$

with all eigenvalues of $\left(\frac{\partial X_i}{\partial x_j}(0)\right)_{i,j\geq \ell+1}$ purely imaginary.

Let $\Gamma_3 \subset \Gamma_2$ be the set defined by: $\xi \in \Gamma_3$ iff

(i) for each value $\bar{\mu} \in B(\xi)$ such that $\xi(\bar{\mu})$ has a saddle-node (Hopf bifurcation) singularity in $p \in M$ then $D\xi(\bar{\mu})(p)$ has one zero eigenvalue (two purely imaginary eigenvalues) while the other eigenvalues form an $(m-1)$-tuple (resp. an $(m-2)$-tuple) in $O_{2,m-1}$ (resp. $O_{2,m-2}$);

(ii) for each value $\bar{\mu} \in B(\xi)$ such that $\xi(\bar{\mu})$ has an orbit of quasi-transversal intersection between W^u_α and W^s_β then there exists C^2 coordinate system near α and β in which we obtain linearization of the vector field $\xi(\bar{\mu})$ near α if α is a singularity of $\xi(\bar{\mu})$ or we obtain a linearization of the Poincaré map near α if α is a periodic orbit of $\xi(\bar{\mu})$. We have the same property for β.

Let $\alpha \in M$ be a hyperbolic critical element of the C^∞ vector field X. Suppose that we have defined the weakest contraction (expansion) at α. Then there exists a unique invariant manifold $W^{ss}_\alpha \subset W^s_\alpha$ (resp. $W^{uu}_\alpha \subset W^u_\alpha$) such that:

a) if α is a singularity then $T_\alpha(W^s_\alpha) = T_\alpha(W^{ss}_\alpha) \oplus T_\alpha(W^{cu}_\alpha \cap W^s_\alpha)$ [resp. $T_\alpha(W^u_\alpha) = T_\alpha(W^{uu}_\alpha) \oplus T_\alpha(W^{cs}_\alpha \cap W^u_\alpha)$];

b) if α is a periodic orbit, $q \in \alpha$, Σ a transversal section to the vector field X at q and P the associated Poincaré map then:

$T_q(W^s_q) = T_q(W^{ss}_q) \oplus T_q(W^{cu}_q \cap W^s_q)$ where $W^s_q = \Sigma \cap W^s_\alpha$,

$W^{cu}_q = \Sigma \cap W^{cu}_\alpha$, $\quad W^{ss}_q = W^{ss}_\alpha \cap \Sigma$

[resp. $T_q(W^u_q) = T_q(W^{uu}_q) \oplus T_q(W^{cs}_q \cap W^u_q)$].

Let α be a non-hyperbolic critical element of the C^∞ vector field X: (i) if α is a saddle-node singularity then the stable set W^s_α is an injectively immersed submanifold with boundary. This boundary is called the strong stable manifold of α and is denoted by W^{ss}_α. Similarly we get the corresponding definition for unstable sets and manifolds; (ii) if α is a Hopf bifurcation then the restriction of the vector field to a center manifold can be either a source or a sink. Suppose it is a sink. In this case the stable set is an injectively immersed submanifold whose dimension is equal to the number of eigenvalues of $DX(\alpha)$ with real negative part plus two. In the stable manifold there exists a unique invariant manifold $W_\alpha^{ss} \subset W_\alpha^s$, tangent at α to the subspace generated by the eigenvalues with negative real part. We can give similar definitions for the case when the restriction of the vector field to a center manifold is a source. (iii) If α is a flip periodic orbit then the restriction of the vector field to a center manifold can be either a source or a sink. Suppose it is a sink. In this case the stable set is an injectively immersed submanifold whose dimension is equal to the number of eigenvalues of $DP(q)$, $q \in \alpha$, with norm smaller than one plus two (one corresponding to the central eigenvalue of $DP(q)$ and the other to the flow direction). In the stable manifold there exists a unique invariant manifold $W_\alpha^{ss} \subset W_\alpha^s$ tangent at q to the subspace generated by the direct sum of the flow direction and the subspace generated by the eigenvalues of $DP(q)$ with norm smaller than one. We can give similar definitions for the case when the restriction of the vector field to the center manifold is a source.

We are only interested in those one parameter families of vector fields were we have elementary bifurcations. With this in mind, we are going to restrict our sets. For this, let us define $WR^\infty(M)$ as the subset of $\mathfrak{X}^\infty(M)$ given by: $WR^\infty(M) = \{X \in \mathfrak{X}^\infty(M) /$ Ω_X is composed by a finite number of critical elements$\}$.

Let $\Gamma_4 \subset \Gamma_3$ be the set defined by: $\xi \in \Gamma_4$ iff

(i) $\xi(\mu) \in WR^\infty(M)$ $\forall \mu \in I$ and $B(\xi) \subset]0,1[$

(ii) if $\bar{\mu} \in B(\xi)$ is such that $\xi(\bar{\mu})$ has a saddle-node (or a Hopf bifurcation) singularity at $p \in M$ then for each critical element α of the vector field $\xi(\bar{\mu})$ we have that W_α^u is transversal to W_p^s, W_p^{ss} and W_α^s is transversal to W_p^u, W_p^{uu} (with the obvious exception $\alpha = p$)

(iii) if $\bar{\mu} \in B(\xi)$ is such that $\xi(\bar{\mu})$ has a flip periodic orbit, $\sigma \subset M$, then for each critical element α of the vector field $\xi(\bar{\mu})$ we have that W^u_α is transversal to W^s_σ, W^{ss}_σ and W^s_α is transversal to W^u_σ, W^{uu}_σ (with the obvious exception $\alpha = \sigma$).

(iv) Let $\bar{\mu} \in B(\xi)$ be such that $\xi(\bar{\mu})$ has an orbit, $\theta \subset M$, of quasi-transversal intersection between W^u_α and W^s_β, $\alpha \neq \beta$. If α is a singularity with real weakest contraction then we suppose that the center manifold $W^c_\alpha = W^{cu}_\alpha \cap W^s_\alpha$ is disjoint of the unstable manifold of the other critical elements of the vector field $\xi(\bar{\mu})$ (similarly when α is a periodic orbit with real weakest contraction). If α is a singularity with complex weakest contraction then we suppose that the center manifold $W^c_\alpha = W^{cu}_\alpha \cap W^s_\alpha$ is transversal to W^u_ρ for all critical element $\rho \neq \alpha$ of the vector field $\xi(\bar{\mu})$. We do similar hypotheses for the critical element β. In all the above situations we suppose that for each critical element $\rho \neq \alpha, \beta$ of the vector field $\xi(\bar{\mu})$, we have that W^u_ρ is transversal to W^{ss}_α and W^s_ρ is transversal to W^{uu}_β. Also, we suppose that there exists a center unstable manifold W^{cu}_α and a center stable manifold W^{cs}_β such that W^{cu}_α is transversal to W^s_β and W^{cs}_β is transversal to W^u_α.

We have now, the following theorem, which is essentially a corollary due to Palis-Takens [P-T] and Newhouse-Palis-Takens [N-P-T].

<u>Theorem</u> ([P-T], [N-P-T])

Let $\xi \in \Gamma_4$, $\bar{\mu} \in B(\xi)$ and $\theta \subset M$ be the unique orbit along which $\xi(\bar{\mu})$ is not Kupka-Smale then:

(1) if θ is a saddle-node or a Hopf bifurcation singularity then ξ is stable at $\bar{\mu}$

(2) if $\mathcal{O}$ is a flip periodic orbit then ξ is stable at $\bar{\mu}$

(3) if $\mathcal{O}$ is an orbit of quasi-transversal intersection between W^u_α and W^s_β where α is a singularity with real weakest contraction and β is a singularity with real weakest expansion then ξ is stable at $\bar{\mu}$.

(I.4) <u>Our results</u>

Let $\xi \in \Gamma_4$ and $\bar{\mu} \in B(\xi)$ be such that $\xi(\bar{\mu})$ has an orbit, $\mathcal{O} \subset M$, of quasi-transversal intersection between W^u_α and W^s_β, $\alpha \neq \beta$, where either

(I) α is a singularity with real weakest contraction and β is a periodic orbit with real weakest expansion.

or

(II) α is a singularity with complex weakest contraction and β is a singularity with real weakest expansion.

In the first case we know, from a result of [H-P-S], that there exists a unique foliation F^{uu}_β of W^u_β with smooth leaves such that W^{uu}_β is one of these leaves and the flow of the vector field $\xi(\bar{\mu})$ send leaves to leaves. These foliation is called the strong unstable foliation of β. We say that the periodic orbit β is u-critical if there exists a critical element $\rho \neq \beta$, of the vector field $\xi(\bar{\mu})$, such that W^s_ρ intersects some leave of F^{uu}_β non-transversally. We say that the u-criticality between W^s_ρ and F^{uu}_β is simple if $W^s_\rho \cap W^u_\beta$ is formed by a unique orbit of the vector field $\xi(\bar{\mu})$ (since W^s_ρ is transversal to W^u_β we have that $\dim(W^s_\rho) = \dim(W^s_\beta)$). We say that the criticality is codimension one if $W^s_\rho \cap W^u_\beta$ is a codimension one submanifold of W^u_β (it is clear that in this case $\dim(W^s_\rho) = n-1$).

Let $\Gamma_5 \subset \Gamma_4$ be the set defined by: $\xi \in \Gamma_5$ iff for each

$\bar{\mu} \in B(\xi)$ such that $\xi(\bar{\mu})$ has a quasi-transversal intersection between W^u_α and W^s_β, $\alpha \neq \beta$, as in Case (I) above, we have:

(a) all the u-criticalities are of quasi-transversal type

(b) suppose that we have a simple criticality between W^s_ρ and F^{uu}_β then:

(i) it is defined the weakest expansion at ρ

(ii) there exists a center-stable manifold for ρ which is locally transversal to F^{uu}_β in a neighborhood of the orbit of intersection of W^s_ρ and W^u_β

(iii) there exists C^s-linearizing coordinates at $(\rho,\bar{\mu})$, $s \geq 2$

(iv) in the case that ρ is a singularity or a periodic orbit with real weakest expansion then ρ is not u-critical

(v) in the case that ρ is a singularity with complex weakest expansion we suppose that $\dim(W^u_\beta) = 3$.

(vi) Let $q \in \beta$ and $\Sigma_q \subset M$ be a transversal section at $q \in \beta$. Let $\pi^{uu}: V_q \cap W^u_\beta \to V_q \cap W^c_\beta$ be the F^{uu}-projection. ($V_q \subset \Sigma_q$ a small neighborhood of q in Σ_q for which we have defined the Poincaré map $P: (V_q,q) \to (\Sigma_q,q)$). Let $\rho_1,\ldots,\rho_r$ be critical elements of the vector field $\xi(\bar{\mu})$ which has simple criticalities with β. Let $D^u_q \subset V_q \cap W^u_q$ be a fundamental domain for P_q. Let us denote the intersections of D^u_q with $W^s_{\rho_i}$ by $\rho_i^1,\ldots,\rho_i^{k_i}$. Let us consider the projections $\ell_i^{s_i} = \pi^{uu}(\rho_i^{s_i})$, $i = 1,2,\ldots,r$, $1 \leq s_i \leq k_i$. Since we have transversality between the manifolds $W^s_{\rho_1},\ldots,W^s_{\rho_r}$ with W^{uu}_β then $\ell_i^{s_i} \neq q$ $\forall\, i,s_i$. Let us denote by $x_i^{s_i}$ the representant of $\ell_i^{s_i}$ in the intersection $D^u_q \cap W^c_\beta$ (i.e. $x_i^{s_i} = P^{n(i,s_i)}(\ell_i^{s_i})$). We assume the generic condition that all the $x_i^{s_i}$ are distincts.

(c) Let $\rho_1,\ldots,\rho_k$ be critical elements such that $W^s_{\rho_i}$ has codimension one u-criticalities with F^{uu}_β. Let $\rho_i^1,\ldots,\rho_i^{k_i}$ be the points of non-transversal intersection between $W^s_{\rho_i} \cap D^u_q$ and F^{uu}_β. Let $\ell_i^{s_i} = \pi^{uu}(\rho_i^{s_i})$ be its π^{uu}-projections $i = 1,\ldots,k$, $s_i = 1,\ldots,k_i$. Since we have transversality between the manifolds $W^s_{\rho_i},\ldots,W^s_{\rho_k}$ with W^{uu}_β then $\ell_i^{s_i} \neq q \ \forall\ i,s_i$. Let us denote by $x_i^{s_i}$ the representant of $\ell_i^{s_i}$ in the intersection $D^u_q \cap W^c_\beta$ (i.e. $x_i^{s_i} = P^{n(i,s_i)}(\ell_i^{s_i})$). We assume the generic condition that all the $x_i^{s_i}$ are distincts.

We note that if there is a u-criticality between W^s_ρ and F^{uu}_β then ρ cannot be a periodic orbit with complex weakest expansion if we want stability of ξ at $\bar{\mu}$ (see Table I and [VS]).

Let $\xi \in \Gamma_5$ and $\bar{\mu} \in B(\xi)$ such that $\xi(\bar{\mu})$ has an orbit of quasi-transversal intersection as in Case (II). Suppose that ρ is a critical element of saddle type of $\xi(\bar{\mu})$ such that: $W^u_\rho \cap W^{ss}_\alpha = \phi$, $W^u_\rho \cap W^c_\alpha = \phi$ and $W^u_\rho \cap W^s_\alpha \neq \phi$. Then, there exists parameter values μ (near $\bar{\mu}$) such that $W^u_{\rho_\mu} \cap W^s_{\beta_\mu} \neq \phi$ and they have a quasi-transversal intersection, clearly this implies that ρ cannot be a periodic orbit with complex weakest contraction. Let $\Gamma_6 \subset \Gamma_5$ be the subset: $\xi \in \Gamma_6$ iff

(1) in the above situation it is defined the weakest contraction at ρ and there exists a center-unstable manifold $W^{cu}_{\rho_\mu}$ which is transversal to $W^s_{\beta_\mu}$;

(2) there exists C^s-linearizing coordinates at $(\rho,\bar{\mu})$, $s \geq 2$;

(3) either the critical element ρ is not s-critical or it is a periodic orbit with s-criticalities of quasi-transversal type and with similar hypothesis as in (b)(vi) or (c) for these criticalities.

Now we state our theorems for $\xi \in \Gamma_6$ and $\bar{\mu} \in B(\xi)$ such that $\xi(\bar{\mu})$ has exactly one orbit of quasi-transversal intersection between W^u_α and W^s_β $(\alpha \neq \beta)$.

<u>Theorem A</u>. Suppose the quasi-transversality is as in Case (I)

(i) if β has not u-criticalities then ξ is stable at $\bar{\mu}$;

(ii) if β has only simple u-criticalities, with critical elements $\rho_1,\ldots,\rho_r$ and all of them are periodic orbits (or singularities) with real weakest expansion then ξ is stable at $\bar{\mu}$;

(iii) if β has only simple u-criticalities, with critical elements $\rho_1,\ldots,\rho_r$, where at least one of them is a singularity with complex weakest expansion then ξ is stable at $\bar{\mu}$.

<u>Theorem B</u>. Suppose the quasi-transversality is as in Case (I). If β has only u-criticalities of codimension one then ξ is stable at $\bar{\mu}$.

<u>Theorem C</u>. Suppose the quasi-transversality is as in Case (II). If $\dim(W^s_\alpha) = 2$ or $\dim(M) = 4$ then ξ is stable at $\bar{\mu}$.

As a corollary of the proofs of Theorems A, B and C we have:

<u>Corollary</u>: For $\xi \in \Gamma_6 \subset \mathfrak{X}^\infty_1(M^n)$, $n \leq 4$, and $\bar{\mu} \in B(\xi)$ such that $\xi(\bar{\mu})$ has exactly one orbit of quasi-transversal intersection between W^u_α and W^s_β $(\alpha \neq \beta)$ as in Cases (I) or (II) we have that ξ is stable at $\bar{\mu}$.

In the second chapter we give the proof of Theorem A and a generalization of it for the case when the critical element ρ is also a u-critical periodic orbit. In Chapter III, IV we give the proof of Theorems B and C respectively.

CHAPTER II

Proof of Theorem A

(II.1) Let $\{X_\mu\} \in \Gamma_6$ and $\mu_o \in B(\{X_\mu\})$ be such that $X = X_{\mu_o}$ has an orbit of quasi-transversal intersection between W^u_p and W^s_σ where p is a singularity with real weakest contraction and σ is a periodic orbit with real weakest expansion.

(II.1.1) Note. Since there are not cycles between the critical elements of the vector field X, it is possible to define a partial order between these elements by: $\alpha \le \beta \Leftrightarrow W^u_\alpha \cap W^s_\beta \neq \phi$. We extend this partial order to a total order between these elements. We have: $\alpha_1 \le \ldots \le \alpha_k \le p \le \sigma \le \beta_1 \le \ldots \le \beta_\ell$. The arguments of the Theorem A for the case where there exists a critical element β of the vector field such that $p \le \beta \le \sigma$ are similar to the case where does not exist such element. We assume this last hypothesis.

(II.1.2) Bifurcations

Let $U(X) \subset \mathfrak{X}^\infty(M^n)$ be a neighborhood of the vector field X such that each $\tilde{X} \in U(X)$ satisfy: $\Omega(\tilde{X}) = \{\alpha_1(\tilde{X}), \alpha_2(\tilde{X}), \ldots, \alpha_k(\tilde{X}), p(\tilde{X}), \sigma(\tilde{X}), \beta_1(\tilde{X}), \ldots, \beta_\ell(\tilde{X})\}$ where each one of these critical elements is hyperbolic and near of the respective critical element of the vector field X. The following proposition is similar to [P-T pg. 396] and its proof is also similar.

Proposition. Let X as below. There exists a neighborhood $U(X) \subset \mathfrak{X}^\infty(M^n)$ of the vector field X such that if $\tilde{X} \in U(X)$ has an orbit of quasi-transversal intersection between $W^u_{\tilde{\alpha}}$ and $W^s_{\tilde{\beta}}$ then,

(a) $W^s_{\tilde{\beta}} \cap W^u_{\tilde{\sigma}} \neq \phi$, (b) $\tilde{\alpha} = \tilde{p}$.

It is clear, from the proposition, that the unique possible bifurcations for vector fields in $U(X)$ are (a) quasi-transversal intersection between $W^u_{\tilde{p}}$ and $W^s_{\tilde{\sigma}}$ or (2) non-transversal intersection between $W^s_{\tilde{\beta}}$ and $W^u_{\tilde{p}}$, where $\tilde{\beta}_j$ is a critical element of saddle type such that $W^s_{\tilde{\beta}_j} \cap W^u_{\tilde{\sigma}} \neq \phi$.

We also note that given a family $\{X_\mu\} \in \Gamma_6$ as above there exists a neighborhood $U(\{X_\mu\}) \subset \mathfrak{X}^\infty_1(M^n)$ of the family $\{X_\mu\}$ such that: any $\{\tilde{X}_\mu\} \in U(\{X_\mu\})$ has one parameter value $\tilde{\mu}_o$ such that $\tilde{X} = \tilde{X}_{\tilde{\mu}_o}$ has an orbit of quasi-transversal intersection between $W^u_{\tilde{p}}$ and $W^s_{\tilde{\sigma}}$. Moreover, all the required hypothesis for $\{X_\mu\}$ in a neighborhood of μ_o are also satisfied for $\{\tilde{X}_\mu\} \in U(\{X_\mu\})$ in a neighborhood of $\tilde{\mu}_o$.

(II.1.3) <u>Remarks</u>

(a) In the sequel, any construction that we do for the family $\{X_\mu\}$ we will suppose constructed to the family $\{\tilde{X}_\mu\}$.

(b) We extend the order defined at (II.1.1) to all the family $\{X_\mu\}$, μ near μ_o.

(c) We do the proof of Theorem A in the case when σ has a unique simple u-criticality, say with ρ, and $\rho = \beta_1$ in the order defined at (II.1.1).

(II.1.4) <u>Definitions and previous lemmas</u>

Let τ be the period of the periodic orbit σ. Let $I_1 = [\mu_1, \mu_2] \subset I$ be a small neighborhood of μ_o in I. $\Sigma_q \subset M$ will denote a codimension one submanifold transversal to the vector field X_μ at $q \in \sigma$ and such that $X_{\mu,\tau}(V_q) \subset \Sigma_q$ for all $\mu \in I_1$, $V_q \subset \Sigma_q$ a small neighborhood of q in Σ_q. We denote by $P_\mu : V_q \to \Sigma_q$ the Poincaré map associated to the vector field X_μ.

Let X be the vector field on $M \times I$ given by $X(x,\mu) = (X_\mu(x),0)$.

Definition. An unstable foliation for $\{X_\mu\}$ or X at (q,μ_o) is a continuous foliation F^u of $V_q \times I$ such that for any $\mu \in I_1$:

(a) the leaves are C^r discs, $1 \le r \le \infty$, varying continuously in the C^r topology and $F^u(q(\mu),\mu) = (W^u_{q(\mu)} \cap V_q) \times \{\mu\}$ ($q(\mu)$ is the hyperbolic fixed point associated to P_μ, near q);

(b) each leave $F^u(y,\mu)$ is contained in $V_q \times \{\mu\}$ for each $(y,\mu) \in V_q \times \{\mu\}$.

(c) F^u is X-invariant, that is: $X_\tau(F^u(y,\mu)) \supset F^u(X_\tau(y,\mu))$ for each $y, X_{\mu,\tau}(y) \in V_q$ [$X_t(y,\mu) = (X_{\mu,t}(y),\mu)$ is the flow associated to X].

A stable foliation for $\{X_\mu\}$ or X at (q,μ_o) is an unstable foliation for $\{-X_\mu\}$ or $-X$ at (q,μ_o).

We know that there exists a small neighborhood $U_p \subset M$ of p and $I_1 \subset I$ of μ_o such that for any $\mu \in I_1$ there exists a hyperbolic singularity $p(\mu) \in U_p$, for the vector field X_μ, where $p(\mu_o) = p$.

Definition. An unstable foliation for $\{X_\mu\}$ or X at (p,μ_o) is a continuous foliation F^u of $U_p \times I_1$ such that for any $\mu \in I_1$:

(a) the leaves are C^r discs, $1 \le r \le \infty$, varying continuously in the C^r-topology and $F^u(p(\mu),\mu) = (W^u_{p(\mu)} \cap U_p) \times \{\mu\}$;

(b) each leave $F^u(y,\mu)$ is contained in $U_p \times \{\mu\}$ for each $(y,\mu) \in U_p \times \{\mu\}$;

(c) F^u is X-invariant, that is: $X_t(F^u(x,\mu)) \supset F^u(X_t(x,\mu))$, for each $t \ge 0$, x and $X_{\mu,t}(x) \in U_p$.

A stable foliation for $\{X_\mu\}$ or X at (p,μ_o) is an unstable foliation for $\{-X_\mu\}$ or $-X$ at (p,μ_o).

It is clear that we have the same definition for any hyperbolic periodic orbit (or singularity) of the vector field X_{μ_o}.

Let us consider, as above, a family $\{X_\mu\}$ and their critical elements in the total order: $\alpha_{1,\mu} \leq \alpha_{2,\mu} \leq \dots \leq \alpha_{k,\mu}$.

<u>Definition</u>. We call a system of unstable foliations $F^u_{\alpha_1}, \dots, F^u_{\alpha_k}$ a compatible system when:

(a) if a leaf F of $F^u_{\alpha_n}$ intersect a leaf $\tilde{F}$ of $F^u_{\alpha_i}$, $\alpha_n \leq \alpha_i \leq \alpha_k$ then $\tilde{F} \subset F$.

(b) The restriction of $F^u_{\alpha_i}$ to a leaf of $F^u_{\alpha_n}$ is a C^2-foliation.

Similarly we define a compatible system of stable foliations for $\beta_1 \leq \beta_2 \leq \dots \leq \beta_\ell$.

<u>Lemma</u>. There exists a compatible system of foliations $F^u_{\alpha_1}, \dots, F^u_{\alpha_k}$.

<u>Proof</u>: See [P-T] pg. 410.

Similarly, there exists a compatible system of foliations $F^s_{\beta_1}, \dots, F^s_{\beta_\ell}$.

(II.1.5) <u>A necessary condition</u>

We want to construct a one parameter family of homeomorphisms $H_\mu : M \to M$ which sends orbits of the vector field X_μ into orbits of the vector field $\tilde{X}_{\rho(\mu)}$ preserving orientation and such that the map $\mu \to H_\mu$, $I \to \text{Homeo}(M)$ is continuous.

Let q, Σ_q be as in (II.1.4). Suppose that we have this one parameter family H_μ such that $H_\mu(\Sigma_q) \subset \Sigma_{\tilde{q}}$, $\forall \mu \in I_1$. It is clear that the restriction $h_\mu = H_\mu\big|_{\Sigma_q}$ must send orbits of P_μ in Σ_q into orbits of $\tilde{P}_{\rho(\mu)}$ in $\Sigma_{\tilde{q}}$. It also must sends $W^u_{p(\mu)} \cap \Sigma_q$ into $W^u_{\tilde{p}(\rho(\mu))} \cap \Sigma_{\tilde{q}}$.

Lemma: [necessary condition for the existence of the family H_μ].

In the above conditions: the homeomorphism h_{μ_o} must send the strong unstable foliation $F^{uu}_{q(\mu_o)} \subset W^u_{q(u_o)} \cap V_q$ onto the strong unstable foliation $F^{uu}_{\tilde{q}(\tilde{\mu}_o)} \subset W^u_{\tilde{q}(\tilde{\mu}_o)} \cap V_{\tilde{q}}$.

Proof: We apply Proposition (2.3) of [dM-P-VS] to conclude that transversal elements to $W^{cs}_{q(\mu)}$ are sent by h_μ, into transversal elements to $W^{cs}_{\tilde{q}(\rho(\mu))}$. In our particular case, we apply this result to $W^u_{p(\mu)} \cap \Sigma_q$ (when $n-u_p = s_\sigma$) or to a foliation of $W^u_{p(\mu)} \cap \Sigma_q$ transversal to $W^{cs}_{q(\mu)}$ (when $n-u_p \le s_\sigma - 1$). Let us see the case $n-u_p = s_\sigma$ ($u_p = \dim(W^u_p)$, $s_\sigma = \dim(W^s_\sigma)$).

Let $(W^u_{p(\mu)} \cap \Sigma_q)_1$ be a connected component of the intersection $W^u_{p(\mu)} \cap \Sigma_q$. Let $\psi_\mu: (V_q, q(\mu)) \to (\mathbb{R}^{n-1}, 0)$ be C^s-coordinates, varying C^s with the parameter wich gives a linearization of P_μ. We suppose that $\psi_\mu(W^u_{q(\mu)} \cap V_q) \subset \{z_{u_\sigma} = \ldots = z_{n-2} = 0\}$, $\psi_\mu(W^{cs}_{q(\mu)} \cap W^u_{q(\mu)} \cap V_q) = \{z_1 = \ldots = z_{n-2} = 0\}$. Take a point $(0,\ldots,0,z^o_{n-1})$, $z^o_{n-1} \neq 0$ and denote by $F^{uu}_q(z^o_{n-1},\mu)$ the leaf of the foliation F^{uu}_q wich contains $\psi^{-1}(0,\ldots,0,z^o_{n-1},\mu)$ ($\psi = (\psi_\mu,\mu)$). We apply the λ-lemma ([dM-P]-1 pg. 80) to find a sequence of the parameter $\mu_n \to \mu_o$ such that for $m \ge m_o$, as big as necessary, $P^m((W^u_{p(\mu_m)} \cap \Sigma_q)_1, \mu_m)$ is ϵ-C^s near of the leaf $F^{uu}_q(z^o_{n-1},\mu_m)$. We obtain the result now as a consequence of the continuous variation (with μ) (a) of the foliation $F^{uu}_{q(\mu)}$; (b) of the homeomorphism h_μ, and by the fact that h_μ must send $W^u_{p(\mu)} \cap \Sigma_q$ onto $W^u_{\tilde{p}(\rho(\mu))} \cap \Sigma_{\tilde{q}}$.

(II.2) Proof of Theorem A in the case that β_1 is a hyperbolic periodic orbit which has a simple u-criticality with σ.

(II.2.1) Let us suppose defined a reparametrization $\rho: [0,1] \to [0,1]$ such that $\rho(0) = 0$, $\rho(1) = 1$, $\rho(\mu_o) = \bar{\mu}_o$.

Following [P-T] we can consider defined homeomorphisms

$$H^s: \bigcup_{\mu\in I_1} \left(\bigcup_{j=1}^{k} (W^s_{\alpha_{j,\mu}} \times \{\mu\})\right) \to \bigcup_{\mu\in \tilde I_1} \left(\bigcup_{j=1}^{k} (W^s_{\tilde\alpha_{j,\mu}} \times \{\mu\})\right)$$

$$H^u: \bigcup_{\mu\in I_1} \left(\bigcup_{i=2}^{\ell} (W^u_{\beta_{i,\mu}} \times \{\mu\})\right) \to \bigcup_{\mu\in \tilde I_1} \left(\bigcup_{i=2}^{\ell} (W^u_{\tilde\beta_{i,\mu}} \times \{\mu\})\right);$$

such that the restrictions $H^s_{j,\mu} = H^s\big|_{W^s_{\alpha_{j,\mu}} \times\{\mu\}}$ and $H^u_{i,\mu} = H^u\big|_{W^u_{\beta_{i,\mu}}\times\{\mu\}}$ are conjugations between $X_{\mu,t}\big|_{W^s_{\alpha_{j,\mu}}\times\{\mu\}}$ and $\tilde X_{\rho(\mu),t}\big|_{W^s_{\tilde\alpha_{j,\rho(\mu)}}\times\{\rho(\mu)\}}$, $X_{\mu,t}\big|_{W^u_{\beta_{i,\mu}}\times\{\mu\}}$ and $\tilde X_{\rho(\mu),t}\big|_{W^u_{\tilde\beta_{i,\rho(\mu)}}\times\{\rho(\mu)\}}$ respectively ($I_1 \subset I$ a small neighborhood of μ_o).

We proceed as in [dM-P] (this time at one parameter) to construct a homeomorphism: $H^u_1: \bigcup_{\mu\in I_1} (W^u_{\beta_{1,\mu}} \times\{\mu\}) \to \bigcup_{\mu\in\tilde I_1} (W^u_{\tilde\beta_{1,\mu}} \times\{\mu\})$ with the following properties:

(1) H^u_1 is compatible with H^u, i.e. if $\bigcup_{\mu\in I_1} (W^s_{\beta_{j,\mu}} \cap W^u_{\beta_{1,\mu}}) \times \{\mu\} \neq \emptyset$ we use $F^s_{\beta_j}$ and H^u to define a homeomorphism in a neighborhood of this set [this kind of construction is, for instance, in [P-T]).

(2) If $H^u_1(x,\mu) = (h^u_{1,\mu}(x), \rho(\mu))$, $x \in W^u_{\beta_{1,\mu}}$ then $h^u_{1,\mu} \circ P_\mu = \tilde P_{\rho(\mu)} \circ h^u_{1,\rho(\mu)}$.

(3) H^u_1 sends leaves of $F^s_{\beta_i} \cap (W^u_{\beta_{1,\mu}} \times\{\mu\})$ into leaves of $F^s_{\tilde\beta_i} \cap (W^u_{\tilde\beta_{1,\rho(\mu)}} \times\{\rho\{\mu\})$.

(4) H^u_1 sends leaves of $F^{uu}_{\beta_{1,\mu}} \subset W^u_{\beta_{1,\mu}} \times \{\mu\}$ into leaves of $F^{uu}_{\beta_{1,\rho(\mu)}} \subset W^u_{\tilde\beta_{1,\rho(\mu)}} \times \{\rho(\mu)\}$. Moreover, if $h^{uu}_{1,\mu}: Q^1_\mu \to \tilde Q^1_{\rho(\mu)}$

is the homeomorphism in the corresponding spaces of leaves $Q^1_\mu = (W^u_{\beta_{1,\mu}} \times\{\mu\} - W^{uu}_{\beta_{1,\mu}} \times\{\mu\})/F^{uu}_{\beta_{1,\mu}}$, which is used to define h^u_1 then $h^{uu}_{1,\mu}$ can be defined in an arbitrary way in a fundamental domain (see [dM-P] pg. 327).

(5) Given $q_1 \in \beta_1$ and Σ_{q_1} a transversal section to the flow $X_{\mu,t}$, invariant by X_{μ,τ_1} (τ_1 = period of the periodic orbit β_1), $\forall \mu \in I_1$, it is possible to construct a one dimensional, invariant, central foliation, F^c_1, such that the leaf through a point $(x,\mu) \in W^u_{q_{1,\mu}} \times \{\mu\}$, $F^c_1(x,\mu)$, is contained in $W^u_{q_{1,\mu}} \times \{\mu\}$, is transversal to leaves of the foliation $F^{uu}_{q_{1,\mu}} \subset W^u_{q_{1,\mu}} \times \{\mu\}$ at $W^u_{q_{1,\mu}} \times \{\mu\}$ and is compatible with the leaves of the foliations $F^s_{\beta_{i,\mu}}$, $i = 2,\ldots,\ell$. [These last foliations were defined in (II.2.1)]. Also, the homeomorphism $H^u_{q_{1,\mu}} = H^u_1\big|_{W^u_{q_{1,\mu}}\times\{\mu\}}$ sends $F^c_1\big|_{W^u_{q_{1,\mu}}\times\{\mu\}}$ into $\tilde{F}^c_1\big|_{W^c_{\tilde{q}_{1,\rho(\mu)}}\times\{\rho(\mu)\}}$.

(6) We can also suppose that there are center-stable manifolds $W^{cs}_{\beta_1} = \bigcup_\beta W^{cs}_{\beta_{1,\mu}} \times \{\mu\}$ and $W^{cs}_{\tilde{\beta}_1} = \bigcup_\mu W^{cs}_{\tilde{\beta}_{1,\mu}} \times \{\mu\}$ such that $((\Sigma_{q_1} \cap W^u_{\beta_{1,\mu}}) \times \{\mu\}) \cap W^{cs}_{\beta_1}$ is a leaf of F^c_1 and $H^u_{q_{1,\mu}}(((\Sigma_{q_1} \cap W^u_{\beta_{1,\mu}})\times\{\mu\} \cap W^{cs}_{\beta_1}) \subset ((\Sigma_{\tilde{q}_{1,\rho(\mu)}} \cap W^u_{\tilde{\beta}_{1,\rho(\mu)}}) \times \{\rho(\mu)\}) \cap W^{cs}_{\tilde{\beta}_1}$.

<u>Note</u>: In the sequel we assume that we have a stable foliation $F^s_{q_1}$ of $W^{cs}_{\beta_1} \cap (\Sigma_{q_1} \times I_1)$ such that its leaves are contained in $W^{cs}_{\beta_{1,\mu}} \times \{\mu\}$ and the leaf through $(q_{1,\mu},\mu)$ is $(W^s_{q_{1,\mu}} \cap \Sigma_{q_1}) \times \{\mu\}$, $\forall \mu \in I_1$. Iterating by the flow we have a stable foliation $F^s_{\beta_1}$.

(II.2.2) Let $W^{cs}_{q(\mu)}$ be a center-stable manifold at $q(\mu)$ in Σ_q and $W^c_{q(\mu)} = W^{cs}_{q(\mu)} \cap W^u_{q(\mu)}$. For each parameter value μ, we take two points $(z(\mu),\mu)$, $(w(\mu),\mu)$ in $W^c_{q(\mu)} \times \{\mu\}$ with $z(\mu)$ and $w(\mu)$ at different sides of $q(\mu)$ in $W^c_{q(\mu)}$, in such way that the maps, $\mu \to z(\mu)$ and $\mu \to w(\mu)$ are C^s-curves and the variation of the points $z(\mu)$, $w(\mu)$ are small with μ.

Let $A_{1,\mu} \subset F^{uu}_q(z(\mu),\mu)) \subset W^u_{q(\mu)} \times \{\mu\}$ and $A_{2,\mu} \subset \subset F^{uu}_q(w(\mu),\mu) \subset W^u_{q(\mu)} \times \{\mu\}$ be closed discs varying continuously with the parameter μ. Let us consider a $(u_\sigma-3)$-dimensional sphere, $S^{u_\sigma-3}_\mu \subset W^{uu}_{q(\mu)} \times \{\mu\}$ and for each point $(x,\mu) \in S^{u_\sigma-3}_\mu$ a one-dimensional leaf $F^c_q(x,\mu) \subset W^u_{q(\mu)} \times \{\mu\}$ transversal to $W^{uu}_{q(\mu)} \times \{\mu\}$ in $W^u_{q(\mu)} \times \{\mu\}$ which intersects $\partial A_{1,\mu}$ in a unique point [resp. intersects $\partial A_{2,\mu}$ at a unique point]. We denote by $C_{q(\mu)} = \bigcup_{(x,\mu)\in S^{u_\sigma-3}_\mu} F^c_q(x,\mu)$ and $C_q = \bigcup_\mu C_{q(\mu)}$.

Now we define a fundamental domain D^u_q for the Poincaré map $p(x,\mu) = (P_\mu(x),\mu)$ in $\bigcup_\mu W^u_{q(\mu)} \times \{\mu\}$ by: D^u_q will be the fundamental domain wich has external boundary $\bigcup_\mu (A_{1,\mu} \cup A_{2,\mu} \cup C_{q(\mu)})$ and internal boundary $P^{-1}(\bigcup_\mu (A_{1,\mu} \cup A_{2,\mu} \cup C_{q(\mu)}))$.

Since we have $W^s_{\beta_1} \cap W^u_\sigma \neq 0$ then $W^s_{\beta_1} \cap V_q$ is not empty and so it is the union of disjoint connected components. We denote the component of $(W^{cs}_{\beta_{1,\mu}} \cap V_q) \times \{\mu\}$ which intersects D^u_q by $\Lambda_{1,\mu} \subset V_q \times \{\mu\}$.

Let us now define in $U_{q_1} \times I_1 \subset \Sigma_{q_1} \times I_1$ a singular center-stable foliation $F^{cs}_{q_1}$ such that, (i) the leaf through $(x,\mu) \in \in U_{q_1} \times I_1$, $F^{cs}_{q_1}(x,\mu)$, is entirely contained in $U_{q_1} \times \{\mu\}$; (ii) the intersection $F^{cs}_{q_1}(x,\mu) \cap W^u_{q_1} \times \{\mu\}$ is a leave of the foliation F^c_1; (iii) the leaves of $F^{cs}_{q_1}$ are compatible with the leaves of the foliation $F^s_{\beta_2},\ldots,F^s_{\beta_\ell}$ (see [dM-P] pg. 332).

The intersection of the saturation by the flow of

$X(x,\mu) = (X_\mu(x),0)$ of these center-stable foliation with $D_q^u \cap$ $\cap\ (V_q \times \{\mu\})$ is as in Figure II.2.2.I.

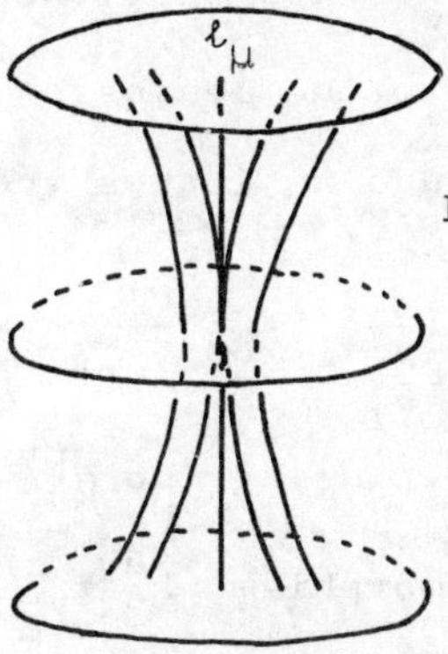

Figure II.2.2.I

Let $(x_o(\mu),\mu) \in D_q^u \cap (W_{\beta_{1,\mu}}^s \times \{\mu\})$ and $\ell_\mu \subset \Lambda_{1,\mu} \cap$ $\cap\ (W_{q(\mu)}^u \times \{\mu\})$ be a curve such that $(x_o(\mu),\mu) \in \ell_\mu$. We know, by hypothesis, that ℓ_μ is transversal to $F_{q(\mu)}^{uu}$ in $W_{q(\mu)}^u \times \{\mu\}$. Let us consider the leaves $F_q^{uu}(x_o(\mu)\pm\varepsilon,\mu) \subset W_{q(\mu)}^u \times \{\mu\}$, $(\varepsilon > 0$ small) then we see the intersection $F_q^{uu}(x_o(\mu)\pm\varepsilon,\mu) \cap$ $\cap\ F_{\beta_{1,\mu}}^{cs}$ $(F_{\beta_{1,\mu}}^{cs} = F_{\beta_1}^{cs} \cap (M \times \{\mu\}))$. It is clear that, for each leaf of $F_{\beta_1}^{cs}$, there is at least one and at most two points in $F_q^{uu}(x_o(\mu)+\varepsilon,\mu) \cup F_q^{uu}(x_o(\mu)-\varepsilon,\mu)$ which belongs to the leave. We use this fact and the definition of $H_{\beta_{1,\mu}}^u$ to construct homeomorphisms $H_\mu^\pm\colon A_\mu^\pm \to A_{\rho(\mu)}^\pm$ ($A_\mu^\pm$ is a disc contained in the intersection $F_q^{uu}(x_o(\mu)\pm\varepsilon,\mu) \cap F_{\beta_{1,\mu}}^{cs}$).

Let us consider a $C^{(o)}$ vector field Y_μ (resp. $\tilde{Y}_{\rho(\mu)}$) in $U(x_o(\mu),\mu) \subset D_q^u \cap (V_q \times \{\mu\})$ ($U(x_o(\mu),\mu)$ small neighborhood of $(x_o(\mu),\mu)$ in $D_q^u \cap (V_q \times \{\mu\})$) which has a singularity at the point $(x_o(\mu),\mu)$ and its orbits (with exception of those through $(x_o(\mu),\mu)$

are the leaves of the foliation $F^{cs}_{\beta_1} \cap U(x_0(\mu),\mu)$.

We proceed as in [dM-P] pg. 326, to construct a one-dimensional foliation in $D^u_q \cap (W^u_{q(\mu)} \times \{\mu\})$, varying continuously with μ, which is compatible with the intersections $(W^s_{\beta_{j,\mu}} \times \{\mu\}) \cap D^u_q$, $j = 2,\ldots,\ell$, transversal to the leaves of $F^{uu}_{q(\mu)}$ (outside of $(x_0(\mu),\mu)$) and which inlcude as leaves the orbits of Y_μ and the leaves of F^c_q.

Using $F^s_{\beta_1} \subset W^{cs}_{\tilde{\beta}_1}$, $F^s_{\beta_1} \subset W^{cs}_{\tilde{\beta}_1}$ and h^{uu}_1 we can define a homeomorphism $h^{uu}_2 : \bigcup_\mu \ell_\mu \to \bigcup_\mu \tilde{\ell}_\mu$. Using the π^{uu}-projections and the map h^{uu}_2 we can define a homeomorphism $h^{uu}_3 : \bigcup_\mu I^o_\mu \to \bigcup_\mu \tilde{I}^o_\mu$ ($I^o_\mu \subset D^u_q \cap$ $\cap (W^c_{q(\mu)} \times \{\mu\})$ small neighborhood of $\pi^{uu}(x_0(\mu),\mu)$).

In a similar way as in [dM-P] pg. 329-330 we construct a homeomorphism $H^u_q : \bigcup_\mu (W^u_{q(\mu)} \times \{\mu\}) \to \bigcup_\mu (W^u_{\tilde{q}(\mu)} \times \{\mu\})$ such that its restriction $H^u_{q(\mu)} = H^u_q | W^u_{q(\mu)} \times \{\mu\} : W^u_{q(\mu)} \times \{\mu\} \to W^u_{\tilde{q}(\rho(\mu))} \times \{\rho(\mu)\}$ satisfies:

(1) $H^u_{q(\mu)}$ is compatible with $H^u_{1,\mu}$ and H^u_μ.

(2) $H^u_{q(\mu)} \circ (P_\mu,\mu) = (P_{\rho(\mu)},\rho(\mu)) \circ H^u_{q(\mu)}$.

(3) $H^u_{q(\mu)}$ sends leaves of $F^s_{\beta_i} \cap (W^u_{q(\mu)} \times \{\mu\})$ into leaves of $F^s_{\tilde{\beta}_i} \cap (W^u_{\tilde{q}(\rho(\mu))} \times \{\rho(\mu)\}$, $i=2,\ldots,\ell$. $H^u_{q(\mu)}$ also sends leaves of $F^{cs}_{\beta_1} \cap (W^u_{q(\mu)} \times \{\mu\})$ into leaves of $F^{cs}_{\tilde{\beta}_1} \cap (W^u_{\tilde{q}(\rho(\mu))} \times \{\rho(\mu)\})$ (in the neighborhood $U(x_0(\mu),\mu)$).

(4) $H^u_{q(\mu)}$ sends leaves of $F^{uu}_q \cap (W^u_{q(\mu)} \times \{\mu\})$ into leaves of $F^{uu}_{\tilde{q}} \cap (W^u_{\tilde{q}(\rho(\mu))} \times \{\rho(\mu)\})$ in such way that the induced homeomorphism $h^{uu}_{3,\mu} : W^c_{q(\mu)} \times \{\mu\} \to W^c_{\tilde{q}(\rho(\mu))} \times \{\rho(\mu)\}$ depends on $h^{uu}_{1,\mu}$.

(5) There exists a one-dimensional invariant, singular center-foliation, compatible with $F^s_{\beta_i} \cap (W^u_{q(\mu)} \times \{\mu\})$ and $F^{cs}_{\beta_1} \cap$

$\cap\ (W^u_{q(\mu)}\times\{\mu\})$ in $W^u_{q(\mu)}\times\{\mu\}$ which is sent, by the map $H^u_{q(\mu)}$, into a foliation with similar properties defined in $W^u_{\tilde{q}(\rho(\mu))}\times\{\rho(\mu)\}$ [denoted by $F^c_{q(\mu)}$ and $F^c_{\tilde{q}(\rho(\mu))}$, respectively].

(II.2.3) Now, we are going to construct partial foliations F^{cu}_p and F^{su}_p in a neighborhood $U_p\times I_1$ of (p,μ_o) in $M\times I_1$. [For F^{cu}_p we take $F^{cu}_p(p(\mu),\mu) = (W^{cu}_{p(\mu)} \cap U_p) \times \{\mu\}$].

We will construct these foliations compatible with the unstable foliation defined in (II.1.4). A typical leaf of F^{su}_p will have dimension equal to the dimension of a typical leaf of F^{cu}_p plus one. The space of leaves of $F^{cu}_p \cap (U_p\times\{\mu\})$ (resp. $F^{su}_p \cap (U_p\times\{\mu\})$) will be the union of two cones in $W^s_{p(\mu)}\times\{\mu\}$ (resp. a sphere in $W^{ss}_{p(\mu)}\times\{\mu\}$). As usual, we wish the leaves of F^{cu}_p to be C^1 embedded discs varying continuously in the C^1 topology (resp. F^{su}_p). Moreover, a leaf of F^{cu}_p (resp. F^{su}_p) through a given point transversely intersects $W^s_{p(\mu)}\times\{\mu\}$ in $U_p\times\{\mu\}$ in a C^1-curve (resp. a two-dimensional C^1-disc).

We introduce these foliations to extend homeomorphisms (which we will define later) to the neighborhood $U_p\times I_1$ of (p,μ_o).

Consider a continuous family of C^2-cylinders C_μ and C^2-discs $D_\mu \subset F_\mu$, $D^*_\mu \subset F^*_\mu$ where F_μ, F^*_μ are leaves of the foliation $F^{ss}_{p(\mu)}$ such that $C_\mu \cup D_\mu \cup D^*_\mu$ is a fundamental domain for X_μ in $W^s_{p(\mu)}$, for each $\mu \in I_1$. We take the cylinders C_μ transversal to $F^{ss}_{p(\mu)}$ and disjoints of the component of $W^u_{\alpha_i(\mu)} \cap W^s_{p(\mu)}$ which does not intersects $W^{ss}_{p(\mu)}$. Moreover, we assume that X_μ is tangent to C_μ along the intersection $C_\mu \cap (D_\mu \cup D^*_\mu)$.

It is possible ([P-T] prop. (3.1)) to construct a one-dimensional foliation F^c_μ of C_μ in such way that its leaves subfoliate leaves of $F^u_{\alpha_i} \cap (U_p\times\{\mu\})$, i=1,...,k. We now raise F^c_μ to a

(u_p+1)-dimensional foliation, $R^{cu}_\mu \subset U_p \times \{\mu\}$, compatible with $F^u_{\alpha_1},\ldots,F^u_{\alpha_k}$.

For $(x,\mu) \in S^{ss}_{p(\mu)}$ we define: $F^{su}_p(x,\mu) = \bigcup_{t\geq 0} X_t(R^{cu}_\mu)$ ($S^{ss}_{p(\mu)} \subset W^{ss}_{p(\mu)} \times \{\mu\}$ a (s_p-2)-dimensional sphere).

Let us now denote by $K_\mu \subset W^s_{p(\mu)} \times \{\mu\}$ the union of two cones $K_{1,\mu}$, $K_{2,\mu}$ whose boundary include the discs D_μ and D^*_μ (see Figure II.2.3.I).

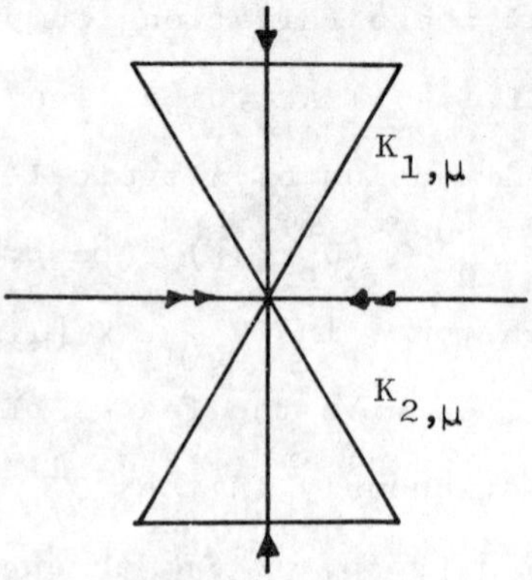

Figure II.2.3.I

Through each point of $\partial D_\mu \cup \partial D^*_\mu$ we have an u_p-dimensional leaf. We extend this foliation to all of K_μ in a compatible way with F^{su}_p and $F^u_{\alpha_1},\ldots,F^u_{\alpha_k}$. If $F^u(x,\mu)$ denotes the leaf through $(x,\mu) \in K_\mu$ then we define:

$$F^{cu}_p(x,\mu) = \bigcup_{t\geq 0} X_t(F^u(x,\mu)).$$

(II.2.4)

<u>Definition</u>. A strong unstable foliation F^{uu} for P ($P = (P_\mu,\mu)$ is the Poincaré map associated to $\Sigma_q \times I_1$, $q \in \sigma$) is a continuous foliation of $V_q \times I_1$ such that:

(a) the leaves are C^k-discs, varying continuously in the C^k-topology, and $F^{uu}(q(\mu),\mu) = (W^{uu}_{q(\mu)} \cap V_q) \times \{\mu\}$;

(b) for each $\mu \in I_1$, $V_q \times \{\mu\}$ is the union of transversal leaves to $W^{cs}_{q(\mu)} \times \{\mu\}$ in $V_q \times \{\mu\}$;

(c) the foliation is P-invariant, i.e. if $F^{uu}(x,\mu)$ is the leaf through $(x,\mu) \in V_q \times \{\mu\}$ then $P(F^{uu}(x,\mu)) \supset F^{uu}(P(x,\mu))$.

Similarly to (II.1.4) we can state the existence of a compatible system of foliations (and partial foliations) $F^u_{\alpha_1}, \ldots, F^u_{\alpha_k}$, F^{cu}_p, F^{su}_p, F^{uu}_σ.

Remarks

(a) The projection $\pi^{uu}: U_q \times I_1 \to \bigcup_\mu W^{cs}_{q(\mu)} \times \{\mu\}$ can be obtained of class C^1 outside a small neighborhood of the intersection of the unstable manifolds with $U_q \times I_1$;

(b) the restriction of this projection to any unstable manifold can be obtained of class C^k.

Let us consider the component $\bigcup_\mu \Lambda_{1,\mu}$ of the intersection $W^{cs}_{\beta_1} \cap (U_q \times I_1)$ (notation from II.2.2). Associated to each point $(x,\mu) \in \Lambda_{1,\mu} \cap (W^u_{q(\mu)} \times \{\mu\})$ we have a leaf of the foliation $F^s_{\beta_{1,\mu}} \subset W^{cs}_{\beta_{1,\mu}} \times \{\mu\}$ such that (x,μ) belongs to this leaf. We denote this leaf by $(F^s_{\beta_{1,\mu}}(x),\mu)$. From λ-lemma [dM-P]-1 pg. 82 we know that the projection $\pi^{uu}(F^s_{\beta_{1,\mu}}(x),\mu)$ is a leaf topologically transversal to $W^c_{q(\mu)} \times \{\mu\}$ in $W^{cs}_{q(\mu)} \times \{\mu\}$. Using these projections we construct a stable foliation $F^s_{q(\mu)}$ of $W^{cs}_{q(\mu)} \times \{\mu\}$, varying continuously with the parameter $\mu \in I_1$. The resultant folation obtained in this way is a $C^{(o)}$ foliation of $W^{cs}_q =$ $= \bigcup_\mu W^{cs}_{q(\mu)} \times \{\mu\}$.

Before proving our theorem, let us recall, the well-known Isotopy Extension Theorem [PA].

Let N be a C^r compact manifold, $r \geq 1$, and $A \subset \mathbb{R}^s$ an open set. Let M be a C^ℓ manifold with $\dim(M) > \dim(N)$, $r \leq \ell$. We indicate by $C^k_A(N\times A, M\times A)$ the set of C^k mappings $f: N\times A \to M\times A$ such that $\pi = \pi' \circ f$ with the C^k-topology $1 \leq k \leq r$. Here π, π' denotes the natural projections $\pi: N\times A \to A$, $\pi': M\times A \to A$. Let $\mathrm{Diff}^k_A(M\times A)$ be the set of C^k diffeomorphisms $\varphi: M\times A \to M\times A$ such that $\pi' = \pi\circ\varphi$, again with the C^k topology.

Isotopy Extension Theorem: Let $i \in C^k_A(N\times A, M\times A)$ be an embedding and A' a compact subset of A. Given neighborhoods U of $i(N\times A)$ in $M\times A$ and V of the identity in $\mathrm{Diff}^k_A(M\times A)$, there exists a neighborhood W of i in $C^k_A(N\times A, M\times A)$ such that for each $j \in W$ there exists $\varphi \in V$ satisfying $\varphi\circ i = j$ restricted to $N\times A'$ and $\varphi(x) = x$ for all $x \notin U$.

For the case of mappings $f: N\times A \to M\times A$, $f(x,a) = (f_a(x), \rho(a))$ with $(x,a) \to f_a(x)$ of class C^k and $\rho: A \to A$ a homeomorphism such that $\rho(a) = a$ $\forall\, a \notin A'$, we can consider $h_\rho: M\times A \to M\times A$, $h_\rho(x,a) = (x, \rho^{-1}(a))$ and the composed map $h_\rho \circ f(x,a) = (f_a(x), a)$. It is clear that $h_\rho \circ f \in C^k_A(N\times A, M\times A)$. Let $i: N\times A \to M\times A$ be the inclusion map and suppose that $h_\rho \circ f \in W_i$ (from the theorem) then there exists $\varphi \in V(id) \subset \mathrm{Diff}^k_A(M\times A)$ such that $\varphi\circ i(x,a) = h_\rho \circ f(x,a)$, i.e. $(\varphi_1(x,a), \varphi_2(x,a)) = (f_a(x), a)$ and so $\varphi_{1,a}(x) = \varphi_1(x,a)$ is an extension of $f_a: N \to M$ varying C^k with $a \in A$. It is clear that the map $\bar{\varphi}: M\times A \to M\times A$ given by $\bar{\varphi}(x,a) = (\varphi_{1,a}(x), \rho(a))$ is an extension of f to all of $M\times A$.

(II.2.5) Now we will construct a homeomorphism $H^{cs}: W^{cs}_q \cap (U_q \times I_1) \to W^{cs}_{\tilde{q}} \cap (U_{\tilde{q}} \times \tilde{I}_1)$ with the followings properties:
(i) $\tilde{P}\circ H^{cs} = H^{cs}\circ P$, (ii) H^{cs} sends leaves of the foliation F^s_q into leaves of the foliation $F^s_{\tilde{q}}$; (iii) H^{cs} sends the inter-

sections $F^u_{\alpha_1} \cap W^{cs}_q \cap (U_q \times I_1), \ldots, F^u_{\alpha_k} \cap W^{cs}_q \cap (U_q \times I_1)$, $F^{cu}_p \cap W^{cs}_q \cap$ $\cap (U_q \times I_1)$, $F^{su}_p \cap W^{cs}_q \cap (U_q \times I_1)$, $(\bigcup_\mu W^u_{p(\mu)} \times \{\mu\}) \cap W^{cs}_q \cap (U_q \times I_1)$ into the corresponding intersections for the vector field $\tilde{X}$.

(II.2.5.(a)) Let us consider first the case $n-u_p = s_\sigma$.

Denote by $\pi^s: U^{cs}_q \to \bigcup_\mu W^c_{q(\mu)} \times \{\mu\}$ the F^s_q-projection $(U^{cs}_q = W^{cs}_q \cap U_q \times I_1)$.

Let $S^s_\mu \subset W^s_{q(\mu)} \times \{\mu\}$ be a $(s_\sigma-2)$-sphere varying continuously with μ and $D^s_\mu \subset W^s_{q(\mu)} \times \{\mu\}$ be a fundamental domain for P_μ such that $\partial D^s_\mu = S^s_\mu \cup P(S^s_\mu)$. Let $N^{cs}_\mu \subset (W^{cs}_{q(\mu)} \cap U_q) \times \{\mu\}$ be a fundamental neighborhood of D^s_μ. Define $(r_\mu, \mu) = (W^u_{p(\mu)} \times \{\mu\}) \cap N^{cs}_\mu$ and $(\lambda_\mu, \mu) = \pi^s(r_\mu, \mu)$.

We first set a C^0 reparametrization $\rho: I_1 \to \tilde{I}_1$ such that if μ_n is a parameter value such that $\pi^{uu}(x_o(\mu_n), \mu_n) =$ $= (P^{n-1}_{\mu_n}(\lambda_{\mu_n}), \mu_n)$, and $\tilde{\mu}_n$ is a parameter value such that $\tilde{\pi}^{uu}(\tilde{x}_o(\tilde{\mu}_n), \tilde{\mu}_n) = (\tilde{P}^{n-1}_{\tilde{\mu}_n}(\tilde{\lambda}_{\tilde{\mu}_n}), \tilde{\mu}_n)$ then: $\rho(\mu_n) = \tilde{\mu}_n$.

We assume, for the homeomorphism $h^{uu}: \bigcup_\mu W^c_{q(\mu)} \times \{\mu\} \to$ $\to \bigcup_\mu W^c_{\tilde{q}(\mu)} \times \{\mu\}$, the property: $h^{uu}(\lambda(\mu), \mu) = (\tilde{\lambda}(\rho(\mu)), \rho(\mu))$. The existence of a homeomorphism with this property is established in the Appendix.

Let us consider the intersections $m_\mu = (W^{cu}_{p(\mu)} \times \{\mu\}) \cap N^{cs}_\mu$, $F^{cu}_p \cap N^{cs}_\mu$ and $F^{su}_p \cap N^{cs}_\mu$. It is clear that m_μ is a C^s-curve (s = class of the linearizing coordinates at p and σ). Let $K_\mu = K^+_\mu \cup K^-_\mu$ be a cone such that its upper face (D^+_μ) and its lower face (D^-_μ) are included in leaves of the intersection $F^s_q \cap N^{cs}_q$ at different sides of the intersection $(W^s_{q(\mu)} \times \{\mu\}) \cap N^{cs}_\mu$.
(See Figure II.2.5.I.)

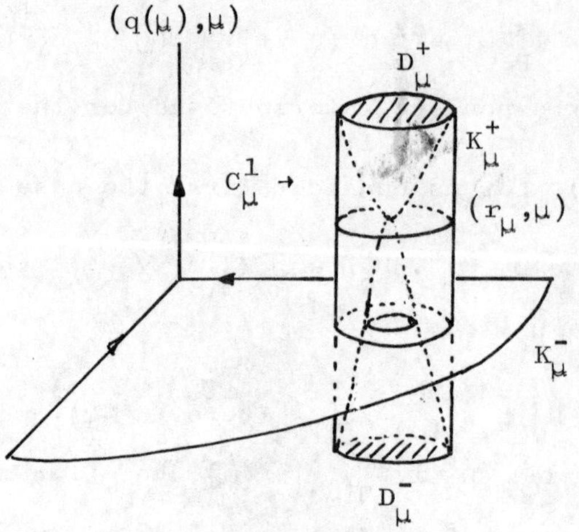

Figure II.2.5.I

This cone is the space of leaves of the intersection $F_p^{cu} \cap N_\mu^{cs}$.

Let $C_\mu^1 \subset N_\mu^{cs}$ be a C^2-cylinder whose intersection with K_μ is the union $D_\mu^+ \cup D_\mu^-$. We denote by $S_\mu \subset C_\mu^1 \cap F^s(r_\mu,\mu)$ a (s_p-2)-sphere which is the space of leaves of the intersection $F_p^{su} \cap W_\mu^{cs}$.

Let $(s_\mu,\mu) = \pi^s(D_\mu^+)$, $(i_\mu,\mu) = \pi^s(D_\mu^-)$. These points are at different sides of $(q(\mu),\mu)$ in $W^c_{q(\mu)} \times \{\mu\}$. Let $(\tilde{s}_{\rho(\mu)},\rho(\mu)) = H^{uu}(s_\mu,\mu)$ and $(\tilde{i}_{\rho(\mu)},\rho(\mu)) = h^{uu}(i_\mu,\mu)$. We construct in $\tilde{N}^{cs}_{\rho(\mu)}$ a cone $\tilde{K}_\mu$ (similar to K_μ) such that $\tilde{\pi}^s(\tilde{D}^+_{\rho(\mu)}) = (\tilde{i}_{\rho(\mu)},\rho(\mu))$.

To define our homeomorphism in $\bigcup_\mu D_\mu^+$ (similarly in $\bigcup_\mu D_\mu^-$) we proceed in a similar way as in [P-T] pg. 413 or [La] pg. 46-59. It is also clear that the definition of a homeomorphism $H_\mu^+ : D_\mu^+ \to \tilde{D}^+_{\rho(\mu)}$ give us an homeomorphism $\hat{H}_\mu^+ : \partial D_\mu^+ \to \partial\tilde{D}^+_{\rho(\mu)}$. Using $\hat{H}_\mu^+$ and the intersection of leaves of F_p^{su} with C_μ^1 we define an homeomorphism $H_\mu^{ss} : S_\mu \to \tilde{S}_{\rho(\mu)}$.

(II.2.5.(b)). Let us denote by T^c the one dimensional foliation determinated by the intersection of F_p^{su} with C_μ^1. Let $(x,\mu) \in S_\mu$ and $H_\mu^{ss}(x,\mu) \in \tilde{S}_{\rho(\mu)}$. Let us consider the respective leaves $F_p^{su}(x,\mu) \cap L_\mu^1$ and $F_{\tilde{p}}^{su}(H_\mu^{ss}(x,\mu)) \cap \tilde{L}^1_{\rho(\mu)}$ (L_μ^1 denotes a solid cylinder with boundary $D_\mu^+ \cup C_\mu^1 \cup D_\mu^-$). These are C^1-surfaces whose boundary is formed by $T^c(x,\mu)$ (resp. $\tilde{T}^c(H_\mu^{ss}(x,\mu))$ and two leaves of the intersection $F_p^{cu} \cap L_\mu^1$ (resp. $F_{\tilde{p}}^{cu} \cap \tilde{L}^1_{\rho(\mu)}$). (See Figure II.2.5.II).

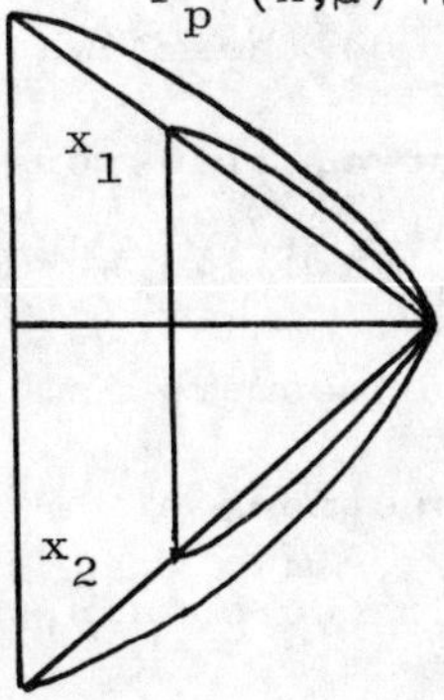

Figure II.2.5.VIII

Let us denote by $\xi(x,\mu)$ this surface, $\xi^+(x,\mu) = K_\mu^+ \cap \xi(x,\mu)$ and $\xi^-(x,\mu) = K_\mu^- \cap \xi(x,\mu)$. The homeomorphism already defined in the space of leaves of F_q^s (II.2.2.(4)) induces homeomorphisms $H^+(x,\mu): \xi^+(x,\mu) \to \xi^+(H_\mu^{ss}(x,\mu)$ and $H^-(x,\mu): \xi^-(x,\mu) \to \xi^-(H_\mu^{ss}(x,\mu))$ (also we use leaves of the foliation F_q^s).

Let now $(x_1,\mu) \in \xi^+(x,\mu)$ and denote by $F_p^{cu}(x_1,\mu)$ the associated leaf of the foliation F_p^{cu}. We consider the leaf $F_{\tilde{p}}^{cu}(H^+(x,\mu))$. Using the homeomorphism defined on the space of leaves of the foliation F_q^s and the intersections of leaves of F_q^s with $F_p^{cu}(x,\mu)$, (resp. $F_{\tilde{q}}^s$ with $F_{\tilde{p}}^{cu}(H^+(x_1,\mu))$ we can define a homeomorphism $H_\mu^{cu}(x_1)$: $F_p^{cu}(x_1,\mu) \cap \xi(x,\mu) \to F_{\tilde{p}}^{cu}(H^+(x_1,\mu)) \cap \tilde{\xi}(H_\mu^{ss}(x,\mu))$.

Varying the point (x_1,μ) on $(\xi^+(x,\mu) \cup \xi^-(x,\mu)-\{(r_\mu,\mu)\})$ and defining $H_\mu^{cu}(r_\mu,\mu) = (\tilde{r}_{\rho(\mu)},\rho(\mu))$ we obtain a homeomorphism in the intersection $F_p^{cu} \cap \xi(x,\mu)$.

To define in all of the surface $\xi(x,\mu)$ we proceed as follows. Let $(x_1,\mu) \in (\xi^+(x,\mu) - \{(r_\mu,\mu)\})$, $(x_2,\mu) \in (\xi^-(x,\mu) - \{(r_\mu,\mu)\})$ and $\ell_{x_2}^{x_1} \subset \xi(x,\mu)$ be a one dimensional C^1-curve, transversal to the intersections $F_q^s \cap \xi(x,\mu)$ in $\xi(x,\mu)$ such that $\partial\ell_{x_2}^{x_1} = \{(x_1,\mu),(x_2,\mu)\}$. Let $(\tilde{x}_1,\rho(\mu)) = H^+(x,\mu)(x_1,\mu)$, $(\tilde{x}_2,\rho(\mu)) = H^-(x,\mu)(x_2,\mu)$ be their images, let $\ell_{\tilde{x}_2}^{\tilde{x}_1} \subset \tilde{\xi}(H_\mu^{ss}(x,\mu))$ be a one dimensional C^1-curve, transversal to the intersections $F_{\tilde{q}}^s \cap \tilde{\xi}(H_\mu^{ss}(x,\mu))$ in $\tilde{\xi}(H_\mu^{ss}(x,\mu))$. We define a homeomorphism from $\ell_{x_2}^{x_1}$ onto $\ell_{\tilde{x}_2}^{\tilde{x}_1}$ using the homeomorphism defined on the space of leaves of F_q^s, the intersections of F_q^s with $\ell_{x_2}^{x_1}$ and the intersections of $F_{\tilde{q}}^s$ with $\ell_{\tilde{x}_2}^{\tilde{x}_1}$. We obtain, in this way, a homeomorphism $\xi(x,\mu) \to \tilde{\xi}(H_\mu^{ss}(x,\mu))$. Varying x (s.t. $(x,\mu) \in S_\mu$) we obtain a homeomorphism $H_\mu^{su}: F_p^{su} \cap L_\mu^1 \to F_{\tilde{p}}^{su} \cap \tilde{L}_{\rho(\mu)}^1$ for each $\mu \in I_1$.

To extend this homeomorphism to L_μ^1 we use (i) H_μ^+, H_μ^-, (ii) $F_p^{cu} \cap L_\mu^1$, $F_{\tilde{p}}^{cu} \cap \tilde{L}_{\rho(\mu)}^1$, (iii) the homeomorphism defined on space of leaves of F_q^s, (iv) the intersections $F_q^s \cap L_\mu^1$, $F_{\tilde{q}}^s \cap \tilde{L}_{\rho(\mu)}^1$.

We remark the following properties of the homeomorphism $H_\mu^1: L_\mu^1 \to \tilde{L}_{\rho(\mu)}^1$ defined as above:

(a) it sends $F_q^s \cap L_\mu^1$ into $F_{\tilde{q}}^s \cap \tilde{L}_{\rho(\mu)}^1$

(b) it sends $F_p^{cu} \cap L_\mu^1$ into $F_{\tilde{p}}^{cu} \cap \tilde{L}_{\rho(\mu)}^1$

(c) it sends $F_p^{su} \cap L_\mu^1$ into $F_{\tilde{p}}^{su} \cap \tilde{L}_{\rho(\mu)}^1$

(d) it sends $m_\mu \cap L_\mu^1$ into $\tilde{m}_{\rho(\mu)} \cap \tilde{L}_{\rho(\mu)}^1$

(e) it varies continuously with the parameter μ

(f) it is compatible with the homeomorphism defined on the space of leaves of F_q^s and with the homeomorphism H^s (II.2.1).

Now we want to extend H_μ^1 to all of $N_\mu^{cs} \subset V_q \cap W_{q(\mu)}^{cs}$.

To do this we consider the intersection $U_\mu^s = L_\mu^1 \cap D_\mu^s$. We have defined a homeomorphism $H_{s,\mu}: U_\mu^s \to \tilde{U}_{\rho(\mu)}^s$ such that its restriction to the boundary of U_μ^s is similar to the homeomorphism H_μ^{ss}. It is clear that proceeding as in [P-T] pg. 413 or [La] pg.63 we can construct a homeomorphism $\tilde{H}_\mu^s: D_\mu^s \to \tilde{D}_{\rho(\mu)}^s$ wich extends $H_{s,\mu}$ and such that conjugates the corresponding Poincaré maps.

To extend $\tilde{H}_\mu^s$ to all N_μ^{cs}, let us first define a one dimensional foliation $T_\mu \subset (N_\mu^{cs} - L_\mu^1)$ with leaves transversal to the intersections $F_q^s \cap N_\mu^{cs}$ and compatible with the foliation $T_\mu^c \subset C_\mu^1$ (that is, the projection map $\pi^c: \bigcup_\mu (N_\mu^{cs} - L_\mu^1) \cup C_\mu^1 \to \bigcup_\mu D_\mu^s$ is continuous). In a similar way we consider a foliation $\tilde{T}_{\rho(\mu)} \subset$ $\subset (\tilde{N}_{\rho(\mu)}^{cs} - \tilde{L}_{\rho(\mu)}^1)$ compatible with $\tilde{T}_{\rho(\mu)}^c \subset \tilde{C}_{\rho(\mu)}^1$. These two foliations are close one to the other one and compatibles with the respective Poincaré maps.

Thus we have defined homeomorphisms $h_{3,\mu}^{uu}$ (II.2.2.(4)) on the space of leaves of $F_q^s \cap (V_q \times \{\mu\})$ and $\tilde{H}_\mu^s$ on the space of leaves of T_μ. By the intersection of these leaves we define a homeomorphism $\tilde{H}_\mu^{cs}: N_\mu^{cs} \to \tilde{N}_{\rho(\mu)}^{cs}$.

To extend $\tilde{H}_\mu^{cs}$ to all of $U_q^{cs} \cap (V_q \times \{\mu\})$ we conjugate the corresponding Poincaré maps. We denote these homeomorphisms by $H_\mu^{cs}: U_q^{cs} \cap (V_q \times \{\mu\}) \to U_{\tilde{q}}^{cs} \cap (V_{\tilde{q}} \times \{\rho(\mu)\})$.

(II.2.5.(c)) <u>Let us now consider the case:</u> $n-u_p \leq s_\sigma - 1$.

Let $\gamma \subset W_p^u \cap W_\sigma^s$ be the orbit of quasi-transversal intersection and $r \in \gamma$. Let us denote by $\xi_\mu : (U_r, r) \to (\mathbb{R}^n, 0)$ the C^∞ diffeomorphism such that $(\xi_\mu)_*(X_\mu) = \frac{\partial}{\partial y_n}$ (i.e. $\xi_\mu(U_r \cap \Sigma_q) \subset \{y_n = 0\}$)

$$\xi_\mu(W^s_{\sigma(\mu)} \cap U_r) = \{y_1 = y_2 = \dots = y_{n-s_\sigma-1} = y_{n-1} = 0\}$$

$$\xi_\mu(W^u_{p(\mu)} \cap U_r) = \{y_{u_p} = y_{u_p+1} = \dots = y_{n-2} = 0,$$
$$y_{n-1} = Q(y_{n-s_\sigma}, \dots, y_{u_p-1}) + \varepsilon(\mu)\}$$

where $\varepsilon : \mathbb{R} \to \mathbb{R}$ is a C^∞ map such that $\varepsilon(\mu_o) = 0$, $\frac{d\varepsilon}{d\mu}(\mu_o) > 0$.

We will assume that:

$$\xi_\mu(W^{cs}_{\sigma(\mu)} \cap U_r) = \{y_1 = \dots = y_{n-s_\sigma-1} = 0\}$$

$$\xi_\mu(W^{cu}_{p(\mu)} \cap U_r) = \{y_{u_p} = \dots = y_{n-2} = 0\}.$$

(This is possible because $W^{cu}_{p(\mu)} \pitchfork W^s_{\sigma(\mu)}$ and $W^{cs}_{\sigma(\mu)} \pitchfork W^u_{p(\mu)}$, $\forall \mu \in I_1$).

Let us denote by $N^u_p(y^o_1, \dots, y^o_{u_p-1}, 0, \dots, 0, y^o_n)$ the normal space to $\{y_{u_p} = \dots = y_{n-2} = y_{n-1} = 0\}$ at the point $(y^o_1, \dots, y^o_{u_p-1}, 0, \dots, 0, y^o_n) = y_o$. It is clear that $N^u_p(y_o) = \{y_1 = \dots = y_{u_p-1} = y_n = 0\}$. Note that the intersection $N^u_p(y_o) \cap W^{cu}_{p(\mu)} = \{y_1 = y_2 = \dots = y_{m-2} = y_n = 0\}$ is a curve and we also have $N^u_p(y_o) \subset W^{cs}_{\sigma(\mu)}$.

We first define a homeomorphism on $U^{us}_\mu = W^{cu}_{p(\mu)} \cap W^{cs}_{q(\mu)} \cap U_r$ preserving $W^u_{p(\mu)} \cap U^{cs}_\mu$, the foliation $\mathcal{F}^s_q \cap U^{us}_\mu$ and a one dimensional foliation $F^c_\mu \subset U^{us}_\mu$ transversal to $F^s_q \cap U^{us}_\mu$ and to $W^u_{p(\mu)} \cap U^{us}_\mu$. Now, for each y_o as above, take a normal $N^u_p(y_o)$ such that $N^u_p(y_o) \cap U^{us}_\mu = F^c_\mu(y_o)$. On each $N^u_p(y_o)$ we define a homeomorphism as in II.2.5.(a) depending continuously on y_o. In this way we define a homeomorphism on a neighborhood of $W^u_{p(\mu)} \cap W^{cs}_{q(\mu)} \cap U_r$. To extend it to all of $W^{cs}_{q(\mu)}$ we proceed similarly as before (II.2.5.(b)). We leave this to the reader.

(II.2.6) Now we will construct a homeomorphism $H_q: V_q \times I_1 \to V_{\tilde{q}} \times \tilde{I}_1$ which will be a conjugation between the associated Poincaré maps $P(x,\mu) = (P_\mu(x),\mu)$ and $\tilde{P}(x,\mu) = (\tilde{P}_\mu(x),\mu)$.

Let us recall (from II.2.2) that we have a homeomorphism $H_q^u: \bigcup_{\mu\in I_1} (W^u_{q(\mu)} \times\{\mu\}) \to \bigcup_{\mu\in\tilde{I}_1} (W^u_{\tilde{q}(\mu)} \times\{\mu\})$ wich sends the level μ into the level $\rho(\mu)$. Also it sends a one-dimensional, central folation F_q^c into one foliation $F_{\tilde{q}}^c$ with similar properties. It also sends leaves of the foliation $F_q^{uu} \cap (W^u_{q(\mu)}\times\{\mu\})$ into leaves of the foliation $F_{\tilde{q}}^{uu} \cap (W^u_{\tilde{q}(\rho(\mu))}\times\{\rho(\mu)\})$.

Construct a C^0 center-stable foliation, F_q^{cs}, of $V_q \times I_1$, **P**-invariant, whose leaf through the point $(x,\mu) \in V_q \times I_1$ is contained in $V_q \times \{\mu\}$, and such that its leaves intersects the set $\bigcup_{\mu\in I_1} (W^u_{q(\mu)}\times\{\mu\})$ in leaves of the foliation F_q^c.

To define the homeomorphism H_q we use:

(i) The homeomorphisms H_q^u and $H_{\tilde{q}}^{cs}$.

(ii) The foliations F_q^{cs}, $F_{\tilde{q}}^{cs}$, F_q^{uu} and $F_{\tilde{q}}^{uu}$.

Since the foliation are P-invariant we must have that this homeomorphism is a conjugation between P and $\tilde{P}$.

It is clear that the restriction map $H_{q(\mu)} = H_q|_{V_q\times\{\mu\}}$ sends the level μ into the level $\rho(\mu)$ and:

(a) apply $F^{cs}_{\beta_1} \cap (V_q\times\{\mu\})$ into $F^{cs}_{\tilde{\beta}_1} \cap (V_{\tilde{q}}\times\{\rho(\mu)\})$; this by the fact that we constructed F_q^{cs} in a compatible way with $F^{cs}_{\beta_1}, F^s_{\beta_2}, \ldots, F^s_{\beta_\ell}$;

(b) apply $F^s_{\beta_1} \cap (V_q\times\{\mu\})$ into $F^s_{\tilde{\beta}_1} \cap (V_{\tilde{q}}\times\{\rho(\mu)\})$; this by the form wich we constructed the homeomorphism H_q^{cs} and by the fact that we used F_q^{uu} and $F_{\tilde{q}}^{uu}$ in the definition of H_q;

(c) apply $\bigcup_{\mu\in I_1} \Lambda_{1,\mu}$ into $\bigcup_{\mu\in\tilde I_1} \tilde\Lambda_{1,\mu}$. The effect of the homeomorphism in this C^s-surface is similar to the effect of H_q^{cs}, that is, off of a small neighborhood of the set $\bigcup_{\mu\in I_1} (W^u_{p(\mu)}\times\{\mu\}) \cap \Lambda_{1,\mu}$, there exists a partial, one dimensional foliation transversal to the leaves of $F^s_{\beta_1} \cap \Lambda_{1,\mu}$ in $\Lambda_{1,\mu}$ (this is only a partial foliation, because we used in the construction of H_q^{cs} (of II.2.5.(c)) the case $n-u_p = s_\sigma$).

(II.2.7) Now we want to extend the homeomorphism defined below to a neighborhood of (p,μ_o). We use the linearization hypothesis to apply a result from [T1] to the vector field $X(x,\mu) = (X_\mu(x),0)$ in a neighborhood of (p,μ_o). We obtain, in this way, C^s-coordinates, $x_1,x_2,\dots,x_n$, μ in a neighborhood $U_p \times I_1$ of (p,μ_o) such that:

$$(1) \qquad X_\mu(x_1,\dots,x_n) = \sum_{i,j=1}^{n} A_{ij}(\mu)x_i \frac{\partial}{\partial x_j}.$$

Let $\Phi\colon U_p\times I_1 \to \mathbb{R}^n\times I_1$ be a coordinate such that $\Phi_*(X)(x_1,\dots,x_n,\mu) = (X_\mu(x_1,\dots,x_n),0)$.

Let $\mathbb{R}^n = \mathbb{R}^{s_p} \times \mathbb{R}^{u_p}$ be the decomposition of $\mathbb{R}^n$ for the singularity $(0,\dots,0) \in \mathbb{R}^n$ of X_μ.

Let $f\colon \mathbb{R}^n\times I_1 \to \mathbb{R}$ be a C^∞ map, such that its restriction $f_\mu(x) = f(x,\mu)$ is a Liapunov function to the vector field X_μ. Let $L_c = \bigcup_{\mu\in I_1} (f_\mu^{-1}(c)) \times \{\mu\}$ and $L_c(\Phi) = \Phi^{-1}(L_c)$.

We assume similar constructions for the vector field $\tilde X(x,\mu) = (\tilde X_\mu(x),0)$.

Recall that in (II.2.5) we defined a homeomorphism $H_r\colon (U_r \cap \Sigma_q)\times I_1 \to (U_{\tilde r} \cap \Sigma_{\tilde q})\times \tilde I_1$. Suppose that there exists $T < 0$ (reparametrizing the vector field if necessary) such that: $\Gamma_T = X_T((U_r \cap \Sigma_q)\times I_1) \subset L_1(\Phi)$. We also assume $\tilde\Gamma_T \subset \tilde L_1(\tilde\Phi)$. We want to define a homeomorphism $\tilde{H}_1: L_1(\Phi) \to \tilde{L}_1(\tilde{\Phi})$ such that (i) it is an extension of $H_T: \Gamma_T \to \tilde{\Gamma}_T$ [wich we define using H_r, X_T and $\tilde{X}_T$]; (ii) it is compatible with H^u defined in (II.2.1). Also, we want that this homeomorphism sends $F^{cu}_p \cap L_1(\Phi)$ into $F^{cu}_{\tilde{p}} \cap \tilde{L}_1(\tilde{\Phi})$, $F^{su}_p \cap L_1(\Phi)$ into $F^{su}_{\tilde{p}} \cap \tilde{L}_1(\tilde{\Phi})$, respectively. The necessity of this last condition follows from the fact: suppose $\pi(1,-1): L_1(\Phi) \to L_{-1}(\Phi)$ denotes the associated Poincaré map. We can define a homeomorphism $\tilde{H}^*_{-1}: [L_{-1}(\Phi) - \bigcup_{\mu \in I_1} W^s_{p(\mu)} \times \{\mu\}] \to [\tilde{L}_{-1}(\tilde{\Phi}) - \bigcup_{\mu \in \tilde{I}_1} W^s_{\tilde{p}(\mu)} \times \{\mu\}]$ by the equation $\tilde{H}^*_{-1}(x,\mu) = \tilde{\pi}(1,-1) \circ \tilde{H}_1 \circ (\pi(1,-1))^{-1}(x,\mu)$. The condition that $\tilde{H}_1$ sends $F^{cu}_p \cap L_1(\Phi)$ into $F^{cu}_{\tilde{p}} \cap \tilde{L}_1(\tilde{\Phi})$ and $F^{su}_p \cap L_1(\Phi)$ into $F^{su}_{\tilde{p}} \cap \tilde{L}_1(\tilde{\Phi})$ is sufficient to guarantee that there exist an extension of $\tilde{H}^*_{-1}$, denoted by $\tilde{H}_{-1}$, to all the set $L_{-1}(\Phi)$.

To construct the homeomorphism $H_p: U_p \times I_1 \to U_{\tilde{p}} \times \tilde{I}_1$ we use, (a) the homeomorphisms $\tilde{H}_1$ and $\tilde{H}_{-1}$, (b) the levels of the Liapunov functions $L_c(\Phi)$, $\tilde{L}_c(\tilde{\Phi})$, $c \in [-1,1]$ and (c) the flows of the vector fields X and $\tilde{X}$. This homeomorphism sends $U_p \times \{\mu\}$ into $U_{\tilde{p}} \times \{\rho(\mu)\}$ and orbits of the vector field X_μ in U_p into orbits of the vector field $\tilde{X}_{\rho(\mu)}$ in $U_{\tilde{p}}$.

The construction of the homeomorphism $\tilde{H}_1$ with all the desired properties is easy and we leave it to the reader.

(II.2.8) Now we want to extend the homeomorphism above to a neighborhood of (β_1,μ_o).

Let $q_1 \in \beta_1$ and Σ_{q_1} be a transversal section to the flow of the vector field X_μ, $\mu \in I_1$, invariant by the diffeomorphism X_{μ,τ_1} (τ_1 = period of β_1). We denote by $U_{q_1} \subset \Sigma_{q_1}$ a neighborhood of q_1. We want to define a homeomorphism $H_{q_1}: U_{q_1} \times I_1 \to U_{\tilde{q}_1} \times \tilde{I}_1$ in a compatible way with the homeomorphisms defined above.

Let $U(\Lambda_1) \subset V_q \times I_1$ (notation as in II.2.6) be a neighborhood of $\bigcup_{\mu \in I_1} \Lambda_{1,\mu}$. Let us assume (reparametrizing if necessary) that there exists $T > 0$ such that $V^1 = X_T(V(\Lambda_1)) \subset U_{q_1} \times I_1$. It is clear that we can define $h^1 \colon V^1 \to \tilde{V}^1$ by the formula $h^1 \circ X_T = \tilde{X}_T \circ H_q$.

We known from the definition of H_q and by the invariance of the foliation $F^{cs}_{\beta_1}$ that h_1 applys the intersection $F^{cs}_{\beta_1} \cap V^1$ into the intersection $F^{cs}_{\tilde{\beta}_1} \cap \tilde{V}^1$ (preserving levels of the map ρ). We also known that for each point $(x,\mu) \in V^1$ we have a unique leaf $F^{uu}_{q_1}(x,\mu) \subset V^1 \cap (U_{q_1} \times \{\mu\})$ which is applied by h_1 into the fiber $F^{uu}_{\tilde{q}_1}(h^1(x,\mu)) \subset \tilde{V}^1 \cap (U_{\tilde{q}_1} \times \{\rho(\mu)\})$. Each leaf of $F^{uu}_{q_1}$ is transversal to $W^{cs}_{q_1} \times \{\mu\}$ in $U_{q_1} \times \{\mu\}$.

Let $D^s_{1,\mu} \subset (W^s_{q_1(\mu)} \cap U_{q_1}) \times \{\mu\}$ be a fundamental domain for the associated Poincaré map and $N^{cs}_{1,\mu} \subset (W^{cs}_{q_1(\mu)} \cap U_{q_1}) \times \{\mu\}$ be one fundamental neighborhood of $D^s_{1,\mu}$ such that $V^1 \cap (W^{cs}_{q_1(\mu)} \times \{\mu\}) \subset \text{Interior}(N^{cs}_{1,\mu})$ in $W^{cs}_{q_1(\mu)} \times \{\mu\}$. We extend continuously the foliation $F^{uu}_{q_1}$ to all the set $N^{cs}_1 = \bigcup_{\mu \in I_1} N^{cs}_{1,\mu}$ in such way that the leaf through $(x,\mu) \in N^{cs}_{1,\mu}$ is contained in $U_{q_1} \times \{\mu\}$, is compatible with the foliations $F^u_{\alpha_i}$ which intersects N^{cs}_1 outside V^1 and is transversal to $W^{cs}_{q_1(\mu)} \times \{\mu\}$ in $U_{q_1} \times \{\mu\}$. Iterating this partial foliation by the Poincaré map we obtain a strong unstable foliation for (q_1,μ_o) [we joint to this foliation the unique strong unstable foliation that we have in $\bigcup_{\mu \in I_1} (W^u_{q_1(\mu)} \cap U_{q_1}) \times \{\mu\}$, see [H-P-S]].

We know that h^1 sends $V^1 \cap W^{cs}_{q_1}$ into $\tilde{V}^1 \cap W^{cs}_{\tilde{q}_1}$ (preserving levels of the map ρ). We want to obtain an extension of the map $h^1\big|_{V^1 \cap W^{cs}_q}$ to all of $U^{cs}_{q_1} = W^{cs}_{q_1} \cap (U_{q_1} \times I_1)$ conjugating the associated Poincaré maps, P and $\tilde{P}$ and preserving the levels of the reparametrization ρ. With this purpose we do the following:

(1) We extend to $(N_1^{cs}-V^1)$ the one dimensional foliation which is the saturation by the diffeomorphism X_T of the one dimensional foliation defined in parts of $\bigcup_{\mu\in I_1} \Lambda_{1,\mu}$ (this last foliation was defined in (II.2.6)). We do this construction in a compatible way with the intersections $F^u_{\alpha_i} \cap N_1^{cs}$ an in such way that the resulting foliation is P-invariant;

(2) We extend the restriction $h^1\big|_{V^1\cap(\bigcup_{\mu\in I_1} D^s_{1,\mu})}$ to all the set $\bigcup_{\mu\in I_1} D^s_{1,\mu}$ in a similar way as in (II.2.5.(b));

(3) We use the fact that h^1 sends the intersection $F^s_{\beta_1} \cap V^1$ into the intersection $F^s_{\tilde{\beta}_1} \cap \tilde{V}^1$ in a compatible way with the homeomorphism h^{uu} of (II.2.1).

With these observations and using $F^c_{\beta_1}$, $F^c_{\tilde{\beta}_1}$ (defined in (1) above) together with $F^s_{\beta_1}$ and $F^s_{\tilde{\beta}_1}$ we can define a homeomorphism $H_1^{cs}: N_1^{cs} \to \tilde{N}^{cs}_{\tilde{q}_1}$ compatible with the maps P_1 and $\tilde{P}_1$.

Conjugating the Poincaré maps we extend the above homeomorphism to a homeomorphism $H_1^{cs}: U^{cs}_{q_1} \to \tilde{U}^{cs}_{\tilde{q}_1}$. To define the homeomorphism H_{q_1}, we proceed in a similar way as in (II.2.6).

(II.2.9) To get an extension of the above homeomorphisms to a neighborhood of (β_2,μ_o), we proveed as in [dM-P] pg. 337-338. The only difference now is that we have to work with a parameter.

Inductively we extend the homeomorphism above to neighborhoods of $(\beta_3,\mu_o),\ldots,(\beta_\ell,\mu_o)$. In this way we construct an equivalence H_μ in all the stable manifolds of the different critical elements of the vector field X_μ, sending these manifolds in the corresponding ones for the vector field $\tilde{X}_{\rho(\mu)}$. Since M^n is the union of the stable manifolds of the critical elements of X_μ, we have that H_μ is a homeomorphism $M^n \to M^n$. (It is clear that this map is continuous in the stable manifolds of $p(\mu), \sigma(\mu), \beta_{1,\mu}, \ldots, \beta_{\ell,\mu}$. The continuity of the map H_μ in the stable manifolds of $\alpha_{i,\mu}$, $1 \le i \le k$ follows from the fact that H_μ preserves the unstable foliations of these critical elements, see [P]).

This conclude the proof of Theorem A in this particular case.

(II.2.10) (I) The above proof also contains the following cases:

(a) the periodic orbit σ is not u-critical

(b) $\dim(W^u_\sigma) = 2$. In this case F^{uu}_q is reduced to points

(c) the case when σ has more than one simple u-criticality.

In this case the construction of the homeomorphism in $\bigcup_{\mu \in I} W^u_{q(\mu)} \times \{\mu\}$ has to be done more carefull than before. The hypothesis that the π^{uu}-projections of the criticalities are differents in $D^u_q \cap$ $\cap\ (W^c_{q(\mu)} \times \{\mu\})$ give us enough "space" to perform the construction in this set.

(II) The case when β_1 has simple u-criticality can be obtained in a similar way as before. In fact, let us suppose that β_{2,μ_o} is a periodic orbit such that $W^s_{\beta_{2,\mu_o}} \cap W^u_{\beta_{1,\mu_o}} \neq \emptyset$ and $\dim(W^s_{\beta_{2,\mu_o}}) = \dim(W^s_{\beta_{1,\mu_o}})$. We know that there exists a foliation F^{uu}_q compatible with $F^{cu}_p, F^{su}_p, F^u_{\alpha_k}, \ldots, F^u_{\alpha_1}$. We use F^{uu}_q and the flow to define a strong unstable foliation $F^{uu}_{q_1}$.

We give now a stable foliation $F^s_{\beta_2} \subset W^{cs}_{\beta_2}$. Using F^s_β and $\pi^{uu}_{q_1}$ we construct a C^o foliation $F^s_{\beta_1} \subset W^{cs}_{\beta_1}$ and using $F^s_{\beta_1}$ and π^{uu}_q we construct $F^s_q \subset W^{cs}_q$.

Using a center-stable foliation, $F^{cs}_{\beta_2}$, we construct a singular center-stable foliation $F^{cs}_{\beta_1}$ and using this last foliation we construct a multi-singular center-stable foliation $F^{cs}_{q_1}$.

After this, we construct homeomorphisms (which conjugate) in $W^u_{\beta_2}$, $W^u_{\beta_1}$, W^u_q with the following properties: (i) they preserve $F^{uu}_{\beta_2}$, $F^{uu}_{\beta_1}$, F^{uu}_q respectively; (ii) they preserve the intersections $F^{cs}_{\beta_2} \cap W^u_{\beta_2}$, $F^{cs}_{\beta_2} \cap W^u_{\beta_1}$, $F^{cs}_{\beta_1} \cap W^u_{\beta_1}$, $F^{cs}_{\beta_1} \cap W^u_q$ and $F^{cs}_q \cap W^u_q$ in their respective domains. After this we proceed as in (II.2.5) to conclude the proof.

It is clear that the reparametrization ρ and the homeomorphism h^{uu} have to satisfy some new conditions in order to get all the required properties for the final homeomorphism. As we will see in the Appendix this is not a problem.

(III) The case when β_1 is a singularity with real weakest expansion is obtained in a similar way.

In the next section we will see the case when β_1 is a singularity with complex weakest expansion.

(II.3) <u>Case when</u> β_1 <u>is a singularity with complex weakest expansion.</u>

(II.3.1) $\dim(M) = 4$.

Let us suppose that $\dim(W^u_p) = \dim(W^s_\sigma) = 2$. Then it is clear that $\dim(W^s_{\beta_1}) = 2$.

If β_j, $j \geq 2$, is a critical element such that $W^s_{\beta_j} \cap W^u_{\beta_2} \neq \phi$ then $\dim(W^s_{\beta_j}) \geq 3$.

We consider defined the same homeomorphisms (H^s, H^u) and the same foliations $(F^u_{\alpha_1}, \ldots, F^u_{\alpha_k}$ and $F^s_{\beta_2}, \ldots, F^s_{\beta_\ell})$ as in (II.2.1). We can also consider defined F^{cu}_p, F^{uu}_q as in (II.2.3) and (II.2.4) respectively.

(II.3.2) Let $S^1_\mu \subset W^1_{\beta_{1,\mu}} \times \{\mu\}$ be a fundamental domain for $W^u_{\beta_{1,\mu}} \times \{\mu\}$. Let j be an index for wich the intersection $S^1_\mu \cap$

$\cap\ (W^s_{\beta_{j,\mu}} \times\{\mu\})$ is a compact set of S^1_μ. Let $U^j_\mu = S^1_\mu \cap F^s_{\beta_j}$. Associated to each point $(x,\mu) \in U^j_\mu$ we define a fiber $I^s(x,\mu) = I^s(x) \times \{\mu\}$; where $I^s(x)$ is a fiber s_{β_1}-dimensional, transversal to $W^u_{\beta_{1,\mu}}$ in M and contained in $F^s_{\beta_j}(x,\mu)$; and such that $U^j_\mu \cap I^s(x,\mu) = \{(x,\mu)\}$. Let j_o be an index for wich the intersection $S^1_\mu \cap (W^s_{\beta_{j_o,\mu}} \times\{\mu\})$ is a non-compact set of S^1_μ. Then there exists an index j_1, $\beta_{j_o} \geq \beta_{j_1}$ such that $W^s_{\beta_{j_o}} \cap W^u_{\beta_{j_1}} \neq \emptyset$ and j_1 is an index as above. Then it is clear that the intersection $(W^s_{\beta_{j_o,\mu}} \times\{\mu\}) \cap (S^1_\mu - \bigcup_{j_1} U^{j_1}_\mu)$ is a compact set (we take the union over all the indexes j_1 as before). In these intersections we can proceed as before. We extend the fibers of I^s to all of S^1_μ. We can assume that each leaf of I^s is of class C^k, $k \geq 2$ and that I^s as a foliation is C^1.

For $(x,\mu) \in S^1_\mu$ we define $F^{us}_{\beta_1}(x,\mu) = \bigcup_{t\leq 0} X_t(I^s(x,\mu))$, where X_t denotes the flow of the vector field $X(x,\mu) = (X_\mu(x),0)$.

(II.3.3) We want to define a homeomorphism $H_q: V_q\times I_1 \to V_{\tilde q}\times\tilde I_1$ conjugating the Poincaré maps $P(x,\mu) = (P_\mu(x),\mu)$ and $\tilde P(x,\mu) = (\tilde P_\mu(x),\mu)$ $(H_q(V_q\times\{\mu\}) \subset V_{\tilde q}\times\{\rho(\mu)\})$, in such wah that we can extend it to neighborhoods $U_{\beta_1} \times I_1$ and $U_p \times I_1$ of (β_1,μ_o) and (p,μ_o) respectively, using for this the flow of the vector fields X, $\tilde X$ and Liapunov functions $f_{\beta_1}: U_{\beta_1}\times I_1 \to \mathbb{R}$, $f_p: U_p\times I_1 \to \mathbb{R}$ (respectively $f_{\tilde\beta_1}, f_{\tilde p}$). To extend to $U_{\beta_1} \times I_1$, we only need to preserve $F^{us}_{\beta_1}$ and, after this, we proceed as in (II.2.7). To extend the homeomorphism to all the manifold, we proceed as in (II.2.9).

(II.3.4) <u>Construction of the homeomorphism</u> $H_q: V_q\times I_1 \to V_{\tilde q}\times\tilde I_1$.

Let us consider a fundamental domain $D^u_q \subset V_q \times I_1$ as in (II.2.2). We denote the component of $(W^s_{\beta_{1,\mu}} \cap V_q) \times \{\mu\}$ which intersects D^u_q by Λ^1_μ. It is clear that $F^{us}_{\beta_1} \cap D^u_{q(\mu)}$ $(D^u_{q(\mu)} = D^u_q \cap V_q \times \{\mu\})$ is the union of spirals in a neighborhood of $\Lambda^1_\mu \cap D^u_{q(\mu)}$. Let us denote this component by E^{us}_μ.

We use the component of $F^{us}_{\beta_1} \cap (V_q \times \{\mu\})$ wich intersects D^u_q and the foliation F^{uu}_q to define a $C^{(o)}$ surface $T_\mu \subset V_q \times \{\mu\}$ which contains all the tangencies between this component and F^{uu}_q. We suppose that $\Lambda^1_\mu \subset T_\mu$. We use the intersections $F^{us}_{\beta_1} \cap T_\mu$ to give a foliation of T_μ by leaves of the same dimension as Λ^1_μ (the principal leaf is Λ^1_μ). We use the projection $\pi^{uu}: V_q \times I_1 \to U^{cs}_q$ $(U^{cs}_q = \bigcup_{\mu \in I_1} (W^{cs}_{q(\mu)} \cap V_q) \times \{\mu\})$ to project these leaves into U^{cs}_q. We use these projections to construct a $C^{(o)}$ foliation $F^s_q \subset U^{cs}_q$.

It is clear that $T_\mu \cap D^u_{q(\mu)}$ is a C^1-curve of tangencies between the spirals of E^{us}_μ and the leaves of $F^{uu}_q \cap D^u_{q(\mu)}$. Let $I^{us}_\mu \subset E^{us}_\mu \cap T_\mu$ be a fundamental domain of spirals. Using the homeomorphism H^u and the foliations $F^s_{\beta_j}$ for j such that $W^s_{\beta_{j,\mu}} \cap W^u_{\beta_{1,\mu}} \neq \emptyset$ we can define a homeomorphism in I^{us}_μ. Using the spirals E^{us}_μ and sending $\Lambda^1_\mu \cap D^u_{q(\mu)}$ into $\tilde{\Lambda}^1_{\rho(\mu)} \cap \tilde{D}^u_{\tilde{q}(\rho(\mu))}$ we define a homeomorphism $h^1_\mu: T_\mu \cap D^u_{q(\mu)} \to \tilde{T}_{\rho(\mu)} \cap \tilde{D}^u_{\tilde{q}(\rho(\mu))}$. We use π^{uu}, $\tilde{\pi}^{uu}$ to define a homeomorphism on $\pi^{uu}(T_\mu \cap D^u_{q(\mu)}) \subset W^c_{q(\mu)} \times \{\mu\}$. Now we extend this homeomorphism to $(W^c_{q(\mu)} \times \{\mu\}) \cap D^u_{q(\mu)}$ and, after this, we conjugate the respective Poincaré maps. Defining $H^{uu}_\mu(q(\mu),\mu) = (\tilde{q}(\rho(\mu)),\rho(\mu))$ we obtain a homeomorphism $h^{uu}_\mu: W^c_{q(\mu)} \times \{\mu\} \to W^c_{\tilde{q}(\rho(\mu))} \times \{\rho(\mu)\}$. Using this homeomorphism on the space of leaves of $F^s_q \subset U^{cs}_q$, we proceed as in (II.2.5) to construct a homeomorphism $H^{cs}_q: U^{cs}_q \to \tilde{U}^{cs}_{\tilde{q}}$.

Now we are going to construct a homeomorphism

$$H^u_q: \bigcup_{\mu \in I_1} (W^u_{q(\mu)} \cap V_q) \times \{\mu\} \to \bigcup_{\mu \in \tilde{I}_1} (W^u_{\tilde{q}(\mu)} \cap V_{\tilde{q}}) \times \{\mu\}.$$

Let $(x_o(\mu),\mu) := \Lambda^1_\mu \cap D^u_{q(\mu)}$ and $x(\mu) > x_o(\mu) > y(\mu)$ with $(y(\mu),\mu) \in E^{us}_\mu(x(\mu),\mu)$. Let R_μ be a small neighborhood of $(x_o(\mu),\mu)$ limited by $F^{uu}_q(x(\mu),\mu)$ and $F^{uu}_q(y(\mu),\mu)$ in $D^u_{q(\mu)}$.

We now extend $T_\mu \cap D^u_{q(\mu)}$ across $D^u_{q(\mu)}$ parallely to $W^c_{q(\mu)} \times \{\mu\}$. Denote this extension by Z_μ. Let Δ^u_μ be a small compact neighborhood of Z_μ containing R_μ in its interior (see Figure II.3.4.I). We fiber Δ^u_μ by an one dimensional foliation L^c_μ such that the restriction of L^c_μ to R_μ is the intersection $F^{us}_{\beta_1} \cap R_\mu$ and Z_μ is a leaf. Now complete L^c_μ to a central foliation $F^c_{q(\mu)}$ of $D^u_{q(\mu)}$ in a compatible way with $F^s_{\beta_2}, \ldots, F^s_{\beta_\ell}$. [It is clear that we must construct the fibers in Δ^u_μ in a compatible way with $F^s_{\beta_j}$ where β_j is such that $W^s_{\beta_j,\mu} \cap W^u_{\beta_1,\mu} \neq \emptyset$].

To define the homeomorphism on $F^{us}_{\beta_1} \cap R_\mu$ we use the fibers of L^c_μ, the leaves of F^{uu}_q and the homeomorphism h^1. To define the homeomorphism on Δ^u_μ we use the homeomorphism h^{uu}_q, the homeomorphism h^1, the fibers of L^c_μ (in Δ^u_μ) and the leaves of F^{uu}_q. To obtain our homeomorphism H^u_q we proceed as in (II.2.2).

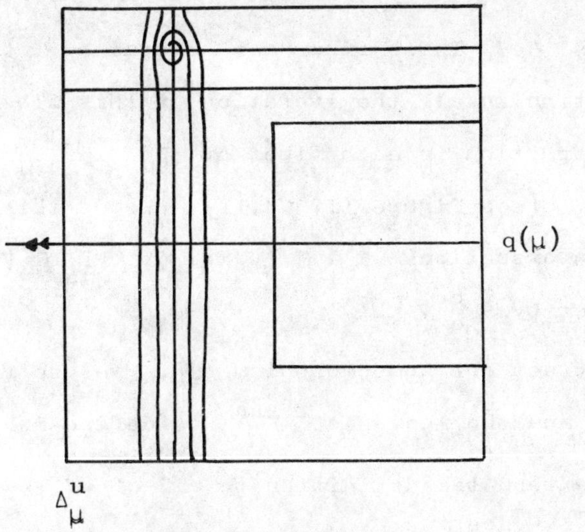

Figure II.3.4.I

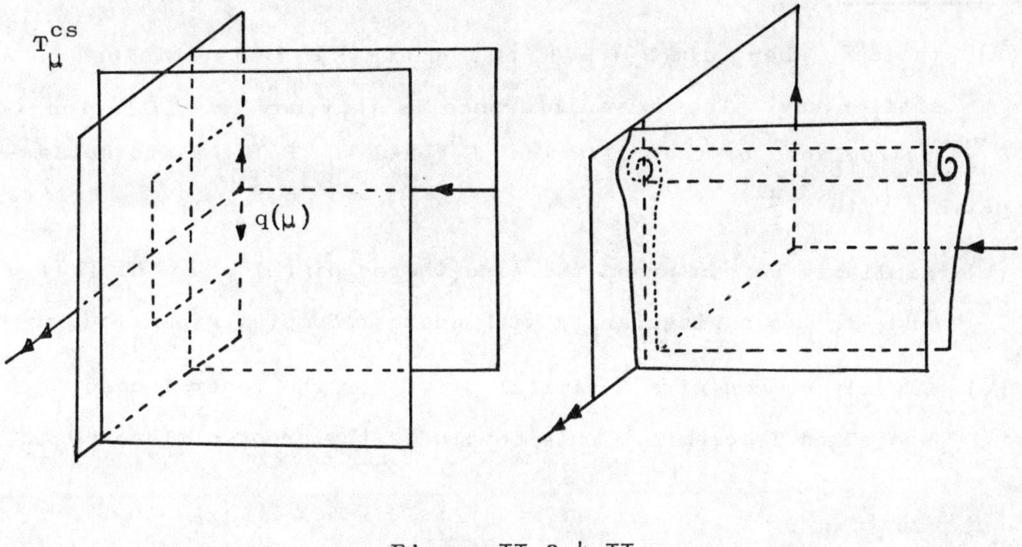

Figure II.3.4.II

Let us denote by ℓ_μ the extension of $T_\mu \cap D^u_{q(\mu)}$ defined above. Now we construct a fiber (T^{cs}_μ) of the same dimension as $W^{cs}_{q(\mu)}$, transversal to $W^u_{q(\mu)} \times \{\mu\}$ and whose intersection with $W^u_{q(\mu)} \times \{\mu\}$ is ℓ_μ. These fiber also contains T_μ. We do a similar construction in all the iteration of this fiber. We realize a similar construction in each fiber of L^c_μ in Δ^u_μ and in each fiber of $F^c_{q(\mu)}$ (see Figure II.3.4.II). We do this in a compatible way with the intersections $F^{us}_{\beta_1} \cap (V_q \times \{\mu\})$, $F^s_{\beta_j} \cap (V_q \times \{\mu\})$. We denote these fiber by F^{cs}_q.

To construct our homeomorphism H^u_q, we use first the homeomorphism H^{cs}_q and the leaves of F^{uu}_q to define in $T^{cs} = \bigcup_{\mu \in I_1} T^{cs}_\mu$. After this we extend the definition to all of $V_q \times I_1$, using H^{cs}_q, F^{uu}_q, H^u_q and F^{cs}_q.

<u>Observations</u>:

(a) the case when $\dim(W^u_p) = 3$ and $\dim(W^u_\sigma) = 2$ is obtained in a similar way. The only difference is that now the foliation F^{cu}_p is replaced by the foliation F^u_p and so F^{uu}_q has to be compatible with F^u_p.

(b) Similarly we can proof the case where $\dim(M) = n$, $\dim(W^u_\sigma) = 3$ and β_1 is a singularity with weakest complex expansion.

(c) Clearly we can give a similar proof for the general case stated in Theorem A. This concludes the proof of Theorem A.

CHAPTER III

Proof of Theorem B

(III.1) Observations

Let $\{X_\mu\} \in \Gamma_6$ and $\mu_0 \in B(\{X_\mu\})$ such that X_{μ_0} has a unique orbit, γ, of quasi-transversal intersection between W^u_p and W^s_σ, where p and σ are as in (II.1). Since we are in the hypothesis of Theorem B we are going to have, through this chapter, only codimension one u-criticalities.

It is clear that, if β has a codimension one u-criticality with σ then $\dim(W^s_\beta) = n-1$, so the weakest expansion at β is real. Let us assume the same order as in (II.1.1) for the critical elements and that $\beta = \beta_1$ has a codimension one u-criticality with σ.

(III.2) Proof of Theorem B

(III.2.1) We proceed as in (II.2.2) to define a fundamental domain $D^u_q \subset \bigcup_{\mu \in I_1} (W^u_{q(\mu)} \cap V_q) \times \{\mu\}$ for the Poincaré map P.

Since we are assuming that $(W^s_{\beta_{1,\mu}} \times\{\mu\}) \cap D^u_q$ is u-critical of codimension one we have the following possibilities:

(a) $(W^s_{\beta_{1,\mu}} \times\{\mu\}) \cap D^u_q$ is a compact subset of $W^u_{q(\mu)} \times \{\mu\}$.

(b) $(W^s_{\beta_{1,\mu}} \times\{\mu\}) \cap D^u_q$ is a non-compact subset of $W^u_{q(\mu)} \times \{\mu\}$.

We can divide case (a) into two subcases:

(a.1) $(W^s_{\beta_{1,\mu}} \times\{\mu\}) \cap \partial D^u_q = \phi$

(a.2) $(W^s_{\beta_{1,\mu}} \times\{\mu\}) \cap \partial D^u_q \neq \phi$.

(II.2.2) Case (a.1)

Let us suppose first that, $\dim(W_\sigma^u) = 3$. In this case $(W^s_{\beta_{1,\mu}} \times\{\mu\}) \cap D^u_q$ is diffeomorphic to a circle.

Since we are assuming that μ_o is the first bifurcation value of the family $\{X_\mu\}$ (i.e. $X_{\mu=0} \in M\text{-}S^\infty(M^n)$, $\mu_o = \inf\{\mu \mid X_\mu \notin M\text{-}S^\infty(M^n)\}$ then $\pi^{uu}((W^s_{\beta_{1,\mu}} \times\{\mu\}) \cap D^u_q) = I^1_\mu$ is contained in one of the components of $(W^c_{q(\mu)}\times\{\mu\} - \{(q(\mu),\mu)\}^{(*)})$. Let us assume that it is contained in the upper component.

Let us denote by $x_o(\mu), x_1(\mu), \ldots, x_\ell(\mu)$ the tangency points between $(W^s_{\beta_{1,\mu}} \times\{\mu\}) \cap D^u_q$ and $F^{uu}_q \cap (W^u_{q(\mu)}\times\{\mu\})$ in $W^u_{q(\mu)} \times \{\mu\}$. We know by hypothesis that these points are of quasi-transversal intersections. Let $(z_i(\mu),\mu) = \pi^{uu}(x_i(\mu))$, $i = 0,\ldots,\ell$. It is clear that it could happen that some of the $(z_i(\mu),\mu)$ are not contained in the intersection $D^u_q \cap (W^c_{q(\mu)}\times\{\mu\})$, but, using the transformation P, there exists some iterated of it which is contained in this intersection. We denote this iterate by $(\tilde{z}_i(\mu),\mu)$. We know that $\tilde{z}_i(\mu) \neq \tilde{z}_j(\mu)$ for $i \neq j$ by hypothesis. This is a generic condition.

(III.2.3) Let us suppose that we have defined a reparametrization $\rho: [0,1] \to [0,1]$ such that $\rho(0) = 0$, $\rho(1) = 1$, $\rho(\mu_o) = \tilde{\mu}_o$.

As in (II.2.1) we can consider defined compatibles stable foliations $F^s_{\beta_\ell},\ldots,F^s_{\beta_2},F^s_{\beta_1}$ and compatibles unstable foliations $F^u_{\alpha_1},\ldots,F^u_{\alpha_k}$. We can also assume defined the homeomorphisms H^s and H^u.

(III.2.4) Now, we are going to define a center singular foliation

---

(*) Here $M\text{-}S^\infty(M)$ denote the intersection $K\text{-}S^\infty(M) \cap WR^\infty(M)$.

$F^c_q \subset D^u_q$, whose leaves are one dimensionals, of class C^1 and (when it is possible) transversal to F^{uu}_q. These foliation will be compatible with $F^s_{\beta_1},\ldots,F^s_{\beta_\ell}$ and will be singular along curves $\ell_0(\mu),\ldots,\ell_\ell(\mu)$ such that $x_i(\mu) \in \ell_i(\mu)$, $i = 0,1,\ldots,\ell$.

Let us consider first the intersection $F^s_{\beta_1} \cap D^u_q$. By hypothesis we know that for each index i there exists a C^1-curve $\ell_i(\mu) \subset D^u_q \cap (W^u_{q(\mu)} \times \{\mu\})$ defined as tangencies between $F^s_{\beta_1}$ and F^{uu}_q in $D^u_q \cap (W^u_{q(\mu)} \times \{\mu\})$. It is clear that $x_i(\mu) \in \ell_i(\mu)$, $i = 0,1,\ldots,\ell$. By hypothesis we know that the projections $J^c_{i,\mu} = \pi^{uu}(\ell_i(\mu))$ satisfies $J^c_{i,\mu} \cap J^c_{j,\mu} = \phi$ for $i \neq j$, $i,j = 0,1,\ldots,\ell$.

Let now $q_1 \in \beta_1$ and $\Sigma_{q_1} \subset M$ be a cross section to the family X_μ, invariant by the map X_{μ,τ_1} (τ_1 = period of β_1). Let $\Psi: \bigcup_{\mu \in I_1} (W^u_{q_1(\mu)} \cap V_{q_1}) \times \{\mu\} \to]v_1, v_2[\times I_1$ ($\delta > 0$ a small number) be a C^s diffeomorphism ($V_{q_1} \subset \Sigma_{q_1}$ a neighborhood of $q_1 \in \beta_1$) such that $\Psi(q_1(\mu),\mu) = (0,\mu)$. To each $(w,\mu) \in [v_1,v_2] \times I_1$ we denote by $F^s_{\beta_1}(w,\mu)$ its associated leaf of the foliation $F^s_{\beta_1}$ wich corresponds to the point $\Psi^{-1}(w,\mu)$. We denote by $S^1(w,\mu)$ the circle $F^s_{\beta_1}(w,\mu) \cap D^u_q$. Let $A(w,\mu): S^1(w,\mu) \to W^c_{q(\mu)} \times \{\mu\}$ be the restriction of π^{uu} to the circle $S^1(w,\mu)$. This is the height function associated to the circle. Let A_μ be the annulus determinated by the circles $S^1(v_1,\mu)$ and $S^1(v_2,\mu)$. From the Appendix of [dM-VS] we know that there exists a diffeomorphism $f_\mu: S^1 \times [v_1,v_2] \to A_\mu$ which is C^∞ in the variable $x \in S^1$ and of class C^1 in $w \in [v_1,v_2]$ such that $f_\mu(S^1 \times \{w\}) = S^1(w,\mu)$. The variation of the family $\{f_\mu\}$ with the parameter μ is C^s. It is possible to define a two parameter family of mappings $f_{w,\mu}: S^1 \to \mathbb{R}$ by the composition $f_{w,\mu}(x) = g_\mu \circ A(w,\mu) \circ f_\mu(x,w)$, where $g_\mu: W^c_{q(\mu)} \times \{\mu\} \to \mathbb{R}$ are C^s-coordinates for each μ. Then we can consider the two parameter family of gradient vector fields $\{G_{w,\mu}\}$ defined by the family $\{f_{w,\mu}\}$. The hypothesis that the u-criticalities are quasi-transversal is equivalent to the fact that the singularities of the gradient vector field $G_{w,\mu}$ are all hyperbolic. That is, each vector field $G_{w,\mu}$ is a Morse-Smale vector field.

Let $v_1 < w_1^1 < w_1^0 < 0 < w_2^0 < w_2^1 < v_2$ and $A_\mu^0 \subset A_\mu^1 \subset A_\mu$ be the annulus determined by $S^1(w_1^0,\mu)$, $S^1(w_2^0,\mu)$; $S^1(w_1^1,\mu)$, $S^1(w_2^1)$ respectively. Let $\eta_\mu : A_\mu \to \mathbb{R}$ be a C^∞ map such that $\eta_\mu(A_\mu^0) = 1$ and $\eta_\mu(\overline{A_\mu \setminus A_\mu^1}) = 0$.

For each μ let $G_\mu(x,w) = (G_{w,\mu}(x),0)$ the associated vector field defined on $S^1 \times [v_1,v_2]$. Let G_μ^* its representant on A_μ using the diffeomorphism f_μ.

For each μ let Y_μ be the parallel vector field in the increasing direction of the height functions $A(w,\mu)$. This vector field varies C^∞ with the variable μ. Let $D_\mu^2 \subset W^u_{q(\mu)} \times \{\mu\}$ be a two dimensional discs wich is a neighborhood of A_μ.

Let us consider the vector field Z_μ defined on D_μ^2 by:

$$Z_\mu(x,w) = \begin{cases} (1-\eta_\mu(x,w))Y_\mu(x,w) + \eta_\mu(x,w)G_\mu^*(x,w) & \text{for } (x,w) \in A_\mu \\ Y_\mu(x,w) & \text{for } (x,w) \in (D_\mu^2 \setminus A_\mu). \end{cases}$$

This is a C^{s-1} vector field in the variable x and $C^{(0)}$ in the variable w. Outside A_μ it is C^∞. The singularities of this vector field are exactly the curves $\ell_i(\mu) \cap A_\mu$, $i=0,1,\dots,\ell$.

In this way we obtain a one dimensional foliation in the set E_μ, $E_\mu = D_\mu^2 - \bigcup_{i=0}^{\ell} \ell_i(\mu) \cap A_\mu$. We note that a leaf through a non-singular point $(x,w) \in S^1(w,\mu) \cap E_\mu$ is totally contained in $S^1(w,\mu)$. We extend this foliation to all $D_q^u \cap (W^u_{q(\mu)} \times \{\mu\} - \bigcup_{i=0}^{\ell} \ell_i(\mu) \cap A_\mu)$ in such way that this extensions is invariant by P, transversal to

$F_q^{uu} \cap D_q^u \cap (W^u_{q(\mu)} \times \{\mu\}$ and compatible with the foliations $F^s_{\beta_2} \cap D^u_q, \dots, F^s_{\beta_\ell} \cap D^u_q$.

(III.2.5) Now we construct a homeomorphism $H^u_q : D^u_q \to D^u_{\tilde{q}}$. With this purpose let us denote by $H^u_{q_{1,\mu}} : W^u_{q_{1,\mu}} \times \{\mu\} \to W^u_{\tilde{q}_{1,\rho(\mu)}} \times \{\rho(\mu)\}$ the homeomorphism induced by $H^u \Big|_{W^u_{\beta_{1,\mu}} \times \{\mu\}} : W^u_{\beta_{1,\mu}} \times \{\mu\} \to W^u_{\tilde{\beta}_{1,\rho(\mu)}} \times \{\rho(\mu)\}$ in the intersection $(W^u_{\beta_{1,\mu}} \cap V_{q_1}) \times \{\mu\}$. We use the homeomorphism $H^u_{q_1} : \bigcup_{\mu\in I_1} W^u_{q_{1,\mu}} \times \{\mu\} \to \bigcup_{\mu\in\tilde{I}_1} W^u_{\tilde{q}_{1,\mu}} \times \{\mu\}$, $H^u_{q_1}\Big|_{W^u_{q_{1,\mu}} \times \{\mu\}} = H^u_{q_{1,\mu}}$ and the foliation $F^s_{\beta_1}$ to define a homeomorphism $h^u_i : \bigcup_{\mu\in I_1} \ell_i(\mu) \to \bigcup_{\mu\in\tilde{I}_1} \tilde{\ell}_i(\mu)$, $i = 0,1,\dots,\ell$, which sends $\ell_i(\mu)$ to $\tilde{\ell}_i(\rho(\mu))$, $x_i(\mu)$ to $\tilde{x}_i(\rho(\mu))$ and $S^1(w^j_k,\mu) \cap \ell_i(\mu)$ to $\tilde{S}^1(\tilde{w}^j_k,\rho(\mu)) \cap \tilde{\ell}_i(\rho(\mu))$, $j = 0,1$, $k = 1,2$, $i = 0,1,\dots,\ell$.

We use the projection π^{uu} and the homeomorphisms h^u_i to induce homeomorphisms $h^c_i : \bigcup_{\mu\in I_1} J^c_{i,\mu} \to \bigcup_{\mu\in\tilde{I}_1} \tilde{J}^c_{i,\mu}$, $J^c_{i,\mu} = \pi^{uu}(\ell_i(\mu))$.

We known, from (II.2.2), that for each $i = 0,1,\dots,\ell$, there exists an integer $n_i \geq 0$ such that $P^{n_i}(J^c_{i,\mu}) \subset D^u_q \cap (W^c_{q(\mu)} \times \{\mu\})$. It is clear that using P^{n_i}, $\tilde{P}^{n_i}$ and h^c_i we can define a homeomorphism $\tilde{h}^c_i : \bigcup_{\mu\in I_1} P^{n_i}(J^c_{i,\mu}) \to \bigcup_{\mu\in\tilde{I}_1} \tilde{P}^{n_i}(\tilde{J}^c_{i,\mu})$. Let us denote by h^c an extension of these homeomorphisms to the set $\bigcup_{\mu\in I_1} D^u_q \cap (W^c_{q(\mu)} \times \{\mu\})$. Now, defining $h^c(q(\mu),\mu) = (\tilde{q}(\rho(\mu)),\rho(\mu))$ and conjugating P and $\tilde{P}$ we can consider defined a homeomorphism $h^c : \bigcup_{\mu\in I_1} W^c_{q(\mu)} \times \{\mu\} \to \bigcup_{\mu\in\tilde{I}_1} W^c_{\tilde{q}(\mu)} \times \{\mu\}$, such that the restriction $h^c_\mu = h^c \Big|_{W^c_{q(\mu)} \times \{\mu\}}$ sends $W^c_{q(\mu)} \times \{\mu\}$ into $W^c_{\tilde{q}(\rho(\mu))} \times \{\rho(\mu)\}$.

If a point $(w,\mu) \in W^u_{q_{1,\mu}} \times \{\mu\}$, we known that its image

$H^u_{q_{1,\mu}}(w,\mu) = (H^u_{1,\mu}(w),\rho(\mu))$. Then we know where we must send the circle $S^1(w,\mu)$, that is, we must send $S^1(w,\mu)$ into the circle $S^1(H^u_{1,\mu}(w),\rho(\mu))$. To do this, we use the homeomorphism h^c_μ and the foliation $F^{uu}_q \cap (W^u_{q(\mu)} \times \{\mu\})$. In this way we obtain a homeomorphism $\dot{H}^u_\mu: A_\mu \to \tilde{A}_{\rho(\mu)}$ such that: (a) it sends $F^s_{\beta_1} \cap A_\mu$ into $F^s_{\tilde{\beta}_1} \cap \tilde{A}_{\rho(\mu)}$ in a compatible way with $H^u_{q_{1,\mu}}$; (b) it sends $F^{uu}_q \cap A_\mu$ into $F^{uu}_{\tilde{q}} \cap \tilde{A}_{\rho(\mu)}$ in a compatible way with h^c_μ and (c) it sends $\ell_i(\mu) \cap A_\mu$ into $\tilde{\ell}_i(\rho(\mu)) \cap \tilde{A}_{\rho(\mu)}$.

Let $B^u_q = D^u_q - \bigcup_{\mu\in I_1} A^1_\mu$. It is clear that, proceeding as in [dM-P] pg. 329-330, we can construct an extension of $\dot{H}^u_\mu\big|_{(A_\mu - A^1_\mu)}$ to all of $B^u_q \cap (W^u_{q(\mu)} \times \{\mu\})$, varying continuously with μ and sending $F^{uu}_q \cap B^u_q \cap (W^u_{q(\mu)} \times \{\mu\})$ into $F^{uu}_{\tilde{q}} \cap B^u_{\tilde{q}} \cap (W^u_{\tilde{q}(\rho(\mu))} \times \{\rho(\mu)\})$ and leaves of the singular foliation F^c_q (define in III.2.4) in B^u_q into leaves of the singular foliation $F^c_{\tilde{q}}$ in $B^u_{\tilde{q}}$ (in a compatible way with the foliations $F^s_{\beta_1}, F^s_{\beta_2}, \ldots, F^s_{\beta_\ell}$ and with the homeomorphism H^u). Let us denote this homeomorphism by $\tilde{H}^u_q$.

It is clear that, using $\dot{H}^u$ and $\tilde{H}^u_q$ we can define a homeomorphism $H^u_q: D^u_q \to \tilde{D}^u_{\tilde{q}}$ such that: (a) it is compatible with the homeomorphism H^u of (III.2.3) and with the foliations $F^s_{\beta_1}, F^s_{\beta_2}, \ldots, F^s_{\beta_\ell}$; (b) it sends a central singular foliation F^c_q of D^u_q into a similar foliation $F^c_{\tilde{q}}$ of $D^u_{\tilde{q}}$ in a compatible may with the orbit structure of the vector field, which define these foliations (we also assume that $(W^c_{q(\mu)} \times \{\mu\}) \cap D^u_q$ is one of the leaves of F^c_q, $\forall \mu \in I_1$).

(III.2.6) Case (a.2)

Being careful (for instance, when the boundary components of the intersection $(W^s_{\beta_{1,\mu}} \times \{\mu\}) \cap D^u_q$ is contained in leaves of F^{uu}_q), the consctruction of the homeomorphism can be done as in (a.1).

(III.2.7) Case (b)

In this case it must exists another critical element β_{μ_o} of the vector field X_{μ_o} such that; β_{μ_o} is not u-critical with σ, $W^s_{\beta_{\mu_o}} \cap W^u_\sigma = \phi$, $\dim(W^s_{\beta_{\mu_o}}) = n-1$ and $W^s_{\beta_{1,\mu_o}} \cap W^u_{\beta_{\mu_o}} \neq \phi$ (see Figure III.2.7.I). It is clear then that $\sigma \leq \beta_{\mu_o} \leq \beta_{1,\mu_o}$.

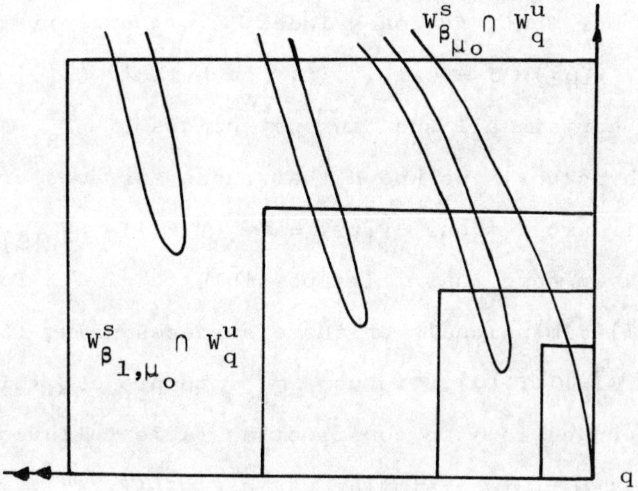

Figure III.2.7.I

Now, we construct a system of compatible stable foliations $F^s_\beta, F^s_{\beta_1}, F^s_{\beta_2}, \ldots, F^s_{\beta_\ell}$ and so a homeomorphism H^u as in (III.2.3).

To obtain a homeomorphism on D^u_q we define first, in a neighborhood of the set $\bigcup_{\mu \in I_1} (W^s_{\beta_\mu} \times \{\mu\}) \cap D^u_q$ and after this we extend it to a neighborhood of $\bigcup_{\mu \in I_1} (W^s_{\beta_{1,\mu}} \times \{\mu\}) \cap D^u_q$. We can do these two constructions in such a way that the foliations F^s_β, $F^s_{\beta_1}$ and F^{uu}_q are preserved by this homeomorphism. The same method of (III.2.5) applys to these constructions. We leave the details to the reader.

(III.3) Now, we want to construct a homeomorphism $H_q : V_q \times I_1 \to V_{\tilde{q}} \times \tilde{I}_1$

wich will be a conjugacy between P and $\tilde{P}$ and in a such way that we can extend it to neighborhoods of (p,μ_o) and (β_1,μ_o) respectively.

Let $N_q^u \subset V_q \times I_1$ be a fundamental neighborhood for the fundamental domain D_q^u.

As in (II.2) we construct foliations F_p^{cu}, F_p^{su} and F_q^{uu}, compatibles with $F_{\alpha_1}^u,\dots,F_{\alpha_k}^u$.

It is clear that, for each index i, there exist a surface $T_i(\mu) \subset N_q^u \cap (V_q \times \{\mu\})$ $i = 0,1,\dots,\ell$, such that $\ell_i(\mu) \subset T_i(\mu)$ and these surfaces contain all the tangency points of $F_{\beta_1}^s \cap N_q^u$ with $F_q^{uu} \cap N_q^u$. By hypothesis we known that these surfaces are all disjoints and their projections $\hat{T}_i(\mu) = \pi^{uu}(T_i(\mu)) \subset W_{q(\mu)}^{cs} \times \{\mu\}$ are also disjoint. We note that it is possible to give a foliation, by leaves $(s_\sigma-1)$-dimensional, of these surfaces using for this the intersections $F_{\beta_1}^s \cap T_i(\mu)$. We use π^{uu} to project this foliation and use these projections to construct a stable foliation $F_q^s \subset$ $\subset (W_{q(\mu)}^{cs} \cap V_q) \times \{\mu\}$ in a similar way as in (II.2.4). Then we proceed as in (II.2.5) to construct a homeomorphism $H_q^{cs}: U_q^{cs} \to U_{\tilde{q}}^{cs}$.

Now we construct the homeomorphism H_q.

<u>Case (a.1)</u>: We first construct a singular center-stable foliation $F_q^{cs} \subset V_q \times I_1$ whose intersections with $W_{q(\mu)}^u \times \{\mu\}$ are leaves of the central-foliation defined in (III.2.4) (see Figure III.3.I). We also suppose that the leaves of F_q^{cs} are transversal to the leaves of F_q^{uu}. We consider the surfaces $T_i(\mu)$ contained in leaves of F_q^{cs}. We construct this foliation in such way that it is invariant by the associated Poincaré map. We note that we decompose leaves of $F_{\beta_1}^s \cap N_q^u$ into leaves of F_q^{cs} wich does not contain points of $\ell_i(\mu)$ and the intesections $F_{\beta_1}^s \cap T_i(\mu)$ (which contains

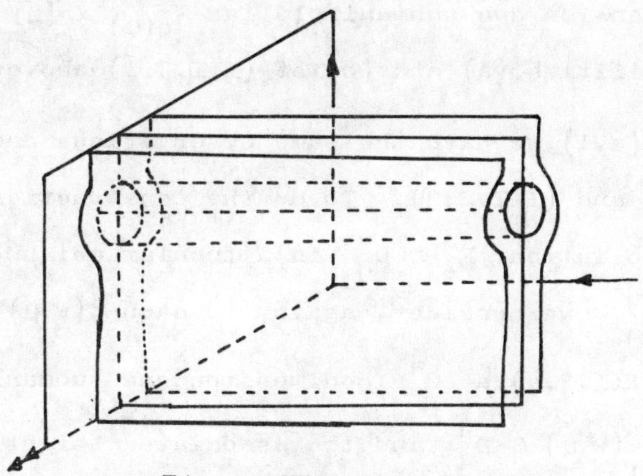

Figure III.2.7.II

points of $\ell_i(\mu)$).

It is clear that we can use F_q^{cs}, F_q^{uu}, the homeomorphisms H_q^u and H_q^{cs} to define our homeomorphism H_q. (We note that H_q, defined in this way, sends $F_{\beta_1}^s \cap T_i(\mu)$ into $F_{\tilde{\beta}_1}^s \cap \tilde{T}_i(\rho(\mu))$).

To extend this homeomorphism to a neighborhood of (p,μ_o) we proceed as in (II.2.7).

To obtain an extension of this homeomorphism to a neighborhood of (β_1,μ_o) we observe that it is $C^{(o)}$ near of inclusion map and that we have constructed it in such way that it sends $F_{\beta_1}^s \cap N_q^u$ into $F_{\tilde{\beta}_1}^s \cap N_{\tilde{q}}^u$.

Proceeding inductively, as in [P-T], we extend this homeomorphism to neighborhoods of $(\beta_2,\mu_o),\ldots,(B_\ell,\mu_o)$.

We obtain, in this way, a proof of Theorem B in the Case (a.1). It is clear that the other cases are similar to this case.

(III.4) Generalizations

Now we analyse the case when $\dim(W_\sigma^u) > 3$ and β_1 is a periodic orbit which has a codimension one u-criticality with σ.

In this case the intersection $((W^u_{q(\mu)} \cap W^s_{\beta_{1,\mu}}) \times \{\mu\}) \cap D^u_q$ is a C^∞ codimension one submanifold in $W^u_{q(\mu)} \times \{\mu\}$. We also have the same possibilities (a) and (b) of (III.2.1) above.

In case (a.1) we have the same observations and constructions as in (III.2.2) and (III.2.3). To do the construction of the singular central foliation $F^c_q \subset D^u_q$ in a tubular neighborhood of $(W^s_{\beta_{1,\mu}} \times\{\mu\}) \cap D^u_q$, we consider, again, to each $(w,\mu) \in W^u_{q_{1,\mu}} \times \{\mu\}$ (notation from III.2.4) a C^∞ codimension one submanifold $S^{u_\sigma -2}(w,\mu) = F^s_{\beta_1}(w,\mu) \cap D^u_q$ and the associated height function $A(w,\mu): S^{u_\sigma -2}(w,\mu) \to W^c_{q(\mu)} \times \{\mu\}$ $(u_\sigma = \dim(W^u_\sigma))$. It is of class C^s. We choose the metric (in D^u_q) in such a way that the gradient of the height function is a Morse-Smale vector field. The decomposition of $S^{s_\sigma -2}(w,\mu)$ into the orbits of this vector field gives a system of curves (except for the singularities). We use these curves in the construction of F^c_q in the tubular neighborhood

$$V_{1,\mu} = \bigcup_{w\in[v_1,v_2]} S^{u_\sigma -2}(w,\mu).$$

To extend this foliation to D^u_q we proceed as in (III.2.4).

To the construction of a homeomorphism $H^u_q: D^u_q \to \tilde{D}^u_{\tilde{q}}$ as in (III.2.5) we use the following facts:

a) Let $N^{u_\sigma -2}$ be a C^∞ manifold diffeomorphic to the intersection $(W^u_{\beta_{1,\mu}} \times\{\mu\}) \cap D^u_q$. Let $f_{w,\mu}: N^{u_\sigma -2} \to \mathbb{R}$ be a two parameter family of associated height functions (as above) then the two parameter family of associated gradient vector fields $\{G_{w,\mu}\}$ is a stable two parameter family of Morse-Smale vector fields (see [P-T]). To construct the equivalence between $\{G_{w,\mu}\}$ and $\{\tilde{G}_{w,\mu}\}$ we use $H^u_{q_{1,\mu}}$ (the homeomorphism induced by $H^u\big|_{W^u_{\beta_{1,\mu}} \times\{\mu\}}$ in the intersection $(W^u_{\beta_{1,\mu}} \cap V_q) \times \{\mu\}$ and the reparametrization ρ [i.e. we construct a two parameter family of homeomorphisms $H_{w,\mu} : N^{u_\sigma - 2} \to$ $\to N^{u_\sigma - 2}$ such that $H_{w,\mu}$ sends orbits of $G_{w,\mu}$ into orbits of $\tilde{G}_{H^u_{q_{1,\mu}}(w,\mu),\rho(\mu)}$. It is clear that this construction varies continuously with (w,μ)]. After of this we translate this construction (using the diffeomorphism between $N^{u_\sigma - 2} \times [v_1, v_2] \times I_1$ and V_1) to a homeomorphism defined in the respective tubular neighborhoods. $V_1 = \bigcup_{\mu \in I_1} V_{1,\mu}$ and $\tilde{V}_1 = \bigcup_{\mu \in \tilde{I}_1} \tilde{V}_{1,\mu}$.

b) We apply several times the Isotopy Extension Theorem to extend the homeomorphism defined on V_1 to all the set D^u_q. It is clear that this last homeomorphism can be obtained in a compatible way with H^u of (III.2.3) and preserving F^c_q and F^{uu}_q.

The rest of the construction is similar to (III.3).

This finish the proof of Theorem B in the general case.

CHAPTER IV

Proof of Theorem C

(IV.1) Observations

Let $\{X_\mu\} \in \Gamma_5$ and $\mu_o \in B(\{X_\mu\})$ be such that X_{μ_o} has and orbit (γ), of quasi-transversal intersection between W^u_p and W^s_q, where p is a singularity with complex weakest contraction and q is a singularity with real weakest expansion.

Let us suppose that $\alpha \neq p$ is a critical elemento of X_{μ_o} such that $W^u_\alpha \cap W^s_p \neq \phi$; then we have the following possibilities:

a) $W^u_\alpha \cap W^c_p \neq \emptyset$, where $W^c_p = W^{cu}_p \cap W^s_p$. Since, by hypothesis, W^u_α is transversal to W^c_p, it follows that $\dim(W^u_\alpha) \geq n-1$ $(n = \dim(M))$;

b) $W^u_\alpha \cap W^{ss}_p \neq \emptyset$; since W^u_α is transversal to W^{ss}_p we have that $\dim(W^u_\alpha) \geq u_p+3$, $u_p = \dim(W^u_p)$;

c) $W^u_\alpha \cap W^{ss}_p = \emptyset = W^u_\alpha \cap W^c_p$; in this case $\dim(W^u_\alpha) \geq u_p+1$.

We recall that we have defined a partial order between the critical elements of the vector field X by the relation $\alpha \leq \beta \Leftrightarrow W^u_\alpha \cap W^s_\beta \neq \emptyset$. We extend this partial order to a total order $\alpha_1 \leq \alpha_2 \leq \ldots \leq \alpha_k \leq p \leq q \leq \beta_1 \leq \beta_2 \leq \ldots \leq \beta_\ell$. We do the same for all the family $\{X_\mu\}$.

(IV.2) Proof of Theorem C in the case that $\dim(W^s_p) = 2$.

(IV.2.1) As in (II.2.1) we consider defined systems of compatible foliations $F^u_{\alpha_1}, \ldots, F^u_{\alpha_k}$; $F^s_{\beta_1}, \ldots, F^s_{\beta_j}$ and homeomorphisms H^s and H^u.

(IV.2.2) Let U_γ be a neighborhood of the closure of the orbit γ. We will construct a homeomorphism $H_\gamma: U_\gamma \times I_1 \to U_{\tilde{\gamma}} \times \tilde{I}_1$ in a compatible way with H^s, H^u such that it sends orbits of $X(x,\mu) = (X_\mu(x), 0)$ in $U_\gamma \times I_1$ into orbits of $\tilde{X}(x,\mu) = (\tilde{X}_\mu(x), 0)$ in $U_{\tilde{\gamma}} \times \tilde{I}_1$.

To do this, let $z_0 \in \gamma$ and $U_{z_0} \subset M^n$ be a small neighborhood of z_0. We know that there exists C^∞ coordinate, $\Phi_\mu: (U_{z_0}, z_0) \to (\mathbb{R}^n, 0)$ such that:

(a) $(\Phi_\mu)_*(X_\mu) = \frac{\partial}{\partial y_n}$

(b) $\Phi_\mu(W^u_{p(\mu)} \cap U_{z_0}) = \{y_1 = y_2 = 0\}$

(c) if $s_q=2$ then $\Phi_\mu(W^s_{q(\mu)} \cap U_{z_0}) = \{y_1 = \varepsilon(\mu),\ y_3 = \ldots = y_{n-1} = 0\}$

if $s_q \geq 3$ then $\Phi_\mu(W^s_{q(\mu)} \cap U_{z_0}) = \{y_1 = Q(y_3, \ldots, y_{s_q}) + \varepsilon(\mu),\ y_{s_q+1} = \ldots =$
$= y_{n-1} = 0\}$

where Q is a non-degenerate homogeneous function and $\varepsilon(\mu)$ is a C^∞ function such that $\frac{d\varepsilon}{d\mu}(\mu_0) \neq 0$ and $\varepsilon(\mu) = 0$.

Let $\Sigma_{z_0} \subset M^n$ be a transversal section to X_μ, $\forall\, \mu \in I_1$ wich corresponds (uniformly in the coordinate μ, reparametrizing X_μ if necessary) in the coordinate Φ_μ to $\{y_n = 0\}$.

To obtain H_γ, we will construct a homeomorphism $H_{z_0} : V_{z_0} \times I_1 \to V_{\bar{z}_0} \times \bar{I}_1$ where $V_{z_0} \subset \Sigma_{z_0}$ is a small neighborhood of z_0, such that using appropriated foliations and Liapunov functions we can obtain H_γ.

To extend to a neighborhood of (p,μ_0) we use a foliation denoted by F^{su}_p, and which is obtained in the following way:

Let $S^1_\mu \subset W^s_{p(\mu)} \times \{\mu\}$ be a fundamental domain for $W^s_{p(\mu)} \times \{\mu\}$ and i be an index such that $\bar{U}^i_\mu = S^1_\mu \cap (W^u_{\alpha_{i,\mu}} \times \{\mu\})$ is a compact set. To each point $(x,\mu) \in \bar{U}^i_\mu$ we define a fiber $I^u(x,\mu) = I^u(x) \times \times \{\mu\}$, where $I^u(x)$ is a u_p-dimensional fiber, transversal to $W^s_{p(\mu)}$ in M and contained in $W^u_{\alpha_{i,\mu}}$ such that $\bar{U}^i_\mu \cap I^u(x,\mu) = \{(x,\mu)\}$. We do the same construction at each point of the intersection $F^u_{\alpha_i} \cap U^i_\mu$ where U^i_μ is a small neibhborhood of $\bar{U}^i_\mu$ in S^1_μ. If i_0 is an index such that $\bar{U}_\mu = S^1_\mu \cap (W^u_{\alpha_{i_0,\mu}} \times \{\mu\})$ is a non-compact set then $\bar{U}^{i_0}_\mu \cap (S^1_\mu - \bigcup_{i_1} U^{i_1}_\mu)$ is a compact set (we take the union over all the indexes i_1 such that $\bar{U}^{i_1}_\mu$ is a compact set). In these intersection we proceed as in the compact case. In this way we obtain a fibration I^u over S^1_μ.

If $(x,\mu) \in S^1_\mu$ we define $F^{su}_p(x,\mu) = \bigcup_{t \geq 0} X_t(I^u(x,\mu))$. Here X_t denotes the flow associated to the vector field. $X(x,\mu) = = (X_\mu(x), 0)$.

Note: Since $\dim(W^u_\alpha) \geq n-1$, for every critical element α such that $W^u_\alpha \cap W^s_p \neq \emptyset$ it follows, from one of the results in the Appendix of [dM-VS] that the foliation F^u_α is always C^1. Then the fibration I^u over S^1_μ over S^1_μ is always C^1.

Let us denote by $L^{su}(x,\mu)$ the intersection $F^{su}_p(x,\mu) \cap \cap (\Sigma_{z_o} \times \{\mu\})$. If our homeomorphism H_{z_o} send leaves of L^{su} into leaves of $\tilde{L}^{su}$ then using a Liapunov function we can extend H_{z_o} to a neighborhood of (p,μ_o).

To extend H_{z_o} to a neighborhood of (q,μ_o), we use partial foliations F^{cs}_q and F^{us}_q which are defined in the same way as the foliation F^{cu}_p and F^{su}_p of (II.2.3). It is clear that if H_{z_o} sends $F^{cs}_q \cap (\Sigma_{z_o} \times \{\mu\})$ into $F^{cs}_{\tilde{q}} \cap (\Sigma_{\tilde{z}_o} \times \{\rho(\mu)\})$ and $F^{us}_q \cap (\Sigma_{z_o} \times \{\mu\})$ into $F^{us}_{\tilde{q}} \cap (\Sigma_{\tilde{z}_o} \times \{\rho(\mu)\})$ then as in (II.2.7) we can extend H_{z_o} to a neighborhood of (q,μ_o).

If H_{z_o} satisfies the two above conditions then we obtain H_γ as we explain before. To extend H_γ to all of $M \times I_1$ we proceed as in (II.2.8) and (II.2.9).

(IV.2.3) Now we construct H_{z_o}.

We will assume that $\Phi_\mu(W^{cs}_{q(\mu)} \cap U_{z_o}) = \{y_{s_q+1} = \ldots = y_{n-1} = 0\}$. Let us define a coordinate $\Phi\colon U_{z_o} \times I_1 \to \mathbb{R}^n \times I_1$ by the equation $\Phi(x,\mu) = (\Phi_\mu(x),\mu)$. Using this coordinate we have that $\Phi(T^{su}_\mu) = \{y_{s_q+1} = \ldots = y_{n-1} = 0\} \times \{\mu\}$ where $T^{su}_\mu = (W^{cs}_{q(\mu)} \cap W^{cu}_{p(\mu)}) \times \{\mu\}$. Let $K^c_\mu \subset T^{su}_\mu$ be a C^k one dimensional foliation, transversal to $\Phi^{-1}(\{y_2 = y_{s_q+1} = \ldots = y_n = 0\} \times \{\mu\})$ in T^{su}_μ and compatible with $W^s_{q(\mu)} \times \{\mu\}$.

Let $S_\mu \subset T^{su}_\mu$ be the surface which contains all the tangencies between the leaves of K^c_μ and $L^{su}|_{T^{su}_\mu}$. We will assume that $(W^u_{p(\mu)} \times \{\mu\}) \cap T^{su}_\mu \subset S_\mu$.

Let θ_μ be the two-dimensional surface $\Phi^{-1}(\{y_3 = \ldots = y_n = 0\} \times \{\mu\})$.

The intersection $\Gamma_\mu = \theta_\mu \cap S_\mu$ is a C^1 curve, transversal to the intersections $L^{su} \cap S_\mu$ in S_μ. Let $I_\mu = [a_\mu^1, a_\mu^2] \subset \Gamma_\mu$ be a fundamental domain of spirals in Γ_μ, that is, each spiral of $L^{su} \cap$ $\cap\, T_\mu^{su}$ intersects the interval I_μ in an unique point, with the only exception of the spiral $L^{su}(a_\mu^1, a_\mu^2)$ for which we have $L^{su}(a_\mu^1, a_\mu^2) \cap I_\mu = \{a_\mu^1, a_\mu^2\}$.

Let i be an index such that the intersection $S_\mu^1 \cap (W^u_{\alpha_{i,\mu}} \times\{\mu\})$ is a compact set and $V_\mu^i \subset]a_\mu^1, a_\mu^2[$ be a small neighborhood of $(W^u_{\alpha_{i,\mu}} \times\{\mu\}) \cap I_\mu$. Using the restriction map $H^s\big|_{W^s_{\alpha_{i,\mu}} \times\{\mu\}}$ and the foliation $F^u_{\alpha_i}$ we define a homeomorphism $H^i_{1,\mu}: V^i_\mu \to \tilde{V}^i_{\rho(\mu)}$.

Let j be an index such that the intersection $S_\mu^1 \cap$ $\cap\, (W^u_{\alpha_{j,\mu}} \times\{\mu\})$ is a non-compact set. Then it is clear that the intersection $(I_\mu - \bigcup_i V_\mu^i) \cap (W^u_{\alpha_{j,\mu}} \times\{\mu\})$ is a compact set (we take the union over all those indexes i for which $S_\mu^1 \cap (W^u_{\alpha_{i,\mu}} \times\{\mu\})$ is a compact set). Let $V_\mu^j \subset]a_\mu^1, a_\mu^2[$ be a small neighborhood of these intersection. Proceeding as before we can construct a homeomorphism $H^j_{1,\mu}: V^j_\mu \to \tilde{V}^j_{\rho(\mu)}$.

We extend these definitions to all of I_μ to obtain a homeomorphism $H_{1,\mu}: I_\mu \to \tilde{I}_{\rho(\mu)}$. Defining $H_{2,\mu}(\Gamma_\mu \cap (W^u_{p(\mu)} \times \{\mu\})) =$ $= \tilde{\Gamma}_{\rho(\mu)} \cap (W^u_{\tilde{p}(\rho(\mu))} \times \{\rho(\mu)\})$ and using the spirals of $L^{su}\big|_{\theta_\mu}$, we define a homeomorphism $H_{2,\mu}: \Gamma_\mu \to \tilde{\Gamma}_{\rho(\mu)}$ sending $(W^s_{q(\mu)} \times\{\mu\}) \cap \Gamma_\mu$ into $(W^s_{\tilde{q}(\rho(\mu))} \times \{\rho(\mu)\}) \cap \tilde{\Gamma}_{\rho(\mu)}$. (This last condition determines the reparametrization ρ.)

Using the above elements we want to construct a homeomorphism $H_3: \bigcup_{\mu\in I_1} S_\mu \to \bigcup_{\mu\in\tilde{I}_1} \tilde{S}_\mu$ such that the restriction $H_{3,\mu} = H_3\big|_{S_\mu}: S_\mu$ $\to \tilde{S}_{\rho(\mu)}$ sends $(W^s_{q(\mu)} \times\{\mu\}) \cap S_\mu$ into $(W^s_{\tilde{q}(\rho(\mu))} \times\{\rho(\mu)\}) \cap \tilde{S}_{\rho(\mu)}$,

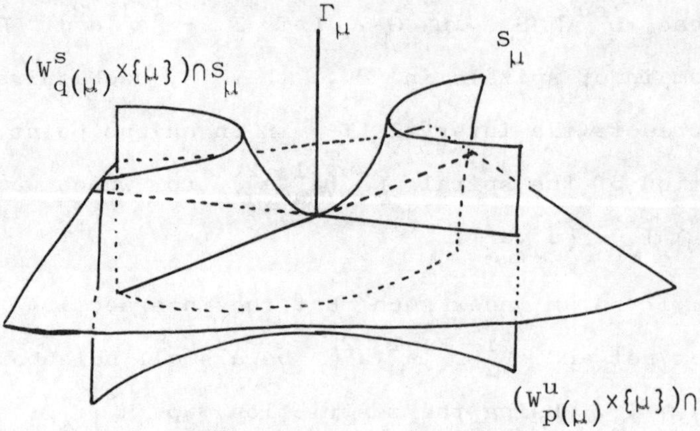

Figure IV.2.3.I

$(W^u_{p(\mu)}\times\{\mu\})\cap S_\mu$ into $(W^u_{\tilde{p}(\rho(\mu))}\times\{\rho(\mu)\})\cap\tilde{S}_{\rho(\mu)}$ and leaves of $L^{su}\cap S_\mu$ into leaves of $\tilde{L}^{su}\cap\tilde{S}_{\rho(\mu)}$ in such way that $H_{3,\mu}|\Gamma_\mu = H_{2,\mu}$ (see Figure IV.2.3.I). This is not hard to do and we left this construction to the reader.

Now we extend $K^c_\mu\subset T^{su}_\mu$ to a one dimensional foliation K^c_μ (same notation) with C^k-leaves, transversal to the surface $\Phi^{-1}(\{y_2 = y_n = 0\}\times\{\mu\})$, contained in $V_{z_0}\times\{\mu\}$ and compatible with the foliations $F^{cs}_q\cap(V_{z_0}\times\{\mu\})$ and $F^{us}_q\cap(V_{z_0}\times\{\mu\})$. We note that these foliation is of class C^k outside a small neighborhood of the intersections $(W^s_{\beta_{j,\mu}}\times\{\mu\})\cap(V_{z_0}\times\{\mu\})$ where j is an index such that $W^s_{\beta_{j,\mu}}\cap W^u_{q(\mu)}\neq\emptyset$. We also note that the restriction of K^c_μ to any leave of $F^s_{\beta_j}$ (j as before) is a C^k foliation.

Let $N_\mu\subset V_{z_0}\times\{\mu\}$ be the surface wich contains all the tangencies between K^c_μ and L^{su} in $(V_{z_0}\times\{\mu\})$. It is clear that $S_\mu\subset N_\mu$ (see Figues IV.2.3.II and IV.2.3.III for the case $u_p+s_q=n$).

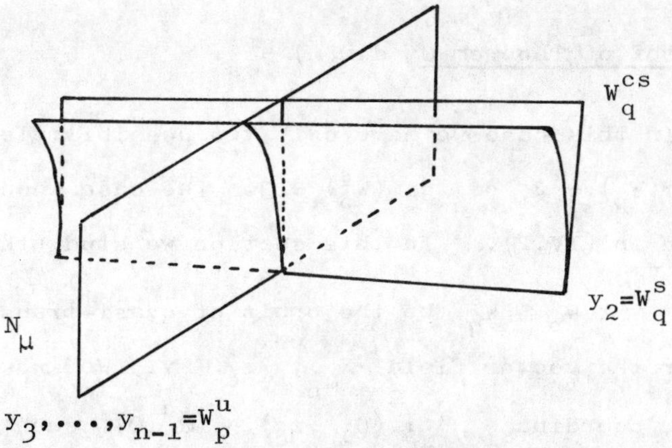

Figure IV.2.3.II

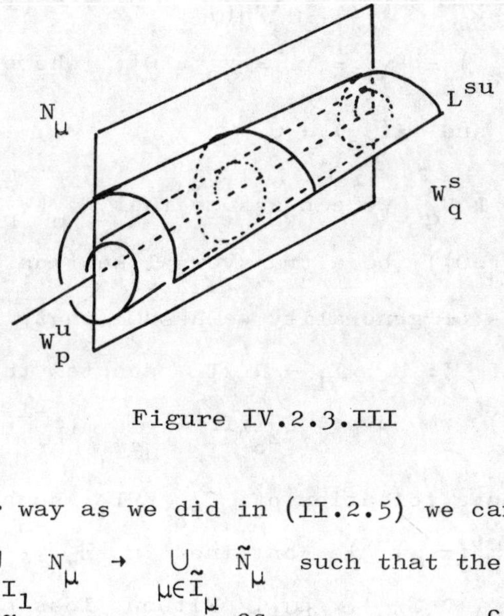

Figure IV.2.3.III

In a similar way as we did in (II.2.5) we can construct a homeomorphism $H_4\colon \bigcup_{\mu\in I_1} N_\mu \to \bigcup_{\mu\in \tilde{I}_\mu} \tilde{N}_\mu$ such that the restriction $H_{4,\mu} = H_4\big|_{N_\mu}\colon N_\mu \to \tilde{N}_{\rho(\mu)}$ sends $F^{cs}_q \cap N_\mu$ into $F^{cs}_{\tilde{q}} \cap \tilde{N}_{\rho(\mu)}$, $F^{us}_q \cap N_\mu$ into $F^{us}_{\tilde{q}} \cap \tilde{N}_{\rho(\mu)}$, $L^{su} \cap N_\mu$ into $\tilde{L}^{su} \cap \tilde{N}_{\rho(\mu)}$ and $(W^u_{p(\mu)} \times \{\mu\}) \cap N_\mu$ into $(W^u_{\tilde{p}(\rho(\mu))} \times \{\rho(\mu)\}) \cap \tilde{N}_{\rho(\mu)}$. Also the restriction map $H_4\big|_{\bigcup_{\mu\in I_1} S_\mu}$ is the map H_3 defined above.

To extend H_4 to the neighborhood $V_{z_o}\times I_1$ we use L^{su}, $\tilde{L}^{su}$, K^c, $\tilde{K}^c$ and the homeomorphisms H_4 and H_2 defined before. We denote this extension by H_{z_o}. This concludes the proof of Theorem C in the case $\dim(W^s_p) = 2$.

(IV.3) <u>Proof of Theorem C</u> $\dim(M) = 4$.

(IV.3.1) In this case we have only two possibilities for $\dim(W^s_p)$; either $\dim(W^s_p) = 2$ or $\dim(W^s_p) = 3$. The case when $\dim(W^s_p) = 2$ was studied in (IV.2). In this section we study the case $\dim(W^s_p) = 3$.

Let $\gamma \subset W^u_p \cap W^s_q$ be the orbit of quasi-transversal intersection for the vector field X_{μ_o}, $z_o \in \gamma$. We know that there exists C^∞ coordinate $\Phi_\mu : (U_{z_o}, z_o) \to (\mathbb{R}^4, 0)$ such that:

(a) $(\Phi_\mu)_*(X_\mu) = \frac{\partial}{\partial y_4}$, (b) $\Phi_\mu(W^s_{q(\mu)} \cap U_{z_o}) = \{y_1 = \varepsilon(\mu)\}$ and (c) $\Phi_\mu(W^u_{p(\mu)} \cap U_{z_o}) = \{y_1 = y_2 = y_3 = 0\}$; where $\varepsilon(\mu)$ is a C^∞ map, $\frac{d\varepsilon}{d\mu}(\mu_o) \neq 0$ and $\varepsilon(\mu_o) = 0$.

Since $W^{cu}_p \pitchfork W^s_q$ we can assume that $\Phi_\mu(W^{cu}_{p(\mu)} \cap U_{z_o}) = \{y_3 = 0\}$. Let $\Sigma_{z_o} = \Phi^{-1}_{\mu_o}(\{y_4 = 0\})$ be a transversal section to the vector field X_{μ_o}. Without loss of generality we assume that $\Sigma_{z_o} = \Phi^{-1}_\mu(\{y_4=0\})$ $\forall\, \mu \in I_1$. Then if $\Phi: U_{z_o} \times I_1 \to \mathbb{R}^4 \times I_1$ denotes the coordinate $\Phi(x,\mu) = (\Phi_\mu(x), \mu)$ we have: $\Sigma_{z_o} \times I_1 = \bigcup_{\mu \in I_1} (\Phi^{-1}_\mu(\{y_4=0\}) \times \{\mu\})$.

Let π^s be a foliation of $\Sigma_{z_o} \times I_1$ such that the leave through (x,μ), $\pi^s(x,\mu)$ is contained in $\Sigma_{z_o} \times \{\mu\}$ and $\pi^s(\varepsilon(\mu),\mu) = (W^s_{q(\mu)} \cap \Sigma_{z_o}) \times \{\mu\}$. Without loss of generality we can suppose that π^s corresponds (in the Φ-coordinate) to the planes $\{(y^o_1, y_2, y_3, 0, \mu) \,/\, y^o_1$ and μ is fixed$\}$.

(IV.3.2) Let us now consider, to each $\mu \in I_1$, linearizing coordinates $\Psi_\mu : (U_p, p(\mu)) \to (\mathbb{R}^4, 0)$ varying smoothly wit the parameter μ. Let us define $\Psi: U_p \times I_1 \to \mathbb{R}^4 \times I_1$ $\Psi(x,\mu) = (\Psi_\mu(x), \mu)$, C^s-coordinates in a small neighborhood of (p, μ_o). We assume, for each μ, that $\Psi_\mu(W^u_{p(\mu)} \cap U_p) \subset \{(0,0,0,x_4)\}$ and $\Psi_\mu(W^s_{p(\mu)} \cap U_p) \subset$ $\subset \{(x_1,x_2,x_3,0)\}$. We consider in the coordinates $(x_1,x_2,x_3,0)$ the cylinder $C_1 = \{(x_1,x_2,x_3,0) / x_1^2+x_2^2 = 1, |x_3| \leq 1\}$ and the discs $B^\pm = \{(x_1,x_2,\pm 1,0) / x_1^2+x_2^2 \leq 1\}$. Then $S_1 = C_1 \cup B^+ \cup B^-$ is a fundamental domain for the vector field $(\Psi_\mu)_*(X_\mu)$. For each point $(x_1^o,x_2^o,x_3^o,0,\mu^o) \in S_1 \times I_1$ we define one unstable leaf, $F^u(x_1^o,x_2^o,x_3^o,0,\mu^o)$ by $F^u(x_1^o,x_2^o,x_3^o,0,\mu^o) = \{(x_1^o,x_2^o,x_3^o,x_4,\mu^o)/|x_4| \leq \delta\}$. Using these leaves we define a stable-unstable foliation F_p^{su} by:

$$F_p^{su}(x_1^o,x_2^o,x_3^o,0,\mu^o) = \bigcup_{t\geq 0} X_t(F^u(x_1^o,x_2^o,x_3^o,0,\mu^o)).$$

The foliation F_p^{su} determines, using Ψ, a foliation F_p^{su} in the set $A = U_p \times I_1 - \bigcup_{\mu\in I_1} W^u_{p(\mu)} \times \{\mu\}$. We assume, for each $x \in (W^c_{p(\mu)} \cap \Psi_\mu^{-1}(S_1)) \times \{\mu\}$ $(W^c_{p(\mu)} = W^s_{p(\mu)} \cap W^{cu}_{p(\mu)})$, that $F_p^{su}(x) \subset W^{cu}_{p(\mu)} \times \{\mu\}$. It is clear that the projection $\pi^{su}: A \to$ $\to \Psi^{-1}(S_1 \times I_1)$ is a C^s-map. Moreover, for each point $x \in (\Psi_\mu^{-1}(S_1) \times \{\mu\} - W^{ss}_{p(\mu)} \times \{\mu\})$ the intersection $(\pi^{su})^{-1}(x) \cap$ $\cap (\Sigma_{z_o} \times \{\mu\})$ is a C^s-spiral in $\Sigma_{z_o} \times \{\mu\}$. Using the coordinate Φ_μ, from (IV.3.1), we can assume that $\pi^{su}(\Phi_\mu^{-1}(\{y_1=y_2=0\}) \times \{\mu\}) \subset$ $\subset W^{ss}_{p(\mu)} \times \{\mu\}$ to each value μ.

(IV.3.3) It is clear that it is possible to define a C^{s-1}-surface $T_\mu \subset \Sigma_{z_o} \times \{\mu\}$ wich contains all the tangencies between the spirals defined by $F_p^{su} \cap (\Sigma_{z_o} \times \{\mu\})$ and the foliation $\pi^s|_{\Sigma_{z_o} \times \{\mu\}}$. Since $\pi^{su}(\Phi_\mu^{-1}(\{y_1=y_2=0\}) \times \{\mu\}) \subset W^{ss}_{p(\mu)} \times \{\mu\}$ to each $\mu \in I_1$ we can also assume that $T_\mu - \Phi_\mu^{-1}(\{y_1=y_2=0\}) \times \{\mu\}$ is the union of two disjoint connected components, which we denote by $T_{1,\mu}$, $T_{2,\mu}$.

Given a point $(x,\mu) \in T_{1,\mu}$, we denote by $E(x,\mu)$ the spiral of $F_p^{su} \cap (\Sigma_{z_o} \times \{\mu\})$ such that $(x,\mu) \in E(x,\mu)$ and by

$R_\mu(x,\mu)$ the first intersection of $E(x,\mu)$ with $T_{2,\mu}$ begining in (x,μ). Also, for points $(x,\mu) \in T_{2,\mu}$ we can define $R_\mu(x,\mu)$ in the same way. Then we have a C^{s-1}-diffeomorphism $R_\mu: T_{1,\mu} \cup T_{2,\mu} \to T_{1,\mu} \cup T_{2,\mu}$ defined using the spirals of $F_p^{su} \cap (\Sigma_{z_o} \times \{\mu\})$.

Let (x,μ) be a point of the curve $\Phi_\mu^{-1}(\{y_1=y_2=0\}) \times \{\mu\})$ different of $(W^u_{p(\mu)} \cap \Sigma_{z_o}) \times \{\mu\}$, $I(x,\mu) \subset T_\mu$ be a C^{s-1} curve transverse to $\Phi_\mu^{-1}(\{y_1=y_2=0\}) \times \{\mu\})$ in T_μ. Then we know the image $R_\mu(I(x,\mu)-\{(x,\mu)\})$ and its closure is a C^{s-1} curve transverse to $\Phi_\mu^{-1}(\{y_1=y_2=0\}) \times \{\mu\}$ in T_μ. Moreover the intersection $\overline{R_\mu(I(x,\mu)-\{(x,\mu)\})} \cap \Phi_\mu^{-1}(\{y_1=y_2=0\}) \times \{\mu\}$ is a unique point (y,μ). Defining $R(x,\mu) = (y,\mu)$ and $R_\mu((W^u_{p(\mu)} \cap \Sigma_{z_o}) \times \{\mu\} = (W^u_{p(\mu)} \cap \Sigma_{z_o}) \times \{\mu\}$ we have a continuous extension of R_μ to all the surface T_μ. We also denote this extension by R_μ. Let $D_\mu(R)$ be a fundamental domain for R_μ in T_μ. We can assume that part of the boundary of $D_\mu(R)$ are leaves of π^s (see Figure IV.3.3.I). Denote by R the map defined on $T = \bigcup_{\mu \in I_1} T_\mu$, $R: T \to T$, $R|_{T_\mu} = R_\mu$.

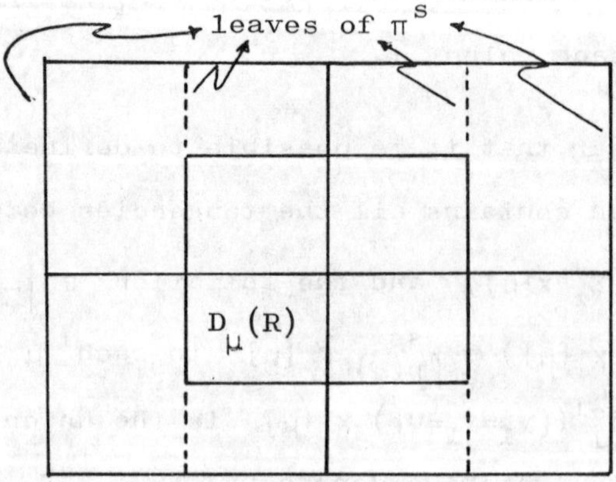

Figure IV.3.3.I

If L^s_μ is the intersection $(W^s_{q(\mu)} \times \{\mu\}) \cap T_\mu$ then L^s_μ is a C^{s-1} curve transversal to $T_\mu \cap (W^{cu}_{p(\mu)} \times \{\mu\})$ in T_μ. By construction we know that $L^s_\mu \subset T_{1,\mu}$ for all μ, $\mu_1 \le \mu < \mu_0$ $(I_1 = [\mu_1, \mu_2])$, $L^s_\mu \subset T_{2,\mu}$ for all μ, $\mu_0 < \mu \le \mu_2$ and $L^s_{\mu_0} \subset \Phi^{-1}_{\mu_o}(\{y_1 = y_2 = 0\}) \times \{\mu_0\})$. If L is the surface $\bigcup_{\mu \in I_1} L^s_\mu$ then it is a C^{s-1} surface in $T = \bigcup_{\mu \in I_1} T_\mu$. We are interested in the representation of the surface L in the fundamental domain $D_R = \bigcup_{\mu \in I_1} D_\mu(R)$. We have that

$$L = \bigcup_{\mu \in [\mu_1, \mu_0]} L^s_\mu \cup \bigcup_{\mu \in [\mu_0, \mu_2]} L^s_\mu \quad \text{and} \quad \bigcup_{\mu \in [\mu_1, \mu_0]} L^s_\mu =$$

$$= \bigcup_{i=0}^{+\infty} \Big(\bigcup_{\mu \in [\bar{\mu}_i, \bar{\mu}_{i+1}]} L^s_\mu \Big) \cup L^s_{\mu_o}, \quad \bigcup_{\mu \in [\mu_o, \mu_2]} L^s_\mu = \bigcup_{i=0}^{+\infty} \Big(\bigcup_{\mu \in [\bar{v}_{i+1}, \bar{v}_i]} L^s_\mu \Big)$$

$\cup L^s_{\mu_o}$ where $\mu_1 = \bar{\mu}_0 < \bar{\mu}_1 < \ldots < \mu_0 < \ldots < \bar{v}_2 < \bar{v}_1 < \bar{v}_0 = \mu_2$ are sequences of parameter values such that $R^{-i}\Big(\bigcup_{\mu \in [\bar{\mu}_i, \bar{\mu}_{i+1}]} L^s_\mu \Big) \subset D_R$ and $R^{-i}\Big(\bigcup_{\mu \in [\bar{v}_{i+1}, \bar{v}_i]} L^s_\mu \Big) \subset D_R$ for $i = 0, 1, 2, \ldots$.

Making a change of the coordinate Φ, if necessary, (of class C^{s-1} in (y_1, y_2, y_3), smooth in the parameter μ) we can assume that $\Phi(T_\mu) \subset \{(y_1, 0, y_3, \mu)\}$ and that the foliation π^s is (yet) in this coordinates represented by the planes $\{(y^o_1, y_2, y_3, \mu) \mid \mu$ is fixed$\}$.

Let us consider the curve $\Phi^{-1}(\{y^o_1, 0, 0, \mu) \mid \mu \in [\mu_1, \mu_2]\} \subset D_R$. Then there exist two sequences of parameter values, $\mu_n(y^o_1)$ and $v_n(y^o_1)$, $\mu_1(y^o_1) < \mu_2(y^o_1) < \ldots < \mu_o < \ldots < v_2(y^o_1) < v_1(y^o_1)$ such that $(y^o_1, 0, 0, \mu_n(y^o_1)) \in \tilde{L}^s_{\mu_n(y^o_1)}$ and $(y^o, 0, 0, v_n(y^o_1)) \in \tilde{L}^s_{v_n(y^o_1)}$ (here $\tilde{L}^s_\mu$ represents the curve L^s_μ in the fundamental domain D_R). Since $F^{\mathbf{su}}_p$ and $W^s_{q(\mu)} \times \{\mu\}$ depends smoothly on the parameter μ, we have that there exist $\lim_{n\to\infty} \tilde{L}^s_{\mu_n(y^o_1)}$ and $\lim_{n\to\infty} \tilde{L}^s_{v_n(y^o_1)}$ and they are equals. We obtain in this way a foliation of $T_{\mu_0} \cap D_R$ invariant by R.

Iterating this foliation and defining $(W^s_{q(\mu_o)} \times \{\mu_o\}) \cap T_{\mu_o}$ as a leaf, we have an invariant foliation of T_{μ_o}. Each leave of this foliation is transversal to $(W^{cu}_{p(\mu_o)} \times \{\mu_o\}) \cap T_{\mu_o}$.

Now we construct a two dimensional surface $S(y^o_1) \subset D_R$ such that $S(y^o_1) \cap D_R = \{(y^o_1,0,0,\mu) / \mu \in [\mu_1,\mu_2]\}$ and $\tilde{L}^s_{\mu_n(y^o_1)} \subset S(y^o_1)$, $\tilde{L}^s_{v_n(y^o_1)} \subset S(y^o_1)$. Of course we do this contruction to each point $(y^o_1,0,0,\mu_1) \in D_R$ and in such way that in the boundary of D_R, these surfaces are $K^o \cup K^1$, where $K^o = \{(1,0,y_3,\mu) / |y_3| \le 1, \mu \in [\mu_1,\mu_2]\}$ and $K^1 = R(K_o)$. Iterating this two dimensional foliation of $T \cap D_R$ and taking $\{(0,0,y_3,\mu) / |y_3| \le 1, \mu \in [\mu_1,\mu_2]\}$ as a leaf, we obtain a two dimensional foliation of T. We can assume that this is a C^{s-1} foliation (outside $\{(0,0,y_3,\mu) / |y_3| \le 1\}$) with C^{s-1} leaves. The intersection of this foliation with T_μ gives a one-dimensional foliation of T_μ invariant by the map R_μ.

We now modify π^s in such a way that its leaves intersects T_μ in leaves of the above foliation and in such way that T_μ is still the surface of tangencies between F^{su}_p and π^s.

(IV.3.4) Let us consider a critical element $\alpha_\mu \le p(\mu)$ such that $W^u_{\alpha_\mu} \cap W^s_{p(\mu)} \neq \phi$. Denote by $S^s_{p(\mu)} \subset W^s_{p(\mu)}$ a two dimensional sphere wich is a fundamental domain for the vector field $X_\mu\big|_{W^s_{p(\mu)}}$.

(I) Suppose α_μ is a critical element of saddle type such that $W^u_{\alpha_\mu} \cap W^c_{p(\mu)} = \phi$ and $W^u_{\alpha_\mu} \cap S^s_{p(\mu)}$ is a compact set. We have two possibilities: (A) $\dim(W^u_{\alpha_\mu} \cap S^s_{p(\mu)}) = 0$ or (B) $\dim(W^u_{\alpha_\mu} \cap S^s_{p(\mu)}) = 1$.

In case (A) we can suppose that the intersection is a point and that there is a center-unstable manifold $W^{cu}_{\alpha_\mu}$ such that $(W^{cu}_{\alpha_\mu} \times \{\mu\}) \cap D_R \cap T_\mu$ is transversal to the above foliation in

$D_R \cap T_\mu$. Yet, in this case we have three possibilities:

(A.1) α_μ is a periodic orbit with real weakest contraction,

(A.2) α_μ is a singularity with real weakest contraction and

(A.3) α_μ is a singularity with complex weakest contraction.

Yet, the case (**A.**1) can be divided into two sub-cases:

(A.1.1) α_μ is a periodic orbit with real weakest contraction and it is not s-critical;

(A.1.2) α_μ is a periodic orbit with real weakest contraction and it is s-critical.

Yet, the case (A.1.2) can be divided into two-subcases:

(A.1.2.1) the s-criticalities are simples or

(A.1.2.2) the s-criticalities are of codimension one.

In case (B) we have that $(W^u_{\alpha_\mu} \times \{\mu\}) \cap T_\mu \cap D_R$ is diffeomorphic to a circle. Since $\{X_\mu\} \in \Gamma_5$ we have that all the non-transversal intersections of $W^u_{\alpha_\mu}$ and $W^s_{q(\mu)}$ are of quasi-transversal type. Let us denote by $\Phi^{-1}((y^\alpha_{1,j}(\mu),0,y^\alpha_{3,j}(\mu),\mu))$ the points of tangency between the circle $S^1_{\alpha_\mu} = (W^u_{\alpha_\mu} \times\{\mu\}) \cap T_\mu \cap D_R$ and the foliation $\pi^s \cap T_\mu \cap D_R$ for $j = 1,2,\ldots,m$. If $\pi^s\colon D_R \to \{(y_1,0,0,\mu)\}$ denotes the π^s-projection, we assume the generic condition:

$\pi^s(\Phi^{-1}((y^\alpha_{1,j}(\mu),0,y^\alpha_{3,j}(\mu),\mu))) \neq \pi^s(\Phi^{-1}(y^\alpha_{1,k}(\mu),0,y^\alpha_{3,k}(\mu),\mu))$ for $k \neq j$.

(It is also clear that we assume that these π^s-projections are different from the π^s-projections of elements of the type (A)).

(II) Suppose α_μ is a critical element of saddle type such that

$W^u_{\alpha_\mu} \cap W^c_{p(\mu)} \neq \emptyset$ and $W^u_{\alpha_\mu} \cap W^c_{p(\mu)} \cap S^s_{p(\mu)}$ is a compact set.

Since $W^u_{\alpha_\mu} \pitchfork W^c_{p(\mu)}$ we have that $\dim(W^u_{\alpha_\mu}) = 3$ and we have the following possibilities: (A) $W^u_{\alpha_\mu} \cap S^s_{p(\mu)}$ is a non-compact set and (B) $W^u_{\alpha_\mu} \cap S^s_{p(\mu)}$ is a compact set.

In case (A) we have that $W^u_{\alpha_\mu} \cap W^s_{\delta_\mu} \neq \phi$ where δ_μ is a critical element of X_μ as in (I). $(\alpha_\mu \leq \delta_\mu \leq p(\mu))$. It is also clear (from the compacticity of $W^u_{\alpha_\mu} \cap W^c_{p(\mu)} \cap S^s_{p(\mu)}$) that we can assume that α_μ is not s-critical with respect of δ_μ. Since $\{X_\mu\} \in \Gamma_5$ we have that the non-transversal intersections between $W^u_{\alpha_\mu}$ and $W^s_{q(\mu)}$ are of quasi-transversal type. Also we assume that the π^s projections of the tangencies between $(W^u_{\alpha_\mu} \times \{\mu\}) \cap T_\mu \cap D_R$ and leaves of the foliation π^s are all different and different from those defined above for others critical elements of type (I).

In case (B) the intersection $(W^u_{\alpha_\mu} \times \{\mu\}) \cap T_\mu \cap D_R$ is a circle $S^1_{\alpha_\mu}$. Also, we assume in this case, that the π^s-projections of the tangencies of $S^1_{\alpha_\mu}$ and leaves of π^s are all different and different from those defined above.

Any other critical element α_μ, for wich we have $W^u_{\alpha_\mu} \cap$ $\cap W^s_{p(\mu)} \neq \phi$ is such that it accumulates in critical elements of type (I) and (II).

(IV.3.5) <u>Construction of the equivalence.</u>

First we deal with the case (II)(B), that is, there exists a critical element α_μ, of saddle type such that $W^u_{\alpha_\mu} \cap W^c_{p(\mu)} \neq \phi$ and so that $W^u_{\alpha_\mu} \cap S^s_{p(\mu)}$ is a compact set.

We can suppose that the critical elements of the vector field X_μ are ordered in the following way: $\alpha_{1,\mu} \leq \ldots \leq \alpha_{k,\mu} \leq \alpha_\mu \leq p(\mu) \leq$ $\leq q(\mu) \leq \beta_{1,\mu} \leq \ldots \leq \beta_{\ell,\mu}$ and that α_μ is the unique critical element of X_μ wich intersects S_1 in a compact set.

(i) as in (IV.2.1) we can consider defined compatible systems of foliations $F^u_{\alpha_1}, \ldots, F^u_{\alpha_k}, F^u_\alpha; F^s_{\beta_1}, \ldots, F^s_{\beta_\ell}$ and homeomorphisms H^s, H^u;

$$H^s: \bigcup_{\mu\in I_1} \left(\bigcup_{i=1}^{k} (W^s_{\alpha_{i,\mu}} \cup W^s_{\alpha_\mu}) \times\{\mu\}\right) \to \bigcup_{\mu\in\tilde{I}_1} \left(\bigcup_{i=1}^{k} (W^s_{\tilde{\alpha}_{i,\mu}} \cup W^s_{\tilde{\alpha}_\mu}) \times\{\mu\}\right)$$

$$H^u: \bigcup_{\mu\in I_1} \left(\bigcup_{j=1}^{\ell} (W^u_{\beta_{j,\mu}} \times \{\mu\})\right) \to \bigcup_{\mu\in\tilde{I}_1} \left(\bigcup_{j=1}^{\ell} (W^u_{\tilde{\beta}_{j,\mu}} \times \{\mu\})\right)$$

whose restrictions $H^s\big|_{W^s_{\alpha_{i,\mu}}\times\{\mu\}} : W^s_{\alpha_{i,\mu}} \times\{\mu\} \to W^s_{\tilde{\alpha}_{i,\rho(\mu)}} \times\{\rho(\mu)\}$

and $H^u\big|_{W^u_{\beta_{j,\mu}}\times\{\mu\}} : W^u_{\beta_{j,\mu}} \times\{\mu\} \to W^u_{\tilde{\beta}_{j,\rho(\mu)}} \times\{\rho(\mu)\}$ are conjugacies between the vector fields $X\big|_{W^s_{\alpha_{i,\mu}}\times\{\mu\}}$ [resp. $X\big|_{W^u_{\beta_{j,\mu}}\times\{\mu\}}$] and $\tilde{X}\big|_{W^s_{\tilde{\alpha}_{i,\rho(\mu)}}\times\{\rho(\mu)\}}$ [resp. $\tilde{X}\big|_{W^u_{\tilde{\beta}_{j,\rho(\mu)}}\times\{\rho(\mu)\}}$] (it is clear that we assume the same for $W^s_{\alpha_\mu} \times \{\mu\}$).

We can assume that the foliations F^u_α and $F^u_{\alpha_j}$ ($F^u_{\alpha_j} \cap F^u_\alpha \neq \emptyset$) are of class C^1. This fact is a consequence of a theorem in the Appendix of [dM-VS] (see that $\dim(W^s_{\alpha_\mu}) = 3$).

(ii) Let us consider the coordinate Ψ defined in (IV.3.2).

We observe, in this coordinate, that the intersection $(W^u_{\alpha_\mu} \cap \Psi^{-1}_\mu(S^1)) \times \{\mu\}$ is a C^s-circle in $\Psi^{-1}(S^1) \times \{\mu\}$. The set $\bigcup_{\mu\in I_1} (W^u_{\alpha_\mu} \cap \Psi^{-1}_\mu(S^1)) \times \{\mu\}$ is a two dimensional surface of class C^s in $\Psi^{-1}(S^1\times I_1)$. It is clear that the intersection $F^u_\alpha \cap (\Psi^{-1}(S^1\times\{\mu\}))$ is a C^1-foliation of a small neighborhood of $(W^u_{\alpha_\mu}\times\{\mu\}) \cap \Psi^{-1}(S^1\times\{\mu\})$. These leaves intersect transversally the set $A^c_\mu = (W^c_{p(\mu)}\times\{\mu\}) \cap \cap \Psi^{-1}(S^1\times\{\mu\})$ in $\Psi^{-1}(S^1\times\{\mu\})$. Wo we can take a small tubular neighborhood V^c_μ of A^c_μ in $\Psi^{-1}(S^1\times\{\mu\})$ and we can define in V^c_μ a C^1-foliation $[F^c]$ with one-dimensional leaves of class C^s transversal to A^c_μ in V^c_μ and compatible with $F^u_\alpha \cap V^c_\mu$.

Now we construct a foliation and a partial foliation of $U_p \times I_1$.

The foliation is F_p^{su} as in (IV.3.2), but this time it must be compatible with F_α^u and with the other unstable foliation of critical elements α_j such that $W^u_{\alpha_{j,\mu}} \cap W^s_{p(\mu)} \neq \phi$. It is clear that this foliation is of class C^1 [that is because we have assumed that α_μ is the unique critical element of X_μ wich intersects the fundamental domain $\Psi_\mu^{-1}(S^1) \subset W^s_{p(\mu)}$ in a compact set. In the general case there are other criticals elements $\alpha_{j,\mu}$ such that $W^u_{\alpha_{j,\mu}} \cap \Psi_\mu^{-1}(S^1)$ is a compact set of dimension zero. For the general case we suppose some linearizing hypothesis for those critical elements of X_μ].

The partial foliation, which we denote by F_p^{uc}, is constructed in the following way. To each point $\Psi^{-1}(x_1^o,x_2^o,0,0,\mu^o) \in \Psi^{-1}(\{x_1,x_2,0,0,\mu \ / \ x_1^2+x_2^2 = 1, \ \mu \in I_1\})$ we associate the leaf F^c, $F^c(\Psi^{-1}(x_1^o,x_2^o,0,0,\mu^o))$, defined above. Denote by $F_1^c(\Psi^{-1}(x_1^o,x_2^o,0,0,\mu^o))$ the leaf $\bigcup_{y\in F^c} F_p^{su}(y)$. Now we define $F_p^{uc}(\Psi^{-1}(x_1^o,x_2^o,0,0,\mu^o))$ by $F_p^{uc}(\Psi^{-1}(x_1^o,x_2^o,0,0,\mu^o)) = \bigcup_{t\geq 0} X_t(F_1^c(\Psi^{-1}(x_1^o,x_2^o,0,0,\mu^o)))$. It is clear that this partial foliation is of class C^1.

As before we denote by $L^{su}(x,\mu)$ the intersection $F_p^{su}(x,\mu) \cap (\Sigma_{z_o} \times \{\mu\})$ for $(x,\mu) \in \Psi^{-1}(S^1 \times \{\mu\})$ and by $L^{uc}(x,\mu)$ the intersection $F_p^{uc}(x,\mu) \cap (\Sigma_{z_o} \times \{\mu\})$ for $(x,\mu) \in \Psi^{-1}(\{(x_1,x_2,0,0,\mu \ / \ x_1^2+x_2^2 = 1, \mu \in I_1\})$.

(iii) As in (IV.3.3) we denote by $T_\mu \subset \Sigma_{z_o} \times \{\mu\}$ the surface wich contains all the tangencies between L^{su} and π^s. As we did in (IV.3.3) we can obtain the foliation π^s in such a way that its intersections with the surface T_μ are invariant by the map R.

Suppose constructed a homeomorphism $H^1\colon \bigcup_{\mu\in I_1} T_\mu \to \bigcup_{\mu\in \tilde{I}_1} \tilde{T}_\mu$

such that its restriction $H^1_\mu = H^1\big|_{T_\mu} : T_\mu \to \tilde{T}_{\rho(\mu)}$ sends:

(a) $\pi^s \cap T_\mu$ into $\tilde{\pi}^s \cap \tilde{T}_{\rho(\mu)}$

(b) $L^{uc} \cap T_\mu$ into $\tilde{L}^{uc} \cap \tilde{T}_{\rho(\mu)}$ and

(c) $\tilde{R} \circ H^1_\mu = H^1_\mu \circ R$.

Then using the spirals of L^{su} and the leaves of π^s we can extend this homeomorphism to the neighborhood $V_{z_0} \times \{\mu\}$ of (z_0,μ) sending $(W^s_{q(\mu)} \cap V_{z_0}) \times \{\mu\}$ into $(W^s_{\tilde{q}(\rho(\mu))} \cap \tilde{V}_{\tilde{z}_0}) \times \{\rho(\mu)\}$ and $(W^{cu}_{p(\mu)} \cap V_{z_0}) \times \{\mu\}$ into $(W^{cu}_{\tilde{p}(\rho(\mu))} \cap \tilde{V}_{\tilde{z}_0}) \times \{\rho(\mu)\}$.

Using Liapunov functions in neighborhoods of (p,μ_0) and (q,μ_0) we can extend this homeomorphism to these neighborhoods. To extend these homeomorphism to a neighborhood of (α_{μ_0},μ_0) we note that we preserved the foliation F^u_α. To extend to the other unstable manifolds of the critical elements $\alpha_{1,\mu},\ldots,\alpha_{k,\mu}$ we preceed in the usual way (see II.2.9).

The construction of the homeomorphism H^1 with the desired properties is easy and we left it to the reader. This concludes the proof of Theorem C in this case.

The construction of the equivalence between $\{X_\mu\}$ and $\{\tilde{X}_\mu\}$ in the other cases is done with sligth modifications of the above construction and we left it to the reader. We observe that the construction of H^1, in such a way that it can be extended to neighborhoods of some determinated critical element (wich intersects $W^s_{p(\mu)}$), depends on an appropriate election of a foliation in this critical element and on a well understanding of the ideas in Chapter II.

<u>Note</u>. We note that, the solution of the general case for bifurca-

tions as in this chapter, depends on the solution of the general case for bifurcations as in Chapter II, because this last kind of bifurcation accumulates in the first one.

APPENDIX

In this appendix we state a lemma related to one parameter families of diffeomorphisms in dimension two. These lemma were used many times in Chapters II, III and IV.

Let $K \subset \mathbb{R}^2$ be a disc. Diff(K) will denote the set of C^∞ diffeomorphisms $f: K \to \mathbb{R}^2$. $P = P(K)$ will be the set of one parameter families of elements of Diff(K). In these two sets we take the usual topologies.

Let $\varphi_\mu : K \to \mathbb{R}^2$ be an element of P such that:

(a) associated to each value $\mu \in I$ there are two hyperbolic fixed points of saddle type s_μ and r_μ of the diffeomorphism φ_μ.

(b) $W^u_{s_\mu} \cap W^s_{r_\mu} \neq \phi$ and $W^u_{s_\mu} \cap W^s_{r_\mu}$ (see Figure A.1).

Let $U \subset K$ (resp. $V \subset K$) be a small neighborhood of $s_{1/2}$

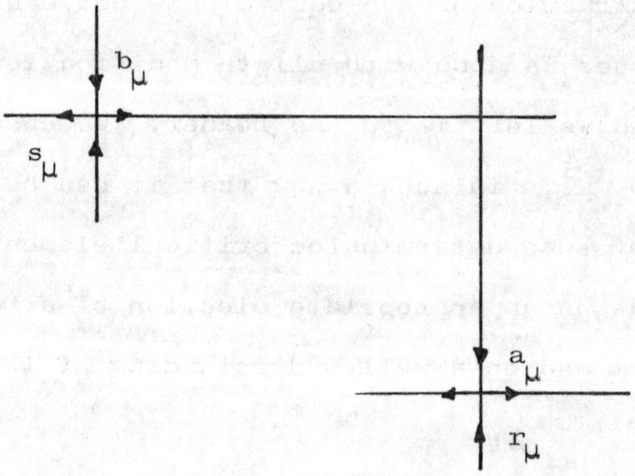

Figure A.1

(resp. $r_{1/2}$) such that $s_\mu \in U$ (resp. $r_\mu \in V$) $\forall \mu \in I$. Denote by b_μ (resp. a_μ) the contraction at s_μ (resp. at r_μ). Without loss of generality we assume that $0 < b_\mu < 1$ (resp. $0 < a_\mu < 1$).

Let Φ be the map $\Phi(x,\mu) = (\varphi_\mu(x),\mu)$. To each value μ let $D^s_{s_\mu} \subset W^s_{s_\mu}$ be a fundamental domain for φ_μ (resp. $D^s_{r_\mu} \subset W^s_{r_\mu}$). Denote by D^s_s the set $\bigcup_{\mu\in I_1} D^s_{s_\mu} \times\{\mu\}$ (resp. $D^s_r = \bigcup_{\mu\in I_1} D^s_{r_\mu} \times \{\mu\}$). This set is a fundamental domain for Φ. Let $\partial D^s_s = \gamma_1 \cup \Phi(\gamma_1) \cup \cup \gamma_2 \cup \Phi(\gamma_2)$.

For each point $(x^1_\mu,\mu) \in \gamma_\mu$ we consider a one dimensional disc of class C^r $\Sigma(x^1_\mu,\mu) \subset U \times \{\mu\}$ transversal to $W^s_{s_\mu} \times \{\mu\}$ in $U \times \{\mu\}$ and varying continuously with the parameter μ (resp. $\Sigma(x^2_\mu,\mu) \subset U \times \{\mu\}$ for each point $(x^2_\mu,\mu) \in \gamma^2$). We foliate now a fundamental neighborhood $N_\mu \subset U \times \{\mu\}$ between $\Sigma(x^1_\mu,\mu)$ and $\Phi(\Sigma(x^1_\mu,\mu))$ with one dimensional leaves of class C^r, transversals to $W^s_{s_\mu} \times \{\mu\}$ in $U \times \{\mu\}$ and varying C^r with the parameter μ. We do this in such a way that the projection along these leaves, $\pi^u_s\colon \bigcup_\mu N_\mu \times\{\mu\} \to D^s_s$ will be a C^r map. Iterating this foliation (by Φ) and defining $F^u_s(s_\mu,\mu) = (W^u_{s_\mu} \cap U) \times \{\mu\}$ we have defined an Φ-invariant foliation F^u_s in $U\times I$. Using the results of the Appendix of [dM-VS] we construct C^1-coordinate $\Psi\colon U\times I_1 \to \mathbb{R}^2\times I$ such that in this coordinate we have for $\Phi^* = \Psi\circ\Phi\circ\Psi^{-1}$: $\Phi^*(x_1,x_2,\mu) = = (\lambda_\mu x_1, b_\mu x_2,\mu)$ and Ψ sends one leaf $F^u_s(x_{\mu^o},\mu^o)$ of F^u_s into the straight line $x_2 = c^{te}$, $\mu = \mu^o$.

Let $(z_\mu,\mu) = (W^u_{s_\mu} \times\{\mu\}) \cap D^s_r$. It is clear that $F^u_s \cap D^s_r$ contains a region $R_s \subset D^s_r$ such that the intersection $R_s \cap (V\times\{\mu\})$ is a interval $(I(z_\mu),\mu)$ with $z_\mu \in$ Interior $(I(z_\mu))$ (see Figure A.2).

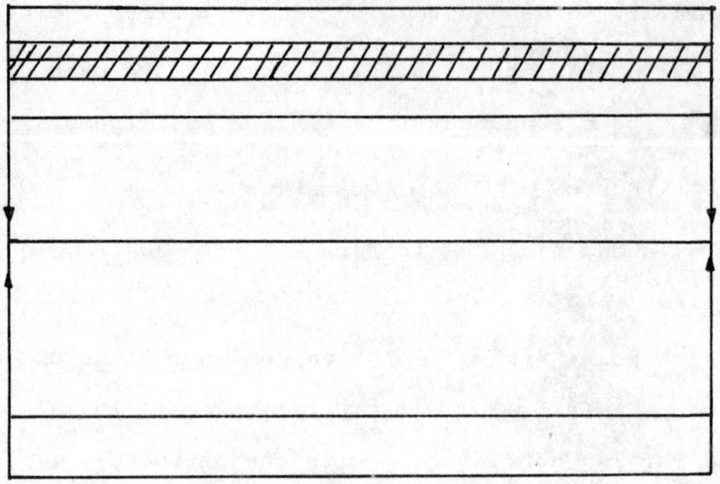

Figure A.2

Let $\partial D_r^s = \delta_1 \cup \Phi(\delta_1) \cup \delta_2 \cup \Phi(\delta_2)$. For each point $(y_\mu^1,\mu) \in \delta_1$ we consider a one dimensional disc of class C^r $\Sigma(y_\mu^1,\mu) \subset V \times \{\mu\}$, transversal to $W_{r_\mu}^s \times \{\mu\}$ in $V \times \{\mu\}$ and varying continuously with the parameter μ (resp. $\Sigma(y_\mu^2,\mu) \subset V \times \{\mu\}$) for each point $(y_\mu^2,\mu) \in \delta_2$). We foliate now a fundamental neighborhood $T_\mu \subset U \times \{\mu\}$ between $\Sigma(y_\mu^1,\mu)$ and $\Phi(\Sigma(y_\mu^1,\mu))$ (resp. between $\Sigma(y_\mu^2,\mu)$ and $\Phi(\Sigma(y_\mu^2,\mu))$) with one dimensional leaves of class C^r, transversals to $W_{r_\mu}^s \times \{\mu\}$ in $V \times \{\mu\}$ and varying continuously with the parameter μ. Trought points of $(I(z_\mu),\mu)$ we use the fibers of the intersection $F_s^u \cap T_\mu$. We do this in such a way that the projection along these leaves, $\pi_r^u\colon \bigcup_\mu T_\mu \times \{\mu\} \to D_r^s$ is a C^1 map. Iterating this foliation (with Φ) and defining $F_r^u(r_\mu,\mu) = (W_{r_\mu}^u \cap V) \times \{\mu\}$ we have one Φ-invariant foliation F_r^u of $V \times I$. Using the results of the Appendix of [dM-VS] we construct C^1 coordinate $\Gamma\colon V \times I \to \mathbb{R}^2 \times I$ such that in this coordinate we have for $\Phi^\Delta = \Gamma \circ \Phi \circ \Gamma^{-1}$: $\Phi^\Delta(y_1,y_2,\mu) = (\rho_\mu u_1, a_\mu y_2,\mu)$ and Γ sends one leaf $F_r^u(y_{\mu^o},\mu^o)$ of

F_r^u into the straight line $y_2 = c^k$, k a constant, $\mu = \mu^o$.

Let $\{\tilde{\varphi}_\mu\}$ be other one parameter family of P C^s-near of $\{\varphi_\mu\}$, $1 \le r \le s \le \infty$. Let us suppose (for the map $\tilde{\Phi}$) similar constructions as we did for the map Φ.

Let now $\chi: I \to \{(0,y_2,\mu)\}$ and $\tilde{\chi}: I \to \{(0,\tilde{y}_2,\mu)\}$, $\chi(\mu) = (0,\varepsilon(\mu),\mu)$, $\tilde{\chi}(\mu) = (0,\tilde{\varepsilon}(\mu),\mu)$ be two C^∞ curves such that

(a) $\varepsilon(0) = \tilde{\varepsilon}(0) = 1$, $\varepsilon(1) = \tilde{\varepsilon}(1) = -1$, $\tilde{\varepsilon}(\tilde{\mu}_o) = 0$ ($\tilde{\mu}_o$ near of μ_o)

(b) $\varepsilon'(\mu_o) < 0$, $\tilde{\varepsilon}'(\tilde{\mu}_o) < 0$ and the two maps are strictly decreasing (see Figure A.3).

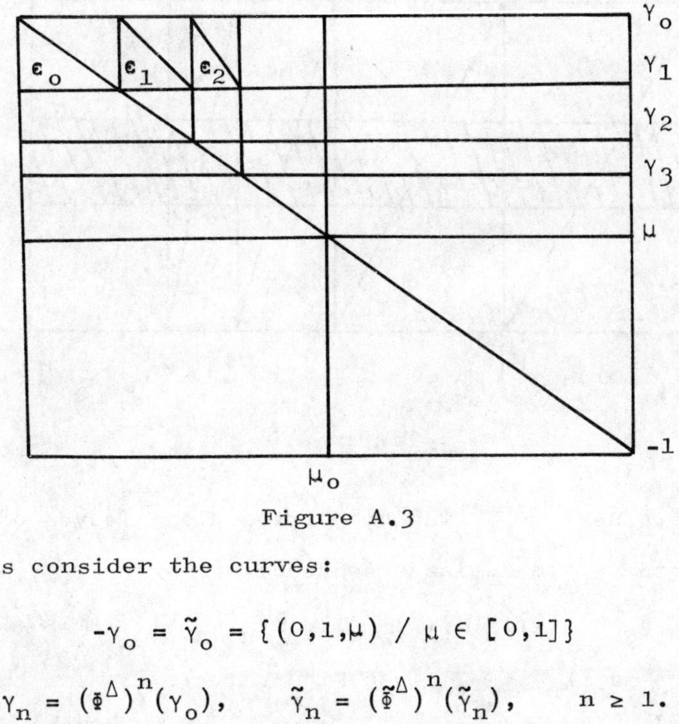

Figure A.3

Let us consider the curves:

$$-\gamma_o = \tilde{\gamma}_o = \{(0,1,\mu) \,/\, \mu \in [0,1]\}$$

$$-\gamma_n = (\Phi^\Delta)^n(\gamma_o), \qquad \tilde{\gamma}_n = (\tilde{\Phi}^\Delta)^n(\tilde{\gamma}_n), \qquad n \ge 1.$$

It is clear that for each n, γ_n (resp. $\tilde{\gamma}_n$) intersects the curve χ (resp. $\tilde{\chi}$) in an unique point: $\gamma_n \cap \chi = \{(0,a^n_{\mu_n},\mu_n)\}$ (resp. $\tilde{\gamma} \cap \tilde{\chi} = \{(0,\tilde{a}^n_{\mu_n},\tilde{\mu}_n)\}$).

Define the curves $\varepsilon_n := (\Phi^{\Delta})^{-n}\{(0,\varepsilon(\mu),\mu) / \mu \in [\mu_n,\mu_{n+1}]\}$ (resp. $\tilde{\varepsilon}_n := (\tilde{\Phi}^{\Delta})^{-n}\{(0,\tilde{\varepsilon}(\mu),\mu) / \mu \in [\tilde{\mu}_n,\tilde{\mu}_{n+1}]\}$) and the curve $\varepsilon_o = \{(0,\varepsilon(\mu),\mu) / \mu \in [0,\mu_1]\}$ (resp. $\tilde{\varepsilon}_o = \{(0,\tilde{\varepsilon}(\mu),\mu) / \mu \in [0,\bar{\mu}_1]\}$).

Consider, for each $n \geq 0$, the intersection $\varepsilon_n \cap \Gamma(R_s)$. Associated to this intersection there are two parameter values, γ_n, λ_n, $0 < \gamma_o < \lambda_o < \mu_1 < \gamma_1 < \lambda_1 < \mu_2 < \ldots < \mu_n < \gamma_n < \lambda_n < \mu_{n+1} < \ldots < \mu_o$ such that $(0,a^n_{\gamma_n}\varepsilon(\gamma_n),\gamma_n)$ and $(0,a^n_{\lambda_n}\varepsilon(\lambda_n),\lambda_n)$ are in the boundary of the set $\Gamma(R_s)$ (similarly for the family $\tilde{\Phi}$, see Figure A.4).

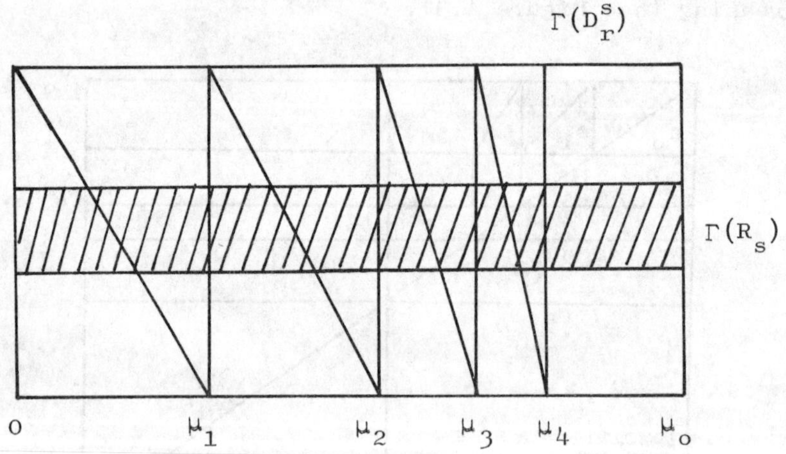

Figure A.4

Now, we use Γ^{-1}, the projection along leaves of F^u_s and the coordinate Ψ to define a sequence of curves $\{\rho_n\}$, $n \geq 0$ in $\{(0,x_2,\mu)\}$, $\rho_n = \{(0,\rho_n(\mu),\mu) / \mu \in [\gamma_n,\lambda_n]\}$. It is clear that we can assume $\rho_n(\lambda_n) = -1$ (see Figure A.5).

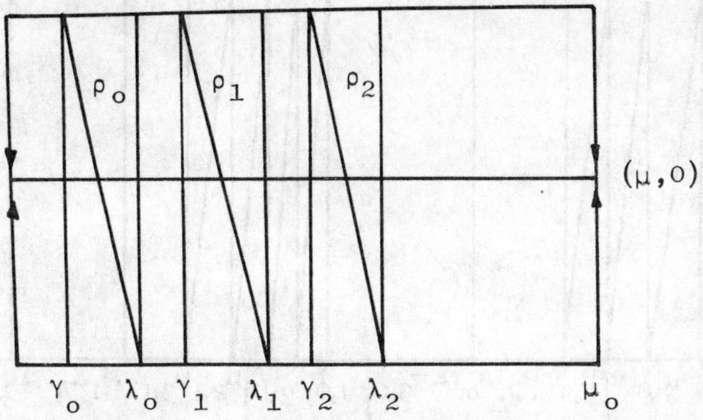

Figure A.5

Let $\theta_n \in]\gamma_n, \lambda_n[$ be the parameter value such that $\rho_n(\theta_n) = 0$ [similarly for the family $\tilde{\Phi}$].

Let $L_o^{\pm} = \{(0, \pm 1, \mu) \ / \ \mu \in [0,1]\} \subset \mathbb{R}_{x_1} \times \mathbb{R}_{x_2} \times I_1$ and $L_m^{\pm} = (\Phi^*)^m (L_o^{\pm})$ (resp. $\tilde{L}_m^{\pm} = (\tilde{\Phi}^*)^m (L_o^{\pm})$), $m \geq 1$.

It is clear that for each pair of integers n, m, $L_m^{\pm}$ intersects the curve ρ_n in an unique point $L_m^+ \cap \rho_n = \{(0, b_{\mu_{m,n}}, \mu_{m,n})\}$ and $L_m^- \cap \rho_n = \{(0, -b_{v_{m,n}}, v_{m,n})\}$ (similarly $\tilde{L}_m^+ \cap \tilde{\rho}_n = \{(0, \tilde{b}_{\tilde{\mu}_{m,n}}, \tilde{\mu}_{m,n})\}$, $\tilde{L}_m^- \cap \tilde{\rho}_n = \{(0, -\tilde{b}_{\tilde{v}_{m,n}}, \tilde{v}_{m,n})\}$) where $\gamma_n < \mu_{1,n} < \mu_{2,n} < \ldots < \theta_n < \ldots < v_{2,n} < v_{1,n} < \lambda_n$. We define now the curves $\rho_n^{m,(+)}$, $\rho_n^{m,(-)}$ by

$$\rho_n^{m,(+)} = (\Phi^*)^{-m}(\{(0, \rho_n(\mu), \mu) \ / \ \mu \in [\mu_{m,n}, \mu_{m+1,n}]\}), \quad m \geq 1,$$

$$\rho_n^{o,(+)} = \{(0, \rho_n(\mu), \mu) \ / \ \mu \in [\gamma_n, \mu_{1,n}]\},$$

$$\rho_n^{m,(-)} = (\Phi^*)^{-m}(\{(0, \rho_n(\mu), \mu) \ / \ \mu \in [v_{m+1,n}, v_{m,n}]\}, \quad m \geq 1$$

and $p_n^{o,(-)} = \{(0, \rho_n(\mu), \mu) \ / \ \mu \in [v_{1,n}, \lambda_n]\}$. [Similar constructions for the family $\tilde{\Phi}$, see Figure A.6].

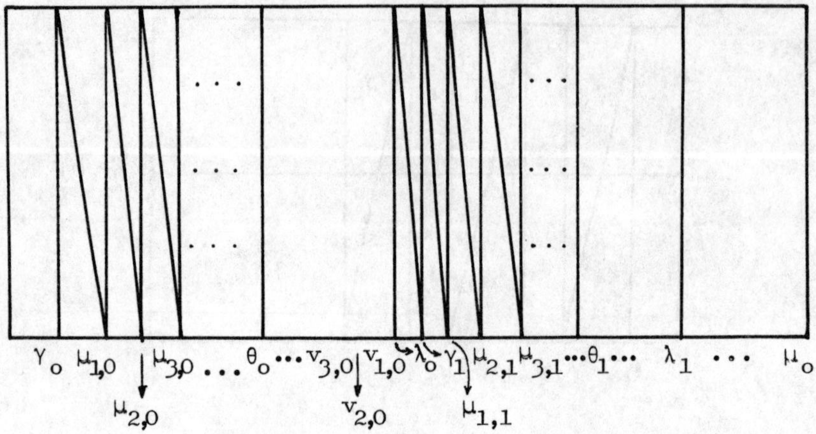

Figure A.6

Denote by $D^s_{s,\Psi}$ the set $\Psi(D^s_s)$. Let $F_s \subset D^s_{s,\Psi}$ be a foliation of class C^1, with leaves transversal to the straight lines $J_\mu = \{(0,x_2,\mu) / b_\mu \le x_2 \le 1\}$ in $D^s_{s,\Psi}$ and such that its leaves intersects the curves $\{\rho_n^{m,(+)}\}$, $\{\rho_n^{m,(-)}\}$ in an unique point for all pair of indexes (m,n). Let us suppose that $L_o^\pm$ and $L_1^\pm$ are leaves of F_s. Denote by $\pi_\mu: J_\mu \to J_1$ the projection along leaves of F_s.

Let $h: J_1 \to \tilde{J}_1$ be a C^2-diffeomorphism such that $h(1) = 1$ and $h(b_1) = \tilde{b}_1$. We want to extend h to a homeomorphism $H: D^s_{s,\Psi} \to \tilde{D}^s_{s,\tilde{\Psi}}$ such that: (a) $H(F_s) = F_{\tilde{s}}$, (b) $H(\rho_n^{m,(+)}) = \tilde{\rho}_n^{m,(+)}$, $H(\rho_n^{m,(-)}) = \tilde{\rho}_n^{m,(-)}$ $\forall$ m,n. (c) $H(J_\mu) = \tilde{J}_{\hat{\rho}(\mu)}$, $\hat{\rho}: I \to I$ a homeomorphism. [$H_\mu = H|_{J_\mu}$ is a diffeomorphism.]

To do this, let $\hat{\rho}: [0,\gamma_o] \to [0,\tilde{\gamma}_o]$ be a diffeomorphism. Define $H(0,x_2,\mu) = \pi^{-1}_{\hat{\rho}(\mu)} \circ h \circ \pi_\mu(0,x_2,\mu)$ to all $\mu \in [0,\gamma_o]$. If $\mu^o \in [\gamma_o,\mu_{1,0}]$ then $J_{\mu^o} \cap \rho_o^{o,(+)} = \{(0,\rho(\mu^o),\mu^o)\}$ so the projection $\pi_{\mu^o}(0,\rho(\mu^o),\mu^o) \in J_1$, and we can consider $h(\pi_{\mu^o}(0,\rho(\mu^o),\mu^o))$ and the leaf $F_{\tilde{s}}(h(\pi_{\mu^o}(0,\rho(\mu^o),\mu^o)))$. By the definition of $F_{\tilde{s}}$ we know that there exists an unique $\hat{\rho}(\mu^o) \in$ $\in [\tilde{\gamma}_o, \tilde{\mu}_{1,o}]$ such that $(0, \tilde{\rho}(\hat{\rho}(\mu^o)), \hat{\rho}(\mu^o)) \in F_{\tilde{s}}(h(\pi_{\mu^o}(0, \rho(\mu^o), \mu^o)))$. We define $H(0, \rho(\mu^o), \mu^o) = (0, \tilde{\rho}(\hat{\rho}(\mu^o)), \hat{\rho}(\mu^o))$.

To each $(0, x_2, \mu^o) \in J_{\mu^o}$, $x_2 \neq \rho(\mu^o)$ we define: $H(0, x_2, \mu^o) = \tilde{\pi}^{-1}_{\hat{\rho}(\mu^o)} \circ h \circ \pi_{\mu^o}(0, x_2, \mu^o)$.

Inductively we can define H on the set $\{(0, x_2, \mu) / b_\mu \le x_2 \le 1, \mu \in \bigcup_{m=1}^{\infty} [\mu_{m,o}, \mu_{m+1,o}]\}$.

To $\{(0, x_2, \theta_o) / b_{\theta_o} \le x_2 \le 1\}$ we define $H(0, x_2, \theta_o) = \tilde{\pi}^{-1}_{\tilde{\theta}_o} \circ h \circ \pi_{\theta_o}(0, x_2, \theta_o)$.

To complete the definition of H in the set $\{(0, x_2, \mu) / b_\mu \le x_2 \le 1$ or $-1 \le x_2 \le -b_\mu$; $\mu \in [\gamma_o, \theta_o]\}$ we define $H(0, x_2, \mu) = \tilde{\pi}^{-1}_{\hat{\rho}(\mu)} \circ h \circ \pi_\mu(0, x_2, \mu)$, $-1 \le x_2 \le -b_\mu$.

The same inductive process can be used to complete the definition of H in all the set $D^s_{s,\Psi}$.

We extend the homeomorphism H to all the set $\{0, x_2, \mu) / -1 \le x_2 \le 1\}$ conjugating Φ^* with $\tilde{\Phi}^*$ and defining $H(0, 0, \mu) = (0, 0, \hat{\rho}(\mu))$.

It is clear that the above homeomorphism sends the curves ρ_n into the curves $\tilde{\rho}_n$, $\forall\, n \ge 0$. It also sends the inverval $L_\mu = \{(0, x_2, \mu) / -1 \le x_2 \le 1\}$ into the corresponding interval $\tilde{L}_{\hat{\rho}(\mu)}$.

Now, we use the above homeomorphism, the coordinate Ψ, the projection $\pi^s_s: \bigcup_{\mu \in I} W^s_{s_\mu} \times \{\mu\} \to \bigcup_{\mu \in I} (I(z_\mu), \mu) \subset V \times I$ and the coordinate Γ to define a homeomorphism $H_2: \Gamma(\bigcup_\mu (I(z_\mu), \mu)) \to$ $\to \tilde{\Gamma}(\bigcup_\mu (I(z_\mu), \mu))$ which sends the intersections $\varepsilon_n \cap \Gamma(\bigcup_\mu (I(z_\mu), \mu))$ and $\{(0, y_2, \mu_o) / a_{\mu_o} \le y_2 \le 1\} \cap$ $\cap\, \Gamma(\bigcup_\mu (I(z_\mu), \mu))$ into the corresponding intersections $\tilde{\varepsilon}_n \cap$

$\cap\ \tilde{\Gamma}(\bigcup_{\mu} (I(\tilde{z}_{\mu}),\mu))$ and $\tilde{\Gamma}(\bigcup_{\mu} (I(\tilde{z}_{\mu}),\mu)) \cap \{(0,\tilde{y}_2,\tilde{\mu}_0) \ / \ \tilde{a}_{\tilde{\mu}_0} \leq \tilde{y}_2 \leq 1\}$.

It is clear that using the image of F_s by the composition $\Gamma \circ \pi_s^s \circ \Psi^{-1}$ we can construct a foliation F_r outside a neighborhood of $\Gamma(\bigcup_{\mu} (I(z_{\mu}),\mu))$ with similar properties as F_s.

As above we can extend H_2 to all the set $\{(0,y_2,\mu) \ / \ -1 \leq y_2 \leq 1, \ \mu \in I\}$ in such a way that it sends the curve $\chi(\mu)$ into the curve $\tilde{\chi}(\mu)$, the intervals $R_{\mu} = \{(0,y_2,\mu) \ / \ -1 \leq y_2 \leq 1\}$ into the corresponding intervals $\tilde{R}_{\hat{\rho}(\mu)}$.

Now we use the coordinates Ψ and Γ to state the following lemma:

<u>Lemma</u>. To families Φ, $\tilde{\Phi}$ and curves χ, $\tilde{\chi}$ as above, it is always possible to construct homeomorphisms $H_s\colon \bigcup_{\mu\in I} W^s_{s_{\mu}} \times\{\mu\} \to \bigcup_{\mu\in I} W^s_{\tilde{s}_{\mu}} \times\{\mu\}$, $H_r\colon \bigcup_{\mu\in I} W^s_{r_{\mu}} \times\{\mu\} \to \bigcup_{\mu\in I} W^s_{\tilde{r}_{\mu}} \times\{\mu\}$ such that

(a) $H_s \circ \Phi = \tilde{\Phi} \circ H_s$, $\quad H_r \circ \Phi = \tilde{\Phi} \circ H_r$

(b) $H_r(\chi(\mu)) \subset \tilde{\chi}(\mu)$

(c) the representation of $\chi(\mu)$ in $\bigcup_{\mu\in I} W^s_{s_{\mu}} \times\{\mu\}$ (using a unstable foliation F^u_s) is sent, by H_s, into the representation of $\tilde{\chi}(\mu)$ in $\bigcup_{\mu\in I} W^s_{\tilde{s}_{\mu}} \times\{\mu\}$.

<u>Observations</u>

(1) The same lemma is true if we consider families $\{\varphi_{\mu}\}$ wich has a finite number $s_{1,\mu}, s_{2,\mu}, \ldots, s_{k,\mu}$ of hyperbolic fixed points such that $W^u_{s_{i,\mu}} \pitchfork W^s_{s_{i+1,\mu}}$, $i = 1,2,\ldots,k-1$.

(2) The same lemma is true if we use $C^{(0)}$ coordinates Ψ and Γ.

BIBLIOGRAPHY

[Be] J.A. Beloqui: Modulo de estabilidade para campos vetoriais em variedades tridimensionais. Tese IMPA.

[Br] P. Brunovsky: On one parameter families of diffeomorphisms I, II, Comment. Mat. Univ. Carolinae 11 595-582 (1970) and 12 765-782 (1971).

[dM-P]-1 W. de Melo - J. Palis: Geometric Theory of Dynamical Systems, An Introduction, Springer-Verlag, 1982.

[dM-P]-2 W. de Melo - J. Palis: Moduli of Stability for Diffeomorphisms, Lecture Notes in Math. 819 Springer-Verlag, Global Theory of Dynamical Systems.

[dM-VS] W. de Melo - S. Van Strien: Diffeomorphisms on Surfaces with a Finite Number of Moduli, preprint, Série A-026, 1984, IMPA.

[dM-P-VS] W. de Melo - J. Palis - S. Van Strien: Characterizing diffeomorphisms with modulus of stability one. Lecture Notes in Math. 898, Springer-Verlag. Dynamical Systems and Turbulence, Warwick 1980.

[H-P-S] M. Hirsch - C. Pugh - M. Shub: Invariant Manifolds, Lecture Notes in Math. 583, Springer-Verlag (1977).

[Ku] I. Kupka: Contributions a la Théorie des Champs génériques, Contribution to Diff. Eq. 2, 457-484 (1963) and 411-420 (1964).

[La] R. Labarca: Estabilidade de famílias a um parâmetro de campos de vetores. Tese IMPA, 1985.

[L] O.E. Lanford III: Bifurcations of periodic solutions into invariant tori. The work of Ruelle and Takens, Lecture Notes in Math. 322 (1983).

[M.J.P.] M.J. Pacifico: Structural stability of vector fields on 3-manifolds with boundary, Jr. of Dif. Eq. 54 nº 3 (1984).

[M] W. de Melo: Moduli of stability of two-dimensional manifolds, Topology 19, 1980, 9-21.

[Ma-P] I.P. Malta - J. Palis: Families of vector fields with finite modulus of stability, Lecture Notes in Math. 898, Dynamical Systems and Turbulence - Warwick 1980.

[N-P-T] S. Newhouse - J. Palis - F. Takens: Bifurcation and stability of families of diffeomorphisms, Publ. I.H.E.S. nº 57, 1983, 5-71.

[PA] R. Palais: Local triviality of the restriction map for embeddings. Comm. Math. Helvet. 34 (1960), 305-312.

[P] J. Palis: A differentiable invariant of topological conjugacies and moduli of stability, Asterisque 51, 1978, 335-346.

[P-T] J. Palis - F. Takens: Stability of parametrized families of gradient vector field, Ann. of Math. 118, 383-421.

[Sm 1] S. Smale: Stable manifolds for differential equations and diffeomorphisms, Ann. Scuola Norm. Sup. Pisa 18, (1963), 97-116.

[So 1] J. Sotomayor: Generic one-parameter families of vector fields on two-dimensional manifolds, Publ. Math. I.H.E.S. vol. 43, (1974), 5-46.

[So 2] J. Sotomayor: Generic bifurcations of Dynamical Systems, Dynamical Systems (M.M. Peixoto ed.), Ac. Press, 1973, 561-582.

[ST 1] S. Sternberg: Local contractions and a theorem of Poincaré, Amer. J. Math. 79, 1957, 809-824.

[ST 2] S. Sternberg: On the structure of local homeomorphisms of Euclidean spaces II, Amer. J. Math. 80, 1958, 623-631.

[TA 1] F. Takens: Moduli and bifurcations, non-transversal intersection of invariant manifolds of vector fields, Conference held at Univ. Fed. de São Carlos (SP), Brazil, in Functional Differential Equations and Bifurcations, (A.F. Izé, ed.), Springer-Verlag 799, 1980, 368-384.

[TA 2] F. Takens: Partially hyperbolic fixed points, Topology 10, (1971), 133-147.

[VS] S. Van Strien: One parameter families of vector fields, Ph.D. Thesis, Preprint, Utrecht, 1982.

Departamento de Matemática y C.C.
Universidad de Santiago de Chile
Casilla 5659 Correo 2
Santiago
Chile

IMPA
Estrada Dona Castorina 110
Jardim Botânico
CEP 22460
Rio de Janeiro - Brasil

R LARBARCA* & M J PACIFICO**

Morse-Smale vector fields on 4– manifolds with boundary

Introduction

Let M be a compact smooth manifold with boundary ∂M. Denote $\mathfrak{X}^r(M,\partial M)$ the space of C^r vector fields on M that are tangent to the boundary of M endowed with the C^r topology. Following the boundaryless situation, we say that a vector field $X \in \mathfrak{X}^r(M,\partial M)$ is C^r structurally stable when it has a neighborhood U such that every $Y \in U$ is topologically equivalent to X, i.e., there exists a homeomorphism h: M↺ that maps orbits of X onto orbits of Y preserving their time induced orientations.

The theory of structural stability in manifolds with boundary cannot be developed just through straightforward modifications of the corresponding boundaryless results. The so far most disclosing example of the novel phenomena made possible by the existence of a boundary is the example constructed in [2] of a C^1 structurally stable vector field in $\mathfrak{X}^\infty(D^3,\partial D^3)$ (where D^3 denotes the three dimensional disk) whose nonwandering set is not hyperbolic. In the boundaryless case, even if the existence of similar examples has not been so far outruled through a theorem, there is at least a sound

*Partially supported by PNUD-CHI-84-004 and CNPq - Brasil.

**Partially supported by CNPq - Brasil.

conjecture (the Stability Conjecture) proposing that C^1 structural stability implies the hyperbolicity of the nonwandering set [3a,7]. In the case of diffeomorphisms on boundaryless manifolds this conjecture has been recently proved by Mañé [3b].

The purpose of this paper is to develop the line of investigations started in [4], namely to define Morse-Smale vector fields on $\mathfrak{X}^r(M,\partial M)$. Obviously, we do not mean the merely formal question of finding a definition resembling enough that of the boundaryless case. We mean characterizing the elements of $\mathfrak{X}^r(M,\partial M)$ that are structurally stable and have a simple nonwandering set, i.e., consisting of finitely many singularities and periodic orbits. For technical reasons we shall restrict our work to the C^∞ case, and even in this space, we shall impose the generic condition of all the singularities and periodic orbits being linearizable. Denote $\mathfrak{X}^\infty_*(M,\partial M)$ this residual subset of $\mathfrak{X}^\infty(M,\partial M)$.

<u>Definition 1</u>. A vector field $X \in \mathfrak{X}^\infty_*(M,\partial M)$ is called Morse-Smale if

(i) $\Omega(X)$ is simple and hyperbolic

(ii) $X/\partial M$ is Morse-Smale

(iii) Let $\sigma_i, \sigma_j \in \Omega(X)$. If $x \in M$ is a nontransversal point of $W^u(\sigma_i)$ with $W^s(\sigma_j)$ then $x \in \partial M$ and either σ_i or σ_j is a singularity of X. Moreover if the α-limit set of x is σ_i and ω-limit set of x is σ_j then it is defined the weakest contraction at σ_i and the weakest expansion at σ_j.

<u>Theorem</u>. If $\dim M = 4$, then $X \in \mathfrak{X}^\infty_*(M,\partial M)$ is Morse-Smale if and only if X is C^∞ structurally stable and its nonwandering set is simple.

The same result in the 3-dimensional case was obtained in [4]. The transition in the dimension of the boundary from 2 to 3, and the related and well known increase in dynamical richness and difficults, is the main new feature of the theorem above. More specifically, in dimension 4 appear certain quasi-transversal saddle connections along the boundary (which are persistent under small perturbations) that pose new interesting problems that look much close to what one will find dealing with the general n-dimensional case. In fact, the 3 and 4-dimensional results are sufficiently substantial to conjecture that the theorem above holds in the n-dimensional case. However in such generality, essential problems remain that the lower dimensional techniques present here don't reach.

The restriction to $\mathfrak{X}_*^\infty(M,\partial M)$ instead of just $\mathfrak{X}^1(M,\partial M)$ is unsatisfactory, but we have reasons to hope that further research will remove it.

We wish to thank M.J. Dias Carneiro and R. Mañé for helpful conversations. We are grateful to IMPA for its very kind hospitality.

2. Preliminaries

Throughout this paper, M is a C^∞ compact 4-dimensional manifold with boundary ∂M.

A point $x \in M$ is a wandering point of X if there is a neighborhood $V \ni x$ in M and a number $t_o > 0$ such that if $|t| > t_o$ then $X_t(V) \cap V = \Phi$. Otherwise x is a nonwandering point of X. Denote by $\Omega(X)$ the set of nonwandering points of X. When $\Omega(X)$ is simple and hyperbolic, we say that $\Omega(X)$ has a cycle if there is a sequence of critical elements $\sigma_1,\ldots,\sigma_{k+1}$ with $\sigma_1 = \sigma_{k+1}$ such that $W^u(\sigma_i) \cap W^s(\sigma_{i+1}) \neq \Phi$, $1 \leq i \leq k$. If $\Omega(X)$ is simple and has no cycles we can define a relation of partial order in the set of critical elements of X as follows: $\sigma_i < \sigma_j$ if and only if $W^u(\sigma_i) \cap W^s(\sigma_j) \neq \Phi$ and $\sigma_i \neq \sigma_j$. We say that the behavior of σ_i with respect to σ_j is one if $\sigma_i < \sigma_j$ and there is no $z \in \Omega(X)$ with $\sigma_i < z < \sigma_j$. We say that the behavior is k if there is a sequence of critical elements $\sigma_i = \gamma_1 < \ldots < \gamma_k = \sigma_j$ such that γ_i has behavior one with respect to γ_{i+1}. We say that $W^u(\sigma_i)$ is transversal to $W^s(\sigma_j)$ if $T_y W^u(\sigma_i) + T_y W^s(\sigma_j) = T_y M$ for each $y \in W^u(\sigma_i) \cap W^s(\sigma_j)$.

Denote by $\mathfrak{X}^\infty_{M-S}(M,\partial M)$ the set of Morse-Smale vector fields in $\mathfrak{X}^\infty_*(M,\partial M)$. Let us make some remarks about the properties that define $\mathfrak{X}^\infty_{M-S}(M,\partial M)$.

Given $X \in \mathfrak{X}^\infty(M,\partial M)$ if $\tilde{Y}$ is a vector field defined on ∂M near $X/\partial M$ it is easy to see that $\tilde{Y}$ can be extended to a vector field Y defined on all of M, Y near X. Therefore, if we are looking for the stable vector fields $X \in \mathfrak{X}^\infty_*(M,\partial M)$ it is necessary that $X/\partial M$ be stable and as $\dim \partial M = 3$ and $\Omega(X)$ is simple then $X/\partial M$ must be a Morse-Smale vector field.

If $X \in \mathfrak{X}^\infty(M,\partial M)$ is such that all its critical elements satisfy (ii) and (iii) from Definition 1 we say that X satisfies the transversality condition modulus ∂M. If $x \in \partial M$ is a nontransversal (in M) intersection point of $W^u(\sigma_o)$ with $W^s(\sigma_1)$ then the dimension of $T_x W^u(\sigma_o) + T_x W^s(\sigma_1)$ is three and so x is a point of quasi transversal intersection of $W^u(\sigma_o)$ with $W^s(\sigma_1)$; moreover, the invariance of ∂M by the flow induced by X implies that $\sigma_o, \sigma_1 \in \partial M$. In this case we say that (σ_o, σ_1) have a quasi-transversal intersection along the boundary of M.

If (σ_o, σ_1) have a quasi-transversal intersection along the boundary of M for $X \in \mathfrak{X}^\infty_*(M,\partial M)$ then, if X is stable, either σ_o or σ_1 must be a singularity of X because if both were closed orbits then X would not be even locally stable in a neighborhood of the closure of an orbit in the intersection $W^u(\sigma_o) \cap W^s(\sigma_1)$ (see [4,6]).

It is also clear that conditions (i) and (ii) are necessary

for the stability of a vector field $X \in \mathfrak{X}_*^\infty(M,\partial M)$ [6,7].

Now we observe that if $X \in \mathfrak{X}_*^\infty(M,\partial M)$ is such that its non-wandering set $\Omega(X)$ is simple, hyperbolic and satisfies (iii) from the Definition 1 then $\Omega(X)$ has no cycles. In fact, if $\sigma_0,\ldots,\sigma_k = \sigma_0$ is a k-cycle in $\Omega(X)$ then, since X satisfy the transversality condition modulus ∂M, there is $i_0 \in \{0,\ldots,k\}$ such that $\sigma_{i_0} \in \partial M$. From the invariance of ∂M by the flow induced by X we conclude that σ_{i_0+1} (or σ_{i_0-1}) belongs to ∂M. Then $(\sigma_{i_0},\sigma_{i_0+1})$ (or $(\sigma_{i_0-1},\sigma_{i_0})$) have a quasi-transversal intersection along ∂M. If $(\sigma_{i_0},\sigma_{i_0+1})$ (or $(\sigma_{i_0-1},\sigma_{i_0})$) is the unique pair of critical elements in the cycle having a quasi-transversal intersection along ∂M and $x \in W^u(\sigma_{i_0}) \cap W^s(\sigma_{i_0+1})$ (or $x \in W^u(\sigma_{i_0-1}) \cap W^s(\sigma_{i_0})$) then since the intersection of $W^u(\sigma_j)$ with $W^s(\sigma_{j+1})$ is transversal for $j \neq i_0$ (or $j \neq i_0-1$), it is not difficult to see that $x \in \Omega(X)$, which is a contradiction. If there is more than one quasi-transversal intersection along ∂M, let $(\sigma_{j_0},\sigma_{j_0+1})$, $j_0 \neq i_0$, be one of them. The transversality condition modulus ∂M implies $W^u(\sigma_{j_0-1}) \cap W^s(\sigma_{j_0+2}) \neq \Phi$ and so $\sigma_0,\sigma_1,\ldots,\sigma_{j_0-1}$, $\sigma_{j_0+2},\ldots,\sigma_k = \sigma_0$ is a (k-2)-cycle in $\Omega(X)$. Repeating the argument for all quasi-transversal intersections along ∂M different from $(\sigma_{i_0},\sigma_{i_0+1})$ we finally get that there is a cycle in $\Omega(X)$ with $(\sigma_{i_0},\sigma_{i_0+1})$ as the unique pair of critical elements having a quasi-transversal intersection along ∂M. Applying the above argument we have the result.

All these remarks together prove the following:

<u>Theorem</u>. Let $X \in \mathfrak{X}_*^\infty(M,\partial M)$ be such that $\Omega(X)$ is simple. If X is structurally stable in $\mathfrak{X}_*^\infty(M,\partial M)$ then X is Morse-Smale.

3. Openess and Stability

The main goal in what follows is to prove the openess of $\mathfrak{X}^\infty_{M-S}(M,\partial M)$ in $\mathfrak{X}^\infty_*(M,\partial M)$ and the stability of its elements.

Let us first give some notations and definitions.

We say that $X \in \mathfrak{X}^\infty_*(M,\partial M)$ has a weakest contracting (expanding resp.) eigenvalue, say λ, at a singularity p if:

(i) $\text{Re}\,\lambda < 0$ (resp. $\text{Re}\,\lambda > 0$)

(ii) λ has multiplicity one

(iii) if γ is an eigenvalue of $DX(p)$ with $\text{Re}\,\gamma < 0$ (resp. $\text{Re}\,\gamma > \text{Re}\,\lambda$), $\gamma \neq \lambda$, $\bar{\lambda}$ then $\text{Re}\,\gamma < \text{Re}\,\lambda$ (resp. $\text{Re}\,\gamma > \text{Re}\,\lambda$).

Similarly, let us suppose that X has a periodic orbit σ. Let Σ be an invariant cross section at $q \in \sigma$ and P the associated Poincaré map. We say that X has a weakest contracting (expanding) eigenvalue at σ, say λ, if:

(i) $|\lambda| < 1$ (resp. $|\lambda| > 1$)

(ii) λ has multiplicity one

(iii) for all eigenvalue γ of $DP(q)$ with $|\gamma| < 1$ (resp. $|\gamma| > 1$), $\gamma \neq \lambda, \bar{\lambda}$, one has $|\gamma| < |\lambda|$ (resp. $|\gamma| > |\lambda|$).

Let $p \in M$ be a hyperbolic singularity of the vector field X. Suppose that it is defined the weakest contraction (resp. expansion) at p. In this case it is possible to define a C^1 center unstable manifold (resp. center stable manifold), not unique, invariant by the flow of X and tangent, at p, to the direct sum of the expansive subspace (resp. contractive subspace), with the subspace associated to the weakest contraction (resp. weakest expansion). We denote this submanifold by $W^{cu}(p)$ (resp. $W^{cs}(p)$). The central manifold at p, $W^c(p)$, is defined as $W^c(p) = W^s(p) \cap W^{cu}(p)$. In the same way if

σ is a hyperbolic periodic orbit of the vector field X such that it is defined the weakest contraction (resp. expansion) at σ then it is possible to define a C^1 center unstable manifold (resp. C^1 center stable manifold), not unique, invariant by the flow of X and tangent, at $q \in \sigma$, to the direct sum of the expansive (resp. contractive) subspace of the Poincaré map with the weakest contractive (resp. expansive) subspace of the Poincaré map and the subspace which corresponds to the flow direction. We denote this submanifold by $W^{cu}(\sigma)$ (resp. $W^{cs}(\sigma)$) and its restriction to the cross section Σ at $q \in \sigma$ by $W^{cu}(q)$ (resp. $W^{cs}(q)$). The central manifold $W^c(q)$ at $q \in \sigma$ is defined as $W^c(q) = W^u(q) \cap W^{cs}(q)$. Moreover, there is a unique invariant manifold $W^{ss}(\sigma) \subset W^s(\sigma)$ (resp. $W^{uu}(\sigma) \subset W^u(\sigma)$ such that:

(i) if σ is a singularity then $T_\sigma W^s(\sigma) = T_\sigma(W^{ss}(\sigma)) \oplus T_\sigma(W^{cu}(\sigma) \cap W^s(\sigma))$ [resp. $T_\sigma(W^u(\sigma)) = T_\sigma W^{uu}(\sigma) \oplus T_\sigma(W^{cu}(\sigma) \cap W^s(\sigma))$.

(ii) if σ is a periodic orbit, $q \in \sigma$, Σ a transversal section to X at $q \in \sigma$ and P is the Poincaré map then $T_q W^s(q) =$ $= T_q W^{ss}(q) \oplus T_q(W^{cu}(q) \cap W^s(q))$ (resp. $T_q W^u(q) = T_q W^{uu}(q) \oplus T_q(W^{cs}(q) \cap W^u(q))$.

Let $\mathfrak{A} \subset \mathfrak{X}^\infty_*(M,\partial M)$ be the set of vector fields $X \in \mathfrak{X}^\infty_{M-S}(M,\partial M)$ such that X has at most one pair of critical elements (σ_0,σ_1) having a quasi-transversal intersection along the boundary of M. Then the stability of (σ_0,σ_1) depends on

(i) if σ_0 and σ_1 are both singularities or one is a singularity and the other a periodic orbit;

(ii) the weakest contraction at σ_0 and the weakest expansion at σ_1 are real or complex numbers. All the different possibilities for σ_0 and σ_1 together yield 16 possibilities, indicated in the table below.

σ_1 \ σ_0		Singularity		Periodic Orbit	
		Real weakest expansion	Complex weakest expansion	Real weakest expansion	Complex weakest expansion
Singularity	Real weakest contraction	1st type	2nd type	3nd type	it is not possible in dimension 4
	Complex weakest contraction	2nd type	it is not possible in dimension 4	4th type	it is not possible in dimension 4
Periodic Orbit	Real weakest contraction	3nd type	4th type	it is not stable [4,6]	it is not possible in dimension 4
	Complex weakest contraction	it is not possible in dimension 4	it is no possible in dimension 4	it is not possible in dimension 4	it is not possible in dimension 4

Let $\mathfrak{X}_i^\infty(M,\partial M) \subset \mathcal{G}$, $0 \le i \le 4$, be defined by $\mathfrak{X}_o^\infty(M,\partial M) = \{X \in \mathcal{G};$ such that X has no quasi-transversal intersections$\}$

$\mathfrak{X}_i^\infty(M,\partial M) = \{X \in \mathcal{G};$ such that X has an unique quasi-transversal intersection along ∂M and this one is of ith type$\}$, $1 \le i \le 4$.

Our purpose is to prove that $\mathfrak{X}_i^\infty(M,\partial M)$, $0 \le i \le 4$, is open in $\mathfrak{X}^\infty(M,\partial M)$ and each of its elements is structurally stable. Before that, let us recall some basic results.

<u>Definition</u>. Let z be a hyperbolic singularity of $X \in \mathfrak{X}^\infty(M,\partial M)$. An unstable foliation for X at z is a C^o foliation $\mathfrak{F}^u(z,X)$ in a neighborhood of $W^u(z)$ satisfying the following properties:

1. The leaves are C^k disks varying continuously in the C^k topology.

2. The foliation is X_t-invariant for all t, that is, $X_t(F^u(x)) \subset F^u(X_t(x))$, where $F^u(x)$ is the leaf through the point x.

3. Each leaf intersects $W^s(z)$ transversally at a unique point.

Observe that $W^u(z)$ is the leaf through z. Similarly we define stable foliation.

If the weakest contraction is defined at z, a central unstable foliation $\mathfrak{F}^{cu}(z)$ in a neighborhood of z is a C^1-foliation invariant by X_t, $t \geq 0$ and its leaves are C^1 embedded disks varying continuously in the C^1 topology. Moreover, a leaf of $\mathfrak{F}^{cu}(z)$ through a given point transversally intersects $W^s(z)$ in a C^1-curve.

<u>Definition</u>. Let γ be a closed orbit of $X \in \mathfrak{X}^\infty(M,\partial M)$, S a cross section for X at $p \in \gamma$ and P the associated Poincaré map. Then P is of class C^∞ and has p as a hyperbolic fixed point. An unstable foliation for P at p is a C^0 foliation $\mathfrak{F}^u(p,P)$ in a neighborhood of $W^u(p,P)$ in S satisfying the following conditions:

1. The leaves are C^k disks varying continuously in the C^k topology.

2. The foliation is P-invariant; namely, $P(F^u(x)) \subset F^u(P(x))$ where $F^u(x)$ is the leaf through the point x.

3. Each leaf intersects $W^s(p,P)$ transversally (in S) at an unique point.

Observe that the leaf through p is $W^u(p,P)$.

<u>Definition</u>. Let γ be a closed orbit of $X \in \mathfrak{X}^\infty(M,\partial M)$. An unstable foliation $\mathfrak{F}^u(\gamma)$ for X at γ is a C^0 foliation in a

neighborhood of $W^u(\gamma)$ such that:

1. The leaves are C^k disks varying continuously in the C^k topology.
2. The foliation is X_t-invariant for all t. Namely, $X_t(F^u(x)) \subset F^u(X_t(x))$ where $F^u(x)$ is the leaf through x.
3. Each leaf intersects $W^s(\gamma)$ transversally at a unique point.
4. $W^u(\gamma) = \bigcup_{t\in\mathbb{R}} X_t(F^u(p))$ where $p \in \gamma$.

Let $\sigma_1, \sigma_2 \in \Omega(X)$ be such that $\sigma_1 < \sigma_2$. We say that the unstable foliation $\mathfrak{F}^u(\sigma_1, X)$ is compatible with the unstable foliation $\mathfrak{F}^u(\sigma_2, X)$ if each leaf of $\mathfrak{F}^u(\sigma_2, X)$ that intersects some leaf of $\mathfrak{F}^u(\sigma_1, X)$ is contained in this leaf and the foliation $\mathfrak{F}^u(\sigma_2, X)$ restricted to a leaf of $\mathfrak{F}^u(\sigma_1, X)$ is a C^k foliation. Similarly we define stable foliations and compatibility of stable foliations.

Let $X \in \mathfrak{X}^\infty_i(M, \partial M)$, $0 \le i \le 4$. Then $\Omega(X) = \{\sigma_j;\ \sigma_j$ is a critical element of X, $0 \le j \le n\}$ is hyperbolic and has no cycles. So we can define a relation of partial order in the set of critical elements of X as follows: $\sigma_i < \sigma_j$ if and only if $W^u(\sigma_i) \cap \cap W^s(\sigma_j) \ne \Phi$ and $\sigma_i \ne \sigma_j$. We extend this partial order to a total one and we assume that $\Omega(X) = \{\alpha_1 \le \ldots \le \alpha_k \le \sigma_o \le \sigma_1 \le \beta_1 \le \ldots \le \beta_\ell\}$ where (σ_o, σ_1) is the pair having the quasi-transversal intersection along the boundary of M. Moreover, we can assume that there is a neighborhood $\mathcal{U} \subset \mathfrak{X}^\infty(M, \partial M)$ of X such that if $Y \in \mathcal{U}$ then $\Omega(X) = \{\alpha_1(Y) \le \ldots \le \alpha_k(Y) \le \sigma_o(Y) \le \sigma_1(Y) \le \beta_1(Y) \le \ldots \le \beta_\ell(Y)\}$ where $\alpha_i(Y)$, $\sigma_o(Y)$, $\sigma_1(Y)$, $\beta_j(Y)$ are the critical elements (all of them hyperbolic) of Y near the corresponding ones of X and $(\sigma_o(Y), \sigma_1(Y))$ is the pair having the quasi-transversal intersection along the boundary of M. We will also assume that:

(i) the unstable manifold $W^u(\alpha_j(Y))$, $1 \le j \le k$, is transversal to the strong stable manifold $W^{ss}(\sigma_o(Y))$,

(ii) the stable manifold $W^s(\beta_j(Y))$, $1 \le j \le \ell$, is transversal to the strong unstable manifold $W^{uu}(\sigma_1(Y))$,

(iii) there is a compatible system of unstable (resp. stable) foliations $\mathfrak{F}^u(\alpha_i(Y))$, $1 \le i \le k$ (resp. $\mathfrak{F}^s(\beta_j(Y))$, $1 \le j \le \ell$),

(iv) all closed orbits of each $Y \in \mathcal{V}$ have the same period w=1 and the same invariant cross section as the corresponding closed orbits of X,

(v) there is $U_o \subset M$ such that for each $Y \in \mathcal{V}$, $\sigma_o(Y) \in U_o$, Y is C^2-linearizable in U_o and the diffeomorphism linearizing Y is C^2-close to the corresponding diffeomorphism for X (see Apendix of [4]). Moreover there is a C^2-Liapunov function $f_Y: U_o \to \mathbb{R}$ for every $Y \in \mathcal{V}$, that is, $Df_Y(x) \cdot Y(x) > 0$ if $x \in U_o \setminus \{\sigma_o(Y)\}$. So the level surfaces $f_Y^{-1}(c)$, $c \in \mathbb{R}$, are transversal to the orbits of Y in $U_o \setminus \{\sigma_o(Y)\}$.

<u>Theorem 1</u>. $\mathfrak{X}_1^\infty(M,\partial M)$ is open in $\mathfrak{X}_*^\infty(M,\partial M)$ and its elements are structurally stable.

<u>Proof</u>: Let $X \in \mathfrak{X}_1^\infty(M,\partial M)$, $\Omega(X) = \alpha_1(X) \le \ldots \le \alpha_k(X) \le \sigma_o(X) \le \le \sigma_1(X) \le \beta_1(X) \le \ldots \le \beta_\ell(X)$, where $(\sigma_o(X),\sigma_1(X))$ is the pair having a quasi-transversal intersection along ∂M. Let $\mathcal{V} \subset \mathfrak{X}_*^\infty(M,\partial M)$ and $U_o \subset M$ be a small neighborhood of X and $\sigma_o(X)$ respectively, $\mathcal{V}$ and U_o as above. Let $\Sigma \subset U_o$ be a cross section for every $Y \in \mathcal{V}$ at $q_o(X) \in W^u(\sigma_o(X)) \cap W^s(\sigma_1(X))$. For each $Y \in \mathcal{V}$ we denote by $q(Y)$ the intersection of $W^u(\sigma_o(Y))$ with Σ.

Let $D_s(\sigma_o(Y))$ be a fundamental domain for $W^s(\sigma_o(Y))$ and a cylinder $C \subset U_o$, $C \cap W^s(\sigma_o(Y)) = D_s(\sigma_o(Y))$, such that C is a cross section for each $Y \in \mathcal{U}$. It is clear that $\mathcal{F}^u(\alpha_i(Y))$, $1 \le i \le k$, induces a foliation in an open subset of C which intersects $D_s(\sigma_0(Y))$ in an open subset $W_0(Y)$ with closure $(W_0(Y)) \cap \partial M = \Phi$. We can extend this foliation to a foliation $\mathcal{F}^u(C,Y)$ of C such that the leaves of $\mathcal{F}^u(C,Y)$ through points of $D_s(\sigma_0(Y)) \setminus \mathrm{int}(W(Y))$ are linear where $W(Y) \subset D_s(\sigma_0(Y))$ is a compact set containing $W_0(Y)$ in its interior. We assume that the linearizing coordinates near $\sigma_0(Y)$ are y_1, y_2, y_3, y_4 (these coordinates depend on Y but this Y-dependence is not expressed in the notation) and, in these coordinates, we have:

$$q(Y) = (0,0,0,1), \qquad \Sigma = \{y_4 = 1\},$$

$$\{y_1 = y_3 = 0\} = W^u(\sigma_0(Y)) \cap \Sigma, \qquad \{y_2 = y_3 = 0\} = W^s(\sigma_1(Y)) \cap \Sigma,$$

$$\{y_3 = 0\} = W^{cu}(\sigma_0(Y)) \cap \Sigma, \qquad \{y_3 = 0\} = W^{cs}(\sigma_1(Y)) \cap \Sigma.$$

Let $\pi_Y\colon \Sigma \setminus \{y_1 = y_3 = 0\} \to C$ be the Poincaré map. Then $\mathcal{F}^{cu}(\Sigma,Y) = \pi_Y^{-1}(\mathcal{F}^u(C,Y))$ is a foliation (with singularity) of Σ whose set of singularities is the y_2-axis. For $x \in \Sigma$, $F^{cu}(x,\Sigma)$ denotes the leaf of $\mathcal{F}^{cu}(\Sigma,Y)$ through x. Observe that $\{y_3 = 0\} = W^{cu}(\sigma_0(Y)) \cap \Sigma$ is a leaf of $\mathcal{F}^{cu}(\Sigma,Y)$ and every leaf $F^{cu}(x,\Sigma)$ at $x \in \Sigma$ is tangent to $\{y_3 = 0\}$, the contact being $\frac{\lambda_2(Y)}{\lambda_1(Y)}$, where $\lambda_1(Y)$ and $\lambda_2(Y)$ are the contracting eigenvalues at $\sigma_0(Y)$, $|\lambda_2(Y)| > |\lambda_1(Y)|$. The foliation $\mathcal{F}^{cu}(\Sigma,Y)$ of Σ looks like the picture below.

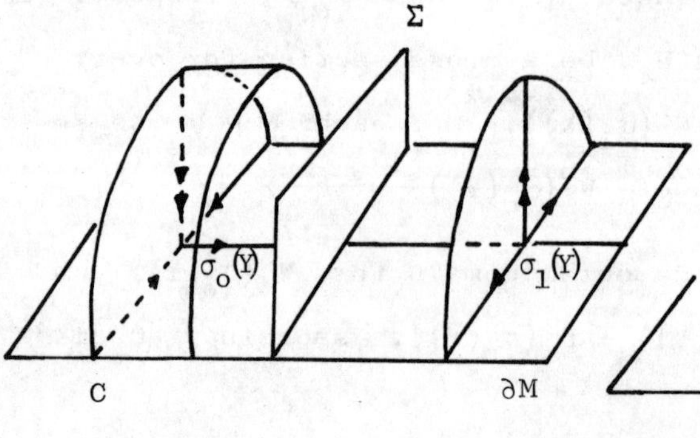

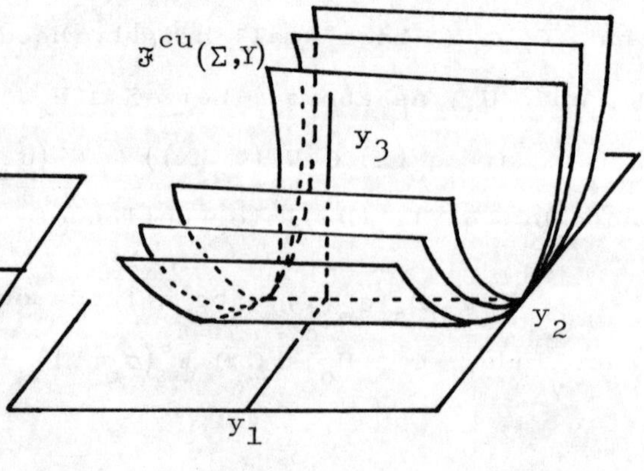

We will modify this foliation $\mathcal{F}^{cu}(\Sigma,Y)$ in order to get a singular foliation $\mathcal{F}^{cu}_S(\Sigma,Y)$ of Σ having $W^s(\sigma_1(Y)) \cap \Sigma$ as space of leaves and satisfying:

(i) the set of singularities of $\mathcal{F}^{cu}(\Sigma,Y)$ coincides with the y_2-axis

(ii) there is a compact set $K_0 \subset C$, $K_0 \cap W^s(\sigma_0(Y)) = D_s(\sigma_0(Y))$, $\bigcup_{x \in W_0(Y)} F^u(x,C) \subset K_0$, the diameter of K_0 going to zero as we approach ∂M, such that if $F^{cu}_S(q(Y))$ is a leaf of $\mathcal{F}^{cu}_S(\Sigma,Y)$ through $q(Y)$ then there is a unique leaf $F^u \in \mathcal{F}^u(C,Y)$ such that $\pi^{-1}(F^u \cap K_0) \subset F^{cu}_S(q(Y))$.

Observe that (ii) will imply that the saturation of $\mathcal{F}^{cu}(\Sigma,Y)$ by the flow induced by Y is compatible with $\mathcal{F}^{cu}(\alpha_i(Y))$, $1 \le i \le k$. The construction of $\mathcal{F}^{cu}_S(\Sigma,Y)$ is as follows:
First observe that the intersection of the leaves of $\mathcal{F}^{cu}(\Sigma,Y)$ with the plane $y_2 = 0$ is a family of curves whose contact with the y_2-axis is $\lambda_2(Y)/\lambda_1(Y)$. We can choose $\beta > 1$, $\beta > \lambda_2(Y)/\lambda_1(Y)$ for every $Y \in \mathcal{V}$ and positive real numbers $M < N$ such that if C_M and C_N are the curves in the plane $y_2 = 0$ given by $y_3 = M|y_1|^\beta$ and $y_3 = N|y_1|^\beta$ respectively then:

(i) for every $x \in W(Y) \subset D_s(\sigma_0(Y))$, $\pi^{-1}(F^u(x,C)) \cap \{y_2=0\}$ is contained in the interior of the region in $\{y_2=0\}$ limited by C_M

(ii) for every $x \in D_s(\sigma_0(Y)) \backslash \overline{W(Y)}$, $\pi^{-1}(F^u(x,C)) \cap \{y_2=0\}$ intersects C_N at a unique point and C_M at a unique point.

Now we consider the singular foliation $\mathcal{F}_S$ of $\{y_2=0\}$ such that the leaves of $\mathcal{F}_S$ in the region interior to C_N coincide with the leaves of $\mathcal{F}^{cu}(\Sigma,Y) \cap \{y_2=0\}$ in this region, the leaves of $\mathcal{F}_S$ in the intersection of the interior to C_N with the exterior to C_M coincide with the intersection of the translation along the y_1-axis of C_N and the leaves of $\mathfrak{F}_S$ in the region exterior to C_N coincide with the translation along the y_1-axis of C_N.

See picture below.

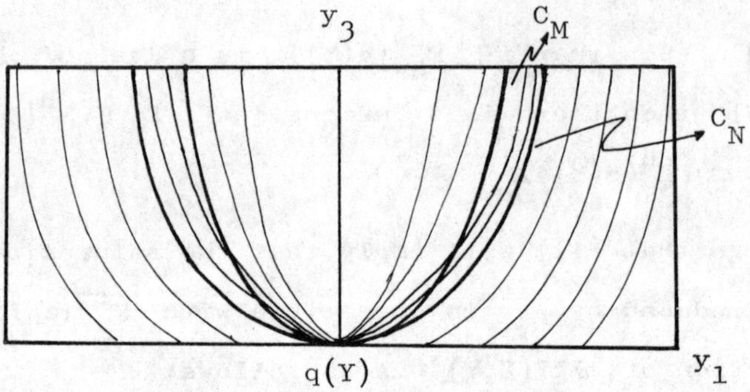

Let R be the region contained in $\{y_2=0\}$ interior to C_M.

Now we raise $\mathfrak{F}_S$ to a singular foliation $\mathfrak{F}_S^{cu}(\Sigma,Y)$ such that each leaf of $\mathfrak{F}_S^{cu}(\Sigma,Y)$ has dimension two and it is transversal to the planes $\{y_3=\text{constant}\}$, the intersections of $\mathfrak{F}_S^{cu}(\Sigma,Y)$ with $\{y_2=0\}$ coincides with $\mathfrak{F}_S$, the intersection of $\mathfrak{F}_S^{cu}(\Sigma,Y)$ with the product of R with the y_2-axis coincides with $\mathfrak{F}_S^{cu}(\Sigma,Y)$.

See picture below.

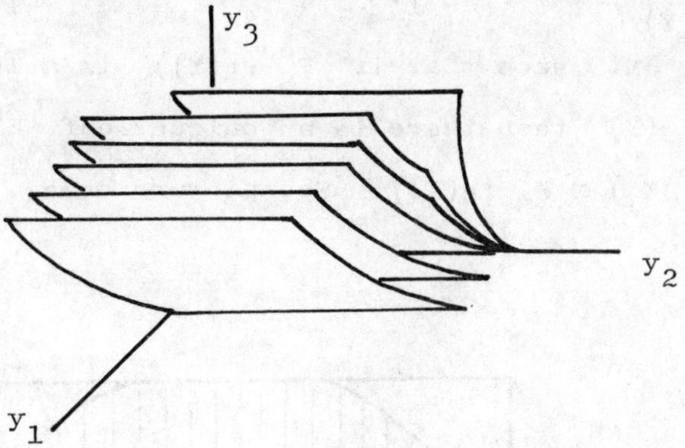

Observe that the y_1-axis is the set of singularities of $\mathfrak{F}_S^{cu}(\Sigma,Y)$ and if $F_S^{cu} \in \mathfrak{F}_S^{cu}(\Sigma,Y)$ then either F_S^{cu} is a differentiable surface or F_S^{cu} has a unique line parallel to $\{y_3=0\}$ along which it is not differentiable. The compact set $K_o \subset C$ satisfying (ii) above is $\overline{\pi_Y(R\times y_2)}$.

Dually, let $D_u(\sigma_1(Y))$ be a fundamental domain for $W^u(\sigma_1(Y))$ and $C \subset U_1$ a cross section for each $Y \in \mathcal{V}$, where U_1 is a fixed neighborhood of $\sigma_1(X)$ such that every $Y \in \mathcal{V}$ is C^2-linearizable in U_1. Let $\mathfrak{F}^s(C,Y)$ be a foliation of C compatible with $\mathfrak{F}^s(\beta_j(Y))$, $1 \le j \le \ell$, and such that the leaves of $\mathfrak{F}^s(C,Y)$ through points outside a compact set $W(Y) \subset D_u(\sigma_1(Y))$ are linear, $W(Y)$ contains in its interior the leaves through points in $\mathfrak{F}^s(\beta_j(Y)) \cap \cap D_u(\sigma_1(Y))$, $1 \le j \le \ell$.

Let $\pi_Y^1\colon \Sigma\backslash\{y_2=y_3=0\} \to C$ be the Poincaré map. Then $\mathfrak{F}^{cs}(\Sigma,Y) = (\pi_Y^1)^{-1}(\mathfrak{F}^s(C,Y))$ is a singular foliation of Σ, whose set of singularities is the y_1-axis.

As before we can modify $\mathfrak{F}^{cs}(\Sigma,Y)$ and get a singular foliation $\mathfrak{F}_S^{cs}(\Sigma,Y)$ having $W^u(\sigma_o(Y)) \cap \Sigma$ as space of leaves and satisfying:

(i) the set of singularities of $\mathfrak{F}_S^{cs}(\Sigma,Y)$ is the y_2-axis

(ii) there is a compact set $K_1 \subset C$, $K_1 \cap W^u(\sigma_1(Y)) = D_u(\sigma_1(Y))$,

$\bigcup_{x \in \text{int } W(Y)} F^s(x,C) \subset K_1$, the diameter of K_1 going to zero as we approach ∂M, such that if $F_S^{cs}(q(Y))$ is a leaf of $\mathfrak{F}_S^{cs}(\Sigma,Y)$ through $q(Y)$ then there is a unique leaf $F^s \in \mathfrak{F}^s(C,Y)$ such that $(\pi_Y^1)^{-1}(F^s \cap K_1) \subset F_S^{cs}(q(Y))$. See picture below.

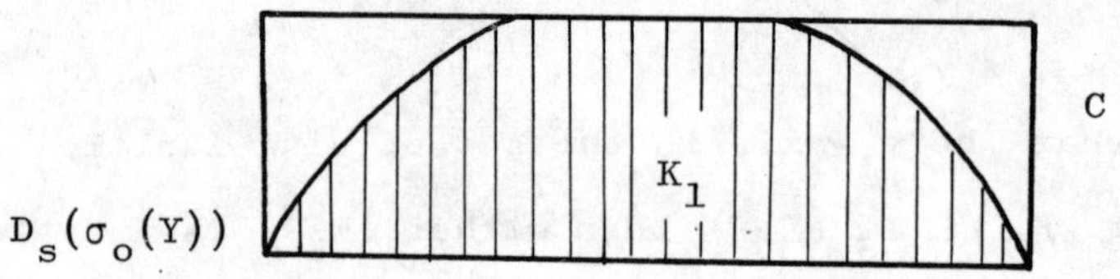

It is possible to get $\mathfrak{F}_S^{cs}(\Sigma,Y)$ and $\mathfrak{F}_S^{cu}(\Sigma,Y)$ in such way that if $F_S^{cs} \in \mathfrak{F}_S^{cs}(\Sigma,Y)$ and $F_S^{cu} \in \mathfrak{F}_S^{cu}(\Sigma,Y)$ then the intersection of $F_S^{cs} \cap F_S^{cu}$ with a plane $\{y_3 = \text{constant} \neq 0\}$ is a unique point.

The constructions above imply that there are neighborhoods $\mathcal{U}$ of X and N of $\sigma_1(X)$ such that for each $Y \in \mathcal{V}$, $W^u(\alpha_i(Y))$ is transversal to $W^s(\beta_j(Y))$ in N for every $1 \leq i \leq k$ and $1 \leq j \leq \ell$. By induction on the behavior of $\sigma_1(X)$ with respect to $\beta_j(X)$ it is possible to prove that for each $j \in \{1,\ldots,\ell\}$ there are neighborhoods $\mathcal{V}_j \subset \mathcal{U}$ of X and N_j of $\beta_j(X)$ such that for every $Y \in \mathcal{V}_j$, $W^u(\alpha_i(Y))$ is transversal to $W^s(\beta_m(Y))$ in N_j, $i \in \{1,\ldots,k\}$, $m \in \{1,\ldots,\ell\}$. Setting $\mathcal{V} = \bigcap_{i=1}^{k} \mathcal{V}_i$ we have $\mathcal{V} \subset \mathfrak{X}_1^\infty(M,\partial M)$ and so we get the openess of $\mathfrak{X}_1^\infty(M,\partial M)$ in $\mathfrak{X}^\infty(M,\partial M)$. The details of this proof can be found in [4].

Now we will prove that $X \in \mathfrak{X}_1^\infty(M,\partial M)$ is structurally stable. Given $X \in \mathfrak{X}_1^\infty(M,\partial M)$, let $Y \in \mathcal{V}$ where $\mathcal{V}$ is a neighborhood of X as above. We will define a homeomorphism $h: M \circlearrowleft$ taking orbits of X into orbits of Y. For this we will use besides the foliations

$\mathcal{F}_S^{cu}(\Sigma,Y)$ and $\mathcal{F}_S^{cs}(\Sigma,Y)$ of Σ, the foliation $\mathcal{F}(\Sigma)$ of Σ whose leaves is the family of planes $\{y_3=\text{constant}\}$. The idea is to define first a homeomorphism $h: \Sigma \circlearrowleft$ compatible with the above foliations. This will imply that the extension of h to the saturation of Σ by the flow induced by X will preserve the intersections of the leaves of $\mathcal{F}^u(C,X)$ with the compact set $K_o \subset C$ defined above. This, together with the levels of Liapunov functions defined in neighborhoods of $\sigma_o(X)$ and $\sigma_1(X)$ respectively will allow us to extend h to a full neighborhood of the quasi-transversal intersection $(\sigma_o(X),\sigma_1(X))$ along the boundary of M. The restriction of h to $W^s(\sigma_o(X))$ (resp. $W^u(\sigma_1(X))$) is compatible with $\mathcal{F}^u(\alpha_i(X))$, $1 \le i \le k$, (resp. $\mathcal{F}^s(\beta_j(X))$, $1 \le j \le \ell$) and coincides with a homeomorphism h^s (resp. h^u) which could be defined a priori in $W^s(\sigma_o(X))$ (resp. $W^u(\sigma_1(X))$) via the compatible system of foliations $\mathcal{F}^u(\alpha_i(X))$, $1 \le i \le k$ (resp. $\mathcal{F}(\beta_j(X))$, $1 \le j \le \ell$). Finally, we complete the definition of h to all of M as usual ([6,7]).

To define such h we proceed as follows.

<u>1st step</u>. Take any diffeomorphism $h: \{y_1=y_2=0\} \subset \Sigma \circlearrowleft$ with $h(q(X)) = q(Y)$.

<u>2nd step</u>. We define, as in [6], a homeomorphism

$$h^s: \bigcup_{1\le i\le k} W^s(\alpha_i(X)) \longmapsto \bigcup_{1\le i\le k} W^s(\alpha_i(Y))$$

compatible with $\mathcal{F}^u(\alpha_i(X))$ and $\mathcal{F}^u(\alpha_i(Y))$, $1 \le i \le k$, conjugating X_t and Y_t, where X_t (resp. Y_t) is the flow induced by X (resp. Y). Dually we define a homeomorphism

$$h^u: \bigcup_{1\le j\le \ell} W^u(\beta_j X)) \longmapsto \bigcup_{1\le j\le \ell} W^u(\beta_j(Y))$$

compatible with $\mathcal{F}^s(\beta_j(X))$ and $\mathcal{F}^s(\beta_j(Y))$, $1 \le j \le \ell$, conjugating X_t and Y_t.

3rd step. We are going to define a homeomorphism $h: \Sigma \circlearrowleft$ compatible with $\mathcal{F}_S^{cu}(\Sigma,X)$, $\mathcal{F}_S^{cs}(\Sigma,X)$ and $\mathcal{F}(\Sigma)$.

Let us define h in the space of leaves of $\mathcal{F}_S^{cu}(\Sigma,X)$. For this it is enough to define h on $\{y_2=0\}$. The leaves of $\mathcal{F}_S^{cu}(\Sigma,X)$ through $q(X)$ bound a region $R(X) \subset \Sigma$ such that $\partial R(X) = F_1^{cu}(q(X)) \cup F_2^{cu}(q(X))$ where $F_i^{cu}(q(X))$, $i = 1,2$ are leaves through $q(X)$. Furthermore if $\pi_X: \Sigma\backslash\{y_2=y_3=0\} \to C$ is the Poincaré map and $K_o(X) \subset C$ is the compact set defined in the construction of $\mathcal{F}_S^{cu}(\Sigma,X)$, we have that if $F_S^{cu}(\Sigma,X) \in \mathcal{F}^{cu}(\Sigma,X)$ is a leaf through a point in the interior of $R(X)$, there is a unique leaf $F^u(C,X) \in \mathcal{F}^u(C,X)$ such that $\pi_X^{-1}(F^u(C,X) \cap K_o(X)) \subset F^{cu}(\Sigma,X)$. Similarly for $\mathcal{F}^{cs}(\Sigma,X)$. Denote $F_i^{cu}(q(Y))$, $i = 1,2$, the leaf in the boundary of $R(Y)$ near $F_i^{cu}(q(X))$, $i = 1,2$. Take any leaf of $\mathcal{F}(\Sigma)$ through $y_3 = 1$ and intersect it with the plane $y_2 = 0$. Denote the interval in this intersection between $F_i^{cu}(q(Y))$, $i = 1,2$, by $I^s(Y)$. It is clear that any leaf of $\mathcal{F}^{cu}(\Sigma,Y)$ through points of $R(Y)$ intersects $I^s(Y)$ at a unique point. Now define a homeomorphism $h^s: I^s(X) \to I^s(Y)$ using the leaves of $\mathcal{F}^u(\alpha_i(X))$, $\mathcal{F}^u(\alpha_i(Y))$ and the homeomorphism h^s from step 2. Now choose any diffeomorphism $h: \{y_3=y_2=0\} \circlearrowleft$, $h(q(X)) = q(Y)$.
In this way we have defined a map h_*^s from the space of leaves of $\mathcal{F}_S^{cu}(\Sigma,X)$ onto the space of leaves of $\mathcal{F}_S^{cu}(\Sigma,Y)$.
Similarly we define a map h_*^u from the space of leaves of $\mathcal{F}_S^{cs}(\Sigma,X)$ onto the space of leaves of $\mathcal{F}_S^{cs}(\Sigma,Y)$.

Finally we define h on Σ in the following way: if $x \in \Sigma$ there are unique $F_S^{cu} \in \mathcal{F}_S^{cu}(\Sigma,X)$, $F_S^{cs} \in \mathcal{F}_S^{cs}(\Sigma,X)$ and $F \in \mathcal{F}(\Sigma)$ such that $x = F_S^{cu} \cap F_S^{cs} \cap F$; we define $h(x) = h_*^s(F_S^{cu}) \cap h_*^u(F_S^{cs}) \cap \cap h_1(F)$, where h_1 is the map defined in (i). Such a map is obviously continuous outside the y_1-axis and the y_2-axis. The continuity of h at these points is a consequence of the continuity of the foliations $\mathcal{F}_S^{cu}(\Sigma,X)$, $\mathcal{F}_S^{cs}(\Sigma,X)$ and $\mathcal{F}(\Sigma)$.

<u>4th step</u>. For $Y \in \mathcal{V}$, let $f_Y\colon U_0 \to \mathbb{R}$ be a Liapunov function for Y. We can suppose that $\Sigma \subset f_Y^{-1}(1)$ and $C \subset f_Y^{-1}(-1)$. We define h on C in the natural way: given $x \in C$, let $y \in \Sigma$ such that $x = X_{-t}(y)$ for some $t \in \mathbb{R}$. We set $h(x)$ as the intersection of the Y-orbit of $h(y)$ with C. Observe that such h defined on C sends leaves of $\mathcal{F}^u(C,X)/K_0(X)$ onto leaves of $\mathcal{F}^u(C,Y)/K_0(Y)$. This allow us to extend h to $W^s(\sigma_0(X))$. Using the levels of f_Y, $f_Y^{-1}(c)$, $-1 \le c \le 1$, and the flow induced by X as a coordinate system of U_0 we extend h to a neighborhood of $\sigma_0(X)$. Similarly, using Liapunov functions for Y near $\sigma_1(Y)$, $Y \in \mathcal{V}$, we extend h to a neighborhood of $\sigma_1(Y)$.

<u>5th step</u>. Here we conclude the construction of the equivalence h between X and Y. So far h is defined on the union of the stable manifolds of $\alpha_i(X)$, $1 \le i \le k$, conjugating X_t and Y_t and preserving the unstable foliations $\mathcal{F}^u(\alpha_i(X))$ and $\mathcal{F}^u(\alpha_i(Y))$, $1 \le i \le k$; h is also defined on the union of the unstable manifolds of $\beta_j(Y)$, $1 \le j \le \ell$, conjugating X_t and Y_t and preserving the stable foliations $\mathcal{F}^s(\beta_j(X))$ and $\mathcal{F}^s(\beta_j(Y))$, $1 \le j \le \ell$. This homeomorphism is also defined on the cross section Σ and on a neighborhood V of the quasi-transversal intersection $(\sigma_0(X),\sigma_1(X))$ saturated by segments of X-orbits. Moreover, reparametrizing the vector fields we can get h in such a way that its restriction to the boundary of V is a conjugacy between X_t and Y_t.

We now extend h to neighborhoods of $\beta_j(X)$, $1 \le j \le \ell$. We proceed by induction. Using the equation $hX_t = Y_th$ we define h on $B = \bigcup_{t\ge 0} X_tC$. Let $D_s(\beta_1(X))$, $D_s(\beta_1(Y))$ be fundamental domains for $W^s(\beta_1(X))$ and $W^s(\beta_1(Y))$. First we extend $h\colon B \to h(B)$

to $h: D_s(\beta_1(X)) \to D_s(\beta_1(Y))$ conjugating X_t and Y_t and being compatible with the unstable foliations of $\alpha_i(X)$, $1 \le i \le k$. This can be done as follows: over $D_s(\beta_1(X))$ we construct an unstable foliation $\mathfrak{F}^u(\beta_1(X))$ defined on $D_s(\beta_1(X))\backslash B$ compatible with the unstable foliations of $\alpha_i(X)$, $1 \le i \le k$. In the same way we construct an unstable foliation $\mathfrak{F}^u(\beta_1(Y))$ defined on $D_s(\beta_1(Y))\backslash h(B)$ compatible with the unstable foliations $\mathfrak{F}^u(\alpha_i(Y))$, $1 \le i \le k$. This is possible because h is differentiable near the boundary of B. Now $\mathfrak{F}^u(\beta_1(X))$ is defined on a neighborhood of $\beta_1(X)$ off B by simply considering the iterates by X_t of the leaves through points of $D_s(\beta_1(X))$ and the leaf $W^u(\beta_1(X))$. Similarly for Y and $\beta_1(Y)$. Since h is differentiable near the boundary of B, we can apply the Isotopy Extension Theorem [5] to extend h to $D_s(\beta_1(X))$. By the equation $hX_t = Y_t h$ we define h on all $W^s(\beta_1(X))$, preserving the foliations $\mathfrak{F}^u(\beta_1(X))$ and $\mathfrak{F}^u(\beta_1(Y))$. Using the complementary foliations $\mathfrak{F}^u(\beta_1(X))$ and $\mathfrak{F}^s(\beta_1(X))$ for X and Y and the conjugacy already defined on their space of leaves we obtain h defined on a full neighborhood of $\beta_1(X)$. The continuity of h on $W^u(\beta_1(X))$ follows from the fact that h preserves $\mathfrak{F}^s(\beta_1(X))$. The induction procedure for the remaining $\beta_j(X)$ is similar. In this way we have defined a homeomorphism h which is a conjugacy on the stable manifolds of $\alpha_i(X)$, $1 \le i \le k$, and $\beta_j(X)$, $1 \le j \le \ell$, and which is an equivalence from a neighborhood V of the quasi-transversal intersection for X to $h(V)$ and, when restricted to the boundary of V, is a conjugacy. Thus we have defined on all of M. By construction h is continuous at the stable manifolds of $\sigma_0(X)$, $\sigma_1(X)$ and $\beta_j(X)$, $1 \le j \le \ell$, $h(\partial M) = \partial M$. That h is continuous at the stable manifolds of $\alpha_i(X)$, $1 \ge i \ge k$, follows from the fact that h sends leaves of $\mathfrak{F}^u(\alpha_i(X))$ to leaves of $\mathfrak{F}^u(\alpha_i(Y))$. The proof of Theorem 1 is complete. ■

Theorem 2. $\mathfrak{X}_2^\infty(M,\partial M)$ is open in $\mathfrak{X}_*^\infty(M,\partial M)$ and its elements are structurally stable.

Proof: Let $X \in \mathfrak{X}_2^\infty(M,\partial M)$ and $(\sigma_o(X),\sigma_1(X))$ be the pair of critical elements having the quasi-transversal intersection along the boundary of M. Since $\dim W^u(\sigma_1(X)) = 1$ we have immediately the openess of $\mathfrak{X}_2^\infty(M,\partial M)$ in $\mathfrak{X}_*^\infty(M,\partial M)$.

We will prove that $X \in \mathfrak{X}_2^\infty(M,\partial M)$ is structurally stable.

Given $X \in \mathfrak{X}_2^\infty(M,\partial M)$, $\Omega(X) = \{\alpha_1(X) \leq \ldots \leq \alpha_k(X) \leq \sigma_o(X) \leq \leq \sigma_1(X) \leq \beta_1(X) \leq \ldots \leq \beta_\ell(X)\}$ let $\mathcal{V}$ be a small neighborhood of X in $\mathfrak{X}_*^\infty(M,\partial M)$ such that if $Y \in \mathcal{V}$ then $\Omega(Y) = \{\alpha_1(Y) \leq \ldots \leq \alpha_k(Y) \leq \leq \sigma_o(Y) \leq \sigma_1(Y) \leq \beta_1(Y) \leq \ldots \leq \beta_\ell(Y)\}$. For every $Y \in \mathcal{V}$ let $\mathcal{F}^u(\alpha_i(Y))$, $1 \leq i \leq k$, $\mathcal{F}^{cu}(\sigma_o(Y))$ be a compatible system of unstable foliations and a center-unstable foliation respectively. We proceed by induction (as in [8]) to define a homeomorphism $h^s\colon \bigcup_{i=1}^{k} W^s(\alpha_i(X)) \to \bigcup_{i=1}^{k} W^s(\alpha_i(Y))$ compatible with the above foliations. Let $x \in W^u(\sigma_o(X)) \cap W^s(\sigma_1(X))$ and Σ be a cross section for every $Y \in \mathcal{V}$ at $x \in \Sigma$. We can assume that $\Sigma \subset f_Y^{-1}(c)$, $c \in \mathbb{R}$, where f_Y is a Liapunov function for every $Y \in \mathcal{V}$, f_Y defined on $U_o \supset \sigma_o(Y)$. In Σ we take coordinates (y_1,y_2,y_3) and we assume $\partial M = \{y_3=0\}$.

Let $C = \{(y_1,y_2,y_3);\ y_1^2+y_2^2 = 1,\ 0 \leq y_3 \leq 1\}$. It is clear that we can suppose that C is transversal to the leaves of the foliation $\mathcal{F}^{cu}(\sigma_o(Y))$ for every $Y \in \mathcal{V}$, that all unstable manifolds which does not intersect $W^s(\sigma_o(X)) \cap \partial M$ intersects

$$C_1 = \{(y_1,y_2,1),\ 0 \leq y_1^2 + y_2^2 \leq 1\}$$

and that the circles $\{(y_1,y_2,0),\ y_1^2+y_2^2 = r,\ 0 < r \leq 1\}$ are transversal to $\mathcal{F}^{cu}(\sigma_o(Y))$, $Y \in \mathcal{V}$.

Let L_X be the two-dimensional surface defined by

$$L_X = \bigcup_{x \in \partial C_1} (F^{cu}(\sigma_o(X))(x) \cap \Sigma)$$

where $F^{cu}(\sigma_o(X))(x)$ is the leaf of $\mathcal{F}^{cu}(\sigma_o(X)) \cap \Sigma$ through x. The planes $\pi_a = \{(y_1,y_2,a)\}$, $a \in (0,1]$ are transversal to the leaves of $\mathcal{F}^{cu}(\sigma_o(Y))$. Now we take a family of cylinders $C_a(X) \subset \Sigma$ such that $C_a(X) \cap \pi_a = \{(y_1,y_2,a);\ y_1^2+y_2^2 = a\}$ and $C_a(X) \cap \{y_3=0\} = \{(y_1,y_2,0);\ y_1^2+y_2^2 = a\}$. Using the homeomorphism h^s and the foliations $\mathcal{F}^u(\alpha_i(Y))$, $1 \le i \le k$, we define a homeomorphism $h_1: C_1(X) \to C_1(Y)$. Let $h_3: \{y_1=y_2=0\} \circlearrowleft$ be any diffeomorphism. Since $\mathcal{F}^{cu}(\sigma_o(Y))$ and the family $C_a(Y)$ give a coordinate system in a neighborhood of $(0,0,0)$ and we have homeomorphisms defined in its space of leaves, we can define a homeomorphism h in a neighborhood of $(0,0,0)$ in Σ setting $h(0,0,0) = (0,0,0)$ and requiring that h preserves the above coordinate system.

To extend h to a neighborhood of $\sigma_i(X)$, $i = 0,1$, we use Liapunov functions. The extension of h to all of M is obtained as in Theorem 1. We leave the details to the reader. ■

The main goal in what follows is to prove the openess of $\mathfrak{X}_i^\infty(M,\partial M)$, $i = 3,4$, and the stability of its elements. The idea of the proof is to define first a singular foliation $\mathcal{F}_S^{cu}(\sigma_o(X))$ associated with the quasi-transversal intersection $(\sigma_o(X),\sigma_1(X))$ of $X \in \mathfrak{X}_i^\infty(M,\partial M)$, $i = 3,4$. A typical leaf of $\mathcal{F}_S^{cu}(\sigma_o(X))$ will be either a central-unstable leaf or the union of central-unstable leaves along a curve in the fundamental domain for $W^s(\sigma_o(X))$. So, the dimension of a typical leaf of $\mathcal{F}_S^{cu}(\sigma_o(X))$ will be either $u+1$ or $u+2$ where u is the dimension of $W^u(\sigma_o(X))$. This foliation induces a singular foliation in a cross section Σ at $x \in W^u(\sigma_o(X)) \cap W^s(\sigma_1(X))$ with leaves either one-dimensional (corresponding to the intersections of the u+1-dimensional leaves of

$\mathcal{F}_S^{cu}(\sigma_o(X))$ with Σ) or two-dimensional (corresponding to the intersection of the (u+2)-dimensional leaves of $\mathcal{F}_S^{cu}(\sigma_o(X))$ with Σ). The two-dimensional ones are the union of one-dimensional leaves which are the intersection of central-unstable leaves with Σ. We will modify this one-dimensional foliation of each two-dimensional leaf as in [4]. The way we make this will allow us to prove the openess of $\mathfrak{X}_i^\infty(M,\partial M)$, $i = 3,4$.

To prove the stability we use this foliation as part of a singular coordinate system which will be preserved by a homeomorphism defined on Σ. This will imply that if $C(\sigma_o(X)) \cup D(\sigma_o(X))$ is a fundamental domain for $W^s(\sigma_o(X))$ such that (i) $C(\sigma_o(X))$ is a cylinder transversal to the restriction of X to $W^s(\sigma_o(X))$, (ii) the intersection $C(\sigma_o(X)) \cap W^{ss}(\sigma_o(X))$ is a fundamental domain $D^{ss}(\sigma_o(X))$ for $W^{ss}(\sigma_o(X))$, and (iii) $D(\sigma_o(X))$ is a disk contained in a leaf of the strong stable foliation associated to $\sigma_o(X)$, then there is a tubular neighborhood V of $C(\sigma_o(X))$ foliated by leaves of dimension two, the intersections of the leaves with $C(\sigma_o(X))$ being an one-dimensional foliation compatible with the unstable system of foliations $\mathcal{F}^u(\alpha_i(X))$, $1 \le i \le k$. Moreover, each leaf is subfoliated by an one-dimensional foliation, each leaf transversal to $C(\sigma_o(X))$. The homeomorphism h defined in the saturation of the cross section by the flow via the conjugacy equation will preserve in each 2-dimensional leaf its one-dimensional foliation restricted to a half cone with vertex at a point of $D^{ss}(\sigma_o(X))$. This, together with the levels of a Liapunov function defined in a neighborhood of $\sigma_o(X)$ allows one to extend the homeomorphism to $W^s(\sigma_o(X))$. The extension to all of M is obtained as usual (see [7,8]).

The Singular Central-Unstable Foliation

We will first construct two partial foliations $\mathcal{F}_1^{cu}(\sigma_o(X))$ and $\mathcal{F}_o^{cu}(\sigma_o(X))$ in a neighborhood $U(\sigma_o(X))$ of $\sigma_o(X)$ in M, compatible with the unstable foliations $\mathcal{F}^u(\alpha_i(X))$, $\alpha_i(X) < \sigma_o(X)$. A typical leaf of $\mathcal{F}_1^{cu}(\sigma_o(X))$ will have dimension of a typical leaf of $\mathcal{F}_o^{cu}(\sigma_o(X))$ plus one and will be a union of leaves of $\mathcal{F}_o^{cu}(\sigma_o(X))$. When the weakest contraction at $\sigma_o(X)$ is real (resp. complex) the space of leaves of $\mathcal{F}_1^{cu}(\sigma_o(X))$ will be a circle in $W^{ss}(\sigma_o(X))$ (resp. in $W^c(\sigma_o(X)) = W^{cu}(\sigma_o(X)) \cap W^s(\sigma_o(X))$ and the space of leaves of $\mathcal{F}_o^{cu}(\sigma_o(X))$ will be a fundamental domain for $W^s(\sigma_o(X))$. The leaves of both such foliations will be C^1 embedded disks varying continuously in the C^1 topology. Moreover, a leaf of $\mathcal{F}_1^{cu}(\sigma_o(X))$ through a given point transversally intersects $W^s(\sigma_o(X))$ (in $U(\sigma_o(X))$) in a two-dimensional disk and a leaf of $\mathcal{F}_o^{cu}(\sigma_o(X))$ through a given point transversally intersects $W^s(\sigma_o(X))$ (in $U(\sigma_o(X))$) in a C^1-curve. We introduce these foliations to extend homeomorphisms (which will be defined later) to $U(\sigma_o(X))$. Suppose first that the weakest contraction at $\sigma_o(X)$ is real. Let $C(\sigma_o(X))$ be a C^2-half cylinder and $D(\sigma_o(X))$ be a C^2-half disk, $D(\sigma_o(X))$ contained in a leaf of the strong stable foliation $\mathcal{F}^{ss}(\sigma_o(X))$ such that $C(\sigma_o(X)) \cup D(\sigma_o(X))$ is a fundamental domain for $W^s(\sigma_o(X))$. The half cylinder is transversal to $\mathcal{F}^{ss}(\sigma_o(X))$, disjoint of the component $W^u(\alpha_i(X)) \cap W^s(\sigma_o(X))$ which does not intersect $W^{ss}(\sigma_o(X))$ and $C \cap W^{ss}(\sigma_o(X))$ is a fundamental domain for $W^{ss}(\sigma_o(X))$. Moreover we assume that X is tangent to $C(\sigma_o(X))$ along $C(\sigma_o(X)) \cap D(\sigma_o(X))$. Let $\mathcal{F}^u(\sigma_o(X))$ be an unstable foliation for $\sigma_o(X)$ compatible with $\mathcal{F}^u(\alpha_i(X))$, $1 \le i \le k$. If $F_o^u(x)$ denotes the leaf of $\mathcal{F}^u(\sigma_o(X))$ at $x \in C(\sigma_o(X)) \cup D(\sigma_o(X))$ we define $F_o^{cu}(x) = \bigcup_{t\geq 0} X_t F_o^u(x)$. The union of these leaves is the central unstable foliation for $\sigma_o(X)$ and it is denoted by $\mathcal{F}_o^{cu}(\sigma_o(X))$. Observe that the space of leaves of this foliation is $C(\sigma_o(X)) \cup \cup D(\sigma_o(X))$ and each leaf has dimension $u+1$.

To define $\mathcal{F}_1^{cu}(\sigma_o(X))$ we proceed as follows: it is possible ([7,8]) to construct an one-dimensional foliation $\mathcal{F}^c(C(\sigma_o(X)))$ of $C(\sigma_o(X))$ in such way that its leaves subfoliate leaves of $\mathcal{F}^u(\alpha_i(X))$, $\alpha_i(X) < \sigma_o(X)$. We now raise $\mathcal{F}^c(C(\sigma_o(X)))$ to a $(u+1)$-dimensional foliations $\mathcal{R}^{cu}$ in the following way: a leaf $R^{cu}(x) \in \mathcal{R}^{cu}$ at $x \in C(\sigma_o(X))$ is, by definition, the union $\bigcup_{Y\in F^c(x)} F_o^u(y)$ where $F^c(x) \in \mathcal{F}^c(C(\sigma_o(X)))$ is the leaf at x and $F_o^u(y)$ is the leave of $\mathcal{F}^u(\sigma_o(X))$ at y. Define $F_1^{cu}(X) = \bigcup_{t\geq 0} X_t\, R^{cu}(x)$. The union of these leaves is denoted by $\mathcal{F}_1^{cu}(\sigma_o(X))$. Observe that $\mathcal{F}_1^{cu}(\sigma_o(X))$ is a partial foliation of a neighborhood of $\sigma_o(X)$ whose space of leaves is $C(\sigma_o(X)) \cap W^{ss}(\sigma_o(X))$ and each of its leaves has dimension $u+2$. Moreover, each leaf of $\mathcal{F}_1^{cu}(\sigma_o(X))$ is a union of leaves of the central unstable foliation defined above.

The leaves of $\mathcal{F}_1^{cu}(\sigma_o(X))$ together with the leaves of $\mathcal{F}_o^{cu}(\sigma_o(X))$ through points of $D(\sigma_o(X))$ determine a singular foliation of a neighborhood $U(\sigma_o(X))$ of $\sigma_o(X)$ which is, by definition, the singular central unstable foliation for $\sigma_o(X)$ and it is denoted by $\mathcal{F}_s^{cu}(\sigma_o(X))$.

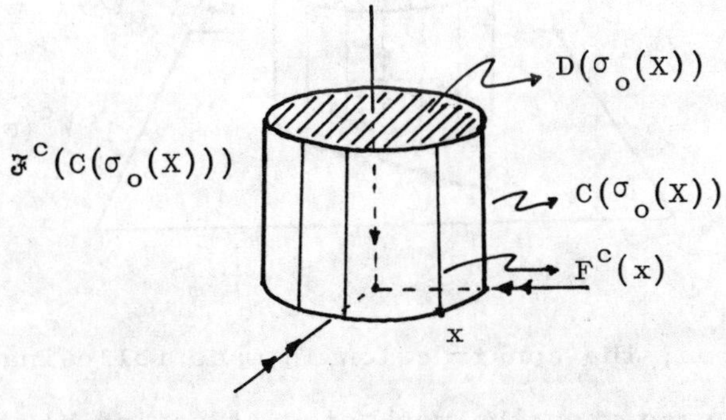

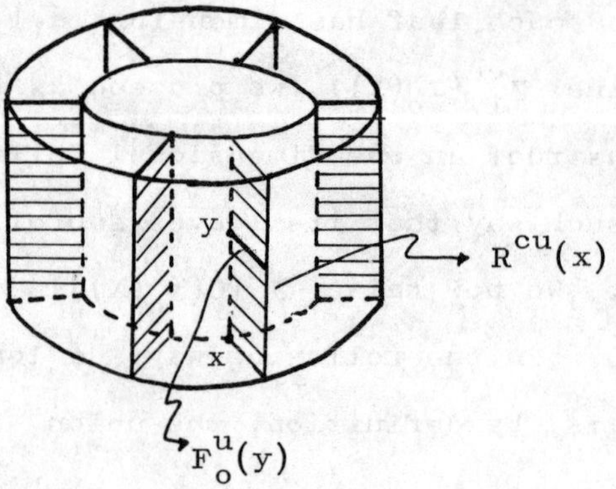

If the weakest contraction at $\sigma_o(X)$ is complex, the singular central unstable foliation $\mathcal{F}_S^{cu}(\sigma_o(X))$ is defined in a similar way. But, in this case, we choose the half cylinder $C(\sigma_o(X))$ and the disk $D(\sigma_o(X))$ satisfying:

(i) $C(\sigma_o(X)) \cup D(\sigma_o(X))$ is a fundamental domain for $W^s(\sigma_o(X))$

(ii) $D(\sigma_o(X))$ is transversal to $W^{ss}(\sigma_o(X))$ in $W^s(\sigma_o(X))$

(iii) $C(\sigma_o(X)) \cap W^c(\sigma_o(X))$ is a fundamental domain for $W^c(\sigma_o(X))$.

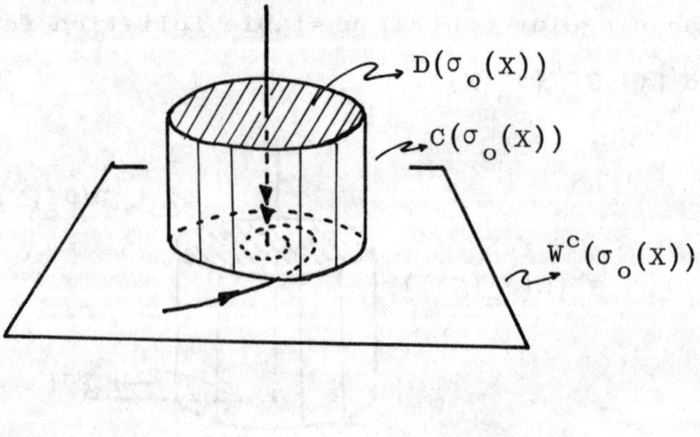

From this point on, the construction is made following the same steps as in the case when the weakest contraction at $\sigma_o(X)$ is real.

We now come to one of the main part of our result: the openess of $\mathfrak{X}_4^\infty(M,\partial M)$ in $\mathfrak{X}_*^\infty(M,\partial M)$ and the stability of its elements.

Let $X \in \mathfrak{X}_4^\infty(M,\partial M)$, $\Omega(X) = \alpha_1(X) \leq \ldots \leq \alpha_k(X) \leq \sigma_o(X) \leq \sigma_1(X) \leq \beta_1(X) \leq \ldots \leq \beta_\ell(X)$ where $(\sigma_o(X),\sigma_1(X))$ corresponds to the quasi-transversal intersection along the boundary of M. Let $\mathcal{V} \subset \mathfrak{X}_*^\infty(M,\partial M)$ be a small neighborhood of X such that for each $Y \in \mathcal{V}$, $\Omega(Y) = \alpha_1(Y) \leq \ldots \leq \alpha_k(Y) \leq \sigma_o(Y) \leq \sigma_1(Y) \leq \beta_1(Y) \leq \ldots \leq \beta_\ell(Y)$, where $\alpha_i(Y)$, $\sigma_o(Y)$, $\sigma_1(Y)$, $\beta_j(Y)$ are the critical elements near the corresponding ones for X, $(\sigma_o(Y),\sigma_1(Y))$ corresponding to the pair having the quasi-transversal intersection along ∂M. Let $\mathcal{F}^u(\sigma_o(Y))$, $\mathcal{F}^u(\alpha_i(Y))$, $1 \leq i \leq k$; $\mathcal{F}^s(\sigma_1(Y))$, and $\mathcal{F}^s(\beta_j(Y))$, $1 \leq j \leq \ell$, be compatible unstable and stable foliations. Taking $\mathcal{V}$ small enough we can suppose that there is $U_o \subset M$ such that for each $Y \in \mathcal{V}$, $\sigma_o(Y) \in U_o$, Y is C^2-linearizable in U_o and the diffeomorphism linearizing Y is C^2-close to the corresponding diffeomorphism for X (see Appendix of [4]). Let $\beta > 1$ be such that $\beta > \dfrac{\lambda(Y)}{a(Y)}$ for every $Y \in \mathcal{V}$ where $-a(Y)$ is the real part of the weakest contraction at $\sigma_o(Y)$ and $\lambda(Y)$ is the other contracting eigenvalue at $\sigma_o(Y)$. We denote by y_1, y_2, y_3, y_4 the linearizing coordinates in U_o (these coordinates depend on Y but this dependence is not expressed in the notation) and, in these coordinates we have

$$\sigma_o(Y) = (0,0,0,0), \quad W^c(\sigma_o(Y)) = \{y_3=y_4=0\}, \quad W^s(\sigma_o(Y)) = \{y_4=0\}$$
$$W^u(\sigma_o(Y)) = \{y_1=y_2=y_3=0\}.$$

When no confusion is possible we will still denote by Y the vector field Y in these coordinates.

Let $\Sigma \subset U_o$ be a cross section for every $Y \in \mathcal{V}$ at $p(X) \in W^u(\sigma_o(X))$. Suppose $p(X) = (0,0,0,1)$ and denote by $P(Y)$ the intersection of $W^u(\sigma_o(Y))$ with Σ. Let $\mathcal{F}_1^{cu}(\Sigma,Y) = \mathcal{F}_1^{cu}(\sigma_o(Y)) \cap \Sigma$

and $\mathcal{F}_o^{cu}(\Sigma,Y) = \mathcal{F}_o^{cu}(\sigma_o(Y)) \cap \Sigma$. The union of $\mathcal{F}_1^{cu}(\Sigma,Y)$ with $\mathcal{F}_o^{cu}(\Sigma,Y)$ is a singular foliation of a neighborhood of $p(Y)$ in Σ which we denote by $\mathcal{F}_S^{cu}(\Sigma,Y)$. Each leaf of $\mathcal{F}_1^{cu}(\Sigma,Y)$ (resp. $\mathcal{F}_o^{cu}(\Sigma,Y)$) is two (resp. one) dimensional. Let $\gamma(Y)$ be the expansion at $\sigma_o(Y)$ and $a(Y) + ib(Y)$ the weakest contraction at $\sigma_o(Y)$, $Y \in \mathcal{U}$. A linear leaf $F_o^{cu} \in \mathcal{F}_o^{cu}(\Sigma,Y)$ can be parametrized by

$$\varphi_Y(s) =$$

$$\left(rs^{-\frac{a(Y)}{\gamma(Y)}} \cos(t(s)b(Y)+w),\ rs^{-\frac{a(Y)}{\gamma(Y)}} \operatorname{sen}(t(s)b(Y)+w),\ s^{-\frac{\lambda(Y)}{\gamma(Y)}},1\right),\quad 0<s\leq 1$$

where $r \in [0,1]$ and $w \in [0,2\pi]$ are fixed, and $t(s) = L_n(s^{-\frac{1}{\gamma(Y)}})$.

If $y = s^{-\frac{a(Y)}{\gamma(Y)}}$ then $0 < y \leq 1$ for $0 < s \leq 1$. Moreover $\varphi_Y(y) = (ry\cos(t(y)b(Y)+w),\ ry\operatorname{sen}(t(y)b(Y)+w),\ y^{\frac{\lambda(Y)}{a(Y)}},1)$, $0 < y \leq 1$, $t(y) = L_n\, y^{\frac{1}{a(Y)}}$. Note that $t(y) \to \infty$ as $y \to 0$. So, F_o^{cu} is a spiral curve around the y_3-axis whose contact with the plane $y_3 = 0$ is $\frac{\lambda(Y)}{a(Y)}$.

A linear leaf $F_1^{cu} \in \mathcal{F}_1^{cu}(\Sigma,Y)$ can be parametrized by

$$\varphi_Y(x,z) = \left(s^{-\frac{a(Y)}{\gamma(Y)}} \cos(t(s)b(Y)+w(z)),\ s^{-\frac{a(Y)}{\gamma(Y)}} \operatorname{sen}(t(s)b(Y)+w(z)),\ s^{\frac{\lambda(z)}{\gamma(Y)}} z,1\right),\quad 0 < s \leq 1,\quad 0 \leq z \leq 1$$

with $t(s) = L_n\, s^{-\frac{1}{\gamma(Y)}}$ and $w(z) \in [0,2\pi]$.

As above we conclude that F_1^{cu} is a spiral surface around the y_3-axis whose contact with the plane $y_3 = 0$ is $\frac{\lambda(Y)}{a(Y)}$. Clearly we have similar equations for non linear leaves. To each $z = z_o \in (0,1]$, $\varphi_Y(s,z_o)$ is a spiral curve in F_1^{cu} spiralling around the y_3-axis whose contact with $\{y_3=0\}$ is also $\frac{\lambda(Y)}{a(Y)}$. The union of such curves determines a singular foliation $\mathcal{F}_o^{cu}(F_1^{cu})$ of

F_1^{cu} and $F_1^{cu} \cap \{y_3=0\}$ is a leaf. We will modify this one dimensional foliation of F_1^{cu} so that the modified one will have $F_1^{cu} \cap \cap \{y_3=0\}$ as space of leaves. Let $C(\Sigma)$ be a solid half cylinder defined by $\partial C(\Sigma) = C_1(\Sigma) \cup D(\Sigma)$ where $D(\Sigma)$ is a disk contained in $\{y_3=1\}$. We can assume that $\partial C(\Sigma)$ is the space of leaves of $\mathcal{F}_0^{cu}(\Sigma,Y)$ and $C(\Sigma) \cap \{y_3=0\} = S$ is the space of leaves of $\mathcal{F}_1^{cu}(\Sigma,Y)$. See picture below.

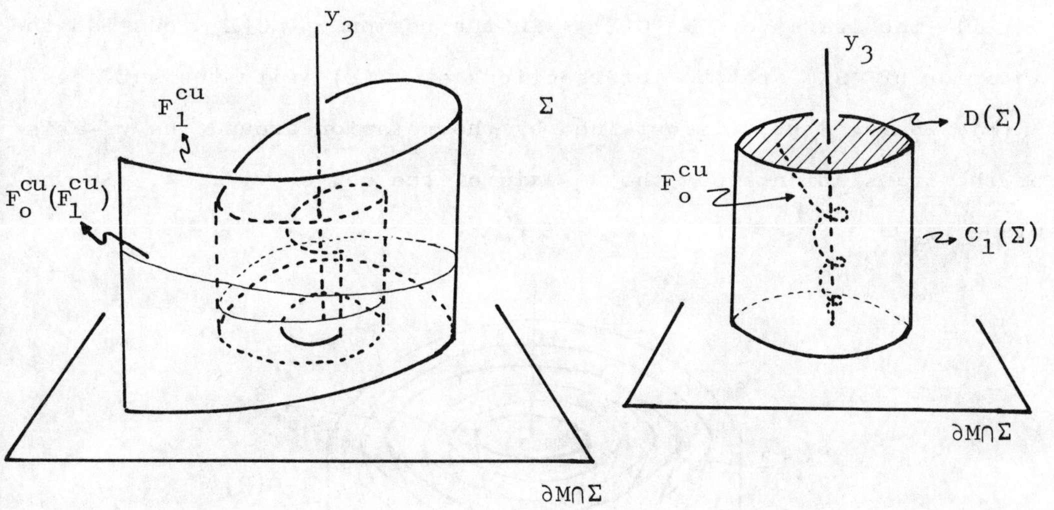

Let us modify the foliation $\mathcal{F}_0^{cu}(F_1^{cu})$ of F_1^{cu}. To do this we proceed as follows.

Let S_M and S_N $(N > M > 0)$ be the surfaces defined by the equations $y_3 = M(y_1^2+y_2^2)^{\beta/2}$, $y_1^2+y_2^2 \le \frac{1}{M^{2/\beta}}$ and $y_3 = N(y_1^2+y_2^2)^{\beta/2}$, $y_1^2+y_2^2 \le \frac{1}{N^{2/\beta}}$, respectively, with $\beta > \frac{\lambda(Y)}{a(Y)}$ for every $Y \in \mathcal{V}$, M, N positive real numbers. The interior of S_M (resp. S_N) is defined by $\{(y_1,y_2,y_3);\ y_3 > M(y_1^2+y_2^2)^{\beta/2}\}$ (resp. $\{(y_1,y_2,y_3);\ y_3 > N(y_1^2+y_2^2)^{\beta/2}\}$). Dually we define the exterior of S_M and S_N respectively. We can choose M and N in such a way that the leaves $F_0^{cu} \in \mathcal{F}_0^{cu}(\Sigma,Y)$ through points at $W^u(\alpha_i(Y)) \cap D(\Sigma)$, $1 \le i \le k$, are contained in the interior of S_N. Observe that any leaf $F_1^{cu} \in \mathcal{F}_1^{cu}(\Sigma,Y)$ intersects both S_M and S_N at C^1 curves C_M and C_N respectively.

Consider the partial foliation $\mathcal{F}_M(C(\Sigma))$ of $C(\Sigma)$ defined by:

(i) the leaves of $\mathcal{F}_M(C(\Sigma))$ in the region of $C(\Sigma)$ given by the intersection of the interior of S_M with the exterior of S_N are the intersection of $C(\Sigma)$ with the translation along the y_3-axis of the surface S_M,

(ii) the leaves of $\mathcal{F}_M(C(\Sigma))$ in the region of $C(\Sigma)$ outside the interior of S_M are the intersections of $C(\Sigma)$ with the surfaces $S_M(r)$ where $S_M(r)$ is obtained by the rotation around the y_3-axis of the translation along the y_2-axis of the curve $y_3 = My_2^{\beta}$ in the plane $y_1 = 0$.

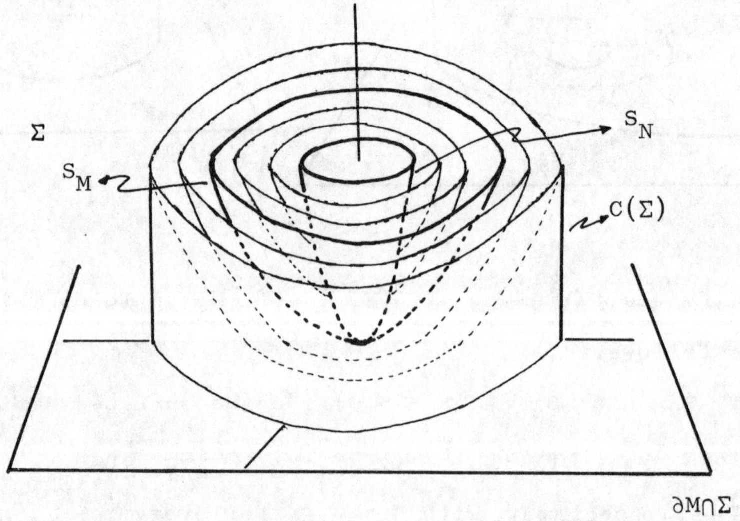

For every $F_1^{cu} \in \mathcal{F}_1^{cu}(\Sigma, Y)$ we consider the singular one-dimensional foliation $\mathcal{F}_{o,S}^{cu}(F_1^{cu})$ of F_1^{cu} defined by:

(i) the leaves of $\mathcal{F}_{o,S}^{cu}(F_1^{cu})$ in the region of F_1^{cu} outside the interior of S_N are the intersections of F_1^{cu} with $\mathcal{F}_M(C(\Sigma))$,

(ii) the leaves of $\mathcal{F}_{o,S}^{cu}(F_1^{cu})$ in the region of F_1^{cu} in the interior of S_N coincide with the restriction of the leaves of

$\mathcal{F}_o^{cu}(F_1^{cu})$ to this region.

So, the space of leaves of $\mathcal{F}_{o,S}^{cu}(F_1^{cu})$ is the union of $F_1^{cu} \cap \{y_3=0\}$ with $F_1^{cu} \cap S_N$. Observe that every $F_{o,S}^{cu}(F_1^{cu}) \in$ $\in \mathcal{F}_{o,S}^{cu}(F_1^{cu})$ is differentiable except possibly at the point corresponding to the intersection of $F_{o,S}^{cu}(F_1^{cu})$ with S_N. The foliation $\mathcal{F}_{o,S}^{cu}(F_1^{cu})$ is the desired modification of $\mathcal{F}_o^{cu}(F_1^{cu})$. In the picture below you can see a typical leaf F_1^{cu} with some leaves of $\mathcal{F}_{o,S}^{cu}(F_1^{cu})$.

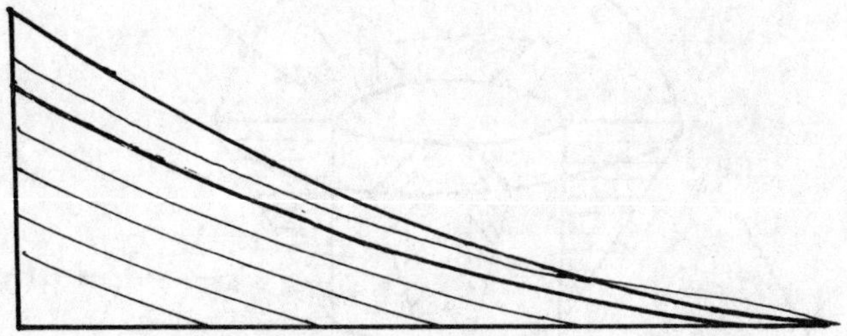

The union $\bigcup \mathcal{F}_{o,S}^{cu}(F_1^{cu})$, $F_1^{cu} \in \mathcal{F}_1^{cu}(\Sigma,Y)$ together with $\mathcal{F}_o^{cu}(\Sigma,Y)$ gives a singular one-dimensional foliation of $C(\Sigma)$ which we denote by $\mathcal{F}_{o,S}^{cu}(C(\Sigma))$, whose space of leaves is $C(\Sigma) \cap \{y_3=0\}$ union $D(\Sigma)$ and the part of S_N below $D(\Sigma)$.

Now we point out a fundamental property of $\mathcal{F}_{o,S}^{cu}(C(\Sigma))$.

Let $V(\sigma_o(Y))$ be a small tubular neighborhood of $C(\sigma_o(Y)) \cup$ $\cup D(\sigma_o(Y))$, where $C(\sigma_o(Y)) \cup D(\sigma_o(Y))$ is a fundamental neighborhood for $W^s(\sigma_o(Y))$ as in the definition of the singular central unstable foliation $\mathcal{F}_S^{cu}(\sigma_o(Y))$. $V(\sigma_o(Y))$ is fibered by the leaves $F_1^{cu}(\sigma_o(Y)) \in \mathcal{F}_1^{cu}(\sigma_o(Y))$ and $F_o^{cu}(\sigma_o(Y)) \in \mathcal{F}_o^{cu}(\sigma_o(Y))$.

Let $\pi_{o,Y}\colon C(\Sigma)\setminus\{p(Y)\} \to V(\sigma_o(Y))$ be the Poincaré map. Then there is a compact set $K(\sigma_o(Y)) \subset V(\sigma_o(Y))$ such that:

(i) $\partial K(\sigma_o(Y)) \supset C(\sigma_o(Y)) \cup D(\sigma_o(Y))$, $K(\sigma_o(Y)) \cap \partial M =$ $= C(\sigma_o(Y)) \cap \partial M$.

(ii) the diameter of $K(\sigma_o(Y))$ goes to zero when we approach ∂M.

(iii) To each $F_1^{cu}(\sigma_o(Y)) \in \mathfrak{F}_1^{cu}(\sigma_o(Y))$ there is a unique $F_1^{cu} \in \mathfrak{F}_1^{cu}(\Sigma,Y)$ such that $\pi_{o,Y}^{-1}(F_1^{cu}(\sigma_o(Y)) \cap K(\sigma_o(Y)) \subset F_1^{cu}$.

(iv) To each $F_o^{cu}(\sigma_o(Y)) \in \mathfrak{F}_o^{cu}(\sigma_o(\sigma_o(Y))$, there is a unique $F_o^{cu} \in \mathfrak{F}_o^{cu}(\Sigma,Y)$ such that $\pi_{o,Y}^{-1}(F_o^u \cap K(\sigma_o(Y))) \subset F_o^{cu}$.

See picture below.

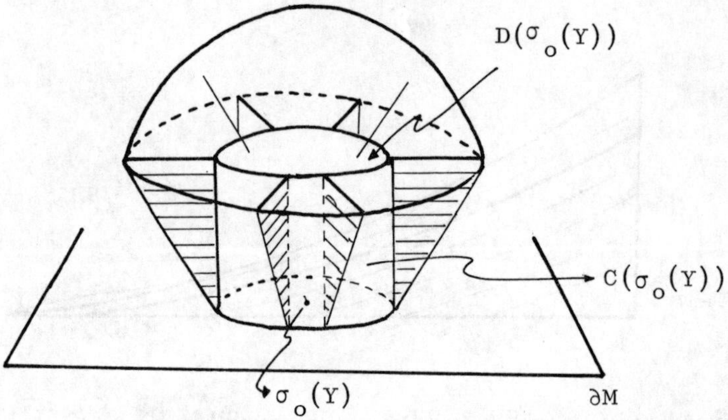

Let S be an invariant cross section at $q(Y) \in \sigma_1(Y)$, P_Y the associated Poincaré map and $D_s(q(Y),P_Y)$ a fundamental domain for $W^s(q(Y),P_Y)$. We denote $r(Y)$ the intersection of $W^u(\sigma_o(Y))$ with $D_s(q(Y),P_Y)$. Consider the Poincaré map $\pi_{1,Y}$: $N(r(Y)) \to \Sigma$ where $N(r(Y))$ is a neighborhood of $r(Y)$ in S. If $\mathfrak{F}^s(q(Y),P_Y)$ is the stable foliation for P_Y compatible with $\mathfrak{F}^s(\beta_j(Y))$, $1 \le j \le \ell$, then $\pi_{1,Y}(\mathfrak{F}^s(q(Y),P_Y)) = \mathfrak{F}^s(p(Y),\Sigma)$ is a C^1 foliation and $\Sigma \cap \partial M$ is a leaf. Next, our goal is to prove that given $X \in \mathfrak{X}_4^\infty(M,\partial M)$, there are neighborhoods $\mathcal{V}$ of X and $V(p(Y))$ of $p(Y)$ in Σ such that for every $Y \in \mathcal{V}$ each leaf of $\mathfrak{F}_{o,S}^{cu}(C(\Sigma))$ in $V(p(Y))$ intersects a leaf of $\mathfrak{F}^s(p(Y),\Sigma)$ at a unique point. We could get this by reasoning that those are transversal foliations except at $p(Y)$. But this is not possible directly because not all leaves of $\mathfrak{F}_{o,S}^{cu}(C(\Sigma))$ are differentiable. To avoid this problem we prove that shrinking $V(p(X))$ we get transversality between $\mathfrak{F}^s(p(X),\Sigma)$ and the restriction of $\mathfrak{F}^{cu}_{o,S}(C(\Sigma))$ to the interior of S_M in $V(p(X))\setminus\{p(X)\}$. Outside S_M, $\mathfrak{F}^{cu}_{o,S}(C(\Sigma))$ is transversal to $\mathfrak{F}^s(p(X),\Sigma)$ in $V(p(X))\setminus V(p(X)) \cap \partial M$. Now we could argue that since the linearization varies continuously with Y in the C^2 topology, the same result remain for Y near enough X. This fails because althoug outside S_M $\mathfrak{F}^{cu}_{o,S}(C(\Sigma))$ is transversal to $\mathfrak{F}^s(p(X),\Sigma)$ in $V(p(X))\setminus V(p(X)) \cap \partial M$, the angle between the leaves of $\mathfrak{F}^{cu}_{o,S}(C(\Sigma))$ and $\mathfrak{F}^s(p(X),\Sigma)$ in $V(p(X))\setminus V(p(X)) \cap \partial M$ goes to zero when we approach $\partial M \cap V(p(X))$. The same occurs with the angle between the leaves of $\mathfrak{F}^{cu}_{o,S}(C(\Sigma))$ in the interior of S_M and $\mathfrak{F}^s(p(X),\Sigma)$ when we get near $p(X)$. This leads us to a careful analysis of $\mathfrak{F}^{cu}_{o,S}(C(\Sigma))$. Roughly speaking, the idea is first to define vector fields $F_N(Y)$, $F_{N,M}(Y)$, $F_M(Y)$, $Y \in \mathcal{U}$ where $\mathcal{U}$ is a small neighborhood of X as before such that the integral curves of $F_N(Y)$, $F_{N,M}(Y)$, $F_M(Y)$ contain the leaves of $\mathfrak{F}^{cu}_{o,S}(C(\Sigma))$ in the interior of S_N, the leaves of $\mathfrak{F}^{cu}_{o,S}(C(\Sigma))$ in the region of $C(\Sigma)$ interior to S_M and exterior to S_N, the leaves of $\mathfrak{F}^{cu}_{o}(C(\Sigma))$ in the exterior of S_M respectively.

Secondly is to prove that the inner product (in the corresponding regions of definition) $F_N(X)(z)\cdot N(X)(z)$, $F_{N,M}(X)(z)\cdot N(X)(z)$ and $F_M(X)(z)\cdot N(X)(z)$ respectively is bounded way from zero, where $N(X)$ is the unitary normal vector to the leaf of $\mathfrak{F}^s(p(X),\Sigma)$ at z. This will imply the existence of neighborhoods $V(p(X))$ of $p(X)$ in Σ and $\mathcal{U}$ of X in $\mathfrak{X}^\infty_*(M,\partial M)$ with the required properties. So, $\pi^{-1}_{1,Y}(\mathfrak{F}^{cu}_{o,S}(C(\Sigma)))$ is a singular one dimensional foliation of $\pi^{-1}_{1,Y}(V(p(X))) = V(r(Y))$, whose fibers are differentiable near the boundary of $V(r(Y))$, each leaf of $\pi^{-1}_{1,Y}(\mathfrak{F}^{cu}_{o,S}(C(\Sigma)))$ intersects a leaf of $\mathfrak{F}^s(q(Y),P_Y)$ at a unique point. We extend this foliation to a P_Y-invariant singular foliation in a neighborhood $V(q(Y))$ of

$q(Y) \in \sigma_1(Y)$ in S in a compatible way with $\mathcal{F}^u(\alpha_i(Y))$, $\alpha_i(Y) < \sigma_1(Y)$, which is denoted by $\mathcal{F}^u_{o,S}(q(Y),P_Y)$.
Note that a leaf F^u_S of $\mathcal{F}^u_{o,S}(q(Y),P_Y)$ is not differentiable if and only if $F^u_S \cap \pi^{-1}_{1,Y}(S_N \setminus \{p(Y)\}) \neq \Phi$ and there is a unique point where F^u_S is not differentiable and this point is this intersection. In particular this will imply that there is a neighborhood $V(\sigma_1(X))$ of $\sigma_1(X)$ such that for each $Y \in \mathcal{U}$, $W^u(\alpha_i(Y))$, $1 \leq i \leq k$, is transversal to $W^s(\beta_j(Y))$, $1 \leq j \leq \ell$, in $V(\sigma_1(X))$. By induction on the behavior of $\sigma_1(X)$ with respect to $\beta_j(Y)$ we prove that for each $j \in \{1,\ldots,\ell\}$ there are neighborhoods $\mathcal{U}_j \subset \mathcal{U}$ of X and $V(\beta_j(X))$ of $\beta_j(X)$ such that for every $Y \in \mathcal{U}_j$, $W^u(\alpha_i(Y))$, $1 \leq i \leq k$, is transversal to $W^s(\beta_m(Y))$ in $V(\beta_m(X))$, $1 \leq m \leq \ell$. Finally, all this together will imply the openess of $\mathfrak{X}^\infty_4(M,\partial M)$ in $\mathfrak{X}^\infty_*(M,\partial M)$. The arguments used to prove all that are very close to the ones in Lemma 2 and Theorem A of [4]. So, we leave the details to the reader.

Now we will prove that if $Y \in \mathcal{U}$, where $\mathcal{U}$ is the neighborhood of X as above then there is a homeomorphism $H: M \circlearrowleft$ taking orbits of X onto orbits of Y preserving their orientations.

<u>1st step</u>. We can assume that there are homeomorphisms

$h^s: \bigcup_{i=1}^{k} W^s(\alpha_i(X)) \to \bigcup_{i=1}^{k} W^s(\alpha_i(Y))$ compatible with the unstable foliations of the critical elements α_i's;

$h^u: \bigcup_{j=1}^{k} W^u(\beta_j(X)) \cup W^u(\sigma_1(X)) \to \bigcup_{j=1}^{k} W^u(\beta_j(Y)) \cup W^u(\sigma_1(Y))$ compatible with the stable foliations of the critical elements β_j's and σ_1, conjugating X_t and Y_t.

<u>2nd step</u>. Using the same notations as above we are going to define a homeomorphism $h: S \circlearrowleft$ compatible with $\mathcal{F}^u_{o,S}(q(X),P_X)$ and $\mathcal{F}^s(q(X),P_X)$.

If $\mathfrak{F}_1^{cu}(\Sigma,X)$ and $\mathfrak{F}_o^{cu}(\Sigma,X)$ are the partial foliations of $V(p(X)) \subset \Sigma$ defined as above, $\pi_{1,Y}^{-1}(\mathfrak{F}_1^{cu}(\Sigma,X)) = \mathfrak{F}_1^{cu}(r(X),S)$ and $\pi_{1,Y}^{-1}(\mathfrak{F}_o^{cu}(\Sigma,X)) = \mathfrak{F}_o^{cu}(r(X),S)$ determine partial foliations of $V(r(X)) \subset S$. Denote by $C_1(S)$ and $D(S)$ the images $\pi_{1,X}^{-1}(C(\Sigma))$ and $\pi_{1,X}^{-1}(D(\Sigma))$ respectively.

Let us assume that $D(S) \subset F^s(q(X),P_X)(x_o)$ where $F^s(q(X),P_X)(x_o)$ is the stable leaf through $x_o \in W^u(\sigma_1(X)) \cap S$. Using the homeomorphism h^s and the unstable foliations which intersect $D(S)$ we can assume defined a homeomorphism $h_o\colon D(S) \subset \subset F^s(q(X),P_X)(x_o) \to D(S) \subset F^s(q(Y),P_Y)(h^u(x_o))$. Using the homeomorphism h^s and the unstable foliations which intersect $W^c(\sigma_o(X))$ we assume defined a homeomorphism $h_1\colon C_1(S) \cap \partial M \to C_1(S) \cap \partial M$.

Consider any homeomorphism $h_2\colon V(r(X)) \cap \partial M \to V(r(Y)) \cap \partial M$ such that h sends the family of intersections $\mathfrak{F}_1^{cu}(r(X)) \cap V(r(X)) \cap \partial M$ onto the family of intersections $\mathfrak{F}_1^{cu}(r(Y)) \cap V(r(Y)) \cap \partial M$ in such way that $h_2/C_1(S) \cap \partial M = h_1$.

Finally we define the homeomorphism h in $V(r(X))$ requiring

1) h preserves the restriction of $\mathfrak{F}^s(q(X),P_X)$ to $V(r(X))$,
2) h sends leaves of $\mathfrak{F}_1^{cu}(r(X),S)$ onto leaves of $\mathfrak{F}_1^{cu}(r(Y),S)$ according with the homeomorphism h_1,
3) h sends leaves of $\mathfrak{F}_o^{cu}(r(X),S)$ onto leaves of $\mathfrak{F}_o^{cu}(r(Y),S)$.

So, we have defined h in the restriction of $\mathfrak{F}^u_{o,S}(q(X),P_X)$ to $V(r(X))$. Using the Isotopy Extension Theorem we extend h to the fundamental domain $D_s(q(X),P_X)$ for $W^s(q(X),P_X)$ in a compatible way with $\mathfrak{F}^u(\alpha_i(X))$, $\alpha_i(X) < \sigma_1(X)$. Near the interior boundary of $D_s(q(X),P_X)$ h is defined by $hP_X = P_Y h$. This homeomorphism induces a map from the space of leaves of $\mathfrak{F}^{cu}_{o,S}(q(X),P_X)/N(X)$ restricted to the leaves through points of $D_s(q(X),P_X)\setminus\{r(X)\}$ where $N(X)$ is a fundamental neighborhood for $D_s(q(X),P_X)$. In this way we have defined a map h from the space of leaves of $\mathcal{F}^{cu}_{o,S}(q(X),P_X)$ onto the space of leaves of $\mathcal{F}^{cu}_{o,S}(q(Y),P_Y)$ restricted to leaves through points of $D_s(q(Y),P_Y)$. Using the equation $hP^n_X = P^n_Y h$ we define h on the space of leaves of $\mathcal{F}^{cu}_{o,S}(q(X),P_X)$. We observe that the restriction of h to $W^s(q(X),P_X)$ is a conjugacy between P_X and P_Y. Thus we have defined homeomorphisms $h\colon W^s(q(X),P_X) \to$ $\to W^s(q(Y),P_Y)$ and $h^u\colon W^u(q(X),P_X) \to W^u(q(Y),P_Y)$ compatible with $\mathcal{F}^{cu}_{o,S}(q(X),P_X)$ and $\mathcal{F}^s(q(X),P_X)$ respectively. By the intersection of the leaves we define $h\colon S \circlearrowleft$. Such a map is obviously continuous outside the P_X-orbit of $r(X)$. The continuity of h at the P_X-orbit of $r(X)$ is a consequence of the continuity of the foliation $\mathcal{F}^{cu}_{o,S}(q(X),P_X)$.

<u>3rd step</u>. We now will extend h to the neighborhood U_o of $\sigma_o(X)$. For this, let $f_X\colon U_o \to \mathbb{R}$ be a Liapunov function for X, $C(\sigma_o(X)) \cup D(\sigma_o(X))$ the fundamental domain for $W^s(\sigma_o(X))$ as before and $\Sigma \subset U_o$ a cross section for every $Y \in \mathcal{U}$ at $p(X) \in$ $\in W^u(\sigma_o(X))$. We suppose that $\Sigma \subset f_X^{-1}(1)$ and $T(\sigma_o(X)) \subset f_X^{-1}(-1)$ where $T(\sigma_o(X))$ is a tubular neighborhood of $C(\sigma_o(X)) \cup D(\sigma_o(X))$ fibered by leaves of $\mathcal{F}^{cu}_1(\sigma_o(X))$ and $\mathcal{F}^{cu}_o(\sigma_o(X))$.

The homeomorphism h defined on S induces, via the projection along the flow, a homeomorphism h defined in a neighborhood $V(p(X))$ in Σ. The way that h was defined implies that $h\colon V(p(X)) \to V(p(Y))$ sends leaves of $\mathcal{F}^{cu}_1(\Sigma,X)$, $\mathcal{F}^{cu}_o(\Sigma,X)$ and $\mathcal{F}^{cu}_{o,S}(V(p(X)))$ into leaves of $\mathcal{F}^{cu}_1(\Sigma,Y)$, $\mathcal{F}^{cu}_o(\Sigma,Y)$ and $\mathcal{F}^{cu}_{o,S}(V(p(Y)))$ respectively.

Define h on $T(\sigma_o(X))$ as: given $x \in T(\sigma_o(X))$, let $y \in \Sigma$ such that $X_t x = y$; define $h(x)$ as the first hit of

$\{Y_t(h(y)),\ t < 0\}$ at $T(\sigma_o(Y))$. Such h sends leaves of $\mathfrak{F}_1^{cu}(\sigma_o(X)) \cap K(X)$, $\mathfrak{F}_o^{cu}(\sigma_o(X)) \cap K(X)$, $\mathfrak{F}_o^{u}(\sigma_o(X)) \cap K(X)$ into leaves of $\mathfrak{F}_1^{cu}(\sigma_o(Y)) \cap K(Y)$, $\mathfrak{F}_o^{cu}(\sigma_o(Y)) \cap K(Y)$ and $\mathfrak{F}_o^{u}(\sigma_o(Y)) \cap$ $\cap K(Y)$ respectively. This implies that we can extend h to $W^s(\sigma_o(X)) \cap T(\sigma_o(X))$. We extend h to the interior of $C(\sigma_o(X)) \cup$ $\cup D(\sigma_o(X))$ using levels of the Liapunov function and the flow.

In this way we have defined a homeomorphism h from a neighborhood of the quasi-transversal intersection $(\sigma_o(X), \sigma_1(X))$ onto a neighborhood of the quasi-transversal intersection $(\sigma_o(Y), \sigma_1(Y))$. Moreover, reparametrizing the vector fields we can get h in such a way that the restriction of h to the boundary of this neighborhood is a conjugacy between X_t and Y_t. To conclude the construction of the equivalence h between X and Y we proceed as in Theorem A. So we have proved the following

<u>Theorem 3</u>. $\mathfrak{X}_4^\infty(M, \partial M)$ is open in $\mathfrak{X}_*^\infty(M, \partial M)$ and its elements are strucutrally stable.

Now we shall prove the openess of $\mathfrak{X}_3^\infty(M, \partial M)$ in $\mathfrak{X}_*^\infty(M, \partial M)$ and the stability of its elements.

Let $X \in \mathfrak{X}_3^\infty(M, \partial M)$, $\Omega(X) = \alpha_1(X) \le \ldots \le \alpha_k(X) \le \sigma_o(X) \le \sigma_1(X)$ $\le \beta_1(X) \le \ldots \le \beta_\ell(X)$ where $(\sigma_o(X), \sigma_1(X))$ corresponds to the quasi-transversal intersection along the boundary of M. Let $\mathcal{V} \subset \mathfrak{X}^\infty(M, \partial M)$ be a small neighborhood of X such that for each $Y \in \mathcal{V}$, $\Omega(X) =$ $= \alpha_1(Y) \le \ldots \le \alpha_k(Y) \le \sigma_o(Y) \le \sigma_1(Y) \le \beta_1(Y) \le \ldots \le \beta_\ell(Y)$, where $\alpha_i(Y)$, $\sigma_o(Y)$, $\sigma_1(Y)$, $\beta_j(Y)$ are the critical elements near the corresponding ones for X, $(\sigma_o(Y), \sigma_1(Y))$ corresponding to the pair having the quasi-transversal intersection along ∂M. Let $\mathfrak{F}^u(\alpha_i(Y))$, $1 \le i \le k$; $\mathfrak{F}^s(\sigma_1(Y))$ and $\mathfrak{F}^s(\beta_j(Y))$, $1 \le j \le \ell$, be the compatible unstable and stable foliations. Taking $\mathcal{V}$ small enough we can suppose that there is $U_o \subset M$ such that for each $Y \in \mathcal{V}$, $\sigma_o(Y) \in U_o$,

Y is C^2-linearizable in U_o and the diffeomorphism linearizing Y is C^2-close to the corresponding diffeomorphism for X (see Appendix of [4]). Let $\beta > 1$ such that $\beta > \max\{\frac{\lambda_2}{\lambda_1}, \frac{\lambda_3}{\lambda_1}, \frac{\lambda_3}{\lambda_2}\}$ for every $Y \in \mathcal{V}$ where $-\lambda_i(Y)$, $1 \le i \le 3$, are the contracting eigenvalues at $\sigma_o(Y)$, $-\lambda_1(Y) < -\lambda_2(Y) < -\lambda_3(Y)$.

1st Case. $\dim W^u(\sigma_o(X)) = 1$ and so $\dim W^s(\sigma_1(X)) = 3$. Here the idea is again to obtain a singular one dimensional foliation of a fundamental domain $D_s(q(X),P_X)$ for $W^s(q(X),P_X)$ where $q(X) \in \sigma_1(X)$ and P_X is the corresponding Poincaré map, which is in a neighborhood of $W^u(\sigma_o(X)) \cap D_s(q(X),P_X)$, a modification of the foliation induced by the singular central unstable foliation $\mathcal{F}_S^{cu}(\sigma_o(X))$ defined before.

We denote by y_1, y_2, y_3, y_4 the linearizing coordinates in U_o (these coordinates depend on Y but this dependence is not expressed in the notation) and, in these coordinates we have $\sigma_o(Y) = (0,0,0,0)$, $W^c(\sigma_o(Y)) = \{y_2=y_3=y_4=0\}$, $W^s(\sigma_o(Y)) = \{y_4=0\}$, $W^u(\sigma_o(Y)) = \{y_1=y_2=y_3=0\}$. When no confusion is possible we will still denote by Y the vector field Y in these coordinates.

Let $\Sigma \subset U_o$ be a cross section for every $Y \in \mathcal{V}$ at $p(X) = (0,0,0,1) \in W^u(\sigma_o(X))$. We denote by $p(Y)$ the intersection of $W^u(\sigma_o(Y))$ with Σ. Let $\mathcal{F}_S^{cu}(\Sigma,Y)$ be the singular foliation of a neighborhood of $p(Y)$ in Σ induced by the intersection of the singular central unstable foliation $\mathcal{F}_S^{cu}(\sigma_o(Y))$ with Σ. The leaves corresponding to the intersection of a typical leaf of $\mathcal{F}_1^{cu}(\sigma_o(Y))$ with Σ have dimension two and are denoted by $\mathcal{F}_1^{cu}(\Sigma,Y)$. The leaves corresponding to the intersection of a typical leaf of $\mathcal{F}_0^{cu}(\sigma_o(Y))$ with Σ have dimension one and are denoted by $\mathcal{F}_0^{cu}(\sigma_o(Y))$.

Let $C(\Sigma)$ be a half solid cylinder defined by $y_2^2+y_3^2 \le 1$,

$y_3 > 0$, $-1 \leq y_1 \leq 1$. Let $D_i(\Sigma) = C(\Sigma) \cap \{y_1 = i\}$, $i = -1,1$. We assume that $\partial C(\Sigma) = C_1(\Sigma)$ is the space of leaves of $\mathfrak{F}_o^{cu}(\Sigma,Y)$ and $C_1(\Sigma) \cap \{y_1=0\} = S$ is the space of leaves of $\mathfrak{F}_1^{cu}(\Sigma,Y)$. The picture below indicates some of the leaves of $\mathfrak{F}_1^{cu}(\Sigma,Y) \cap C(\Sigma)$, $i = 0,1$.

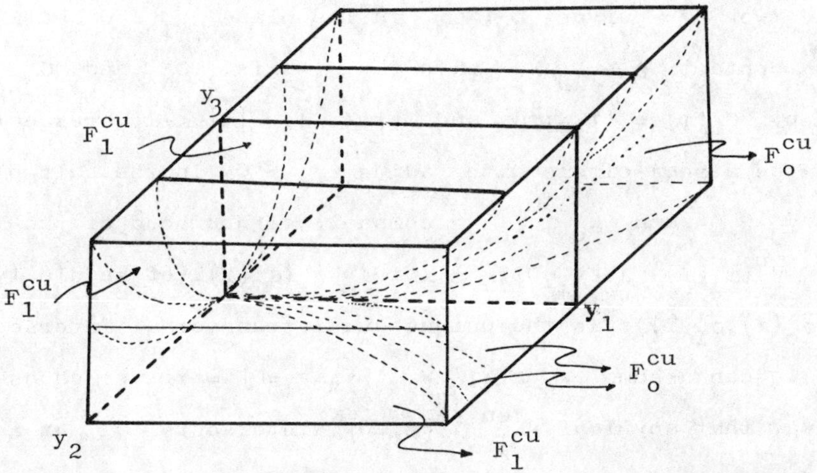

From the construction of the singular central unstable foliation we have that each $F_1^{cu} \in \mathfrak{F}_1^{cu}(\sigma_o(Y))$ is a union of center-unstable leaves $F_o^{cu}(x)$ with x varying along a curve. This implies that a leaf $F_1^{cu} \in \mathfrak{F}_1^{cu}(\Sigma,Y)$ is foliated by an one dimensional foliation corresponding to the intersections $\bigcup_{t\geq 0} Y_t F_o^{cu}(x) \cap \Sigma$. We denote this foliation by $\mathfrak{F}_o^{cu}(F_1^{cu})$. The picture below indicates some of the leaves of $\mathfrak{F}_o^{cu}(F_1^{cu})$.

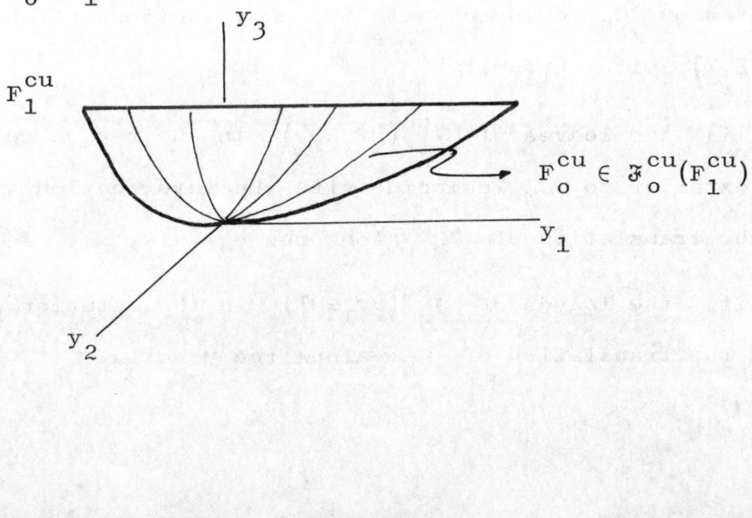

We now will start the modifications of these one dimensional foliations. We will do this in steps.

<u>1st step</u>. We first modify the intersections of $\mathcal{F}_1^{cu}(\Sigma,Y)$ with the plane $y_1 = 0$. It is easy to see that this intersection is a union of C^1 curves all of them with contact λ_3/λ_2 with the y_2-axis. We choose two C^1 curves C_N, C_M in the plane $y_1 = 0$ both of them with contact $\beta > \lambda_3/\lambda_2$ with the y_2-axis, C_N and C_M parametrized by $C_N(t) = (t, Nt^\beta)$ and (t, Mt^β), $|t| \leq 1$, respectively. Let S_{r_0} be a semi-circle with radius $r_0 > 0$ in the half plane $y_1 = 0$, $y_3 \geq 0$ and $W_0 \subset S_0$ a compact neighborhood of the intersections $W^u(\alpha_i(Y)) \cap \{y_1=0\}$, $1 \leq i \leq k$ ($\alpha_i(Y)$ of saddle type). Since $(\sigma_0(Y), \sigma_1(Y))$ is the unique quasi-transversal intersection for Y, we can assume $\bar{W}_0 \subset W$, $\bar{W} \cap \{y_3=y_1=0\} = \Phi$. We choose $M > N$ in such way that any leaf $F_1^{cu} \cap \{y_1=0\}$ intersects C_N at a unique point and C_M at a unique point. On the other hand we can choose r_0 small enough such that the leaves through points of $\bar{W}_0$ intersect C_N and C_M in the exterior of S_{r_0}. Shrinking $\mathcal{U}$, we can take $r_0 = r_0(Y)$ for every $Y \in \mathcal{U}$.

We now consider the one dimensional foliation $\mathcal{F}_0^{cu}(\{y_1=0\})$ of $\{y_1=0\}$ given by:

(i) the leaves of $\mathcal{F}_0^{cu}(\{y_1=0\})$ in the interior of the region limited by C_N coincide with the intersections of the leaves of $\mathcal{F}_1^{cu}(\Sigma,Y)$ with $\{y_1=0\}$,

(ii) the leaves of $\mathcal{F}_0^{cu}(\{y_1=0\})$ in the region interior to C_M and exterior to C_N coincide with the intersection with this region of the translation of C_M along the y_3-axis,

(iii) the leaves of $\mathcal{F}_0^{cu}(\{y_1=0\})$ in the exterior of C_M coincide with the translation of C_M along the y_2-axis.

The picture below shows some of the leaves of $\mathcal{F}_o^{cu}(\{y_1=0\})$.

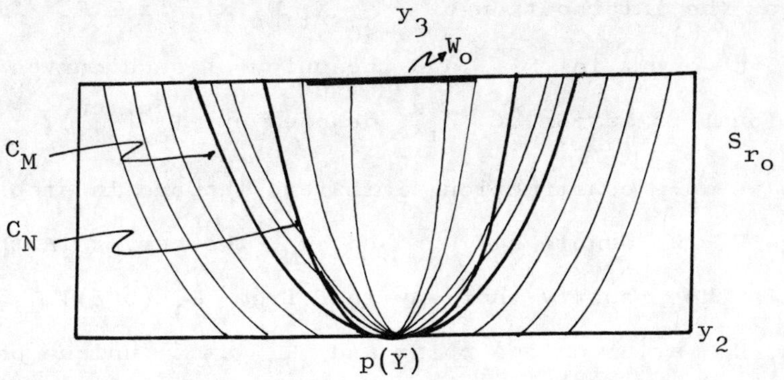

$\underline{2^{nd}\ \text{step}}$. Let $C(\sigma_o(Y)) \cup D(\sigma_o(Y))$ be the fundamental domain for $W^s(\sigma_o(Y))$ as in the construction of the singular central unstable foliation $\mathcal{F}_S^{cu}(\sigma_o(Y))$. See picture below.

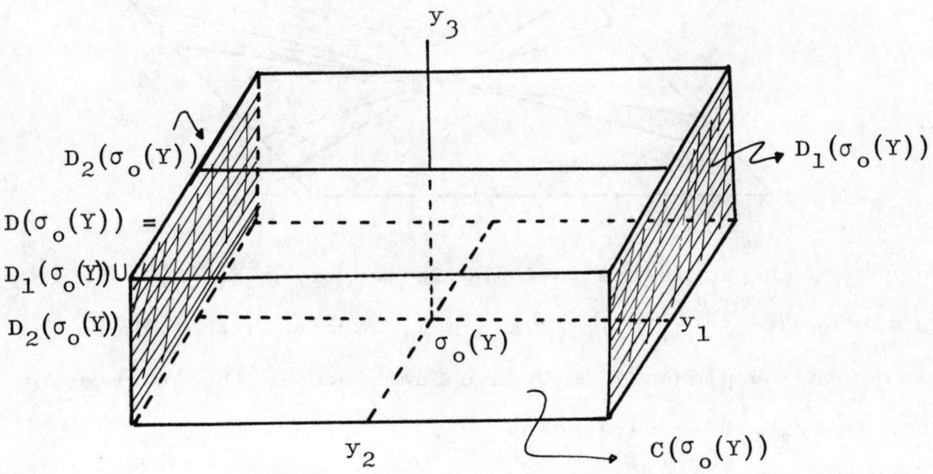

It is possible to construct a singular one dimensional foliation $\mathcal{F}_S^c(D(\sigma_o(Y))$ of $D(\sigma_o(Y))$ compatible with $\mathcal{F}^u(\alpha_i(Y))$, $1 \le i \le k$. Raise $\mathcal{F}_S^c(D(\sigma_o(Y)))$ to a two dimensional foliation $\mathcal{R}^{cu}$ in the following way: a leaf $R^{cu} \in \mathcal{R}^{cu}$ is, by definition, the union $\bigcup_{x \in F_S^c} F_o^u(x)$ where F_S^c is a leaf of $\mathcal{F}_S^c(D(\sigma_o(Y)))$ and $F_o^u(x)$ is the unstable leaf at x. The intersection of $\bigcup_{t \ge 0} X_t\, \mathcal{R}^{cu}$ with $C(\Sigma)$ induces a partial singular two dimensional foliation of $C(\Sigma)$ whose space of leaves is $D(\Sigma) \cap \{y_3=0\}$. We denote this foliation by $\mathcal{F}_1^{cu}(D(\Sigma))$.

Observe that the intersection of $\bigcup_{t\geq 0} X_t F_o^u(x)$, $x \in F_S^c$ fixed, with $C(\Sigma)$, is a C^1-curve in F_1^{cu} and the union of such curves gives an one dimensional foliation of F_1^{cu} denoted by $\mathcal{F}_o^{cu}(F_1^{cu})$.

Now we do a construction similar to the one in step 1 in the plane $y_3 = 0$. We denote by R_N and R_M the curves in the plane $y_3 = 0$ with the property that any leaf from $\mathcal{F}_1^{cu}(D(\Sigma)) \cap \{y_3=0\}$ intersects R_N at a unique point and R_M at a unique point. Since the y_1-axis corresponds to the direction of the weakest contraction at $\sigma_o(Y)$, R_N and R_M can be chosen as straightlines.

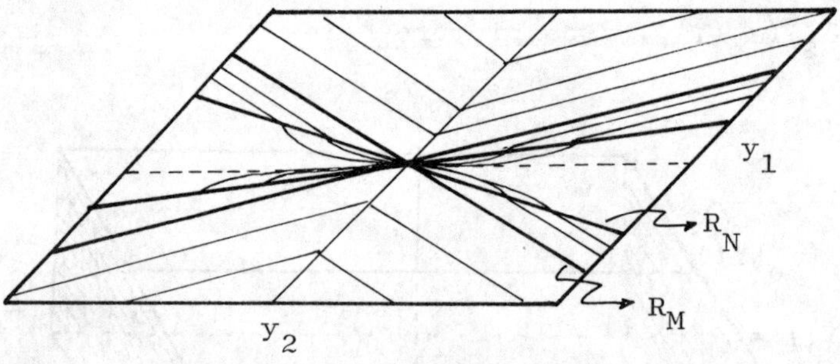

Moreover, we can assume that the leaves of $\mathcal{F}_1^{cu}(D(\Sigma))$ corresponding to leaves of $\mathcal{F}^u(\alpha_i(Y))$, $1 \leq i \leq k$, are such that their intersection with the plane $y_3 = 0$ are contained in the interior of the cone in $y_3 = 0$ generated by R_N.

3rd <u>step</u>. Now we shall modify the one-dimensional foliation $\mathcal{F}_o^{cu}(F_1^{cu})$ of F_1^{cu} for every $F_1^{cu} \in \mathcal{F}_1^{cu}(D(\Sigma))$ such that the intersection of $F_1^{cu} \cap \{y_3=0\}$ with $C(\Sigma)$ is contained in the interior of the cone in the plane $y_3 = 0$ generated by R_N. It is easy to see that each F_1^{cu} as above can be parametrized by

$$\varphi(y,w) = (y, y^{\lambda_2/\lambda_1} a(w),\ y^{\lambda_3/\lambda_1} w), \quad 0 \leq y \leq 1, \quad 0 \leq w \leq 1$$

and the leaves of $\mathcal{F}_o^{cu}(F(_1^{cu})$ are the image under φ of the horizontal lines (y,w_o), $0 \le y \le 1$.

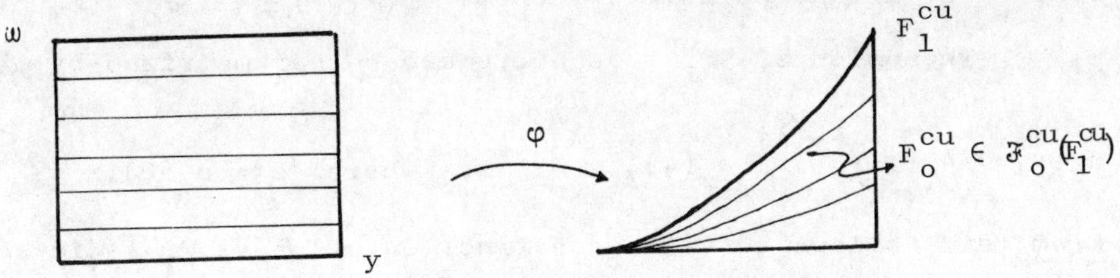

Moreover $a(w)$ is differentiable with respect to w and φ is a local diffeomorphism outside the w-axis. Let L_1 and L_2 be two lines in the square $Q = \{(y,w), 0 \le y,w \le 1\}$ through the origin such that the angle between L_1 (resp. L_2) and the y-axis is θ_1 (resp. θ_2), $\theta_1 < \theta_2$.

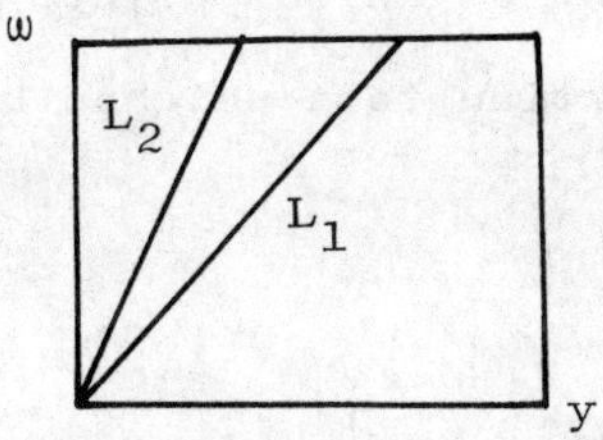

Consider the foliation $\mathfrak{F}(Q)$ of Q as below

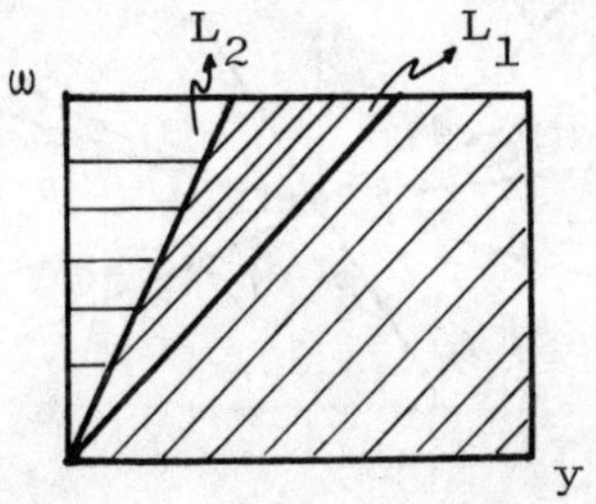

Observe that the pre-image under φ of the horizontal lines in F_1^{cu} is the family of hyperboles $w = \dfrac{c}{y^{\lambda_3/\lambda_1}}$, $0 \le c \le 1$. So, each leaf of $\mathfrak{F}(Q)$ intersects a hyperbole of this family at a unique point.

Then $\varphi(\mathfrak{F}(Q))$ determines a singular foliation of F_1^{cu} with the property that each leaf of $\varphi(\mathfrak{F}(Q))$ intersects a horizontal line at a unique point. We denote $\varphi(\mathfrak{F}(Q))$ by $\mathfrak{F}^{cu}_{o,S}(F_1^{cu})$.

The union of F_1^{cu} as above can be parametrized by

$$\varphi_z(y,w) = (y, y^{\lambda_2/\lambda_1} a_z(w), y^{\lambda_3/\lambda_1} w) \quad \text{where} \quad z = a_z(0).$$

From this follows that the construction of $\mathfrak{F}^{cu}_{o,S}(F_1^{cu})$ is such that $\mathfrak{F}^{cu}_{o,S}(F_1^{cu})$ varies continuously with z, and so, with F_1^{cu}.

<u>4th step</u>. Let T be the topological cone in $C(\Sigma)$ defined by the union $\bigcup_{z\in I} F_1^{cu}(z)$ where $F_1^{cu}(z) \in \mathfrak{F}_1^{cu}(D(\Sigma))$ is the leaf through z and I is the intersection of $D(\Sigma)$ with the cone generated by R_N in the plane $y_3 = 0$. So far we have defined a singular one dimensional foliation in the interior of T such that each leaf intersects a horizontal plane at a unique point. Now we will extend this one dimensional foliation to a neighborhood of $p(Y)$.

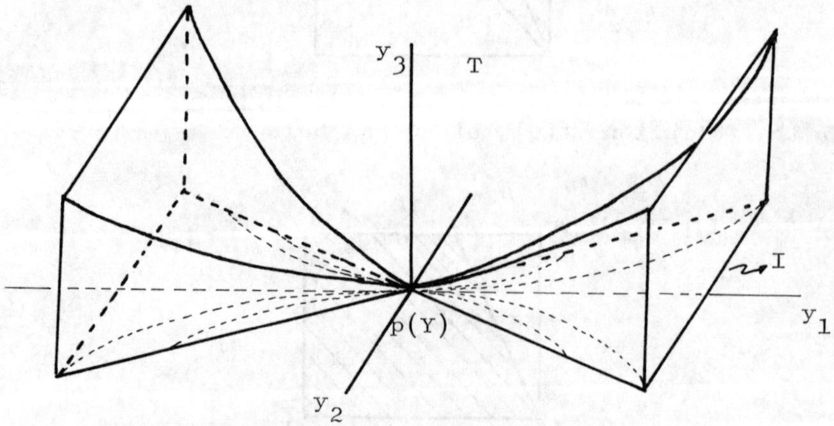

Consider the topological half solid cylinder $\bar{C}(Y)$ in Σ such that $\partial\bar{C}(Y) \cap D(\Sigma) = T \cap D(\Sigma)$.

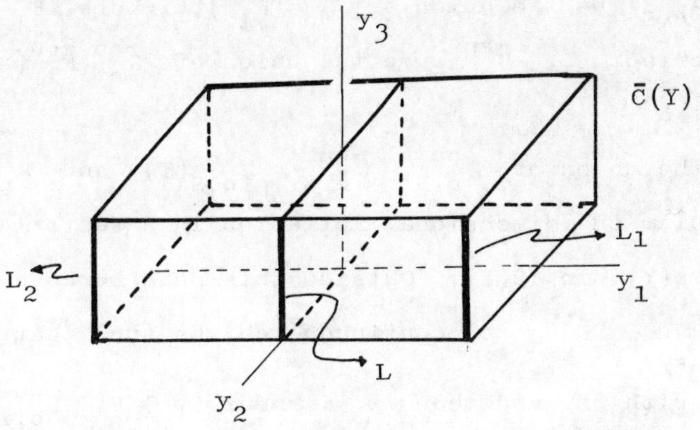

We can suppose that the intersection of $\partial\bar{C}(Y)$ with the curve C_N in the plane $y_1 = 0$, C_N as in step 1, are the corners of $\partial\bar{C}(Y) \cap \{y_1=0\}$ outside the plane $\{y_3=0\}$. Let L_1, L_2 and L be the segments in the boundary of $\bar{C}(Y)$ as in the picture above. Let F_1 and F_2 be the leaves of $\mathfrak{F}_1^{cu}(D(\Sigma))$ such that $F_i \cap \partial T \supset L_i$. To each $x \in L$, let $F_1^{cu}(x)$ be the leaf of $\mathfrak{F}_1^{cu}(\Sigma,Y)$ through x. From the choice of C_N it follows that there is a unique $\bar{x} \in F_1^{cu}(x) \cap C_N$. Let $L_{\bar{x}} = F_1^{cu}(x) \cap P_{\bar{x}}$, where $P_{\bar{x}}$ is the plane parallel to $y_3 = 0$ through $\bar{x}$ and let $y_i(\bar{x})$ be the intersection of $L_{\bar{x}}$ with F_i, $i = 1,2$.

We denote by $\tilde{F}_1^{cu}(\bar{x})$ the intersection of $F_1^{cu}(x)$ with the region of Σ below the plane $P_{\bar{x}}$.

It is not difficult to see that the union $\bigcup_{\bar{x}\in C_N} y_i(\bar{x})$ is a C^0 curve C_i in F_i with the property that every leaf $\in \mathfrak{F}_o^{cu}(F_i)$ intersects C_i at a unique point and every plane parallel to $y_3 = 0$ intersects C_i at a unique point. We now perform in F_i a construction similar to the one in step 1.

We remark that the constructions in step 3 can be performed in such way that the singular one dimensional foliation $\mathfrak{F}_{o,S}^{cu}(T)$ defined in T is a continuous foliation.

Let $\mathcal{F}^{cu}_{o,S}$ be the union of $\mathcal{F}^{cu}_o(F^{cu}_1)$, where $F^{cu}_1 \in \mathcal{F}^{cu}_1(\Sigma,Y)$ does not intersect L, $\tilde{\mathcal{F}}^{cu}_{o,S}$ be the union of $\mathcal{F}^{cu}_o(\tilde{F}^{cu}_1(\bar{x}))$, $\tilde{F}^{cu}_1(\bar{x})$ as above.

Then the union of $\mathcal{F}^{cu}_{o,S}$, $\tilde{\mathcal{F}}^{cu}_{o,S}$, $\mathcal{F}^{cu}_{o,S}(T)$ and $\mathcal{F}^{cu}_o(\{y_1=0\})$ gives a singular one dimensional foliation in a sectorial neighborhood V of p(Y) in $\bar{C}(Y)$. Outside this neighborhood we take any family $\bigcup_{x\in\{y_3=0\}\setminus V} F^{cu}_{o,x}$ of continuous curves such that $F^{cu}_{o,x}$ varies continuously with x and when x is near $y \in V$, $F^{cu}_{o,x}$ is C^o-close to $F^{cu}_{o,y}$, where $F^{cu}_{o,y}$ is the leaf of $\mathcal{F}^{cu}_{o,S} \cup \tilde{\mathcal{F}}^{cu}_{o,S}(T) \cup \mathcal{F}^{cu}_{o,S}(T)$ through y.

In this way we have defined a singular one dimensional foliation in $\bar{C}(Y)$. We can suppose that the leaves along the boundary of $\bar{C}(Y)$ are C^1-curves. Remark that the diameter of the leaves through points in $V\setminus T$ tends to zero when we approach p(Y).

As before, a fundamental property of the foliation obtained above is that there is a small tubular neighborhood $V(\sigma_o(Y))$ of $C(\sigma_o(Y)) \cup D(\sigma_o(Y))$, where $C(\sigma_o(Y)) \cup D(\sigma_o(Y))$ is a fundamental domain for $W^s(\sigma_o(Y))$ as in the definition of the singular central unstable foliation $\mathcal{F}^{cu}_S(\sigma_o(Y))$, fibred by the leaves $F^{cu}_1(\sigma_o(Y)) \in \mathcal{F}^{cu}_1(\sigma_o(Y))$ and $F^{cu}_o(\sigma_o(Y)) \in \mathcal{F}^{cu}_o(\sigma_o(Y))$ and a compact set $K(\sigma_o(Y)) \subset V(\sigma_o(Y))$ such that:

(i) $\partial K(\sigma_o(Y)) \supset C(\sigma_o(Y)) \cup D(\sigma_o(Y))$, $K(\sigma_o(Y)) \cap \partial M = C(\sigma_o(Y)) \cap \partial M$

(ii) the diameter of $K(\sigma_o(Y))$ goes to zero when we approach ∂M

(iii) let $\pi_{o,Y}: \bar{C}(Y)\setminus\{p(Y)\} \to V(\sigma_o(Y))$ be the Poincaré map. To each $F^{cu}_1(\sigma_o(Y)) \in \mathcal{F}^{cu}_1(\sigma_o(Y))$ there is a unique $F^{cu}_1 \in \mathcal{F}^{cu}_1(\Sigma,Y)$ such that $\pi^{-1}_{o,Y}(F^{cu}_1(\sigma_o(Y)) \cap K(\sigma_o(Y)) \subset F^{cu}_1$

(iv) to each $F^{cu}_o(\sigma_o(Y)) \in \mathcal{F}^{cu}_o(\sigma_o(Y))$ there is a unique $F^{cu}_o \in \mathcal{F}^{cu}_o(\Sigma,Y)$ such that $\pi^{-1}_{o,Y}(F^u_o \cap K(\sigma_o(Y)) \subset F^u_o$.

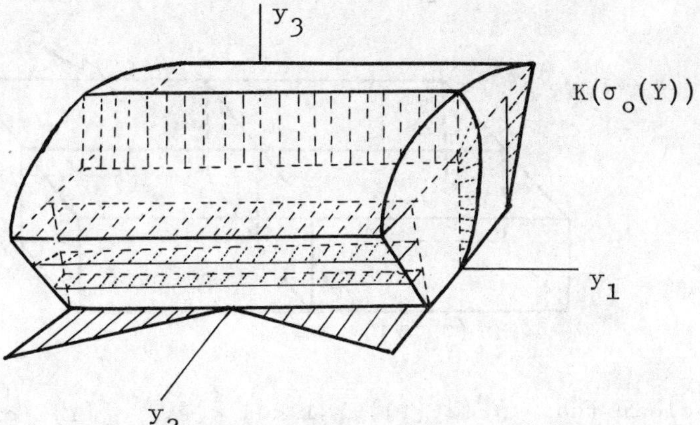

From this point on, the arguments to prove the openess of $\mathfrak{X}_3^\infty(M,\partial M)$ and the stability of $X \in \mathfrak{X}_3^\infty(M,\partial M)$ are very close to the ones in Theorems 1, 2 and 3. So, we leave the details to the reader.

<u>2nd Case</u>. $\dim W^u(\sigma_o(Y)) = 2$ and $\dim W^s(\sigma_1(Y)) = 3$.

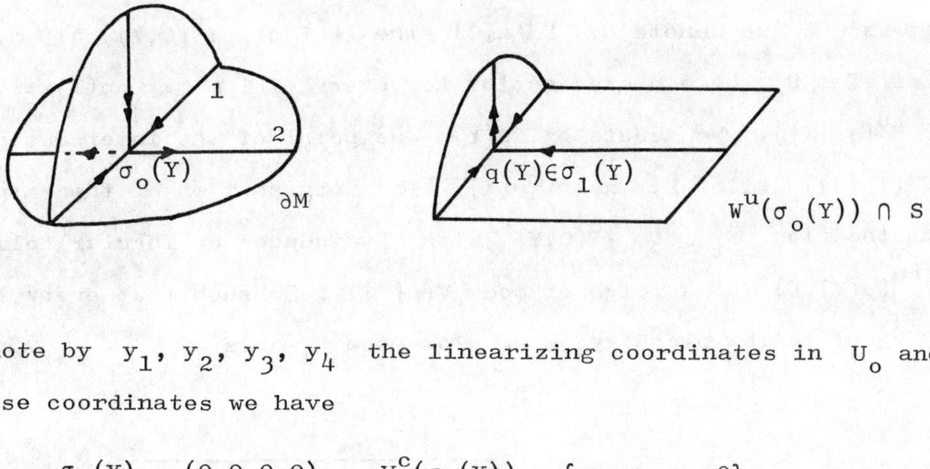

We denote by y_1, y_2, y_3, y_4 the linearizing coordinates in U_o and in these coordinates we have

$$\sigma_o(Y) = (0,0,0,0), \quad W^c(\sigma_o(Y)) = \{y_1=y_3=y_4=0\},$$

$$W^u(\sigma_o(Y)) = \{y_2=y_3=0\}.$$

When no confusion is possible we still denote by Y the vector field Y in these coordinates.

Let $D_s(\sigma_o(Y)) \subset U_o$ be a fundamental domain for $W^s(\sigma_o(Y))$ and $C \subset U_o$ a cross section for each $Y \in \mathcal{V}$, C as in the picture below.

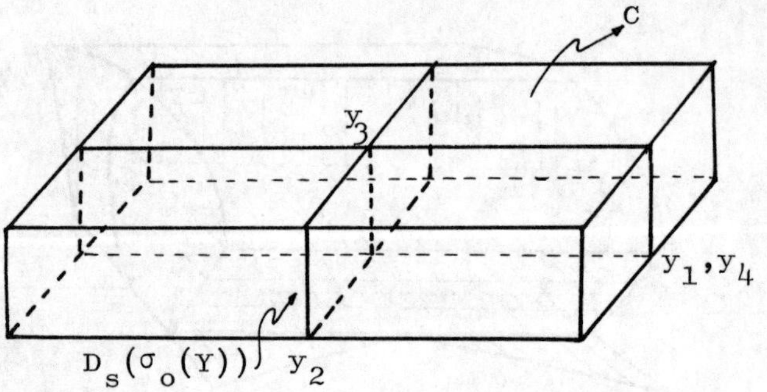

It is clear that $\mathfrak{F}^u(\alpha_i(Y))$, $1 \le i \le k$, $\alpha_i(Y)$ of saddle type, induces a foliation in an open subset of C which intersects $D_s(\sigma_o(Y))$ in an open set $W_o(Y)$ with $\bar{W}_o(Y) \cap \partial M = \Phi$. We can extend this foliation to a foliation $\mathfrak{F}^u(C,Y)$ of C such that the leaves of $\mathfrak{F}^u(C,Y)$ through points of $D_s(\sigma_o(Y)) \setminus \text{int}\, W(Y)$ are linear where $W(Y) \subset D_s(\sigma_o(Y))$ is a compact set containing $W_o(Y)$ in its interior. If $x \in C$ we denote by $F_Y^u(x,C)$ the leaf of $\mathfrak{F}^u(C,Y)$ at x.
Let $\Sigma \subset U_o$ be a cross section for every $Y \in \mathcal{V}$ at $p(X) = (0,0,0,1) \in W^u(\sigma_o(X))$. We denote by $p(Y)$ one point of the intersection of $W^u(\sigma_o(Y))$ with Σ, near $p(X)$. The intersection of the saturation by the flow Y_t of $\mathfrak{F}^u(C,Y)$ with Σ induces a singular foliation $\mathfrak{F}^{cu}(p(Y),\Sigma)$ of a neighborhood $V(p(Y)) \subset \Sigma$ such that every leaf is tangent to the plane $y_3 = 0$ along the y_1-axis.

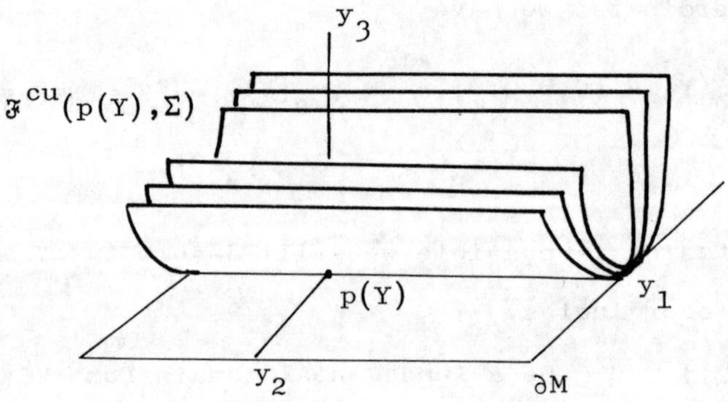

We will modify this foliation to get a new one, $\mathcal{F}_S^{cu}(p(Y),\Sigma)$, in such way that its space of leaves includes the y_2-axis and satisfying:

(i) the set of singularities of $\mathcal{F}_S^{cu}(p(Y),\Sigma)$ is the y_1-axis

(ii) there is a compact set $K \subset C$, $K \cap W^s(\sigma_o(Y)) = D_s(\sigma_o(Y))$, $\bigcup_{x \in W_o(Y)} F_Y^u(x,C) \subset K$, the diameter of K going to zero as we approach ∂M, such that if $F_S^{cu} \in \mathcal{F}_S^{cu}(p(Y),\Sigma)$ is a leaf through $p(Y)$, then there is a unique $F^u \in \mathcal{F}^u(C,Y)$ such that $\pi_{o,Y}^{-1}(F^u \cap K) \subset F_S^{cu}$, where $\pi_{o,Y}: V(p(Y))\backslash\{y_2=y_3=0\} \to C$ is the Poincaré map.

To get this foliation we proceed as in Theorem 1 and so, we leave the details to the reader.

Let $\mathcal{F}(p(Y),Y)$ be the two-dimensional foliation of $V(p(Y))$ whose leaves are planes parallel to the plane $y_1 = 0$. Then the intersection of $\mathcal{F}_S^{cu}(p(Y),\Sigma)$ with $\mathcal{F}(p(Y),Y)$ gives a singular one dimensional foliation $\mathcal{F}_{o,S}^{cu}(p(Y),\Sigma)$.

Let S be an invariant cross section at $q(Y) \in \sigma_1(Y)$, P_Y the associated Poincaré map and $D_s(q(Y),P_Y)$ a fundamental domain for $W^s(q(Y),P_Y)$. We denote $r(Y)$ the intersection of the orbit of $p(Y)$ with $D_s(q(Y),P_Y)$. Consider the Poincaré map $\pi_{1,Y}: N(r(Y)) \to V(p(Y))$ where $N(q(Y))$ is a neighborhood of $r(Y)$ in S. If $\mathcal{F}^s(q(Y),P_Y)$ is the stable foliation for P_Y compatible with $\mathcal{F}^s(\beta_j(Y))$, $1 \le j \le \ell$, then $\pi_{1,Y}(\mathcal{F}^s(q(Y),P_Y) = \mathcal{F}^s(p(Y),\Sigma)$ is a C^1-foliation of $V(p(Y))$ and $\Sigma \cap \partial M$ is a leaf. As in Theorem 1 we prove that given $X \in \mathfrak{X}_3^\infty(M,\partial M)$ there are neighborhoods $\mathcal{V}$ of X and $V(p(X))$ of $p(X)$ in Σ such that for every $Y \in \mathcal{V}$, each leaf of $\mathcal{F}_{o,S}^{cu}(p(Y),\Sigma)$ intersects a leaf of $\mathcal{F}^s(p(Y),\Sigma)$ at a unique point. So, $\pi_{1,Y}^{-1}(\mathcal{F}_{o,S}^{cu}(p(Y),\Sigma))$ gives a singular one dimensional foliation of a neighborhood of $W^u(\sigma_o(Y)) \cap D_s(q(Y),P_Y)$ such that each leaf of this foliation intersects a leaf of $\mathcal{F}^s(q(Y),P_Y)$ at a unique point. We extend this foliation to a singular one dimensional invariant foliation for $W^s(q(Y),P_Y)$ compatible with $\mathcal{F}^u(\alpha_i(Y))$, $1 \le i \le k$. We denote this one dimensional foliation by $\mathcal{F}^{cu}_{o,S}(q(Y),P_Y)$.

We use the above foliations to prove, in this case, the openess of $\mathfrak{X}^\infty_3(M,\partial M)$.

To define a homeomorphism $h: M \circlearrowleft$ taking orbits of X onto orbits of Y, where $X,Y \in \mathcal{V}$ are near enough we proceed as follows:

(i) we suppose that there are homeomorphisms

$$h^s: \bigcup_{i=1}^{k} W^s(\alpha_i(X)) \to \bigcup_{i=1}^{k} W^s(\alpha_i(Y))$$

and

$$h^u: \bigcup_{j=1}^{\ell} W^u(\beta_j(X)) \cup W^u(\sigma_1(X)) \to \bigcup_{j=1}^{\ell} W^u(\beta_j(Y)) \cup W^u(\sigma_1(Y))$$

compatible with $\mathcal{F}^u(\alpha_i(X))$, $1 \le i \le k$, and $\mathcal{F}^s(\beta_j(X))$, $\mathcal{F}^s(\sigma_1(X))$ respectively.

(ii) We define $h: V(p(X)) \to V(p(Y))$, $V(p(X)) \subset \Sigma$, $V(p(Y)) \subset \Sigma$ in a compatible way with h^s, h^u and such that h sends leaves of $\mathcal{F}^s(p(X),\Sigma)$ into leaves of $\mathcal{F}^s(p(Y),\Sigma)$, leaves of $\mathcal{F}(p(X),\Sigma)$ into leaves of $\mathcal{F}(p(Y),\Sigma)$, and leaves of $\mathcal{F}^{cu}_S(p(X),\Sigma)$ into leaves of $\mathcal{F}^{cu}_S(p(Y),\Sigma)$. So, h sends leaves of $\mathcal{F}^{cu}_{o,S}(p(X),\Sigma)$ into leaves of $\mathcal{F}^{cu}_{o,S}(p(Y),\Sigma)$.

(iii) h defined from $V(p(X)) \subset \Sigma$ onto $V(p(Y)) \subset \Sigma$ induces a homeomorphism from a neighborhood of $W^u(\sigma_o(X)) \cap D_s(q(X),P_X)$ onto a neighborhood of $W^u(\sigma_o(Y)) \cap$ $\cap D_s(q(Y),P_Y)$, compatible with the restriction of $\mathcal{F}^{cu}_{o,S}(q(X),P_X)$ to $W^u(\sigma_o(X)) \cap D_s(q(X),P_X)$. We use the Isotopy Extension Theorem to extend h to $D_s(q(X),P_X)$. We can do this in such way that near the interior boundary of $D_s(q(X),P_X)$, h is defined by $hP_X = P_Y h$. Requiring that h sends leaves of $\mathcal{F}^{cu}_{o,S}(q(X),P_X)$ into leaves of $\mathcal{F}^{cu}_{o,S}(q(Y),P_Y)$ and leaves of $\mathcal{F}^s(q(X),P_X)$ into leaves of $\mathcal{F}^s(q(Y),P_Y)$, we define h on a fundamental neighborhood for $D_s(q(X),P_X)$. Using the conjugacy equation $h^n P_X = P_Y h^n$ we extend h to S, in a compatible way with $\mathcal{F}^s(q(X),P_X)$ and $\mathcal{F}^{cu}_{o,S}(q(X),P_X)$.

(iv) h can be extended to a full neighborhood of $\sigma_o(X)$ by the same reasons as in Theorem 1.

Finally the extension of h to all of M is obtained as in Theorem 1.

<u>3rd Case</u>. $\dim W^u(\sigma_o(Y)) = 2$ and $\dim W^s(\sigma_1(Y)) = 2$.

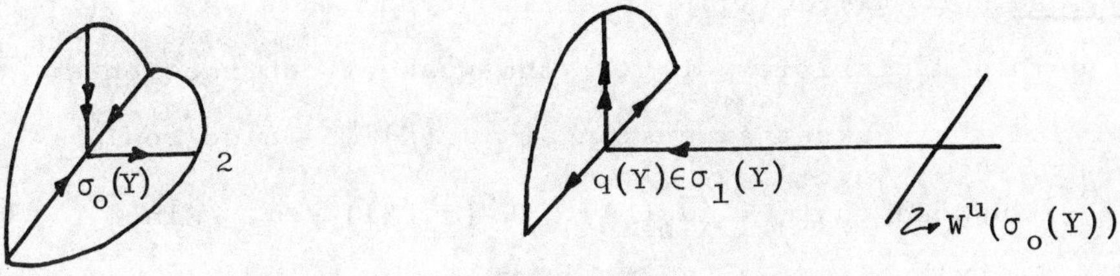

In this case we construct, as in the 2nd Case, a singular two dimensional foliation $\mathcal{F}^{cu}_S(q(Y),P_Y)$ for $W^s(q(Y),P_Y)$ compatible with $\mathcal{F}^u(\alpha_i(Y))$, $1 \le i \le k$, and such that each leaf of $\mathcal{F}^{cu}_S(q(Y),P_Y)$ intersects a leaf of $\mathcal{F}^s(q(Y),P_Y)$ at a unique point. These foliations are used to prove the openess of $\mathfrak{X}^\infty_3(M,\partial M)$ in this case.

To construct the homeomorphism h between X and Y we also proceed as in the 2nd Case. The difference is that now we use just $\mathcal{F}^{cu}_S(q(X),P_X)$ and $\mathcal{F}^s(q(X),P_X)$ to get h defined on neighborhoods of $W^u(\sigma_o(X)) \cap D_s(q(X),P_X)$. We leave the details to the reader. So we have proved the following

<u>Theorem 4</u>. $\mathfrak{X}^\infty_3(M,\partial M$ is open in $\mathfrak{X}^\infty_*(M,\partial M)$ and its elements are

structurally stable.

Thus we proved that a vector field $X \in \mathfrak{X}^{\infty}_{M-S}(M,\partial M)$ having a unique quasi-transversal intersection along the boundary of M is structurally stable.

We indicate now how to proceed when we have more than one quasi-transversal intersection along ∂M. Suppose we have two quasi-transversal intersections $(\sigma_o(X),\sigma_1(X))$ and $(\bar{\sigma}_o(X),\bar{\sigma}_1(X))$ along ∂M. If $\sigma_o(X) \neq \bar{\sigma}_o(X)$ and $\sigma_1(X) \neq \bar{\sigma}_1(X)$ then the construction of the compatible families is independent. If $\sigma_o(X) \neq \bar{\sigma}_o(X)$, $\sigma_1(X) = \bar{\sigma}_1(X)$ and $W^u(\bar{\sigma}_o(X)) \cap W^s(\sigma_o(X)) = \Phi$ then again the construction of the compatible families is independent. Let us analyse the case $W^u(\bar{\sigma}_o(X)) \cap W^s(\sigma_o(X)) \neq \Phi$.

<u>1st Case</u>. $(\sigma_o(X),\sigma_1(X))$ is of 1st type, that is, $\sigma_o(X)$ and $\sigma_1(X)$ are both singularities of X, the weakest contraction at $\sigma_o(X)$ is real and the weakest expansion at $\sigma_1(X)$ is also real.

Suppose that $W^u(\bar{\sigma}_o(X)) \cap W^c(\sigma_o(X)) \neq \Phi$. Since $\dim W^c(\sigma_o(X)) = 1$ and $X/\partial M$ is Morse-Smale we have that $W^u(\bar{\sigma}_o(X))/\partial M$ has dimension three.

(i) $\bar{\sigma}_o$ is a periodic orbit. This implies that the ω-limit set of $W^c(\sigma_1(X))\backslash\sigma_1(X)$ is not a periodic orbit unless this periodic orbit is a sink. Otherwise we would have a quasi-transversal intersection between two periodic orbits and so, X would not be Morse-Smale. Anyway if β is contained in the ω-limit set of $W^u(\sigma_1(X))\backslash\sigma_1(X)$ then $\dim W^s(\beta) = 3$. If β is a singularity, we extend any homeomorphism defined in a neighborhood of $\sigma_1(X)$ to a neighborhood of β just using Liapunov functions. If β is a periodic orbit then β is a sink and it is easy to extend the homeomorphism defined on a neighborhood of $\sigma_1(X)$ to a neighborhood of β.

Thus, if Σ is a transversal section to Y, $Y \in \mathcal{U}$, through $p(Y) \in W^u(\sigma_o(X)) \cap W^s(\sigma_1(X))$, we have to construct a homeomorphism $h: V(p(X)) \to V(p(Y))$, $V(p(X))$, $V(p(Y)) \subset \Sigma$, $p(Y) = W^u(\sigma_o(Y)) \cap \Sigma$ in such way that it can be extended to a neighborhood of $\bar{\sigma}_o(X)$. For this we proceed as follows: Let $\mathcal{F}^u(\bar{\sigma}_o(X))$ be the unstable foliation for $\bar{\sigma}_o(X)$ and suppose that there is a homeomorphism from the space of leaves of $\mathcal{F}^u(\bar{\sigma}_o(X))$ to the space of leaves of $\mathcal{F}^u(\bar{\sigma}_o(Y))$, preserving the leaves. The intersection of $\mathcal{F}^u(\sigma_o(Y))$ with Σ, where Σ is as in Theorem 1, looks like in the picture below.

$$\{y_3 = 0\} \subset W^u(\bar{\sigma}_o(X))$$
$$\{y_2 = 0 = y_3\} = W^u(\sigma_o(X)) \cap \Sigma$$
$$\{y_1 = 0 = y_3\} = W^u(\sigma_o(X)) \cap \Sigma$$

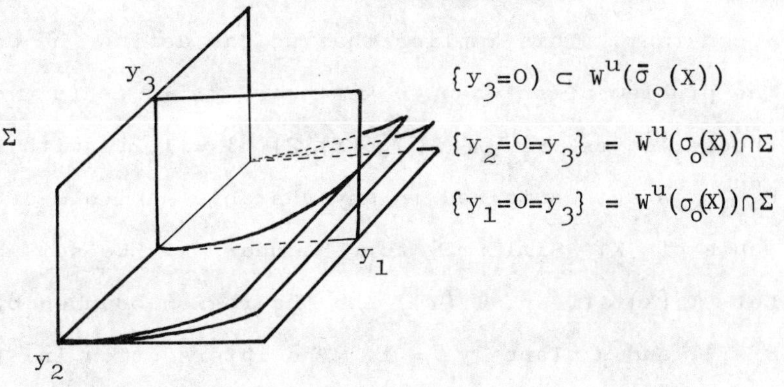

It is clear that now we can not modify this foliation as we did in Theorem 1. This is because the leaves approaching ∂M can be contained in unstable manifolds of critical elements of X. But in the plane $y_2 = 0$ we still have freedom to modify $\mathcal{F}^{cs}(\sigma_1(X)) \cap \Sigma$. And we do this as we did in Theorem 1. Let $\mathcal{F}_S^{cs}(p(X), \Sigma)$ be this foliation of Σ. The picture below shows some of the leaves of this foliation of Σ.

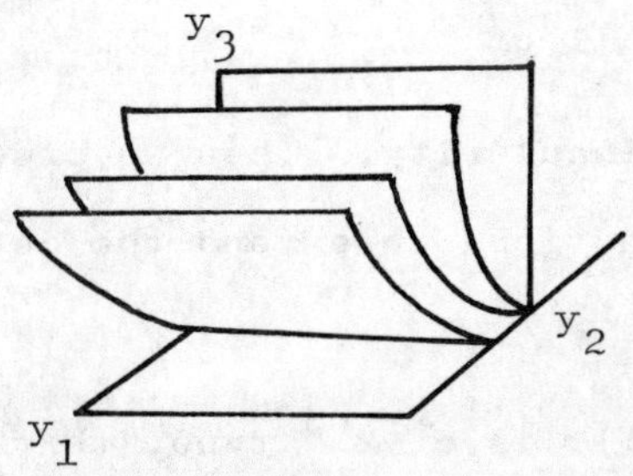

The intersection of $\mathcal{F}_S^{cs}(p(X),\Sigma)$ with y_2 = constant looks like the picture below.

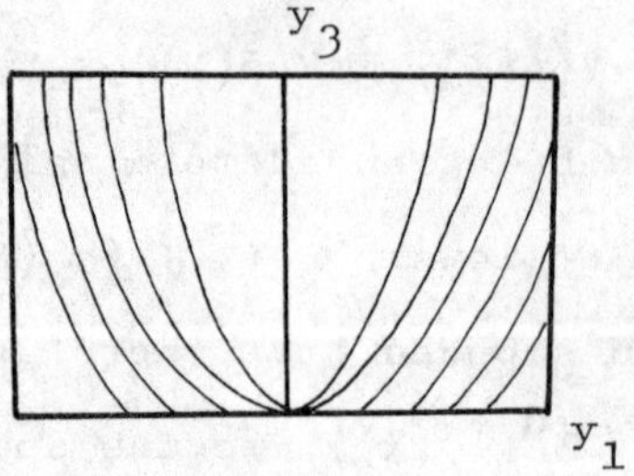

It is clear that the intersections of $\mathcal{F}^{cu}(\sigma_o(X))$ and $\mathcal{F}^{cs}(\sigma_1(X))$ with a plane y_2 = constant gives a coordinate system in y_2 = = constant. This implies that we can define h on y_2 = constant. The problem arise when we get near $y_2 = 0$. To avoid this, we choose leaves $F_{X,i}^{cu} \in \mathcal{F}^{cu}(p(X),\Sigma)$ $i = 1,2$, with the condition that $F_{X,i}^{cu}$ is not contained in any unstable manifold of the critical elements of X. Similarly for Y near X. We set $h(F_{X,i}^{cu}) = F_{Y,i}^{cu}$. Let $T_1(X)$ (resp. $T_1(Y)$) be the region bounded by $F_{X,i}^{cu}$ (resp. $F_{Y,i}^{cu}$) and a plane $y_3 = 1$. The intersection in $T_1(X)$ (resp. $T_1(Y)$) of leaves of $\mathcal{F}^{cu}(p(X),\Sigma)$, $\mathcal{F}^{cs}(p(X),\Sigma)$ and the planes y_3 = constant gives us a coordinate system in $T_1(X)$ (resp. $T_1(Y)$). So, outside $T_1(X)$ we define h using as a coordinate system $\mathcal{F}^{cu}(p(X),\Sigma)$, $\mathcal{F}^{cs}(p(X),\Sigma)$ and the planes y_2 = constant. We can do this in a compatible way with the homeomorphism already defined in the space of leaves of $\mathcal{F}^u(\sigma_o(X))$. Finally we extend h to $h\colon T_1(X) \to T_1(Y)$, using as a coordinate system the one defined above. As in Theorem 1 we extend such h to all of M.

(ii) $\bar{\sigma}_o(X)$ is a singularity. Then we proceed as in Theorem 1 and we use Liapunov functions to extend the definition to a neighborhood of $\bar{\sigma}_o(X)$.

<u>2nd Case</u>. $(\sigma_o(X),\sigma_1(X))$ is of 2nd type, that is, $\sigma_o(X)$ and $\sigma_1(X)$ are both singularities of X, the weakest contraction (resp. weakest

expansion) at $\sigma_0(X)$ (resp. $\sigma_1(X)$) is complex (resp. real). So $\dim W^u(\sigma_1(X)$ is one and if Σ is a cross section to every Y nearby X at $p(X) \in W^u(\sigma_0(X)) \cap W^s(\sigma_1(X))$, our problem is to give a definition of the homeomorphism h on Σ in such way that we can extend it to a neighborhood of $\bar{\sigma}_0(X)$.

(i) $\bar{\sigma}_0(X)$ is a singularity and $\dim W^u(\bar{\sigma}_0(X)) = 3$.
In this case, using a Liapunov function defined on a neighborhood of $\bar{\sigma}_0(X)$ we extend h defined on Σ as in Theorem 2 to a neighborhood of $\bar{\sigma}_0(X)$.

(ii) $\bar{\sigma}_0(X)$ is a singularity, $\dim W^u(\bar{\sigma}_0(X)) = 2$.
Let $U_0 \ni \sigma_0(X)$ be a neighborhood of $\sigma_0(X)$ such that every $Y \in \mathcal{U}$ is C^2-linearizable in U_0. We denote y_1, y_2, y_3, y_4 the linearizing coordinates in U_0 and, in these coordinates we have:

$\sigma_0(Y) = (0,0,0,0)$, $W^c(\sigma_0(Y)) = \{y_3=y_4=0\}$, $W^s(\sigma_0(Y)) = \{y_4=0\}$.

Let $C(Y) \cup D(Y)$ be a fundamental domain for $W^s(\sigma_0(X))$ as in the picture:

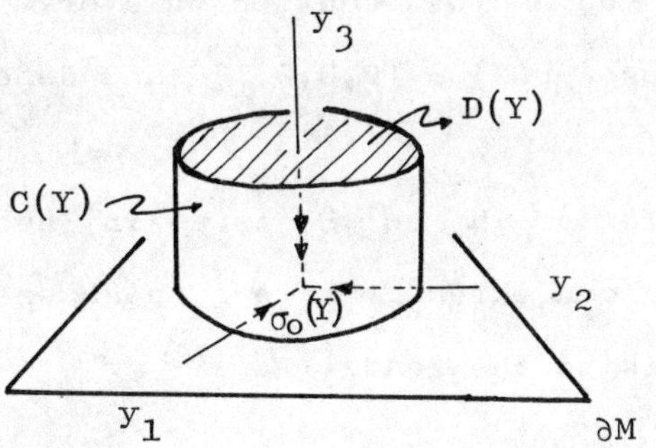

The intersection of $\mathcal{F}^u(\bar{\sigma}_0(Y))$ with $C(Y)$ is a family of curves through $W^u(\bar{\sigma}_0(Y)) \cap W^s(\sigma_0(Y))$, all of them with contact $\frac{\bar{\lambda}_2(Y)}{\bar{\lambda}_1(Y)}$ where $\bar{\lambda}_i(Y)$ are the contracting eigenvalues at $\bar{\sigma}_0(Y)$, $\bar{\lambda}_1(Y) < \bar{\lambda}_2(Y)$.

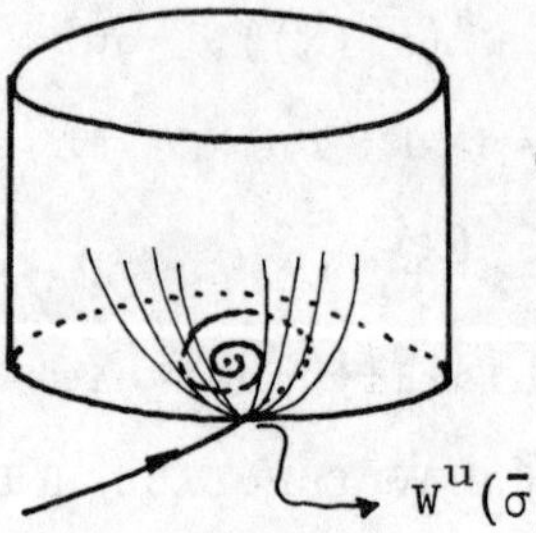

We can modify this family of curves as before and get a singular one dimensional foliation in a neighborhood of $W^u(\bar{\sigma}_o(Y)) \cap W^s(\sigma_o(Y))$ in $C(Y)$.

We extend this foliation to a singular one dimensional foliation $\mathcal{F}^c(C(\sigma_o(Y)))$ compatible with $\mathcal{F}^u(\alpha_i(Y))$, $1 \le i \le k$. Now we construct a singular central unstable foliation for $\sigma_o(Y)$, $\mathcal{F}^{cu}_S(\sigma_o(Y))$, pasting together the unstable leaves of $\mathcal{F}^u(\sigma_o(Y))$ along each leaf of $\mathcal{F}^c(C(\sigma_o(Y)))$.

Let $\Sigma \subset U_o$ be a cross section to every $Y \in \mathcal{V}$ at $p(X) \in$ $\in W^u(\sigma_o(X))$. Suppose $p(X) = (0,0,0,1)$ and denote by $p(Y) =$ $= W^u(\sigma_o(Y)) \cap \Sigma$.

The definition of h on Σ is as in Theorem 2 and it is easy to see that we can extend h to a neighborhood of $\bar{\sigma}_o(X)$. We leave the details to the reader.

(iii) $\bar{\sigma}_o(X)$ is a periodic orbit.

Let $\beta_1(X)$; $W^u(\sigma_1(X)) \cap W^s(\beta_1(X)) \neq \phi$. Then either $\beta_1(X)$ is an attracting periodic orbit or an attracting singularity. So, as before, we can extend any homeomorphism h defined on a neighborhood of $\sigma_1(X)$ to a neighborhood of $\beta_1(X)$. Thus we have to worry just with the extension of h to a neighborhood of $\bar{\sigma}_o(X)$.

<u>1st possibility</u>. $\dim W^u(\bar\sigma_o(X)) = 2$, $W^u(\bar\sigma_o(X)) \subset \partial M$.

In this case we consider a singular central unstable foliation $\mathcal{F}^{cu}(\bar\sigma_o(X))$ compatible with $\mathcal{F}^u(\alpha_i(X))$, $1 \le i \le k$, $\alpha_i(X) < \bar\sigma_o(X)$; and a center unstable foliation $\mathcal{F}^{cu}(\sigma_o(X))$, compatible with $\mathcal{F}^u(\alpha_i(X))$, $1 \le i \le k$, $\alpha_i(X) < \sigma_o(X)$. Let $C(p(X)) \subset \Sigma$ be a cylinder containing $p(X)$ in the interior of $C(p(X)) \cap \partial M$, where Σ is a cross section for every $Y \in \mathcal{V}$ at $p(X) \in W^u(\sigma_o(X))$. If $x \in W^u(\bar\sigma_o(X)) \cap C(p(X))$, the intersection of $\mathcal{F}^{cu}(\bar\sigma_o(X))$ with $C(p(X))$ induces a singular one dimensional foliation in a neighborhood $V(x)$ of x in $C(p(X))$ as below:

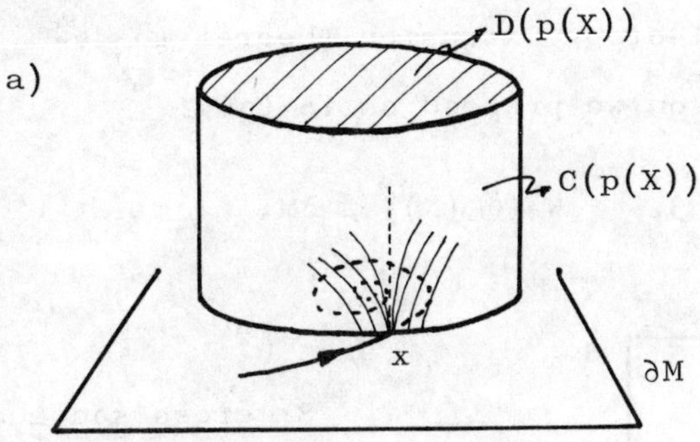

$W^c(\bar\sigma_o(X)) \pitchfork \partial M$

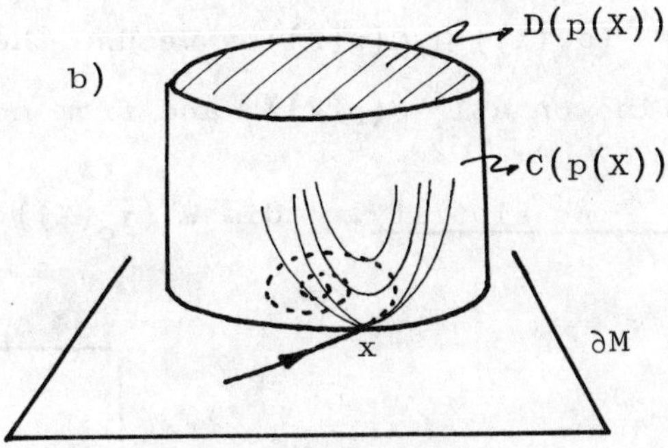

$W^c(\bar\sigma_o(X)) \subset \partial M$

In (a) we define h in a small neighborhood of x preserving the intersections of $\mathcal{F}^{cu}(\bar\sigma_o(X))$ with $C(p(X))$ and the intersection of parallel planes to ∂M in $C(p(X))$ in such way that we can use the Isotopy Extension Theorem to get h defined from $C(p(X))$ onto $C(p(Y))$.

If $h^s\colon \bigcup_{1\le i\le k} W^s(\alpha_i(X)) \to \bigcup_{1\le i\le k} W^s(\alpha_i(Y))$ is the homeomorphism already defined in a compatible way with the unstable foliations $\mathcal{F}^u(\alpha_i(X))$, $1 \le i \le k$, then h^s induces a homeomorphism from an open subset of $D(p(X))$ to an open subset of $D(p(Y))$. We extend this homeomorphism to all of $D(p(X))$ using the Isotopy Extension Theorem. Observe that this can be done in such way that the definition of h on $C(p(X))$ and on $D(p(X))$ concide along $C(p(X)) \cap D(p(X))$.

From this point on, the construction of h is as in Theorem 2. Since such h preserves $\mathfrak{F}^{cu}(\bar{\sigma}_o(X))$ and an one dimensional foliation in Σ transversal to $\mathfrak{F}^{cu}(\bar{\sigma}_o(X))$ it follows that h can be extend to a neighborhood of $\bar{\sigma}_o(X)$. The extension of h to all of M is obtained as usual. We leave the details to the reader (see [1]).

In (b) we define h in a small neighborhood of x preserving the intersection of $\mathfrak{F}^{cu}(\bar{\sigma}_o(X))$ with $C(p(X))$ and an one dimensional foliation of $C(p(X))$ transversal to the leaves of $\mathfrak{F}^{cu}(\bar{\sigma}_o(X)) \cap C(p(X))$. We use the Isotopy Extension Theorem to define on all $C(p(X))$ and from now on we proceed as in (a).

<u>2nd possibility</u>. dim $W^u(\bar{\sigma}_o(X)) = 3$, $W^u(\bar{\sigma}_o(X)) \not\subset \partial M$.

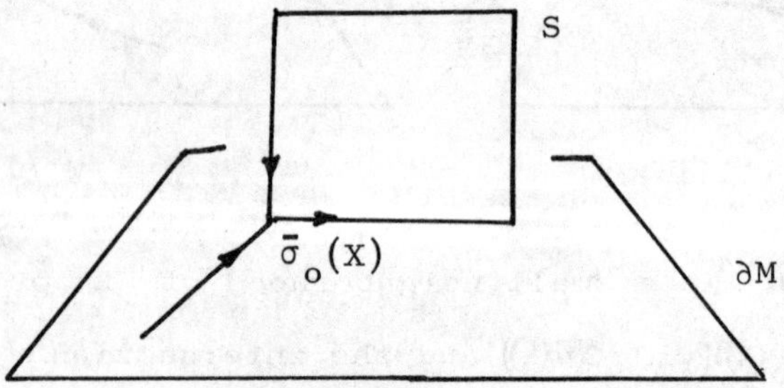

S: cross section at $\bar{q}(X) \in \bar{\sigma}_o(X)$

In this case we consider $\mathfrak{F}^u(\bar{\sigma}_o(X))$ and we proceed as before, but now we require that h preserves the intersections of $\mathfrak{F}^u(\bar{\sigma}_o(X))$ with $C(p(X))$ where $C(p(X))$ is as above.

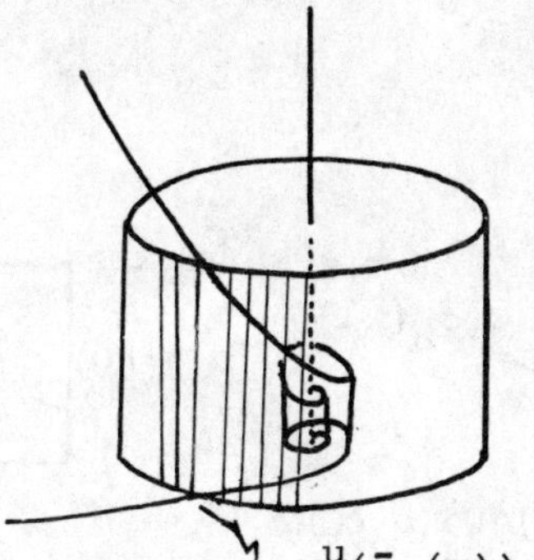

3rd possibility. $\dim W^u(\bar\sigma_o(X)) = 3$, $W^u(\bar\sigma_o(X)) \subset \partial M$.

In this case, if $C(p(X)) \subset \Sigma$ is the cylinder as above we have that there is a segment $I \subset \partial(C(p(X))) \cap \partial M$ such that $I \subset W^s(\bar\sigma_o(X))$, $\partial I = \{x_1, x_2\}$ where x_i is a transversal intersection point of $W^u(\alpha_i(X))$ with $W^s(\sigma_o(X))$. Observe that since $X/\partial M$ is Morse-Smale then $W^u(\alpha_i(X)) \cap W^s(\bar\sigma_o(X)) \neq \Phi$.

We define h on a neighborhood of I in $C(p(X))$ in a compatible way with $\mathfrak{F}^u(\alpha_i(X))$, $1 \le i \le k$. We use the Isotopy Extension Theorem to extend h to $C(p(X))$. This implies that we can extend h to a neighborhood of $\alpha_i(X)$, $i = 1,2$ and to a neighborhood of $\bar\sigma_o(X)$.

To complete the definition of h we proceed as in Theorem 2.

3rd Case. $(\sigma_o(X), \sigma_1(X))$ is of 3rd type, that is $\sigma_o(X)$ is a singularity with real weakest contraction and $\sigma_1(X)$ is a periodic orbit with real weakest expansion. So $\bar\sigma_o(X)$ is a singularity of X.

(i) $\dim W^u(\sigma_o(X)) = 1$ and $\dim W^u(\bar\sigma_o(X)) = 2$. To solve this case we use a construction similar to the one in Theorem 4. We consider a central unstable foliation $\mathfrak{F}^{cu}(\bar\sigma_o(X))$ for $\bar\sigma_o(X)$. The intersection of this foliation with a fundamental domain $C(\sigma_o(X))$ for $W^s(\sigma_o(X))$ induces a singular one dimensional foliation in a neighborhood of $W^u(\bar{\sigma}_o(X)) \cap W^s(\sigma_o(X))$ in $C(\sigma_o(X))$, which looks like the picture below.

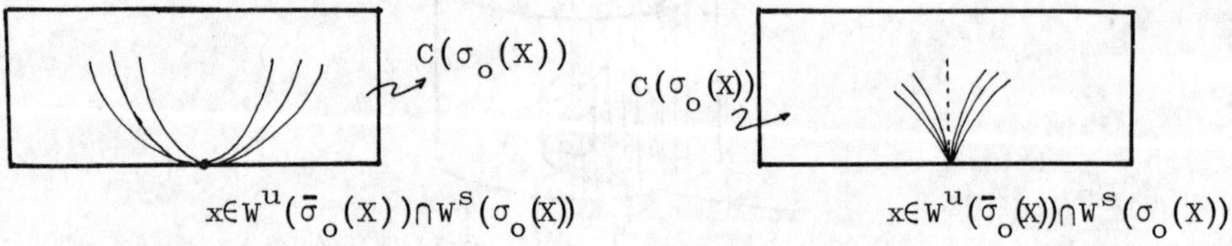

$W^{cu}(\bar{\sigma}_o(X)) \subset \partial M$ $\qquad\qquad$ $W^{cu}(\bar{\sigma}_o(X)) \pitchfork \partial M$

We modify this foliation as before and we get a singular one dimensional foliation of a neighborhood of x in $C(\sigma_o(X))$. We extend this foliation to a singular one dimensional foliation $\mathfrak{F}^c(C(\sigma_o(X))$ compatible with $\mathfrak{F}^u(\alpha_i(X))$, $1 \le i \le k$. From this point on we proceed as in Theorem 4.

(ii) $\dim W^u(\sigma_o(X)) = 1$ and $\dim W^u(\bar{\sigma}_o(X)) = 3$. In this case we extend h defined on Σ (as in Theorem 4) to a neighborhood of $\bar{\sigma}_o(X)$ using a Liapunov function for $\bar{\sigma}_o(X)$.

(iii) $\dim W^u(\sigma_o(X)) = 2$, $\dim W^s(\sigma_1(X)) = 3$.

Then $\bar{\sigma}_o(X)$ must be a repellor in ∂M. So, we use a Liapunov function for $\bar{\sigma}_o(X)$ to extend h to a neighborhood of $\bar{\sigma}_o(X)$.

(iv) $\dim W^u(\sigma_o(X)) = 2$ and $\dim W^s(\sigma_1(X)) = 2$.

Here we proceed as in (iii).

Remark. We proved all the theorems assuming that $W^{cu}(\sigma_o(X))$ is not transversal to $W^s(\sigma_1(X))$ and $W^{cs}(\sigma_1(X))$ is not transversal to $W^u(\sigma_o(X))$. If one of the above hypothesis fails we can either proceed as before to get the results or simplify the construction of the homeomorphism in a neighborhood of the quasi-transversal intersection along ∂M using techniques similar to the ones in ([1,4 and 8]). We leave the details to the reader.

The theorem is proved. ■

REFERENCES

[1] R. Labarca, Stability of Parametrized Families of Vector Fields, this volume.

[2] R. Labarca and M.J. Pacifico, Stability of Singular Horseshoes, Topology 25, 1986.

[3a] R. Mañé, An ergodic closing lemma, Ann. of Math. 166, 1982.

[3b] R. Mañé, The Characterization of Structural Stability, to appear.

[4] M.J. Pacifico, Structural Stability of Vector Fields on 3-Manifolds with Boundary, J. Diff. Equations 54, 1984.

[5] R. Palais, Local triviality of the restriction map for embeddings, Comm. Math. Helvet. 34, 1960.

[6] J. Palis, A Differentiable Invariant of Topological Conjugacies and Moduli of Stability, Astérisque 51, 1978.

[7] J. Palis and S. Smale, Structural Stability Theorems, Proc. Sympos. Pure Math. 14, 1970.

[8] J. Palis and F. Takens, Stability of parametrized families of gradient vector fields, Annals Math., 118, 1983.

R. Labarca

Departamento de Matematica
Universidad de Santiago de Chile
Casilla 5659, Correo 2,
Santiago

Chile

M.J. Pacifico

Instituto de Matemática
Universidade Federal do
Rio de Janeiro
Caixa Postal 68530,
CEP 21910
Rio de Janeiro, RJ, Brasil

and

Instituto de Matemática
Universidade Federal do Rio de Janeiro
Caixa Postal 68530,
CEP 21910

Rio de Janeiro, RJ, Brasil

J LLIBRE

Structure of the set of periods for the Lorenz map

Abstract. We show that the structure of the set of periods for a class of maps of the interval into itself with a single discontinuity (Lorenz maps) is also given by the Sarkovskii's ordering.

1. Introduction

In the study of the geometrical model of the Lorenz attractor an one-dimensional map plays an important role, we refer to such map as the Lorenz map (see [GH], [Sp] and [T]), although it is different from the one-dimensional map presented by Lorenz (see [L]).

In this note we are interested in the structure of the set of periods of the Lorenz map.

Let $I = [-1,1]$. We shall say that a map $f: I \setminus \{0\} \to I$ is a Lorenz map if

L1) f is odd (i.e. $f(-x) = -f(x)$),

L2) f is once continuously differentiable, and $f' > 1$

L3) $f(-1) < 0$,

L4) f has a single discontinuity at 0, and $\lim_{x \to 0^+} f(x) = -1$, $\lim_{x \to 0^-} f(x) = 1$.

We extend a Lorenz map to all the interval I defining

L5) $f(0) = 1$.

In what follows an _extended_ _Lorenz_ _map_ will be a map f from I into itself satisfying the above five properties. In fact, our results will hold for a Lorenz map (extended or not) satisfying L2') instead of L2), where

L2') f is strictly increasing on the intervals $[-1,0)$ and $(0,1]$.

For $x \in I$, we say that x is _periodic_ if there exists a positive integer n such that $f^n(x) = x$. The _period_ of x is the smallest integer satisfying the above relation. Let $P(f)$ denote the set of periods of f.

We consider Sarkovskii's ordering $\gg$ on the set of positive integers, defined as follows

$$3 \gg 5 \gg 7 \gg \dots \gg 2\cdot 3 \gg 2\cdot 5 \gg 2\cdot 7 \gg \dots \gg 4\cdot 3 \gg 4\cdot 5 \gg 4\cdot 7 \gg \dots \gg 4 \gg 2 \gg 1.$$

That is, first the odd integers ≥ 3, then the powers of 2 times the odd integers, and then the powers of 2 backwards.

The structure of the set of periods for a continuous map from the interval into itself is given by the following theorem.

THEOREM (Sarkovskii [Sa,St,BGMY,CE]). _Let_ f _be a continuous map from the interval_ I _into itself_. _If_ $n \in P(f)$ _and_ $n \gg k$, _then_ $k \in P(f)$.

Our main result is to show that the set of periods for a Lorenz map has the same structure that the set of periods for a continuous map from the interval into itself.

THEOREM A. _Let_ f _be a Lorenz map_. _If_ $n \in P(f)$ _and_ $n \gg k$ _with_ $k > 1$, _then_ $k \in P(f)$.

The proof of Theorem A will be given in Section 4. The scheme of the proof is similar to the proof of Sarkovskii's theorem given by Collet and Eckmann, see [CE].

2. Orbits and itineraries

Let f be a Lorenz map. We shall denote f^0 = identity, $f^1 = f$, $f^2 = f\circ f$, $f^k = f\circ k^{-1}$ for $k > 2$. If $x \in I$ then the orbit of x is the set $\{f^k(x) : k=0,1,2,\ldots\}$. If x is a periodic point of period n then its orbit is the set $\{f^k(x) : k=0,1,\ldots,n-1\}$, we refer to such an orbit as a periodic orbit of period n.

LEMMA 1. Let f be a Lorenz map and $x \in I\setminus\{0\}$. Then the following hold.

(a) $f^k(-x) = -f^k(x)$ for every positive integer k.

(b) If x is periodic, then $-x$ is periodic with the same period.

(c) If $f^n(x) = -x$ and $f^k(x) \neq -x$ for $k=1,2,\ldots,n-1$, then x is a periodic point of period $2n$.

Proof: (a) Since f is odd we have $f^k(-x) = f^{k-1}(f(-x)) =$ $= f^{k-1}(-f(x)) = f^{k-2}f(-f(x)) = f^{k-2}(-f^2(x)) = \ldots = -f^k(x)$.

(b) From (a) and since x is a periodic point of period n we have $f^n(-x) = -f^n(x) = -x$ and $f^k(-x) = -f^k(x) \neq -x$ for $k=1,2,\ldots,n-1$.

(c) From (a) it follows $f^{2n}(x) = f^n(-x) = -f^n(x) = x$. It is clear that $f^n(x) \neq x$. For $k = n,n+1,\ldots,2n$ we have $f^k(x) = f^{k-n}(f^n(x)) =$ $= f^{k-n}(-x) \neq x$, and for $k = 1,2,\ldots,n-1$ we have $f^k(x) \neq x$ (if not there exists a positive integer m such that $n < km < 2n$ and $f^{km}(x) = x$, but this is a contradiction).

Q.E.D.

A periodic orbit in the hypotheses of Lemma 1.(c) is called symmetric. Of course, any symmetric orbit has even period. A periodic orbit which is not symmetric is called asymmetric. By Lemma 1.(b) any asymmetric periodic orbit $\{x_1,x_2,\ldots,x_n\}$ has a twin orbit $\{-x_1,-x_2,\ldots,-x_n\}$ (see [B]).

A sequence $\underline{I}$ formed of symbols L, C, R is called <u>admissible</u> if either $\underline{I}$ is an infinite sequence of L's and R's or if $\underline{I}$ is a finite (or empty) sequence of L's and R's, followed by C.

Let f be an extended Lorenz map. We associate with $x \in I$ a finite or infinite sequence of the symbols L, C, R called its <u>itinerary</u> $\underline{I}(x)$, as follows:

I1) $\underline{I}(x)$ is either an infinite sequence of L's and R's, or a finite (or empty) sequence of L's and R's, followed by C. The j-th element of $\underline{I}(x)$ will be denoted $I_j(x)$, for $j = 0,1,\ldots$.

I2) If $f^j(x) \neq 0$, for $j = 0,1,2,\ldots$, then $I_j(x)$ is equal to L or R according to $f^j(x)$ is negative or positive.

I3) If $f^n(x) = 0$, for some n, then letting k denote the smallest such n, we set $I_k(x) = C$ and $I_j(x)$ is equal to L or R according to $f^j(x)$ is negative or positive, for $j = 0,1,\ldots,k-1$.

It is clear that every itinerary is an admissible sequence.

Let $\underline{A} = A_0A_1\ldots$ be an admissible sequence finite or not, then we define $\underline{A}' = \underline{A}_0'\underline{A}_1'\ldots$, where $L' = R$, $R' = L$ and $C' = C$. So, from Lemma 1.(a) it follows that $\underline{I}(-x) = \underline{I}'(x)$.

The <u>shift</u> <u>operator</u> S is defined by $S\underline{A} = A_1A_2\ldots$ if $\underline{A} = A_0A_1A_2\ldots$. We write S^k for the k-fold iterated of S. If $I = C$, then S is not defined. Note that $S(\underline{I}(x)) = \underline{I}(f(x))$ unless $x = 0$.

We shall introduce an <u>ordering</u> $<$ between different admissible sequences $\underline{A}$ and $\underline{B}$. First, we shall say $L < C < R$. Let k be the first index for which $A_k \neq B_k$. We say that $\underline{A} < \underline{B}$ if $A_k < B_k$. We shall use the notation $\underline{A} \leq \underline{B}$, $\underline{A} \geq \underline{B}$ and $\underline{A} > \underline{B}$ in the standard way. It is easy to see that $<$ is a complete ordering.

LEMMA 2. *Let* f *be an extended Lorenz map and* suppose that $x, y \in I$. *Then the following hold*.

(a) *If* $\underline{I}(x) < \underline{I}(y)$ *then* $x < y$.

(b) *If* $x < y$ *then* $\underline{I}(x) \leq \underline{I}(y)$.

Proof: By induction on the smallest k such that $I_k(x) < I_k(y)$. It is clear that the statement is true for $k = 0$. Suppose that $k > 0$ and that we know the result for $k-1$. Since $I_k(x) < I_k(y)$ and $k > 0$ we have $\underline{I}(f(x)) = S\underline{I}(x) < S\underline{I}(y) = \underline{I}(f(y))$. By the induction hypothesis $f(x) < f(y)$. Since $I_0(x) = I_0(y)$ and $k > 0$ we have either $x, y \in [-1, 0)$, or $x, y \in (0, 1]$. So, by L2'), $f(x) <$ $< f(y)$ implies $x < y$.

(b) Since $<$ is a complete ordering, if $\underline{I}(x) \not\leq \underline{I}(y)$ it follows that $\underline{I}(x) > \underline{I}(y)$. By (a) $x > y$, which is a contradiction. So, $\underline{I}(x) \leq \underline{I}(y)$.

Q.E.D.

From Lemma 2 it follows that the order of the real line is reflected in the order of the itineraries.

3. Maximal sequences and min-max

Let $\underline{A}$ and $\underline{B}$ be two sequences formed of symbols L, R. We shall write $\underline{AB}$ for the concatenation of $\underline{A}$ and $\underline{B}$, and $\underline{A}^n = \underline{A} \dots \underline{A}$ (n times) and $\underline{A}^\infty = \underline{AA} \dots$ indefinitely.

We shall say that an admissible sequence $\underline{B}$ is *periodic of period* n if there exists $\underline{A} = A_0 \dots A_{n-1}$ a finite sequence of symbols L, R such that $\underline{B} = \underline{A}^\infty$ and $S^k(\underline{B}) \neq \underline{B}$ for $k = 1, 2, \dots, n-1$.

An admissible sequence $\underline{B}$ is called *symmetric periodic of period* $2n$ if there exists $\underline{A} = A_0 \dots A_{n-1}$ a finite sequence of symbols L, R such that $\underline{B} = (\underline{AA}')^\infty$ and $S^k(\underline{B}) \neq \underline{B}'$ for $k = 1, 2, \dots, n-1$ (see [B]).

Let f be an extended Lorenz map and x be a periodic point of period n. Now, consider the union S of the periodic orbits associated to x and $-x$, perhaps they are the same, see Lemma 1. Each of the points $y \in S$ has a periodic itinerary. In some sense they are equivalent because they represent the same symmetric periodic orbit of the union of an asymmetric periodic orbit with its twin. We want to single out the maximal itinerary $\underline{I}(y)$ where $y \in S$. By Lemma 2, it is $\underline{I}(y)$ with $y = \max S$. Then all the shifts $S^k(\underline{I}(y))$ and $S^k(\underline{I}'(y))$ are less than or equal to $\underline{I}(y)$. This motivates the following definition of maximal sequence.

An admissible sequence $\underline{A}$ is called <u>maximal</u> if $S^k(A) \leq A$ and $S^k(\underline{A}') \leq \underline{A}$ for $k = 1,2,\ldots,|A|-1$, when $\underline{A}$ is finite and for $k = 1,2,\ldots$ when $\underline{A}$ is infinite. Here, $|\underline{A}|$ denotes the cardinality of $\underline{A}$.

Now, we consider the periodic sequences of L's and R's and we ask for their "first" appearance with respect to the ordering $<$. More precisely, consider all periodic sequences of L's and R's which have period $n \geq 2$. This is, for a fixed n, a finite set of sequences (at most 2^n). Let $\underline{A}^{\infty}_{1,n} < \underline{A}^{\infty}_{2,n} < \ldots < \underline{A}^{\infty}_{i(n),n}$ be an enumeration of the maximal sequence in this set. We call $\underline{P}_n = \underline{A}_{1,n}$ the <u>min-max</u> of period n.

LEMMA 3. <u>The only maximal sequences not starting with</u> RR... <u>are</u> C, $(RL)^{\infty}$ <u>and</u> RC.

<u>Proof</u>: It is clear that the unique maximal sequences not starting with R... is C. A maximal sequence starting with R... but not with RR... must start with RL... or RC. But a maximal sequence $\underline{A}$ starting with RL... can not have two consecutive L's or R's (if not there exists a positive integer k such that $S^k(\underline{A}')$ or

$S^k(\underline{A})$ starts with RR... and $\underline{A}$ would not be maximal). Therefore $\underline{A} = (RL)^\infty$.

Q.E.D.

Now, we shall define the R* product. This product will allow us to describe the min-max sequences $\underline{P}_n$. We define $R*\underline{A}$ as follows:

P1) If $\underline{A}$ is an infinite or finite sequence of L's and R's, then $R*\underline{A} = B_0A_0B_1A_1\ldots$ where $B_0 = R$, $B_i = A_i$ if $A_{i-1} \neq A_i$, and $B_i = A_i'$ if $A_{i-1} = A_i$, for $i = 1,2,\ldots$ when $\underline{A}$ is infinite and for $i = 1,2,\ldots,|\underline{A}|-1$ when $\underline{A}$ is finite.

P2) If $\underline{A} = C$ then $R*A = RC$. If $\underline{A} = A_0A_1\ldots A_{n-1}C$ is a finite admissible sequence, then $R*A = B_0A_0B_1A_1\ldots B_{n-1}A_{n-1}B_nC$, where $B_0 = R$, $B_i = A_i$ if $A_{i-1} \neq A_i$, and $B_i = A_i'$ if $A_{i-1} = A_i$, for $i = 1,2,\ldots,n-1$; and $B_n = A_{n-1}'$.

THEOREM 4. (a) $\underline{A}$ is maximal if and only if $R*\underline{A}$ is maximal.
(b) If $\underline{A} < \underline{D}$ then $R*\underline{A} < R*\underline{D}$.

Proof: (a) Assume that $\underline{A}$ is maximal. If $\underline{A}$ does not start with RR... then $R*\underline{A}$ is maximal, because the sequences $R*L^\infty = (RL)^\infty$, $R*C = RC$, $R*(RL)^\infty = (RRLL)^\infty$ and $R*RC = RRLC$ are maximal (see Lemma 3).

Suppose $\underline{A} = RR\ldots$. Then $R*\underline{A} = RRLR\ldots$. Consider $S^{2m+1}(R*\underline{A}) = A_mB_{m+1}A_{m+1}\ldots$ for $m = 0,1,2,\ldots$. The only possibilities for $A_mB_{m+1}A_{m+1}$ are LRL, LRR, RLL and RLR. So $S^{2m+1}(R*\underline{A}) < R*\underline{A}$.

Now, consider $S^{2m}(R*A)$ for $m = 1,2,\ldots$, and suppose that $S^{2m}(R*\underline{A}) \neq R*\underline{A}$. Since $B_0A_0B_1A_1 = RRLR$ we have either $B_mA_mB_{m+1}A_{m+1}C < B_0A_0B_1A_1C$, or $B_mA_m\ldots B_{m+j-1}A_{m+j-1} = B_0A_0\ldots B_{j-1}A_{j-1}$ with $j = j(m) \geq 2$. In the first case we are done. Assume that we are in the second case. If $A_{m+j} = C$ then $A_j = R$ (because $\underline{A}$ is maximal) and we have $A_{m+j-1}A'_{m+j-1}C = LRC < A_{j-1}B_jA_j\ldots = LRR\ldots$ or $A_{m+j-1}A'_{m+j-1}C = RLC < A_{j-1}B_jA_j\ldots = RLR\ldots$, according to $A_{m+j-1} = A_{j-1}$ is equal to L or R. So $S^{2m}(R*\underline{A}) < R*\underline{A}$. If $A_{m+j} \neq C$ then four cases are possible, either $A_{m+j-1} = A_{m+j}$ and $A_{j-1} = A_j$, or $A_{m+j-1} = A_{m+j}$ and $A_{j-1} \neq A_j$, or $A_{m+j-1} \neq A_{m+j}$ and $A_{j-1} = A_j$, or $A_{m+j-1} \neq A_{m+j}$ and $A_{j-1} \neq A_j$. In any case it follows that $B_{m+j} = B_j$. So we can assume $B_mA_m\ldots B_{m+j-1}A_{m+j-1}B_{m+j} = B_0A_0\ldots B_{j-1}A_{j-1}B_j$ and $A_{m+j} \neq A_j$. Therefore, since $\underline{A}$ is maximal we have $A_{m+j} < A_j$. Hence $S^{2m}(R*\underline{A}) < R*A$. In short, $R*\underline{A}$ is maximal.

Assume that $\underline{A}$ is not maximal. If $\underline{A} = L^iR\ldots$ with $i > 0$, then $R*\underline{A} = (RL)^iRR\ldots$ is not maximal. Suppose $\underline{A} = R^iL\ldots$ with $i > 0$. Let m be the smallest positive integer such that $S^m\underline{A} > \underline{A}$. Then $S^m\underline{A} = R^jL\ldots$ with $j > i$ and $S^{m-1}\underline{A} = LR^jL\ldots$. So $S^{2m}(R*\underline{A}) = RR(LR)^{j-1}LL\ldots > R*\underline{A} = RR(LR)^{i-1}LL\ldots$ and $R*\underline{A}$ is not maximal. Hence (a) is proved.

(b) Since $\underline{A} < \underline{D}$ we have either $A_0 < D_0$ or $A_0A_1\ldots A_{i-1} = D_0D_1\ldots D_{i-1}$ and $A_i < D_i$ with $i > 0$. In the first case, $R*\underline{A} = RA_0\ldots < R*\underline{D} = RD_0\ldots$. Assume that we are in the second case. Set $R*\underline{A} = B_0A_0\ldots B_{i-1}A_{i-1}B_iA_i\ldots$ and $R*\underline{D} = E_0D_0\ldots E_{i-1}D_{i-1}E_iD_i\ldots$. We have three possibiliites $A_i = L < D_i = R$, $A_i = L < D_i = C$, and $A_i = C < D_i = R$. We consider the first case, the other two cases are easy variants. Assume $A_{i-1} = D_{i-1} = L$ then $A_{i-1}B_iA_i = LRL$ and $D_{i-1}E_iD_i = LRR$. Therefore, $R*\underline{A} < R*\underline{D}$. Now, if $A_{i-1} = D_{i-1} = R$ we have $A_{i-1}B_iA_i = RLL$ and $D_{i-1}E_iD_i = RLR$. Again, $R*\underline{A} < R*\underline{D}$.

Q.E.D.

We shall denote $R^{*0}*\underline{A} = \underline{A}$, $R^{*1}*\underline{A} = R*\underline{A}$, $R^{*2}*\underline{A} = R*(R*\underline{A})$, $R^{*k}*\underline{A} = R*(R^{*(k-1)}*\underline{A})$ for $k > 2$.

For $n = 2,3,\ldots$ we define $\underline{Q}_n$ as follows:

$$\underline{Q}_n^\infty = R^{*i} * (R(RL)^{(k-1)/2})^\infty = (R^{*i} * R(RL)^{(k-1)/2})^\infty,$$

if $n = 2^i k$ with $i \geq 0$ and $k \geq 3$ odd; and

$$\underline{Q}_n^\infty = R^{*i} * L^\infty = (R^{*i} * L)^\infty,$$

if $n = 2^i$ with $i > 0$.

LEMMA 5. <u>The sequences</u> $\underline{Q}_n^\infty \to (RL)^\infty$ <u>are maximal</u>.

<u>Proof</u>: Since $(R(RL)^{(k-1)/2})^\infty$ and L^∞ are maximal, by Theorem 4.(a), the lemma follows.

Q.E.D.

LEMMA 6. <u>Suppose</u> $n \geq 3$. <u>Then the following hold.</u>

(a) $\underline{P}_n$ <u>starts with</u> RRL...

(b) $\underline{P}_n^\infty$ <u>does not have three consecutive</u> L's or R's.

(c) $\underline{P}_n$ <u>ends with</u> L.

<u>Proof</u>: (a) By Lemma 5 we have $\underline{P}_n^\infty \leq \underline{Q}_n^\infty$. Since $\underline{Q}_n = RRL\ldots$ for $n \geq 3$, from Lemma 3, it follows that $\underline{P}_n = RRL\ldots$.

(b) From (a), since $\underline{P}_n^\infty$ is maximal, it follows that $\underline{P}_n$ does not have three consecutive L's or R's.

(c) By (a) and (b) we have that $\underline{P}_n$ ends with L.

Q.E.D.

THEOREM 7. <u>If</u> $n \geq 3$ <u>is odd, then</u> $\underline{P}_n = \underline{Q}_n = R(RL)^{(n-1)/2}$.

<u>Proof</u>: Suppose $\underline{P}_n \neq \underline{Q}_n$. Since $\underline{P}_n$ is min-max we have $\underline{P}_n^\infty < \underline{Q}_n^\infty$. So, from Lemma 6.(a) it follows $P_0 P_1 \ldots P_{2k-1} = Q_0 Q_1 \ldots Q_{2k-1}$ and $P_{2k} = L < Q_{2k} = R$ with $k \geq 2$. Then, from Lemma 6.(b) we have $\underline{P}_n = R(RL)^{k-1}(LR)\ldots$. Now, we consider three cases.

Case 1: $\underline{P}_n = \ldots(LR)(RL)\ldots$. It is not possible that $\underline{P}_n^\infty = \ldots(LR)(RL)^j(RR)\ldots$ with $j \geq 1$, because $R(RL)^j(RR)\ldots > \underline{P}_n^\infty$.

So, $\underline{P}_n = \ldots(LR)(RL)^j(LR)\ldots$ with $j \geq 1$, and this case is reduced to Case 2.

Case 2: $\underline{P}_n = \ldots(LR)\ldots$. By Lemma 6.(b) we have $\underline{P}_n = \ldots(LR)^j(RL)$ or $\underline{P}_n = \ldots(LR)^j(LL)$ with $j \geq 1$. If the first is true, using Case 1 we have that $\underline{P}_n = \ldots(LR)^j(RL)^i(LR)\ldots$ and we are, again, in Case 2. In short, since $\underline{P}_n$ does not end with (LR) (see Lemma 6.(c)), we have that $\underline{P}_n = \ldots(LR)^j(LL)$ with $j \geq 1$. Then this case is reduced to Case 3.

Case 3: $\underline{P}_n = \ldots(LR)(LL)\ldots$. In fact, from above we must have $\underline{P}_n = \ldots(RL)(LR)^j(LL)\ldots$ with $j \geq 1$. Therefore, $\underline{P}'_n = \ldots R(RL)^j RR\ldots$ is a contradiction, because $R(RL)^j RR \ldots > \underline{Q}_n^\infty > \underline{P}_n^\infty$.

Hence, $\underline{P}_n = \underline{Q}_n$.

Q.E.D.

LEMMA 8. <u>Suppose</u> $n \geq 6$ <u>even</u>. <u>Then the following hold.</u>

(a) $\underline{P}_n = RRLRLL\ldots$.

(b) <u>The sequences</u> LLRLL, RRLRR, LLRLRL <u>and</u> RRLRLR <u>do not appear into</u> $\underline{P}_n$.

(c) $\underline{P}_2 = \underline{Q}_2 = RL$ <u>and</u> $\underline{P}_4 = \underline{Q}_4 = RRLL$.

<u>Proof</u>: By Lemma 6.(a) we have that $\underline{P}_n$ starts with RRL... . Assume $\underline{P}_n = RRLL\ldots$. $\underline{P}_n$ is different from $(RRLL)^{n/4}$ or $(RRLL)^{(n-2)/4}RR$ because in the firs case $\underline{P}_n$ would have period 4 and in the second one would end with R (see Lemma 6.(c)). So, it must appear at least one of the sequences LLRL or RRLR into $\underline{P}_n$. This is a contradiction because $(LLRL)'\ldots = RRLR\ldots > \underline{P}_n^\infty = RRLL\ldots$. Hence $\underline{P}_n = RRLR\ldots$.

Suppose $k \geq 3$ odd and $i \geq 1$ such that either $2^i k \geq 6$ or $2^i \geq 6$. By Lemma 5, $\underline{Q}_{2^{i-1}k}$ and $\underline{Q}_{2^{i-1}}$ are maximal. Therefore, since either $2^{i-1}k \geq 3$ or $2^{i-1} \geq 3$, by Lemma 6.(a) we have

$\underline{Q}_2 i\text{-}1_k$ and $\underline{Q}_2 i\text{-}1$ starts with RRL... . So $\underline{Q}_2 i_k = R * \underline{Q}_2 i\text{-}1_k$ and $\underline{Q}_2 i = R * \underline{Q}_2 i\text{-}1$ start with RRLRLL... . Then, from $\underline{P}_n^\infty = RRLR... \leq$ $\leq \underline{Q}_n^\infty = RRLRLL...$ it follows (a).

Since (LLRLL)′ ... = RRLRR... > $\underline{P}_n^\infty$ = RRLRLL... and (LLRLRL)′... = RRLRLR... > $\underline{P}_n^\infty$ = RRLRLL... , it follows (b).

To prove $\underline{P}_2 = RL$ is immediate. From (a) and (c) of Lemma 6, it follows that $\underline{P}_4 = RRLL$.

Q.E.D.

THEOREM 9. <u>For</u> $n \geq 2$ <u>even there exists a unique</u> $\underline{A}$ <u>such that</u> $\underline{P}_n = R*\underline{A}$ <u>and</u> $\underline{A}^\infty$ <u>is maximal or</u> $A = L$.

<u>Proof</u>: By Lemma 8.(c), $\underline{P}_2 = RL = R*L$ and $\underline{P}_4 = RRLL = R*RL$, so the theorem follows from $n = 2$ and $n = 4$. Assume $n \geq 6$. We shall prove that $\underline{P}_n = R*A = RA_0B_1A_1...B_iA_i...B_{\frac{n}{2}-1}A_{\frac{n}{2}-1}$ with

$$B_j = A_j \text{ if } A_{j-1} \neq A_j, \text{ and } B_j = A'_j \text{ if } A_{j-1} = A_j, \quad (1)$$

for $j = 1,2,...,\frac{n}{2}-1$.

By Lemma 8.(a), $\underline{P}_n$ = RRLRLL... . Therefore, $\underline{P}_n$ satisfies (1) for $j = 1,2$. Suppose that (1) is true for $j = 1,2,...,i-1$ with $i < \frac{n}{2} - 1$. Now, assume that

$$B_i = A'_i \text{ if } A_{i-1} \neq A_i, \text{ or } B_i = A_i \text{ if } A_{i-1} = A_i. \quad (2)$$

Then $A_{i-1}B_iA_i$ is one of the following sequences LLL, LLR, RRL and RRR. By Lemma 6.(b) the first and the last are no possible. Suppose $A_{i-1}B_iA_i$ = LLR. By Lemma 6.(b) $B_{i-1} = R$. Then (1) implies $A_{i-2} = L$. By Lemma 8.(b), the sequence LLRLL can not appear into $\underline{P}_n$, so $B_{i-2} = R$, and from (1) $A_{i-3} = L$. Again, by Lemma 8.(b) the sequence LLRLRL can not appear into $\underline{P}_n$, so $B_{i-3} = R$. Repeating these arguments we obtain $\underline{P}_n = (RL)^j LR...$ for some $j > 0$, and this is a contradiction with Lemma 8.(a).

Now, assume $A_{i-1}B_iA_i = RRL$. In a similar way we would obtain $\underline{P}_n = (LR)^jRL\ldots$ for some $j > 0$, and again this would be a contradiction with Lemma 8.(a). So, (2) is not true and there is $\underline{A}$ such that $\underline{P}_n = R*\underline{A}$. From Theorem 5.(a) it follows that $\underline{A}^\infty$ is maximal.

Q.E.D.

COROLLARY 10. (a) $\underline{P}_{2^i} = \underline{Q}_{2^i} = R^{*i}*L$ *for* $i = 1,2,\ldots$.
(b) $\underline{P}_{2^ik} = \underline{Q}_{2^ik} = R^{*i}*R(RL)^{(k-1)/2}$ *with* $i \geq 0$ *and* $k \geq 3$ *odd*.

Proof: (a) By Theorem 9, $\underline{P}_{2^i} = R^{*i}*A$ with $A \in \{L,R\}$. Then, by Theorem 4.(b) we have $A = L$.
(b) By Theorem 9, $\underline{P}_{2^ik} = R^{*i}*\underline{A}$. Then, from Theorem 4.(b), it follows that $\underline{A} = \underline{P}_k$. Hence, by Theorem 7, (b) follows.

Q.E.D.

THEOREM 11. *Let* n *and* k *be two positive integers with* $k > 1$. *If* $n > k$ *then* $\underline{P}_n^\infty > \underline{P}_k^\infty$.

Proof: From Theorem 7, for the odd integers ≥ 3 it is clear that $\underline{P}_3^\infty > \underline{P}_5^\infty > \underline{P}_7^\infty > \ldots$. Then, by Theorem 4.(b) we have

$$\underline{P}_{2^i3}^\infty > \underline{P}_{2^i5}^\infty > \underline{P}_{2^i}^\infty > \ldots , \tag{3}$$

for $i = 1,2,\ldots$. If $k \geq 3$ is odd and $n \geq 6$ is even, then (by Lemma 8.(a) and Theorem 7) we have $\underline{P}_k^\infty > \underline{P}_n^\infty$. Therefore

$$\underline{P}_3^\infty > \underline{P}_5^\infty > \underline{P}_7^\infty > \ldots > \underline{P}_{2\cdot3}^\infty > \underline{P}_{2\cdot5}^\infty > \underline{P}_{2\cdot7}^\infty > \ldots . \tag{4}$$

From Theorem 4.(b), (3) and (4) it follows that

$$\underline{P}_{2\cdot3}^\infty > \underline{P}_{2\cdot5}^\infty > \underline{P}_{2\cdot7}^\infty > \ldots > \underline{P}_{2^23}^\infty > \underline{P}_{2^25}^\infty > \underline{P}_{2^27}^\infty > \ldots ,$$

and so on.

Let i and j be positive integers and $k \geq 3$ odd. If $j \leq i$ then (by Theorem 4.(b)) $\underline{P}_{2^i}^\infty = R^{*i}*L^\infty < R^{*i}*\underline{P}_k^\infty =$

$= \underline{P}_{2}^{\infty} i_k \leq \underline{P}_{2}^{\infty} j_k$. When $j > i$ we have (by Theorem 4.(b)) $\underline{P}_{2}^{\infty} i = R^{*i} * L^{\infty} < R^{*j} * L^{\infty} < R^{*j} * \underline{P}_{k}^{\infty} = \underline{P}_{2}^{\infty} j_k$. This completes the proof of the theorem in all cases.

Q.E.D.

4. Proof of Theorem A

LEMMA 12. *Let* f *be an extended Lorenz map and assume* $\underline{I}(1) = \underline{D}C$. *Given* $s \geq 1$, *for* $x < 0$ *sufficiently near to* 0, $\underline{I}(f(x)) = (\underline{D}L)^s \ldots$.

Proof: Let $p = |\underline{D}| + 1$. Then $f^p(0) = 0$. There is an interval $(w_0, 1]$ on which f^{p-1} is strictly monotone. Therefore f^p maps some interval $(w,1]$ into $(w_0,1]$ provided w is sufficiently near to 1. Thus the itinerary of $x \in (w,1)$ starts $\underline{D}L\ldots$. Iterating s times in this fashion we can find a $w < 1$ such that $f^{kp}(w,1] \subset (w_0,1]$ for $k = 0,1,\ldots,s$. Thus the itinerary of $y \in (w,1)$ starts with $(\underline{D}L)^s \ldots$.

Q.E.D.

LEMMA 13. *Let* f *be an extended Lorenz map. If* $\underline{I}(x)$ *is a periodic itinerary with period* p, *then* $f^j(x)$ *converges towards a periodic orbit of period* p *of* f *as* $j \to \infty$. *One of the points in the periodic orbit has* $\underline{I}(x)$ *as itinerary*.

Proof: We define sets $J_0, J_1, \ldots, J_{p-1}$ as follows. For every r, $r = 0,1,\ldots,p-1$, J_r is the smallest closed subinterval of I containing all the points $f^{r+pn}(x)$, when $n \geq 0$. Then the interior of J_r does not contain 0 and hence $f|_{J_r}$ is a homeomorphism into $J_{r+1 \pmod p}$. Thus f^p is a homeomorphism of J_0 into itself which preserves orientation. Then if $x \leq f^p(x)$ it follows that $x \leq f^p(x) \leq f^{2p}(x) \leq \ldots$ and hence this subsequence converges to a limit in J_0. This limit belongs to a periodic orbit whose period q divides p. If $x \geq f^p(x)$ the inequalities $x \geq f^p(x) \geq f^{2p}(x) \geq \ldots$ will lead to the same conclusion.

If $f^j(x)$ converges to a periodic orbit of period q which does not contain 0, then the period p of the periodic itinerary $\underline{I}(x)$ must be a divisor of q. So, $q = p$.

Suppose $f^j(x)$ converges to a periodic orbit of period q which contain 0. Since $\underline{I}(x)$ is periodic, if $\underline{I}(1) = \underline{D}C$, by Lemma 12, we have $\underline{I}(x) = (\underline{D}L)^\infty$ because $f^{r+qn}(x) \to 0^-$ when $n \to \infty$ for some $r \in \{0,1,\ldots,p-1\}$ (recall that $f(0) = 1$). Hence $q = p$.

Q.E.D.

We define the <u>extended itinerary</u> $\underline{I}_E(x)$ as follows:

EI1) $\underline{I}_E(x) = \underline{I}(x)$ if $\underline{I}(x)$ is infinite,

EI2) $\underline{I}_E(x) = \underline{I}(x)\underline{I}(1)$ if $\underline{I}(x)$ is finite and $\underline{I}(1)$ is infinite.

EI3) $\underline{I}_E(x) = \underline{I}(x)(\underline{I}(1))^\infty$ if both $\underline{I}(x)$ and $\underline{I}(1)$ are finite.

If $\underline{A}$ is admissible and

$$S^k\underline{A} < \underline{I}(1) \quad \text{if} \quad \underline{I}(1) \text{ is infinite,}$$

$$S^k\underline{A} < (\underline{D}L)^\infty \quad \text{if} \quad \underline{I}(1) = \underline{D}C,$$

for $k = 0,1,2,\ldots$ when $\underline{A}$ is infinite or $k = 0,1,2,\ldots,$ $|\underline{A}| - 1$ when $\underline{A}$ is finite, then we shall say that $\underline{A}$ is <u>dominated</u> by $\underline{I}(1)$, and we shall use the notation $\underline{A} \leftarrow \underline{I}(1)$.

LEMMA 14. <u>Let f be an extended Lorenz map. Assume $\underline{A}$ is an admissible sequence satisfying $\underline{A} \leftarrow \underline{I}(1)$. Then the sets $L_{\underline{A}} = \{x \in I : \underline{I}(x) < \underline{A}\}$ and $R_{\underline{A}} = \{x \in I : \underline{I}(x) > \underline{A}\}$ are open.</u>

<u>Proof</u>: We shall prove that $R_{\underline{A}}$ is open. The case $L_{\underline{A}}$ is similar. Assume $y \in R_{\underline{A}}$. Let n be the first index for which $I_n(y) > A_n$.

1. Suppose $\underline{I}(y)$ is infinite. Then $I_n(y) = R$. By continuity of f^n in a neighborhood of y we can preserve the equalities $I_j(x) = A_j$ when $j < n$ and $I_n(x) = R$ for x sufficiently near to y.

So $y \in \mathrm{Int}(R_{\underline{A}})$.

2. If $\underline{I}(y)$ is finite and $I_n(y) \neq C$, then we can apply the above argument to obtain $y \in \mathrm{Int}(R_{\underline{A}})$.

3. Suppose $\underline{I}(y)$ is finite, $I_n(y) = C$ and $\underline{I}(1)$ infinite. Then we have $\underline{I}(y) = \underline{B}C$ and $\underline{A} = \underline{B}L\hat{\underline{A}}$. The extended itinerary of y is $\underline{I}_E(y) = \underline{B}C\underline{I}(1)$. Since $\underline{A} \leftarrow \underline{I}(1)$, we have $\hat{\underline{A}} < \underline{I}(1)$. Let m denote the first index for which $\hat{A}_m < I_m(1)$. Since $\underline{I}(1)$ is infinite, $I_m(1) = R$. Therefore, there is an $x_m < 1$ such that for $z \in (x_m, 1]$ we have $I_j(z) = \hat{A}_j$ for $j < m$ and $I_m(z) = R$. Since 1 is the image of 0, there is some interval $(w, o]$ mapped into $(x_m, 1]$. So, for y' sufficiently near to y, we have

$$\underline{I}(y') = \underline{B}L\hat{A}_0 \ldots \hat{A}_{m-1}R \ldots \quad \text{if} \quad y' < y,$$

$$\underline{I}(y') = \underline{B}R \ldots \quad \text{if} \quad y' > y.$$

Thus $\underline{I}(y') > \underline{A}$. Hence $y \in \mathrm{int}(R_{\underline{A}})$.

4. Suppose $\underline{I}(y)$ is finite, $I_n(y) = C$ and $\underline{I}(1)$ is finite. Then we have $\underline{I}(1) = \underline{B}C$, $\underline{A} = \underline{B}L\hat{\underline{A}}$ and $\underline{I}_E(y) = \underline{B}C(\underline{I}(1))^\infty$. By Lemma 12, for y' sufficiently near to y we have

$$\underline{I}(y') = \underline{B}L(\underline{D}L)^s\underline{X} \quad \text{if} \quad y' < y,$$

$$\underline{I}(y') = \underline{B}R \ldots \quad \text{if} \quad y' > y.$$

Since $\underline{A} \leftarrow \underline{I}(1)$ we have $\hat{\underline{A}} < (\underline{D}L)^\infty$, i.e., $\hat{\underline{A}} = (\underline{D}L)^{s-1}\underline{E}$ with $\underline{E} < \underline{D}L\underline{X}$, $\underline{E} \neq \underline{D}\ldots$ for some s. Therefore, $\underline{I}(y') > \underline{A}$ if $y' < y$. So, $y \in \mathrm{Int}(R_{\underline{A}})$.

Q.E.D.

THEOREM 15. <u>Let</u> f <u>be an extended Lorenz map, and assume</u> $\underline{A}$ <u>is an admissible sequence satisfying</u> $\underline{I}(-1) \leq \underline{A} \leftarrow \underline{I}(1)$. <u>Then there is an</u> $x \in I$ <u>such that</u> $\underline{I}(x) = \underline{A}$.

<u>Proof</u>: By Lemma 14, $L_{\underline{A}}$ and $R_{\underline{A}}$ are open. Therefore $L^c_{\underline{A}} = I \backslash L_{\underline{A}}$

and $R_{\underline{A}}^{c} = I \backslash R_{\underline{A}}$ are closed. Since $\underline{I}(-1) \le \underline{A} \le \underline{I}(1)$, neither $L_{\underline{A}}^{c}$ nor $R_{\underline{A}}^{c}$ are empty. Furthermore, $L_{\underline{A}}^{c} \cup R_{\underline{A}}^{c} = I$. Since I is connected, $L_{A}^{c} \cap R_{A}^{c}$ is not empty.

Q.E.D.

THEOREM A. Let f be a Lorenz map. If $n \in P(f)$ and $n \gg k$ with $k > 1$, then $k \in P(f)$.

Proof: Assume that x is a periodic point of the Lorenz map f of period n. Now, we extend f to all the interval I defining $f(0) = 1$. Then $\underline{I}(x)$ is periodic because the periodic orbit does not contain 0. By Lemma 13, the period of $\underline{I}(x)$ is n. Without loss of generality we may assume $\underline{I}(x)$ is maximal. So $\underline{I}(x) \ge \underline{P}_n^{\infty}$. By Lemma 2, $\underline{I}(x) \le I(1)$. Thus $\underline{P}_n^{\infty} \le \underline{I}(1)$.

Assume $k \ll n$ with $k > 1$ then $\underline{P}_k^{\infty} < \underline{P}_n^{\infty}$ by Theorem 11. Since $\underline{P}_k^{\infty}$ is maximal, if $\underline{I}(1)$ is infinite then $\underline{P}_k^{\infty} \leftarrow \underline{I}(1)$. Now, suppose $\underline{I}(1)$ is finite and equals $\underline{D}C$. Since $\underline{I}(x)$ is periodic, we have $\underline{I}(x) < \underline{I}(1)$. Therefore, by Lemma 2, $x < 1$. By Lemma 12, if $y < 0$ is sufficiently near to 0 we have $\underline{I}(f(y)) = (\underline{D}L)^{n}\ldots$ and $x < f(y) < 1$. So, by Lemma 2, $\underline{I}(x) \le \underline{I}(f(y))$. Since $\underline{I}(x)$ is periodic of period n, it follows that $\underline{I}(x) \le (\underline{D}L)^{\infty}$. Hence, $\underline{P}_k^{\infty} < (\underline{D}L)^{\infty}$ and, by maximality, $\underline{P}_k^{\infty} \leftarrow \underline{I}(1)$.

From $\underline{I}(-1) = L\ldots$ and $\underline{P}_k^{\infty} = R\ldots$ we have $\underline{I}(-1) < \underline{P}_k^{\infty}$. Therefore, by Theorem 15, there is a $y \in I$ with $\underline{I}(y) = \underline{P}_k^{\infty}$. It follows, by Lemma 13, that there is a $z \in I$ with period k, i.e. $k \in P(f)$.

Q.E.D.

This work was partially supported by CAYCIT nº 3534/83.C3.

5. References

[BGMY] L. Block, J. Guckenheimer, M. Misiurewicz and L.S. Young, Periodic points and topological entropy of one dimensional maps, Lect. Notes in Math. 819, 18-39, Springer-Verlag, 1980.

[B] B. Branner, Iterations by odd functions with two extrema, J. of Math. Anal. and Appl. 105 (1985), 276-297.

[CE] P. Collet and J.P. Eckmann, Iterated maps on the interval as dynamical systems, Progress on Physics, Vol. I, Birkhäuser, 1980.

[GH] J. Guckenheimer and P. Holmes, Nonlinear oscillations, dynamical systems, and bifurcations of vector fields, Applied Math. Sciences 42, Springer-Verlag, 1983.

[L] E.N. Lorenz, Deterministic non-periodic flow, J. Atmos. Sci. 20 (1963), 130-141.

[Sa] A.N. Sarkovskii, Coexistence of cycles of a continuous map of a line into itself, Ukr. Mat. Z 16 (1964), 61-71.

[Sp] C. Sparrow, The Lorenza equations, Appl. Math. Sc. 41, Springer-Verlag, 1982.

[St] P. Stefan, A theorem of Sarkovskii on the existence of periodic orbits of continuous endomorphism of the real line, Comm. Math. Phys. 54 (1977), 237-248.

[T] F. Takens, Implicit differential equations: some open problems, Lecture Notes in Math. 535, 237-253, Springer-Verlag, 1976.

Secció de Matemàtiques
Universitat Autònoma de Barcelona
Bellaterra, Barcelona, Spain

A LINS NETO

Complex codimension one foliations leaving a compact submanifold invariant

§1. Introduction

Let $\mathcal{F}$ be a codimension one singular foliation on a complex manifold V of dimension $n+1$, $n \geq 1$. Such a foliation is given by a covering $\{U_\alpha\}_{\alpha \in I}$ of V by open sets and a collection of holomorphic integrable 1-forms $\{\omega_\alpha\}_{\alpha \in I}$, satisfying the following properties:

a) For each $\alpha \in I$, the set of singularities of ω_α in U_α, $S_\alpha = \{p \in U_\alpha \mid \omega_\alpha(p) = 0\}$ is an analytic subvariety of codimension at least 2.

b) If $\alpha, \beta \in I$ are such that $U_\alpha \cap U_\beta \neq \emptyset$, then there exists a holomorphic function $g_{\alpha\beta} \in \mathcal{O}^*(U_\alpha \cap U_\beta)$ such that $\omega_\alpha = g_{\alpha\beta} \cdot \omega_\beta$.

We say that ω_α represents $\mathcal{F}$ in U_α.

The set of singularities of $\mathcal{F}$ is defined by $S(\mathcal{F}) = \bigcup_{\alpha \in I} S_\alpha$. From the definition, clearly $S(\mathcal{F})$ is a union of analytic subvarieties of V, each of them with codimension at least 2. Moreover, since for each α, ω_α is integrable, the differential equation $\omega_\alpha = 0$ defines a codimension 1 foliation on $U_\alpha - S_\alpha$. From condition b) these foliations glue together and define a global codimension 1 foliation on $V-S(\mathcal{F})$, whose leaves we call the <u>regular</u> leaves of $\mathcal{F}$.

We are interested in the case where $\mathfrak{F}$ leaves invariant a compact connected codimension one submanifold $M \subset V$. In this case $M - S(\mathfrak{F})$ is a regular leaf of $\mathfrak{F}$ (since it is connected and invariant).

Let $\varphi = (x,y): U \to \mathbb{C}^n \times \mathbb{C}$ be a local chart on U, where $U \cap M \neq \phi$, and $\varphi(M \cap U) \subset \{y=0\}$. Let ω be an integrable 1-form on U which represents $\mathfrak{F}$ in U. Since M is invariant by $\mathfrak{F}$, ω can be written as

$$(1) \qquad \omega = y\mu + f(x,y)\,dy$$

where $\mu = \sum_{j=1}^{n} \mu_j(x,y)\,dx_j$ and $f(x,0) \not\equiv 0$. It follows that $S(\mathfrak{F}) \cap \cap M \cap U = \{(x,0) \mid f(x,0) = 0\}$. Hence $S(\mathfrak{F}) \cap M$ has codimension 1 in M (if not empty). Let $S(\mathfrak{F}) \cap M = S_1 \cup \ldots \cup S_k$, where S_j's are the components of $S(\mathfrak{F}) \cap M$. Let us define a normal index for each S_j.

First observe that if $\varphi = (x,y)$ and $\omega = y\mu + f(x,y)\,dy$ are as above, then the 1-form in $M \cap U$, $\eta = \frac{1}{f(x,0)} \sum_{j=1}^{n} \mu_j(x,0)\,dx_j$ is closed. In fact, from the integrability condition $\omega \wedge d\omega = 0$, we get

$$y\mu \wedge d\mu + f\,dy \wedge d\mu + \mu \wedge df \wedge dy = 0,$$

and so, by doing $y = 0$,

$$dy \wedge (f\,d\mu - df \wedge \mu)_{y=0} = 0.$$

As it is easy to see this equation implies that $d(\frac{\mu}{f}\Big|_{U \cap M}) = = d\eta = 0$.

Now let $p \in S_j - \bigcup_{i \neq j} S_i$ be a non singular point of S_j. Take $\varphi = (x,y)$ and ω as above in a neighborhood of p. Define the normal index of S_j as

$$(2) \qquad i(S_j, \mathfrak{F}, M) = -\mathrm{Res}(\eta, S_j).$$

As usual this residue is defined by taking a 1-dimensional embedded disk D transversal to S_j at p and then

$$\mathrm{Res}(\eta, S_j) = \frac{1}{2\pi i} \int_\gamma \eta$$

where γ is a generator of $H_1(D-\{p\}, \mathbb{Z})$.

If $\tilde{\omega} = y\tilde{\mu} + \tilde{f}\, dy$ is another form which represents $\mathcal{F}$ in U, then $\tilde{\omega} = g\omega$, where g does not vanishes, therefore $\tilde{\mu}/\tilde{f} = \mu/f$. Hence $\mathrm{Res}(\eta, S_j)$ does not depend on the form which represents $\mathcal{F}$ locally. It can be checked also that if we change variables in (1) by a local diffeomorphism $x = g(u,v)$, $y = vh(u,v)$, sending $\{v=0\} \mapsto \{y=0\}$, then the residue does not change. It follows that $i(S_j, \mathcal{F}, M)$ is well defined and depends only on $\mathcal{F}$ in a neighborhood of S_j.

Now let L be the normal line bundle of M in V and $L^k = L \oplus \ldots \oplus L$ (k times). We denote by c_k the kth Chern class of this bundle: if ∇ is any connection in L and Θ its curvature form, $c_k = (\frac{i}{2\pi})^k \Theta^k = (\frac{i}{2\pi})^k \Theta \wedge \ldots \wedge \Theta$ (k times).

Theorem 1. Let V, $\mathcal{F}$ and M be as before and φ be a C^∞ closed (2n-2)-form on M. Then

$$\int_M c_1 \wedge \varphi = \sum_{j=1}^{k} i(S_j, \mathcal{F}, M) \int_{S_j} \varphi.$$

In particular we have:

a) $$\int_M c_n = \sum_{j=1}^{k} i(S_j, \mathcal{F}, M) \int_{S_j} c_{n-1}$$

b) $[c_1] = [\sum_{j=1}^{k} i(S_j, \mathcal{F}, M)\Omega_j]$, where $[\eta]$ denotes the cohomology class of a closed form in the De Rham cohomology group and $[\Omega_j]$ is the Poincaré dual of the divisor associated to S_j, j=1,...,k.

In b) we are considering each S_j as an irreducible subvariety of M.

Remark 1. This result was communicated to me by J.P. Brasselet past year, in the form b). Here we will give another proof.

We will prove also the following results:

<u>Theorem 2</u>. Let M be a compact connected complex manifold of dimension n and $S = S_1 \cup \dots \cup S_k$ be a subvariety, where $S_1, \dots, S_k$ are the irreducible components of S. For each $j = 1, \dots, k$, let $[\Omega_j] \in H^2(M, \mathbb{C})$ be the Poincaré dual of S_j. Suppose that there are non zero complex numbers $\lambda_1, \dots, \lambda_k$ such that the class $c = [\sum_{j=1}^{k} \lambda_j \Omega_j] \in H^2(M, \mathbb{Z})$. Then there exist a manifold V and a singular codimension one foliation $\mathfrak{F}$ on V such that:

a) V is a line bundle over M, say $V \xrightarrow{\pi} M$. The first Chern class of V is c.

b) The zero section $M_0 \approx M$ is invariant by $\mathfrak{F}$.

c) The set of singularities of $\mathfrak{F}$ in M is S. In $\pi^{-1}(M-S)$, $\mathfrak{F}$ is transverse to the fibers of π.

d) For each $j = 1, \dots, k$, $i(S_j, \mathfrak{F}, M) = \lambda_j$.

e) If $\lambda_1, \dots, \lambda_k$ are integers then V is the line bundle associated to the divisor $\sum_{j=1}^{k} \lambda_j S_j$.

In the special case where M is a compact Kähler manifold we can prove more.

<u>Corollary</u>. Let M be a compact connected Kähler manifold and L be a line bundle over M. Let $S = S_1 \cup \dots \cup S_k \subset M$ and $[\Omega_1], \dots, [\Omega_k]$ be as in Theorem 2. Suppose that $\lambda_1, \dots, \lambda_k \in \mathbb{C} - \{0\}$ are such that

$$[\sum_{j=1}^{k} \lambda_j \Omega_j] = [c_1(L)]$$

where $[c_1(L)] \in H^2(M, \mathbb{Z})$ is the first Chern class of L. Then there exists a singular foliation on L satisfying b), c) and d) of Theorem 2.

<u>Remark 2</u>. In the special case where M is compact Riemann surface, Theorem 1 was proved by Camacho and Sad in [1]. Theorem 2 is a kind

of generalization of [2], where we construct singular foliations in a 2-dimensional manifold which leaves invariant a Riemann surface M and the holonomy of M-S is any given group of germs of diffeomorphisms at $0 \in \mathbb{C}$ with k-1 generators, where $S \cap M$ has k points in this case. The foliations that will be constructed in Theorem 2, will be so that the holonomy of the leaf M-S will be linear and abelian. In this sense the results of [2] are more general than Theorem 2, since there the holonomy need not be abelian. On the other hand Theorem 2 is more general in the sense that the dimension of M can be greater than 1. An interesting problem which arises is to obtain in Theorem 2 any holonomy compactible with $\pi_1(M-S)$, as in the results of [2].

I would like to thank A. Douady who indicated to me the corollary of Theorem 2 and to P. Sad for helpfull conversations.

More details about Theorem 2 and the corollary are given in Remark 4 at the end of the paper.

§2. Proof of Theorem 1

2.1 - Linear approximation of $\mathfrak{F}$

Let V, $\mathfrak{F}$ and M be as in §1. We will associate to $\mathfrak{F}$ a foliation $\mathfrak{L}$ on the normal bundle $L \xrightarrow{\pi} M$ of M in V with the following properties:

a) $\mathfrak{L}$ leaves invariant the zero section M_0, which we identify with M via the natural inclusion $M \hookrightarrow M_0 \subset L$.

b) The set of singularities of $\mathfrak{L}$ in M is contained in $S = S_1 \cup \ldots \cup S_k$. We observe that some of the componentes of S (for which $i(S_j,\mathfrak{F},M) = 0$) could disappear in the process.

c) For each $j = 1,\ldots,k$, $i(S_j,\mathfrak{F},M) = i(S_j,\mathfrak{L},M)$.

After this construction, Theorem 1 for $\mathfrak{F}$ and $\mathfrak{L}$ will be equivalent.

Let $\{U_\alpha, \varphi_\alpha\}_{\alpha \in I}$ be a covering of M by local charts and $\{\omega_\alpha\}_{\alpha \in I}$ be a collection of integrable 1-forms such that ω_α represents $\mathfrak{F}$ in U_α. We assume also that $\varphi_\alpha = (x_\alpha, y_\alpha): U_\alpha \to \mathbb{C}^n \times \mathbb{C}$ is such that $M \cap U_\alpha = W_\alpha \subset \{y_\alpha = 0\}$. Since M is invariant we can write

$$\omega_\alpha = y_\alpha \mu_\alpha + f_\alpha(x_\alpha, y_\alpha) dy_\alpha$$

where $\mu_\alpha = \sum_{j=1}^{n} \mu_j(x_\alpha, y_\alpha) dx_j$, $x_\alpha = (x_1, \ldots, x_n)$. Define

$$\tilde{\omega}_\alpha = y_\alpha \tilde{\mu}_\alpha + \tilde{f}_\alpha(x_\alpha) dy_\alpha \tag{3}$$

where $\tilde{\mu}_\alpha = \sum_{j=1}^{n} \mu_j(x_\alpha, 0) dx_j$ and $\tilde{f}_\alpha(x_\alpha) = f_\alpha(x_\alpha, 0)$.

Clearly $\tilde{\omega}_\alpha$ is an integrable 1-form in $W_\alpha \times \mathbb{C}$ (since $\frac{\tilde{\mu}}{\tilde{f}} = \left.\frac{\mu}{f}\right|_{W_\alpha}$ is a closed 1-form in W_α). Let $L = TV|_M/TM$ be the normal bundle of M. For each $\alpha \in I$ let $L_\alpha = L|_{W_\alpha} = TV|_{W_\alpha}/TW_\alpha$ and $\tilde{\varphi}_\alpha : L_\alpha \to x_\alpha(W_\alpha) \times \mathbb{C}$ be defined by

$$\tilde{\varphi}_\alpha(q, [v]) = (x_\alpha(q), Dy_\alpha(q) \cdot v),$$

where $q \in W_\alpha$, $v \in T_qV$ and $[v]$ denotes its equivalent class in T_qV/T_qM. It is not difficult to prove that $\tilde{\varphi}_\alpha$ is well defined and that the set $\{\tilde{\varphi}_\alpha, L_\alpha\}_{\alpha \in I}$ defines an analytic atlas on L.

Let $\alpha, \beta \in I$ are such that $U_\alpha \cap U_\beta \neq \emptyset$ and $\varphi_\beta \circ \varphi_\alpha^{-1}(x_\alpha, y_\alpha) = (x_\beta(x_\alpha, y_\alpha), y_\beta(x_\alpha, y_\alpha))$, where $y_\beta(x_\alpha, 0) = 0$. It is easy to check that

$$\tilde{\varphi}_\beta \circ \tilde{\varphi}_\alpha^{-1}(x_\alpha, y_\alpha) = (x_\beta(x_\alpha, 0), D_2 y_\beta(x_\alpha, 0), y_\alpha)$$

where $D_2 y_\beta$ denotes the partial derivative of y_β with respect to the second variable.

Now, suppose we apply the changes of variables $\varphi_\beta \circ \varphi_\alpha^{-1}$ and $\tilde{\varphi}_\beta \circ \tilde{\varphi}_\alpha^{-1}$ on ω_β and $\tilde{\omega}_\beta$ respectively, obtaining expressions for these forms in the coordinate systems $(U_\alpha \cap U_\beta, \varphi_\alpha)$ and $((W_\alpha \cap W_\beta)\times\mathbb{C}, \tilde{\varphi}_\alpha)$, say ω_1 and ω_2. It can be proved that $\tilde{\omega}_1 = \omega_2$, so that $\tilde{\omega}_\beta$ is a well defined 1-form in L_β. On the other hand, if $\omega_\alpha = g_{\alpha\beta}\omega_\beta$ on $U_\alpha \cap U_\beta$, then $\tilde{\omega}_\alpha = \tilde{g}_{\alpha\beta}\tilde{\omega}_\beta$, where $\tilde{g}_{\alpha\beta} = g_{\alpha\beta}\big|_{W_\alpha \cap W_\beta}$. This implies that the collection $\{\tilde{\omega}_\alpha\}_{\alpha\in I}$ defines a foliation $\tilde{\mathfrak{L}}$ on L. Moreover the singular sets of $\mathfrak{F}$ and $\mathfrak{L}$ in M are the same.

Observe that the singular set of $\tilde{\omega}_\alpha$ could have codimension 1 for some α's. In this case $\tilde{\omega}_\alpha$ must be divided by the equation of the codimension 1 components of $\tilde{S}_\alpha = \{q \in W_\alpha\times\mathbb{C} \mid \tilde{\omega}_\alpha(q) = 0\}$. After this division the singular set of $\tilde{\mathfrak{L}}$ in M could have less components than the singular set of $\mathfrak{F}$ in M. We call $\mathfrak{L}$ the foliation obtained from $\tilde{\mathfrak{L}}$ after this division.

Set S_j be one of the components of S, $q \in S_j$ and $(U_\alpha, \varphi_\alpha)$ be a local chart around q as before. Let $\omega_\alpha = y_\alpha\mu_\alpha + f_\alpha dy_\alpha$ and $\tilde{\omega}_\alpha = y_\alpha\tilde{\mu}_\alpha + \tilde{f}_\alpha dy_\alpha$ represent $\mathfrak{F}$ and $\mathfrak{L}$ on U_α and $W_\alpha\times\mathbb{C}$ respectively. Then clearly $\dfrac{\mu_\alpha}{f_\alpha}\Big|_{W_\alpha} = \dfrac{\tilde{\mu}_\alpha}{\tilde{f}_\alpha}$ and so $i(S_j, \mathfrak{F}, M) = i(S_j, \mathfrak{L}, M)$.

Remark 3. We observe that the meromorphic form $\theta_\alpha = \dfrac{dy_\alpha}{y_\alpha} + \dfrac{\tilde{\mu}_\alpha}{\tilde{f}_\alpha}$ on $W_\alpha\times\mathbb{C}$ can be extended to a meromorphic form on L. In fact, if $W_\alpha \cap W_\beta \neq \emptyset$, then $\tilde{\omega}_\alpha = \tilde{g}_{\alpha\beta}\tilde{\omega}_\beta$ on $(W_\alpha \cap W_\beta)\times\mathbb{C}$. If the change of variables of L is $y_\alpha = h_{\alpha\beta}(x)\cdot y_\beta$, $h_{\beta\alpha} = h_{\alpha\beta}^{-1}$, then

$$\tilde{g}_{\alpha\beta}\tilde{\omega}_\beta = \tilde{g}_{\alpha\beta}\cdot(y_\beta\tilde{\mu}_\beta + \tilde{f}_\beta dy_\beta) = y_\alpha(\tilde{g}_{\alpha\beta}h_{\beta\alpha}\tilde{\mu}_\beta f_\alpha + \tilde{f}_\beta\tilde{g}_{\alpha\beta}dh_{\beta\alpha}) + $$
$$+ \tilde{g}_{\alpha\beta}\tilde{f}_\beta h_{\beta\alpha}dy_\alpha = y_\alpha\tilde{\mu}_\alpha + \tilde{f}_\alpha dy_\alpha .$$

Hence, $\tilde{f}_\alpha = \tilde{g}_{\alpha\beta}h_{\beta\alpha}\tilde{f}_\beta$ and $\tilde{\mu}_\alpha = \tilde{g}_{\alpha\beta}h_{\beta\alpha}\tilde{\mu}_\beta + \tilde{g}_{\alpha\beta}\tilde{f}_\beta dh_{\beta\alpha}$. It follows that

$$\frac{dy_\alpha}{y_\alpha} + \frac{\tilde{\mu}_\alpha}{\tilde{f}_\alpha} = \frac{dh_{\alpha\beta}}{h_{\alpha\beta}} + \frac{dy_\beta}{y_\beta} + \frac{\tilde{\mu}_\beta}{\tilde{f}_\beta} + \frac{dh_{\beta\alpha}}{h_{\beta\alpha}} = \frac{dy_\beta}{y_\beta} + \frac{\tilde{\mu}_\beta}{\tilde{f}_\beta} .$$

This implies that $\theta_\alpha = \theta_\beta$ on $(W_\alpha \cap W_\beta) \times \mathbb{C}$, which proves the assertion.

2.2 - Proof of Theorem 1

Let ∇ be a connection on L and Θ be its curvature form. The idea of the proof is to construct a C^∞ 1-form η on $M\text{-}S(\mathcal{L})$ such that $\Theta\big|_{M-S} = d\eta$. Let $\{W_\alpha \times \mathbb{C}, \tilde{\varphi}_\alpha\}_{\alpha \in I}$ and $\{\tilde{\omega}_\alpha\}_{\alpha \in I}$ be as before. Let $p \in W_\alpha - S(\mathcal{L})$ and U be a simply connected neighborhood of p in $W_\alpha - S(\mathcal{L})$. Let $\sigma: U \to L$ be a non zero solution of $\tilde{\omega}_\alpha = 0$, that is a local section which can be written as $\sigma(x) = \xi(x) \cdot 1_\alpha$ (1_α is the local section $y_\alpha = 1$), where ξ satisfies

(4) $$\xi(x)\tilde{\mu}_\alpha + \tilde{f}_\alpha \cdot d\xi = 0.$$

Such a ξ exists because the 1-form $\tilde{\mu}_\alpha / \tilde{f}_\alpha$ is closed and U is simply connected. Let η_U be the 1-form defined by

$$\nabla\sigma = \eta_U \otimes \sigma.$$

By definition $d\eta_U = \Theta\big|_U$. If σ_1 is another non zero solution of $\tilde{\omega}_\alpha = 0$, then $\sigma_1 = a \cdot \sigma$, where $a \neq 0$ is a constant, because (4) is a linear equation in ξ. Now, $\nabla\sigma_1 = \nabla(a\cdot\sigma) = a\nabla\sigma = a\,\eta_U \otimes \sigma = \eta_U \otimes \sigma_1$. Therefore η_U does not depend on the solution σ of $\tilde{\omega}_\alpha = 0$. If $\beta \in I$ is such that $W_\beta \cap W_\alpha \neq \emptyset$ and $S(\mathcal{L}) \cap W_\beta = \emptyset$, then $\tilde{\omega}_\beta\big|_{W_\alpha \cap W_\beta} = \tilde{g}_{\beta\alpha} \cdot \tilde{\omega}_\alpha\big|_{W_\beta \cap W_\alpha}$, where $\tilde{g}_{\beta\alpha}$ does not vanishes. It follows that if σ_β is a solution of $\tilde{\omega}_\beta = 0$ on $V \subset W_\beta - S(\mathcal{L})$, then $\sigma_\beta\big|_{V \cap W_\alpha}$ is a solution of $\tilde{\omega}_\alpha = 0$. This implies that the collection $\{\eta_U\}_U$ defines a global well defined 1-form η on $M\text{-}S$. From the definition of Θ, clearly $\Theta\big|_{M-S} = d\eta$.

We observe that for each $\alpha \in I$, $\eta\big|_{W_\alpha} = -\tilde{\mu}_\alpha / \tilde{f}_\alpha + \gamma_\alpha$, where γ_α is a C^∞ 1-form on W_α. In fact, if σ_α is a local non zero solution of $\tilde{\omega}_\alpha = 0$, then $\sigma_\alpha = \xi \cdot 1_\alpha$, where 1_α is the section given by $y_\alpha = 1$ and $\frac{d\xi}{\xi} = -\tilde{\mu}_\alpha/\tilde{f}_\alpha$. Therefore

$$\nabla\sigma_\alpha = d\xi \otimes 1_\alpha + \xi\nabla 1_\alpha = d\xi \otimes 1_\alpha + \xi\gamma_\alpha \otimes 1_\alpha = \left(\frac{d\xi}{\xi} + \gamma_\alpha\right) \otimes \sigma_\alpha .$$

Hence $\eta|_{W_\alpha} = \frac{d\xi}{\xi} + \gamma_\alpha = -\tilde{\mu}_\alpha/\tilde{f}_\alpha + \gamma_\alpha$. We set $-\tilde{\mu}_\alpha/\tilde{f}_\alpha = \delta_\alpha$, so that $\eta|_{W_\alpha} = \delta_\alpha + \gamma_\alpha$.

Now the divisor S is defined by the collection $\{\tilde{f}_\alpha\}_{\alpha\in I}$. Let us suppose that $S = \sum_{j=1}^{k} \ell_j S_j$ $(\ell \geq 1,\ j=1,\ldots,k)$, so that $\tilde{f}_\alpha = f_{1\alpha}^{\ell_1} \ldots f_{k\alpha}^{\ell_k}$, where the collection $\{f_{j\alpha}\}_{\alpha\in I}$ defines the divisor S_j. Since the 1-form $\delta_\alpha = -\tilde{\mu}_\alpha/\tilde{f}_\alpha$ is closed with poles on $f_{1\alpha}^{\ell_1} \ldots f_{k\alpha}^{\ell_k} = 0$, we can write

$$(5) \qquad \delta_\alpha = \sum_{j=1}^{k} \lambda_{j\alpha} \frac{df_{j\alpha}}{f_{j\alpha}} + d\left(\frac{h_\alpha}{f_1^{\ell_1-1} \ldots f_k^{\ell_k-1}}\right)$$

where $\lambda_{1\alpha},\ldots,\lambda_{n\alpha} \in \mathbb{C}$ and h_α is holomorphic (cf. [3] p.37). When $W_\alpha \cap S_j \neq \phi$ we have $\lambda_{j\alpha} = \mathrm{Res}(\delta_\alpha, S_j) = i(S_j,\mathcal{L},M)$. It follows that we can take $\lambda_{j\alpha} = i(S_j,\mathcal{L},M)$ for all $\alpha \in I$. We set $i(S_j,\mathcal{L},M) = \lambda_j$.

Let us prove the result in the case where $\ell_1 = \ldots = \ell_k = 1$. In this case

$$(5') \qquad \delta_\alpha = \sum_{j=1}^{k} \lambda_j \frac{df_{j\alpha}}{f_{j\alpha}} + d\, h_\alpha ,$$

where h_α is holomorphic. Let φ be a C^∞ closed $(2n-2)$-form on M. Let $\langle\ ,\ \rangle$ be a hermitian metric on the line bundle over M, defined by the divisor $\{\tilde{f}_\alpha\}_{\alpha\in I}$. Let σ be the section on this bundle defined by $\sigma|_{W_\alpha} = \tilde{f}_\alpha$. We set $S(\varepsilon) = \{p \in M \mid \|\sigma(p)\|_p \leq \varepsilon\}$. Since $\sigma(p) = 0$ if and only if $p \in S$, it follows that if $\varepsilon > 0$ is small then $S(\varepsilon)$ is C^∞ submanifold of M with boundary $\partial S(\varepsilon) = \{p \in M \mid \|\sigma(p)\|_p = \varepsilon\}$. Write

$$\int_M c_1\wedge\varphi = \frac{i}{2\pi}\int_M \Theta\wedge\varphi = \int_{S(\epsilon)} c_1\wedge\varphi + \frac{i}{2\pi}\int_{M-S(\epsilon)} d\eta\wedge\varphi =$$

$$= \int_{S(\epsilon)} c_1\wedge\varphi + \frac{i}{2\pi}\int_{M-S(\epsilon)} d(\eta\wedge\varphi) = \int_{S(\epsilon)} c_1\wedge\varphi - \frac{i}{2\pi}\int_{\partial S(\epsilon)} \eta\wedge\varphi .$$

Now, since $\bigcap_{\epsilon>0} S(\epsilon) = S$ we have $\lim_{\epsilon\to 0}\int_{S(\epsilon)} c_1\wedge\varphi = 0$. Therefore it is enough to prove that

$$\text{(6)} \qquad \lim_{\epsilon\to 0}\int_{\partial S(\epsilon)} \eta\wedge\varphi = \sum_{j=1}^{k} 2\pi i\,\lambda_j\int_{S_j} \varphi .$$

At this point it is convenient to observe the following fact: let S_j^* be the set of non singular points of $S_j - \bigcup_{\ell\neq j} S_\ell$. Then $\int_{S_j}\varphi = \int_{S_j^*}\varphi$ (cf. [4]). Let $V^* = V-A$, where $A = S - \bigcup_{j=1}^{k} S_j^*$ is the singular set of S. We can consider a locally finite covering $\{W_\alpha\}_{\alpha\in I}$ of V^*, with the following properties:

i) If $W_\alpha \cap S_j^* \neq \phi$ for some $j\in\{1,\dots,k\}$, then $W_\alpha\cap S_i^* = \phi$ for $i\neq j$. This is possible because $S_i^*\cap S_j^* = \phi$.

ii) If $W_\alpha\cap S_j^*\neq\phi$ then $\pi^{-1}(W_\ell)$ is the domain of a local chart $\psi_\alpha = (x_1, z, y_\alpha)\colon \pi^{-1}(W_\alpha)\to \mathbb{C}\times\mathbb{C}^{n-1}\times\mathbb{C}$, where $\pi^{-1}(W_\alpha)\cap M\subset\{y_\alpha=0\}$ and $\pi^{-1}(W_\alpha)\cap S_j^*\subset\{x_1=y_\alpha=0\}$. Moreover, since $\mathcal{L}\big|_{\pi^{-1}(W_\alpha)}$ is represented by $\tilde\omega_\alpha = y_\alpha\tilde\mu_\alpha + \tilde f_\alpha dy$, where $\tilde f_\alpha = 0$ defines $\pi^{-1}(W_\alpha)\cap S_j^*$, we can suppose that $\tilde f_\alpha(x_1,z) = x_1$ and $\partial S(\epsilon)\cap W_\alpha = \{(x_1,z)\mid |x_1|=\epsilon\}$.

We consider also a partition of the unity $\{\rho_\alpha\}_{\alpha\in I}$ subordinated to the covering $\{W_\alpha\}_{\alpha\in I}$ of V^*. Let $\varphi_\alpha = \rho_\alpha\cdot\varphi$, so that $\sum_\alpha \varphi_\alpha=\varphi$ and $\int_{\partial S(\epsilon)}\eta\wedge\varphi = \sum_{\alpha\in I}\int_{\partial S(\epsilon)\cap W_\alpha}\eta\wedge\varphi_\alpha$. In order to prove the result it is sufficient to prove

$$\text{I-}\quad \lim_{\epsilon\to 0}\int_{\partial S(\epsilon)\cap W_\alpha} \eta\wedge\varphi_\alpha = 0 \quad \text{if} \quad W_\alpha \cap S = \emptyset$$

$$\text{II-}\quad \lim_{\epsilon\to 0}\int_{\partial S(\epsilon)\cap W_\alpha} \eta\wedge\varphi_\alpha = 2\pi i\,\lambda_j \int_{S_j\cap W_\alpha} \varphi_\alpha \quad \text{if} \quad W_\alpha \cap S_j^* \neq \emptyset .$$

Equality I is clear. Let us prove II. We know that $\eta\big|_{W_\alpha} = \delta_\alpha + \gamma_\alpha$ where γ_α is C^∞ and δ_α is as in (5'). Since $\lim_{\epsilon\to 0} \mathrm{vol}(\partial S(\epsilon)\cap W_\alpha) = 0$ we have $\lim_{\epsilon\to 0}\int_{\partial S(\epsilon)\cap W_\alpha} \gamma_\alpha\wedge\varphi_\alpha = 0$ and so $\lim_{\epsilon\to 0}\int_{\partial S(\epsilon)\cap W_\alpha} \eta\wedge\varphi_\alpha = \lim_{\epsilon\to 0}\int_{\partial S(\epsilon)\cap W_\alpha} \delta_\alpha\wedge\varphi_\alpha$. Now, from the assumption $\ell_1 = \ldots = \ell_k = 1$ we have

$$\delta_\alpha = \lambda_j \frac{dx_1}{x_1} + \sum_{m\neq j} \lambda_m \frac{df_{\alpha m}}{f_{\alpha m}} + dh_\alpha = \lambda_j \frac{dx_1}{x_1} + \theta_\alpha$$

where θ_α is holomorphic because $f_{\alpha m}$ does not vanishes on W_α for $m\neq j$. Therefore

$$\lim_{\epsilon\to 0}\int_{\partial S(\epsilon)\cap W_\alpha} \delta_\alpha\wedge\varphi_\alpha = \lim_{\epsilon\to 0} \lambda_j \int_{\partial S(\epsilon)\cap W_\alpha} \frac{dx_1}{x_1}\wedge\varphi_\alpha .$$

Observe that $\partial S(\epsilon)\cap W_\alpha = \{(x_1,z) \mid |x_1| = \epsilon\}$ can be parametrized by $\gamma_\epsilon(\theta,z) = (\epsilon e^{i\theta}, z)$, where $\theta\in\mathbb{R}$ and $z = (x_2,\ldots,x_n)\in D_\alpha = S_j^*\cap W_\alpha$. Write $\varphi_\alpha = dx_1\wedge\varphi_1 + d\bar{x}_1\wedge\varphi_2 + a(x_1,z)\Omega$, where φ_1 and φ_2 are $(2n-3)$-forms and $\Omega = d\bar{x}_2\wedge dx_2\wedge\ldots\wedge d\bar{x}_n\wedge dx_n$. Of course $\varphi_\alpha\big|_{D_\alpha} = a(0,z)\Omega$. On the other hand $\gamma_\epsilon^*(\frac{dx_1}{x_1}\wedge\varphi) = i\,a(\epsilon e^{i\theta},z)\,d\theta\wedge\Omega$. Therefore

$$\lim_{\epsilon\to 0}\int_{\partial S(\epsilon)\cap W_\alpha} \frac{dx_1}{x_1}\wedge\varphi_\alpha = \lim_{\epsilon\to 0}\int_{D_\alpha}\left[\int_0^{2\pi} i\,a(\epsilon e^{i\theta},z)\,d\theta\right]\Omega =$$

$$= 2\pi i\int_{D_\alpha} a(0,z)\Omega = 2\pi i\int_{D_\alpha}\varphi_\alpha .$$

This proves the result in the case where $\ell_1 = \ldots = \ell_k = 1$. Let us consider the case where some of the ℓ_j's are bigger than 1, say $\ell_j > 1$ for $1 \le j \le m$ and $\ell_j = 1$ for $m < j \le k$. In this case we can write

$$\delta_\alpha = \sum_{j=1}^{k} \lambda_j \frac{df_{j\alpha}}{f_{j\alpha}} + d\left(\frac{h_\alpha}{f_{1\alpha}^{r_1} \cdots f_{m\alpha}^{r_m}}\right)$$

where $r_j = \ell_j - 1$, $1 \le j \le m$. We observe that if $W_\alpha \cap S_j^* \ne \phi$ then $h_\alpha \not\equiv 0$ on $W_\alpha \cap S_j^*$. This follows from the fact that the set of singularities of the form

$$f_{1\alpha}^{\ell_1} \cdots f_{k\alpha}^{\ell_k} dy_\alpha - f_{1\alpha}^{\ell_1} \cdots f_{k\alpha}^{\ell_k} \delta_\alpha$$

in W_α is $\{f_{j\alpha} = 0\}$.

Let $g_\alpha = h_\alpha / f_{1\alpha}^{r_1} \cdots f_{m\alpha}^{r_m}$. We assert that $g_\beta - g_\alpha$ has no poles on $W_\alpha \cap W_\beta$.

<u>Proof</u>: We have seen in Remark 3 that $\delta_\beta - \delta_\alpha = dh_{\beta\alpha}/h_{\beta\alpha}$ on $W_\alpha \cap W_\beta$, where $h_{\beta\alpha}$ does not vanishes and has no poles on $W_\alpha \cap W_\beta$. This implies that

$$d(g_\beta - g_\alpha) = dh_{\beta\alpha}/h_{\beta\alpha} + \sum_{j=1}^{k} \lambda_j \, dg_{j\alpha\beta}/g_{j\alpha\beta}$$

where $g_{j\alpha\beta} = f_{j\alpha}/f_{j\beta}$ does not vanishes and has no poles on $W_\alpha \cap W_\beta$, because the collection $\{f_{j\alpha}\}_{\alpha \in I}$ defines the divisor S_j. It follows that $d(g_\beta - g_\alpha)$ has no poles on $W_\alpha \cap W_\beta$ and so the same is true for $g_\beta - g_\alpha$. This proves the assertion.

Now, let us consider the C^∞ function $g = \sum_{\alpha \in I} \rho_\alpha g_\alpha$. Since φ is closed we have

$$(*) \qquad \sum_{\alpha \in I} \int_{\partial S(\epsilon) \cap W_\alpha} dg \wedge \varphi_\alpha = \int_{\partial S(\epsilon)} dg \wedge \varphi = 0.$$

From the first part of the proof we have

$$\int_M c_1 \wedge \varphi - \sum_{j=1}^{k} \lambda_j \int_{S_j} \varphi = \lim_{\epsilon \to 0} \sum_{\alpha \in I} \int_{\partial S(\epsilon) \cap W_\alpha} dg_\alpha \wedge \varphi_\alpha .$$

On the other hand, from (*) we have

$$\sum_{\alpha\in I} \int_{\partial S(\varepsilon)\cap W_\alpha} dg_\alpha \wedge \varphi_\alpha = \sum_{\alpha\in I} \int_{\partial S(\varepsilon)W_\alpha)} (dg_\alpha - dg) \wedge \varphi_\alpha .$$

But,

$$dg_\alpha - dg = d(g_\alpha - \sum_{\beta\in I} \rho_\beta g_\beta) = d(\sum_{\beta\in I} \rho_\beta (g_\alpha - g_\beta)).$$

Since for every $\beta \in I$, such that $W_\alpha \cap W_\beta \neq \phi$, $g_\alpha - g_\beta$ has no poles, it follows that $dg_\alpha - dg$ has no poles on W_α. Therefore

$$\lim_{\varepsilon\to 0} \sum_{\alpha\in I} \int_{\partial S(\varepsilon)\cap W_\alpha} (dg_\alpha - dg) \wedge \varphi_\alpha = \sum_{\alpha\in I} \lim_{\varepsilon\to 0} \int_{\partial S(\varepsilon)\cap W_\alpha} (dg_\alpha - dg) \wedge \varphi_\alpha = 0.$$

This ends the proof of Theorem 1.

§3. Constructions

3.1 - Proof of Theorem 2

First of all let us consider the divisor $S = S_1 \cup \ldots \cup S_k \subset M$ and a covering $\mathcal{U} = \{W_\alpha\}_{\alpha\in I}$ of M by open sets such that for each $j = 1,\ldots,k$ the divisor S_j is defined by a collection of holomorphic functions $\{f_{j\alpha} : W_\alpha \to \mathbb{C}\}_{\alpha\in I}$, where $f_{j\alpha}/f_{j\beta}$ does not vanishes and has no poles on $W_\alpha \cap W_\beta$. We suppose also that W_α, $W_\alpha \cap W_\beta$ and $W_\alpha \cap W_\beta \cap W_\gamma$ are diffeomorphic to polydisks for any $\alpha,\beta,\gamma \in I$ such that $W_\alpha \cap W_\beta \cap W_\gamma \neq \phi$. We set $g_{j\alpha\beta} = f_{j\alpha}/f_{j\beta}$, $j=1,\ldots,k$.

The idea is to show that there exist a line bundle $L \xrightarrow{\pi} M$ and a meromorphic closed 1-form θ on L such that for each $\alpha \in I$, the expression of θ on $\pi^{-1}(W_\alpha) \approx W_\alpha \times \mathbb{C}$ is

$$(7) \qquad \theta_\alpha = \frac{dy_\alpha}{y_\alpha} - \sum_{j=1}^{k} \lambda_j \frac{df_{j\alpha}}{f_{j\alpha}}$$

where y_α is the vertical coordinate on $W_\alpha \times \mathbb{C}$. If we can do this, then the holomorphic 1-form on $\pi^{-1}(W_\alpha)$

$$\omega_\alpha = y_\alpha f_{\alpha 1} \cdots f_{\alpha k} \theta_\alpha = f_{1\alpha} \cdots f_{k\alpha}\, dy_\alpha - y_\alpha \left(f_{1\alpha} \cdots f_{k\alpha} \sum_{j=1}^{k} \lambda_j \frac{df_{j\alpha}}{f_{j\alpha}}\right)$$

is integrable and defines a singular foliation on $\pi^{-1}(W_\alpha)$, such that $\{y_\alpha = 0\} = M_0 \cap \pi^{-1}(W_\alpha)$ is invariant and whose singular set on $\{y_\alpha = 0\}$ is the divisor $\{f_{1\alpha} \cdots f_{k\alpha} = 0\} = S \cap W_\alpha$. Since θ is defined globally, such foliations glue together and we obtain a singular foliation $\mathfrak{F}$ on L which satisfies b), c) and d).

Let us suppose the problem solved, so that we have found the transition functions $h_{\alpha\beta} : W_\alpha \cap W_\beta \to \mathbb{C}^*$ of the bundle L. Since the change of variables of the bundle are given by $y_\alpha = h_{\alpha\beta} y_\beta$, the form $\theta_\alpha |_{W_\alpha \cap W_\beta \times \mathbb{C}}$ is transformed into

$$\theta_\alpha^* = \frac{dh_{\alpha\beta}}{h_{\alpha\beta}} + \frac{dy_\beta}{y_\beta} - \sum_{j=1}^{k} \lambda_j \frac{df_{j\alpha}}{f_{j\alpha}} .$$

Now, θ is globally defined and so $\theta_\alpha^* = \theta_\beta |_{W_\alpha \cap W_\beta \times \mathbb{C}}$, which is equivalent to

$$(8) \qquad \frac{dh_{\alpha\beta}}{h_{\alpha\beta}} = \sum_{j=1}^{k} \lambda_j \left(\frac{df_{j\alpha}}{f_{j\alpha}} - \frac{df_{j\beta}}{f_{j\beta}}\right) = \sum_{j=1}^{k} \lambda_j \frac{dg_{j\alpha\beta}}{g_{j\alpha\beta}} .$$

Observe that if the λ_j's are integers then (8) has a solution

$$h_{\alpha\beta} = g_{1\alpha\beta}^{\lambda_1} \cdots g_{k\alpha\beta}^{\lambda_k}$$

which corresponds to the transition function of the bundle associated to the divisor $\sum_{j=1}^{k} \lambda_j S_j$. On the other hand, if the λ_j's are not integers then every solution of (8) is of the form

$$h_{\alpha\beta} = \exp\left(c_{\alpha\beta} + \sum_{j=1}^{k} \lambda_j \ell_{j\alpha\beta}\right)$$

where for each $j = 1, \ldots, k$, $\ell_{j\alpha\beta}$ is a branch of the logarithm of $g_{j\alpha\beta}$. Since $g_{j\alpha\beta} \in \mathcal{O}^*(W_\alpha \cap W_\beta)$ and $W_\alpha \cap W_\beta$ is simply connected, we can define such branches of the logarithm. We suppose that these branches are fixed from now on.

Now observe that the collection $\{h_{\alpha\beta}\}$ defines the transition functions of a line bundle if and only if for any $\alpha,\beta,\gamma \in I$ such that $W_\alpha \cap W_\beta \cap W_\gamma \neq \emptyset$ the cocycle condition

$$h_{\alpha\beta}\cdot h_{\beta\gamma}\cdot h_{\gamma\alpha} = 1$$

is satisfied. This condition is equivalent to

$$(9) \qquad c_{\alpha\beta}+c_{\beta\gamma}+c_{\gamma\alpha} + \sum_{j=1}^{k} \lambda_j \, (\ell_{j\alpha\beta}+\ell_{j\beta\gamma}+\ell_{j\gamma\alpha}) \in 2\pi i\,\mathbb{Z}.$$

Since $\ell_{j\alpha\beta}$ is a branch of the logarithm of $g_{j\alpha\beta} = f_{j\alpha}/f_{j\beta}$ we have

$$\exp(\ell_{j\alpha\beta}+\ell_{j\beta\gamma}+\ell_{j\gamma\alpha}) = \frac{f_{j\alpha}}{f_{j\beta}} \cdot \frac{f_{j\beta}}{f_{j\gamma}} \cdot \frac{f_{j\gamma}}{f_{j\alpha}} = 1.$$

Hence $\ell_{j\alpha\beta} + \ell_{j\beta\gamma} + \ell_{j\gamma\alpha}$ is a constant of the form $2\pi i\, k_{j\alpha\beta\gamma}$, on $W_\alpha \cap W_\beta \cap W_\gamma$, where $k_{j\alpha\beta\gamma} \in \mathbb{Z}$. The collection $\{k_{j\alpha\beta\gamma}\}$ defines a cocycle on $\check{H}^2(\mathcal{U},\mathbb{Z}) \approx H^2(M,\mathbb{Z})$, which corresponds to the fundamental class of S_j (cf. [4] p.141). Now the hypothesis of the theorem says that the class of the cocycle $\{\sum_{j=1}^{k} \lambda_j \, k_{j\alpha\beta\gamma}\}$ in $\check{H}^2(\mathcal{U},\mathbb{C})$ is integer. Hence there exist cocycles $\{m_{\alpha\beta\gamma}\} \in Z^2(\mathcal{U},\mathbb{Z})$ and $\{d_{\alpha\beta}\} \in$ $\in Z^1(\mathcal{U},\mathbb{C})$ such that

$$\{m_{\alpha\beta\gamma}\} = \{\sum_{j=1}^{k} \lambda_j \, k_{j\alpha\beta\gamma}\} + \delta\{d_{\alpha\beta}\} =$$

$$= \{\sum_{j=1}^{k} \lambda_j \, k_{j\alpha\beta\gamma} + d_{\alpha\beta} + d_{\beta\gamma} + d_{\gamma\alpha}\}.$$

Now it is easy to see that the collection of functions

$$\{h_{\alpha\beta} = \exp(2\pi i\, d_{\alpha\beta} + \sum_{j=1}^{k} \lambda_j \, \ell_{j\alpha\beta})\}$$

defines a line bundle $L \xrightarrow{\pi} M$ where the forms θ_α as in (7) glue together. This ends the proof of Theorem 2.

3.2 - Proof of the corollary

Let M be a compact Kähler manifold. We consider first the case where the divisor S is empty. Let us recall briefly the method of suspension in the case we are interested. Let $\varphi: \pi_1(M,x_o) \to \mathrm{Diff}(\mathbb{C})$ be a representation, where in our case we suppose that for each $[\gamma] \in \pi_1(M,x_o)$, $\varphi([\gamma]): \mathbb{C} \to \mathbb{C}$ is $\mathbb{C}$-linear. In this case for each $[\gamma] \in \pi_1(M,x_o)$ we have $\varphi([\gamma])(z) = a([\gamma])\cdot z$, where $a([\gamma]) \in \mathbb{C}^*$. Since φ is a representation, the map

$$[\gamma] \in \pi_1(M,x_o) \longmapsto a([\gamma]) \in \mathbb{C}^*$$

is an abelian representation, that is $a([\gamma_1 * \gamma_2]) = a([\gamma_1])\cdot a([\gamma_2]) = a([\gamma_2 * \gamma_1])$. It follows that the map a can be considered as a homomorphism

$$a: H_1(M,\mathbb{Z}) \to \mathbb{C}^*.$$

The suspension of φ (or a) is a non singular foliation $\mathfrak{F}(\varphi)$ on a holomorphic line bundle $L(\varphi) \xrightarrow{\pi} M$, defined as follows: let $\hat{M} \xrightarrow{P} M$ be the universal covering of M and $\hat{\mathfrak{F}}$ be the foliation on $\hat{M}\times\mathbb{C}$ whose leaves are the horizontals $\hat{M}\times\{c\}$, $c \in \mathbb{C}$. Let $\hat{\varphi}: \pi_1(M,x_o) \to \mathrm{Diff}(\hat{M}\times\mathbb{C})$ be the action defined by

(10) $$\hat{\varphi}([\gamma])(\hat{x},z) = (f_{[\gamma]^{-1}}(\hat{x}),\ a([\gamma])\cdot z),$$

where $f_{[\gamma]^{-1}}$ denotes the covering automorphism associated to the loop $[\gamma]^{-1} \in \pi_1(M,x_o)$. It can be easily proved that every orbit of $\hat{\varphi}$ is discrete and so the space obtained by identifying two points of $\hat{M}\times\mathbb{C}$ in the same orbit is a manifold. We denote this manifold by $L(\varphi)$. Let $\hat{M}\times\mathbb{C} \xrightarrow{P} L(\varphi)$ be the projection of the equivalence relation which defines $L(\varphi)$. From (10) it follows that for any $[\gamma] \in \pi_1(M,x_o)$, the transformation $\hat{\varphi}([\gamma])$ sends leaves of $\hat{\mathfrak{F}}$ onto leaves of $\hat{\mathfrak{F}}$. Therefore there exists a foliation $\mathfrak{F}(\varphi)$ on $L(\varphi)$ such that $P^*(\mathfrak{F}(\varphi)) = \hat{\mathfrak{F}}$. Moreover, it can be proved easily that there exists a map $\pi: L(\varphi) \to M$ such that the following diagram commutes

$$\begin{array}{ccc} \hat{M}\times\mathbb{C} & \xrightarrow{p_1} & \hat{M} \\ \Big\downarrow P & & \Big\downarrow p \\ L(\varphi) & \xrightarrow{\pi} & M \end{array}$$

This map π defines a bundle structure on $L(\varphi)$, which in our case is $\mathbb{C}$-linear from (10) and so $L(\varphi) \xrightarrow{\pi} M$ is a line bundle. The foliation $\mathfrak{F}(\varphi)$ is transverse to the fibers of this bundle and the zero section $M_o \subset L(\varphi)$ is a leaf (from (10)). It can be proved also that the holonomy of this leaf on the fiber $\pi^{-1}(x_o)$ is given by the representation φ. Moreover, if F is a leaf of $\mathfrak{F}(\varphi)$ then $\pi|_F: F \to M$ is a covering map (see [5] for the details).

From the construction, it can be proved that $\mathfrak{F}(\varphi)$ satisfies also the following property:

(11) Let $U \subset M$ be a connected open set such that on $\pi^{-1}(U)$ is defined a chart of the bundle $\psi: \pi^{-1}(U) \to U\times\mathbb{C}$. Let $\mathfrak{F}_* = \psi_*(\mathfrak{F}(\varphi)\big|_{\pi^{-1}(U)})$. If F is a leaf of $\mathfrak{F}_*$ then all the other leaves of $\mathfrak{F}_*$ are of the form $c\cdot F$, where $c \in \mathbb{C}$. We will call <u>linear</u> a foliation which satisfies this property.

When U is simply connected, a leaf F of $\mathfrak{F}_*$ is the graph of a holomorphic function $f: U \to \mathbb{C}$. In this case condition (11) means that all the leaves of $\mathfrak{F}_*$ are of the form $F_c = c\cdot F = \{(x, c\cdot f(x));\ x \in U\}$. In particular, if M is simply connected then all the leaves of $\mathfrak{F}(\varphi)$ are diffeomorphic to M and $L(\varphi)$ is holomorphically equivalent to the trivial bundle $M\times\mathbb{C}$.

Now, let $L \xrightarrow{\pi} M$ be a line bundle and suppose that there exists a linear foliation $\mathfrak{F}$ on L (that is $\mathfrak{F}$ is transverse to the fibers and satisfies (11)). In this case the zero section $M_o \subset L$ is a leaf of $\mathfrak{F}$ and it can be proved that its holonomy is linear. If $\varphi\colon \pi_1(M_o,x_o) \to \mathrm{Diff}(\pi^{-1}(x_o)) \approx \mathrm{Diff}(\mathbb{C})$ is this holonomy, then the bundle obtained from φ by suspension is holomorphically equivalent to L (cf. [5]). We observe that in this case the first Chern class of L is zero (this follows from Theorem 1). The proof of the corollary will be based on the following result:

__Lemma__. If M is Kähler then any line bundle $L \xrightarrow{\pi} M$, with Chern class zero, can be obtained by suspension of some linear representation.

__Proof__: We will use the following result, which is proved in [4] (p. 313):

Let M be a compact Kähler manifold and $L \xrightarrow{\pi} M$ be a holomorphic line bundle with Chern class zero. Then there exists a covering $\mathfrak{U} = \{U_\alpha\}_{\alpha\in I}$ of M and a collection $\{\psi_\alpha\colon \pi^{-1}(U_\alpha) \to U_\alpha\times\mathbb{C}\}_{\alpha\in I}$ of bundle charts such that if $U_\alpha \cap U_\beta \neq \emptyset$ then the transition map $\psi_\alpha\circ\psi_\beta^{-1}\colon (U_\alpha\cap U_\beta)\times\mathbb{C} \to (U_\alpha\cap U_\beta)\times\mathbb{C}$ is of the form

(12) $$\psi_\alpha\circ\psi_\beta^{-1}(x,z) = (x, c_{\alpha\beta}\cdot z),$$

where $c_{\alpha\beta} \in \mathbb{C}^*$ and $c_{\alpha\beta}c_{\beta\gamma}c_{\gamma\alpha} = 1$ if $U_\alpha \cap U_\beta \cap U_\gamma = 1$.

We can define a linear foliation $\mathcal{F}$ on L as follows: let $\mathcal{F}_\alpha$ be the foliation on $U_\alpha\times\mathbb{C}$, whose leaves are the horizontals $U_\alpha\times\{c\}$, $c \in \mathbb{C}$. Clearly $\mathcal{F}_\alpha$ satisfies (11) and is transverse to the fibers $\{x\}\times\mathbb{C}$, $x \in U_\alpha$. From (12), it follows that if $U_\alpha \cap U_\beta \neq$ $\neq \emptyset$, then the foliations $\mathcal{F}_\alpha$ and $\mathcal{F}_\beta$ glue together along $\pi^{-1}(U_\alpha\cap U_\beta)$. This clearly defines a global linear foliation $\mathcal{F}$ on L.

Now, the holonomy of the leaf $M_o \subset L$ is linear and the bundle obtained by the suspension of this holonomy is holomorphically equivalent to L. This proves the lemma and the corollary in the case $S = \emptyset$.

Let us suppose that $\emptyset \neq S = S_1 \cup\ldots\cup S_k \subset M$. Let $\lambda_1,\ldots,\lambda_k \in \mathbb{C}$ be such that $[\sum_{j=1}^{k} \lambda_j\Omega_j] = [c_1(L)]$, where L is some line bundle over M. We will use the following well known fact (cf.[4]): let L_1 be a line bundle over M with the same Chern class as L. Then there exists a line bundle L_0 over L, with Chern class 0 and such that $L = L_1 \otimes L_0$. This follows from the group structure on the set of line bundles over M.

From Theorem 2 there exist a line bundle $L_1 \xrightarrow{\pi_1} M$, and a singular foliation $\mathcal{F}_1$ on L_1 which satisfies b), c) and d) of Theorem 2. From Theorem 1 we have

$$[c_1(L_0)] = [\sum_{j=1}^{k} \lambda_j \Omega_j] = [c_1(L)].$$

Hence there exists a line bundle $L_0 \xrightarrow{\pi_0} M$ such that $L = L_1 \otimes L_0$. Let $\mathcal{F}_0$ be a non singular foliation on L_0 given by the lemma. Let us consider a covering $\mathcal{U} = \{U_\alpha\}_{\alpha \in I}$ of M by open sets with the following properties:

i) For each $\alpha \in I$, there exist bundle charts $\psi_\alpha^0: \pi_0^{-1}(U_\alpha) \to U_\alpha \times \mathbb{C}$ and $\psi_\alpha^1: \pi_1^{-1}(U_\alpha) \to U_\alpha \times \mathbb{C}$ of L_0 and L_1 respectively, where the transition functions of $\mathcal{U}$ with respect to L_0 are constants as in the lemma.

ii) The foliation $\mathcal{F}_0\big|_{\pi_0^{-1}(U_\alpha)}$ is represented in $U_\alpha \times \mathbb{C}$ by the linear foliation whose leaves are the horizontals $U_\alpha \times \{z\}$, $z \in \mathbb{C}$.

iii) The foliation $\mathcal{F}_1\big|_{\pi_1^{-1}(U_\alpha)}$ is represented in $U_\alpha \times \mathbb{C}$ by the closed meromorphic 1-form $\theta_\alpha = \frac{dy_\alpha}{y_\alpha} - \sum_{j=1}^{k} \lambda_j \frac{df_{j\alpha}}{f_{j\alpha}}$, where $\{f_{j\alpha}\}_{\alpha \in I}$, $j=1,\dots,k$, are as in the proof of Theorem 2.

If we let $\psi_\alpha^0 = (x, z_\alpha): \pi_0^{-1}(U_\alpha) \to U_\alpha \times \mathbb{C}$, then the foliation $\mathcal{F}_0$ is represented in $U_\alpha \times \mathbb{C}$ by the closed 1-form dz_α, that is, its leaves are the solutions of $dz_\alpha = 0$. From condition i), the meromorphic 1-form dz_α / z_α is globally defined, because $z_\beta = c_{\beta\alpha} z_\alpha$ on $U_\alpha \cap U_\beta$ and so $dz_\beta / z_\beta = dz_\alpha / z_\alpha$.

From the definition, the transition functions of the bundle $L = L_1 \otimes L_o$ are $g_{\alpha\beta} = c_{\alpha\beta} \cdot h_{\alpha\beta}$, where $\{h_{\alpha\beta} \mid U_\alpha \cap U_\beta \neq \phi\}$ are the transition functions of L_1. If we set

$$\eta_\alpha = \frac{dw_\alpha}{w_\alpha} - \sum_{j=1}^{k} \lambda_j \frac{df_{j\alpha}}{f_{j\alpha}}$$

where $w_\alpha = g_{\alpha\beta} \cdot w_\beta = c_{\alpha\beta} \cdot h_{\alpha\beta} \cdot w_\beta$, then it is not difficult to see that the collection of 1-forms $\{\eta_\alpha\}_{\alpha \in I}$, defines a globally well defined closed meromorphic 1-form on L. The foliation $\mathcal{F}$ on L, obtained from this form satisfies b), c) and d) of Theorem 2, and this proves the corollary.

Remark 4. From the construction of the corollary we can compute the dimension of the space of linear foliations on L which have a fixed singular set $S = S_1 \cup \ldots \cup S_k \subset M$ and fixed normal indexes, say $\lambda_1, \ldots, \lambda_k$. In fact this set can be parametrized by the set of closed holomorphic 1-forms on M, as we shall see below. First of all we give a definition in order to simplify the notations.

Definition. Let $L_1 \xrightarrow{\pi_1} M$ and $L_2 \xrightarrow{\pi_2} M$ be line bundles over M, and $\mathcal{F}_1, \mathcal{F}_2$ be linear foliations on L_1 and L_2 respectively. We define the tensor product, $\mathcal{F}_1 \otimes \mathcal{F}_2$, of $\mathcal{F}_1$ and $\mathcal{F}_2$ as the linear foliation on $L_1 \otimes L_2$ given as follows: let $\{U_\alpha\}_{\alpha \in I}$ be a covering of M by open sets such that $\pi_j^{-1}(U_\alpha) \simeq U_\alpha \times \mathbb{C}$, $j = 1,2$. We take bundle coordinates (x, y_α^j) on $\pi_j^{-1}(U_\alpha)$, $j = 1,2$. Let $\{g_{\alpha\beta}^j \mid U_\alpha \cap U_\beta \neq \phi, \alpha,\beta \in I\}$ be the collection of transition functions of the bundle L_j, $j=1,2$. As we have seen $\mathcal{F}_j$ can be given by a meromorphic close 1-form θ^j on L_j, whose expression on $\pi_j^{-1}(U_\alpha)$ is of the form

$$(13) \qquad \theta_\alpha^j = \frac{dy_\alpha^j}{y_\alpha^j} - \mu_\alpha^j$$

where μ_α^j is a closed meromorphic 1-form depending only on x.

Let $\theta_\alpha = \frac{dy_\alpha}{y_\alpha} - (\mu^1_\alpha + \mu^2_\alpha)$, where (x,y_α) are bundle coordinates of $L_1 \otimes L_2 \xrightarrow{\pi} M$, on $\pi^{-1}(U_\alpha) \simeq U_\alpha \times \mathbb{C}$. It is not difficult to see that the set $\{\theta_\alpha \mid \alpha \in I\}$ defines a meromorphic closed 1-form θ on $L_1 \otimes L_2$, where $\theta\big|_{\pi^{-1}(U_\alpha)} = \theta_\alpha$. This follows from the fact that $\mu^j_\beta - \mu^j_\alpha = \frac{dg^j_{\beta\alpha}}{g^j_{\beta\alpha}}$ if $U_\alpha \cap U_\beta \neq \phi$, $j = 1,2$. As before this form defines a linear foliation on $L_1 \otimes L_2$, which we denote by $\mathcal{F}_1 \otimes \mathcal{F}_2$.

On the other hand, if one has a linear foliation $\mathcal{F}_1$ on $L_1 \xrightarrow{\pi_1} M$ given by a closed form as in (13), then we can define a linear foliation $\mathcal{F}_1^{-1}$ on the bundle $L_1^{-1} \xrightarrow{\pi} M$, whose transition functions are $(g_{\alpha\beta})^{-1}$, $U_\alpha \cap U_\beta \neq \phi$. This foliation is given by the meromorphic closed 1-form θ on L_1^{-1}, such that

$$\theta\Big|_{\pi^{-1}(U_\alpha)} = \frac{dy_\alpha}{y_\alpha} + \mu^1_\alpha$$

where (x,y_α) are the bundle coordinates on $\pi^{-1}(U_\alpha) \simeq U_\alpha \times \mathbb{C}$.

We have the following properties:

(a) Let $S^1 = S^1_1 \cup \ldots \cup S^1_{k_1}$ and $S^2 = S^2_1 \cup \ldots \cup S^2_{k_2}$ be the sets of singularities of $\mathcal{F}_1$ and $\mathcal{F}_2$ respectively. Suppose that $i(S^j_m, \mathcal{F}_j, M) = \lambda^j_m \in \mathbb{C}^*$, $m \in \{1,\ldots,k_j\}$, $j=1,2$. Then the set of singularities, S of $\mathcal{F}_1 \otimes \mathcal{F}_2$, is contained in $S^1 \cup S^2$. Moreover if $i \in \{1,\ldots,k_1\}$ then

$$i(S^1_i, \mathcal{F}_1 \otimes \mathcal{F}_2, M) = \lambda^1_i + \sum_{j=1}^{k_2} \delta(S^1_i, S^2_j) \cdot \lambda^2_j$$

where $\delta(S^1_i, S^2_j) = 0$ if $S^1_i \neq S^2_j$ and $\delta(S^1_i, S^2_j) = 1$ if $S^1_i = S^2_j$. Analogously

$$i(S^2_i, \mathcal{F}_1 \otimes \mathcal{F}_2, M) = \lambda^2_i + \sum_{j=1}^{k_1} \delta(S^2_i, S^1_j) \cdot \lambda^2_j .$$

In particular, if $S^1_i \neq S^2_j$ for any $i \in \{1,\ldots,k_1\}$ and $j \in \{1,\ldots,k_2\}$, then $S = S_1 \cup S_2$ and $i(S^1_k, \mathcal{F}_1 \otimes \mathcal{F}_2, M) = \lambda^1_k$ and

$i(S_k^2, \mathcal{F}_1 \otimes \mathcal{F}_2, M) = \lambda_k^2$, for any k.

(b) Let $S = S_1 \cup \ldots \cup S_k$ be the set of singularities of $\mathcal{F}$, a linear foliation on $L \xrightarrow{\pi} M$. Suppose $i(S_j, \mathcal{F}, M) = \lambda_j$, $j=1,\ldots,k$. Then the set of singularities of $\mathcal{F}^{-1}$ is S and $i(S_j, \mathcal{F}^{-1}, M) = -\lambda_j$, $j=1,\ldots,k$.

(c) Let $\mathcal{F}$ be a linear foliation on $L \xrightarrow{\pi} M$. Then $\mathcal{F} \otimes \mathcal{F}^{-1}$ is equivalent to the product foliation on $L \otimes L^{-1} \simeq M\times\mathbb{C}$.

We leave the proof of the above properties for the reader.

Now suppose we have a linear foliation $\mathcal{F}_0$ defined in a line bundle $L \xrightarrow{\pi} M$, with singular set $S = S_1 \cup \ldots \cup S_k \subset M$ and normal indexes $i(S_j, \mathcal{F}_0, M) = \lambda_j$. Suppose also that all singularities of $\mathcal{F}_0$ are of order 1, in the sense that in (4) of §2.2 we have $\ell_1 = \ldots = \ell_k = 1$. Let $\mathcal{F}_1$ be another linear foliation on L with the same singular set, same indexes and singularities of order 1. Let $\mathcal{F}(\mathcal{F}_1) = \mathcal{F}_1 \otimes \mathcal{F}_0^{-1}$. Then $\mathcal{F}(\mathcal{F}_1)$ is a linear foliation on $L \otimes L^{-1} \simeq M\times\mathbb{C}$, without singularities, since from the assumptions their singularities cancel one another. On the other hand, this foliation is given by a meromorphic 1-form θ, on $M\times\mathbb{C}$, which locally looks like

$$\theta_\alpha = \frac{dy_\alpha}{y_\alpha} - \eta_\alpha$$

where η_α is closed and has no poles, since $\mathcal{F}(\mathcal{F}_1)$ has no singularities. Now, since $y_\beta = y_\alpha$ on $U_\alpha \cap U_\beta$ $(L \otimes L^{-1} = M\times\mathbb{C})$, we can write in fact

$$\theta_\alpha = \frac{dy}{y} - \eta_\alpha$$

where y is the vertical coordinate in $M\times\mathbb{C}$. It follows that the collection $\{\eta_\alpha\}_{\alpha\in I}$ defines a closed holomorphic 1-form in M, say η. In this case $\theta = \frac{dy}{y} - \eta$ and the foliation $\mathcal{F}(\mathcal{F}_1)$ is defined by the linear equation

(14) $$dy = y\eta .$$

This defines a map $\mathfrak{F}_1 \mapsto \eta$ = closed holomorphic 1-form on M. On the other hand, given a closed holomorphic 1-form η on M, we consider the foliation $\mathfrak{F}_1'$, whose leaves are solutions of (14) and set $\mathfrak{F}_1 = \mathfrak{F}_0 \otimes \mathfrak{F}_1'$, which is clearly a foliation on L. This shows that the correspondence $\mathfrak{F}_1 \mapsto \eta$ is in fact a bijection, and so the set of singular foliations with the same singular set, same normal indexes as $\mathfrak{F}_0$ and with singularities of order 1, can be parametrized by the set of holomorphic closed 1-forms on M.

References

[1] C. Camacho and P. Sad - Invariant varieties through singularities of holomorphic vector fields - Ann. of Math. 115 (1982), 579-595.

[2] Alcides Lins Neto - Construction of singular holomorphic vector fields and foliations in dimension two - To appear in the Journal of Diff. Geometry.

[3] D. Cerveau and J.F. Mattei - Formes intégrables holomorphes singulières. Astérisque 97.

[4] P. Griffiths and J. Harris - Principles of algebraic geometry - Wiley (1978).

[5] C. Camacho and A. Lins Neto - Geometric theory of foliations - Birkhauser (1985).

Instituto de Matemática Pura e Aplicada (IMPA)
Estrada Dona Castorina 110
20460 - Rio de Janeiro, RJ - Brazil

R MAÑÉ

On a theorem of Klingenberg

I - Introduction

Let M be a complete Riemannian manifold and let UM be its unit tangent bundle endowed with its standard Riemannian structure. The geodesic flow $\varphi_t\colon UM \circlearrowleft$, $t \in \mathbb{R}$, is defined by $\varphi_t(p,v) = (\gamma(t),\dot{\gamma}(t))$ where $\gamma\colon \mathbb{R} \to M$ is the geodesic with initial condition $\gamma(0) = p$, $\dot{\gamma}(0) = v$. This and all the geodesics in this work will be parametrized by arc lenght. When the curvature of M is $\leq -a < 0$, its geodesic flow is an Anosov flow, i.e., there exists a continuous splitting $T(UM) = E^s \oplus E^u \oplus E^{\varphi}$, where E^{φ} is the one dimensional subbundle of $T(UM)$ tangent to the orbits of φ, such that there exist constants $C > 0$, $0 < \lambda < 1$ satisfying

$$\|(D\varphi_t)/E^s_\theta\| \leq C\lambda^t$$

$$\|(D\varphi_{-t})/E^u_\theta\| \leq C\lambda^t$$

for all $\theta \in UM$ and $t \geq 0$.

This strong property was discovered by Hadamard [6]. Later, it was successfully exploited by many authors; most notably by Hopf [8] (in the two-dimensional case) and Anosov [1] (in the general n-dimensional case), whose works are the foundation of the nowadays remarkably rich theory of the geodesic flow of negatively curved manifolds.

On the other hand there are manifolds whose geodesic flow is of Anosov type and whose curvature is strictly positive in certain regions. Examples of this phenomena and a study of sufficient conditions that grant the Anosov property to the geodesic flow were developed by Eberlein [3].

Riemannian manifolds whose geodesic flow is Anosov share several of the main properties of negatively curved manifolds. Some of them, like (in the compact case) the ergodicity of the geodesic flow and the density of the closed geodesics, because they follow from a theorem that holds for any Anosov volume preserving flow. But there also other shared properties, of a more strictly geometric nature. Here is a list of the main ones in the case when M is compact:

I) M has no conjugate points.

II) There exist $C_o > 0$ and $\mu > 0$ such that every perpendicular Jacobi vector field J with $J(0) = 0$ satisfies $\|J(t)\| \geq C_o \|\dot{J}(0)\| \sinh \mu t$ for all t.

III) If $\tilde{M}$ is the universal covering of M, then two different geodesics $\gamma_i : \mathbb{R} \to \tilde{M}$, $i = 1,2$ with $\gamma_1(0) = \gamma_2(0)$, satisfy

$$\lim_{t \to \pm\infty} d(\gamma_1(t), \gamma_2(t)) = +\infty.$$

IV) Given a geodesic $\gamma: \mathbb{R} \to \tilde{M}$ and a point $p \in \tilde{M}$ there exists a unique geodesic γ_1 with $\gamma_1(0) = p$ and ω-asymptotic with γ i.e. satisfying

$$\sup_{t \geq 0} d(\gamma_1(t), \gamma(\mathbb{R}^+)) < \infty.$$

In this case there exists a such that

$$\lim_{t \to +\infty} d(\gamma_1(t), \gamma(t+a)) = 0.$$

In a similar way, but replacing $t \geq 0$ and $t \to +\infty$ by $t \leq 0$ and $t \to -\infty$, we define α-asymptotic geodesics.

V) If two geodesics of $\tilde{M}$ are α and ω-asymptotic then they coincide up to a translation of the parametrization.

VI) Given two geodesics γ_1 and γ_2 such that γ_1 is not ω-asymptotic with $t \to \gamma_2(-t)$, there exists a geodesic γ in $\tilde{M}$ that is ω-asymptotic with γ_1 and α-asymptotic with γ_2. Moreover γ is unique up to reparametrizations by a translation of the parameter.

Of these properties the crucial one is the expanding property of the Jacobi vector fields (property II); that in particular implies that there are no conjugate points. The other properties follow from (II) as reasonably easy corollaries adapting the proofs of the negatively curved case. That property was proved by Klingenberg in [9] assuming M to be compact.

The objective of this paper is to extend this result to complete non compact manifolds with curvature bounded from below. The rest of the list of properties then follow by strict repetition of their proofs in the compact case. Our proof will contain a proof for the compact case different from that given by Klingenberg in [9].

When M is compact we shall show that if the derivative of the geodesic flow leaves invariant a continuous Lagrangian subbundle, then M has no conjugate points. When the geodesic flow is Anosov, both the stable and unstable subbundles are Lagrangian.

To state our results let us first recall the basic definitions and properties of the unit tangent bundle and the geodesic flow.

If $\theta = (p,v) \in UM$, denote $N(\theta)$ the space of vectors $w \in T_pM$ such that $\langle w,v \rangle = 0$. Let $\pi\colon UM \to M$ be the canonical projection and define $S(\theta) = (D_\theta\pi)^{-1}N(\theta)$. $S(\theta)$ is orthogonal to E_θ^φ and $(D_\theta\pi)E_\theta^\varphi$ is the one dimensional space generated by v. Moreover $S(\theta)$ is invariant under $D\varphi$, i.e., $(D\varphi)S(\theta) = S(\varphi_t(\theta))$ for all θ and t. The vertical subspace of $S(\theta)$ is defined as $V(\theta) = (D_\theta\pi)^{-1}(\{0\})$. The horizontal subspace $H(\theta)$ is the orthogonal complement of $V(\theta)$ in $S(\theta)$. The map $(D_\theta\pi)/H(\theta)$ is an isometry between $H(\theta)$ and $N(\theta)$. Moreover there exists an isometry $J_\theta : S(\theta)\circlearrowleft$ such that $J_\theta^2 = -I$, $J_\theta V(\theta) = H(\theta)$, $J_\theta H(\theta) = V(\theta)$ and the symplectic form $\Omega_\theta : S(\theta)\times S(\theta) \to \mathbb{R}$ defined by $\Omega_\theta(v,w) = \langle v, J_\theta w\rangle$ is invariant under $D\varphi$, i.e., $\Omega_\theta(v,w) = \Omega_{\varphi_t(\theta)}((D\varphi_t)v,(D\varphi_t)w)$ for all θ, t, w and v. For proofs of this basic properties see [4]. Denoting $n = \dim M$, it is clear that $\dim H(\theta) = \dim V(\theta) = \dim N(\theta) = n-1$. A subspace $E \subset S(\theta)$ is Lagrangian if $\dim E = n-1$ and $\Omega_\theta(v,w) = 0$ for all $v \in E$, $w \in E$ (or, what is the same, if $S(\theta)$ is the orthogonal sum of E and $J_\theta E$. The angle between two subspaces $E \subset S(\theta)$, $F \subset S(\theta)$ with $\dim E = \dim F = n-1$, is defined by $\alpha(E,F) = \infty$ if $E \cap F^\perp \neq \{0\}$ and, when $E \cap F^\perp = \{0\}$, by $\alpha(E,F) = \|L\|$ where $L\colon F \to F^\perp$ is the unique linear map such that E is the graph of L, i.e., $E = \{x+Lx \mid x\in F\}$.

When the geodesic flow is Anosov, E_θ^s and E_θ^u are Lagrangian subspaces of $S(\theta)$. To prove this property let us first show that $E_\theta^s \subset S(\theta)$, $E_\theta^u \subset S(\theta)$ for all θ. Suppose that $v \in T_\theta(UM)$ and $v \notin S(\theta)$. Let w be its orthogonal projection on E_θ^φ. Then, since $(D\varphi_t)S(\theta) = S(\varphi_t(\theta))$ it follows that $(D\varphi_t)w$ is the orthogonal projection of $(D\varphi_t)v$ on $E_{\varphi_t(\theta)}^\varphi$. Moreover $\|(D\varphi_t)w\| = \|w\|$. Then $\|(D\varphi_t)v\| \geq \|(D\varphi_t)w\| \geq \|w\|$. This means that a vector not contained in $S(\theta)$ cannot be contracted by $(D\varphi_t)$. Hence $E_\theta^s \subset S(\theta)$. Moreover, if x and y belong to E_θ^s, we have $\Omega_\theta(x,y) = \Omega_{\varphi_t(\theta)}((D\varphi_t)x,(D\varphi_t)y)$ for all t. Taking the limit for $t \to +\infty$

we obtain $\Omega_\theta(x,y) = 0$. Hence E^s_θ is a Lagrangian subspace of $S(\theta)$. A similar argument proves the same property for E^u_θ.

<u>Theorem A</u>. If M is a complete Riemannian manifold whose geodesic flow is Anosov and with curvature $\geq -c$, $c > 0$, then M has no conjugate points and for all $\theta \in UM$ the stable and unstable spaces E^s_θ, E^u_θ, are Lagrangian subspaces of $S(\theta)$ satisfying

$$\alpha(E^s_\theta, H(\theta)) \leq \sqrt{c}$$

$$\alpha(E^u_\theta, H(\theta)) \leq \sqrt{c}.$$

<u>Corollary</u>. Under the assumptions of Theorem A, there exist constants $C_0 > 0$ and $\mu > 1$ such that every perpendicular Jacobi vector field J, with $J(0) = 0$, satisfies

$$\|J(t)\| \geq C_0 \|\dot{J}(0)\| \sinh \mu t$$

for all $t \geq 0$.

The proof of Theorem A and its corollary will be given in the next section.

For the statement of the next theorem we need the following definition: we say that the geodesic flow of a complete Riemannian manifold M has an invariant Lagrangian subbundle if there exists a continuous subbundle F of $T(UM)$ such that for all $\theta \in UM$ the fiber F_θ is a Lagrangian subspace of $S(\theta)$ and $F_{\varphi_t(\theta)} = (D\varphi_t)F_\theta$ for all $t \in R$.

<u>Theorem B</u>. A complete, finite volume Riemannian manifold whose geodesic flow has an invariant Lagrangian subbundle, has no conjugate points.

The proof of Theorem B is completely different from that of Theorem A and can be read independently. It uses an argument involving a Maslov index, that we shall introduce through a definition slightly different from the standard one exposed by Arnold in [1], that makes possible its calculation in our case as a direct application of the Ricatti equation.

II - Proof of Theorem A

In this section we shall prove Theorem A using the following propositions that will be proved after completing the proof of Theorem A.

Proposition II.1. Let M be a Riemannian manifold and $\gamma: [0,a] \to M$ a geodesic arc. If there exists a Lagrangian subspace $E \subset S(\gamma(0),\dot{\gamma}(0))$ such that $V(\gamma(t),\dot{\gamma}(t)) \cap (D\varphi_t)E = \{0\}$ for all $0 \le t \le a$ then the geodesic arc γ doesn't contain conjugate points.

Proposition II.2. Let M be a complete Riemannian manifold with curvature $\ge -c$, $c \ge 0$, and let $\gamma: \mathbb{R} \to M$ be a geodesic such that there exists a Lagrangian subspace $E \subset S(\gamma(0),\dot{\gamma}(0))$ such that $E \cap V(\gamma(0),\dot{\gamma}(0)) \ne \{0\}$ and constants $C > 0$, $0 < \lambda < 1$ satisfying

$$\|(D\varphi_t)/(D\varphi_T)E\| \le C\lambda^t$$

for all $t \ge 0$, $T \ge 0$. Then there exists in $\gamma(\mathbb{R}^+)$ points conjugate to $\gamma(0)$ and moreover there exists $\tau = \tau(c,\lambda,C)$ such that every interval $[a,b] \subset \mathbb{R}^+$ with $b-a > \tau$ contains a point t such that

$$V(\gamma(t),\dot{\gamma}(t)) \cap (D\varphi_t)E \ne \{0\}.$$

Proposition II.2 also holds reversing the time, i.e., if $\|(D\varphi_t)/(D\varphi_T)E\| \le C\lambda^{|t|}$ for all $t \le 0$, $T \le 0$ then there exist in $\gamma((-\infty,0])$ points conjugate to $\gamma(0)$ and every interval $[a,b] \subset$ $\subset (-\infty,0]$ with $b-a > \tau$ contains a point t such that

$V(\gamma(t),\dot\gamma(t)) \cap (D\varphi_t)E \neq \{0\}$.

To prove Theorem A let us first recall the formalism through which the dynamics of $D\varphi$ is related to Jacobi vector fields and Lagrangian subspaces to symmetric linear maps. Given $\theta \in UM$ define a linear map $L_\theta: S(\theta) \to N(\theta)\times N(\theta)$ by $L_\theta(v) = ((D_\theta\pi)v, 0)$ if $v \in H(\theta)$ and $L_\theta v = (0, (D_\theta\pi)J_\theta v)$ if $v \in V(\theta)$. By the properties explained in the Introduction L_θ is an isometry. Identifying $S(\theta)$ with $N(\theta)\times N(\theta)$ through this isometry the vertical subspace becomes $\{0\}\times N(\theta)$ and the horizontal becomes $N(\theta)\times\{0\}$. Any subspace $E \subset S(\theta)$ with $\dim E = n-1$ (where $\dim M = n$) such that $E \cap (\{0\}\times N(\theta)) = \{0\}$ can be written as the graph of a linear map $T: N(\theta) \circlearrowleft$ i.e. $E = \{(v,Tv) \mid v \in N(\theta)\}$ and E is Lagrangian if and only if T is symmetric. Moreover for every $(p,v) = \theta \in UM$ we have the curvature symmetric map $K(\theta): N(\theta) \circlearrowleft$ given by $K(\theta)w = R_p(v,w)v$. If E is a Lagrangian subspace of $S(\theta)$ such that $(D\varphi_t)E \cap V(\varphi_t(\theta)) = \{0\}$ for all t in an interval $(-\epsilon,\epsilon)$, then, as explained above, we can write $(D\varphi_t)E = \text{graph } U(t)$ where $U(t): N(\varphi_t(\theta)) \circlearrowleft$ is a symmetric map. This family of maps satisfies the Ricatti equation

$$\dot U(t) + U^2(t) + K(t) = 0 \qquad (1)$$

where $K(t) = K(\varphi_t(\theta))$. Finally, if $v \in S(\theta)$, we can write, using the identifications of $S(\varphi_t(\theta))$ with $N(\varphi_t(\theta)) \times N(\varphi_t(\theta))$, $(D\varphi_t)v = (J(t), Y(t))$, where $J(t)$ and $Y(t)$ belong to $N(\varphi_t(\theta))$ for all t. Then $Y(t) = \dot J(t)$ and $J(t)$ satisfies the Jacobi equation

$$\ddot J(t) = -K(t)J(t).$$

Now suppose that M satisfies the hypothesis of Theorem A. Define Λ as the set of points $\theta \in UM$ such that

$$E^s_{\varphi_t(\theta)} \cap V(\varphi_t(\theta)) = \{0\}$$

for all t. If $\theta \in \Lambda$ we can write

$$E^s_{\varphi_t(\theta)} = \text{graph } U(t)$$

where $U(t): N(\varphi_t(\theta)) \circlearrowleft$ is a symmetric linear map that satisfies the Ricatti equation (1) for all $t \in \mathbb{R}$. By a result of Green, [5], a symmetric solution of (1) defined for all $t \in \mathbb{R}$ satisfies

$$\sup_t \|U(t)\| \leq \left(\sup_{v \neq 0, t} (\langle -K(t)v, v\rangle / \|v\|^2) \right)^{1/2}$$

Hence, in our case:

$$\sup_t \|U(t)\| \leq \sqrt{c}.$$

Therefore, $\theta \in \Lambda$ implies

$$\sup_t \alpha(E^s_{\varphi_t(\theta)}, H(\theta)) \leq \sqrt{c}. \tag{2}$$

This proves that Λ is closed. Now let us study the complement Λ^c of Λ. We claim that there exists $\tau > 0$ such that if $\theta \in \Lambda^c$ then every interval $[a,b]$ with $b-a > \tau$ contains a point t such that $E^s_{\varphi_t(\theta)} \cap V(\varphi_t(\theta)) \neq \{0\}$. Let $C > 0$ and $0 < \lambda < 1$ be the contraction constants of the Anosov flow φ. Let $\tau = \tau(c,\lambda,C) > 0$ be given by II.2. Suppose that $\theta \in \Lambda^c$. Hence $E^s_{\varphi_t(\theta)} \cap V(\varphi_t(\theta)) \neq$ $\neq \{0\}$ for some $t \in \mathbb{R}$. Suppose that θ doesn't satisfy the property of the claim for $\tau = \tau(c,\lambda,C)$. Then there exists an interval $[a,b]$, with $b-a > \tau$, such that $E^s_{\varphi_t(\theta)} \cap V(\varphi_t(\theta)) = \{0\}$ for all $t \in [a,b]$. But then $E^s_{\varphi_t(\theta)} \cap V(\varphi_t(\theta)) = \{0\}$ for all $t < a$ because if $E^s_{\varphi_{\bar{t}}(\theta)} \cap V(\varphi_{\bar{t}}(\theta)) \neq \{0\}$ for some $\bar{t} \leq a$ then, by II.2 (applied with the origin of the parametrization placed at $\bar{t}$) there exists $t \in [a,b]$ satisfying $E^s_{\varphi_t(\theta)} \cap V(\varphi_t(\theta)) \neq \{0\}$. This contradiction shows that $E^s_{\varphi_t(\theta)} \cap V(\varphi_t(\theta)) = \{0\}$ for all $t < a$. Then, by II.1, there no conjugate points in $\gamma/(-\infty,a]$. But then

$E^u_{\varphi_t(\theta)} \cap V(\varphi_t(\theta)) = \{0\}$ for all $t \leq a$ because if $E^u_{\varphi_{\bar t}(\theta)} \cap$ $\cap V(\varphi_{\bar t}(\theta)) \neq \{0\}$ for some $\bar t \leq a$, then, applying II.2 (reversing time and placing the origin of the parametrization at $\bar t$) we obtain points in $(-\infty, \bar t]$ conjugate with $\bar t$. This contradiction shows that $E^u_{\varphi_t(\theta)} \cap V(\varphi_t(\theta)) = \{0\}$ for all $t \leq a$. But then $E^u_{\varphi_t(\theta)} \cap$ $\cap V(\varphi_t(\theta)) = \{0\}$ for all $t \in R$ because if this property fails at $\bar t \geq a$, applying II.2 (reversing time and placing the origin at $\bar t$) we should have in any interval $[c,d] \subset (-\infty, \bar t]$ with $d-c > \tau$ points t where $E^s_{\varphi_t(\theta)} \neq \{0\}$. We have thus proved $E^s_{\varphi_t(\theta)} \cap V(\varphi_t(\theta)) =$ $= \{0\}$ for all t. By II.1 it follows that γ doesn't contain conjugate points. But $E^s_{\varphi_r(\theta)} \cap V(\varphi_r(\theta)) \neq \{0\}$ and this, by II.2, implies the existence of conjugate points in $[r, +\infty)$. This contradiction completes the proof of the claim.

Now observe that the claim easily implies that Λ^c is closed. Since Λ is also closed one of them must be empty. When M is not compact there exists $\theta \in UM$ such that the geodesic $\gamma: \mathbb{R} \to M$ with initial condition $(\gamma(0), \dot\gamma(0)) = \theta$ is a ray i.e. $\gamma/\mathbb{R}^+$ doesn't contain conjugate points. In particular, by II.2 $E^s_{\varphi_t(\theta)} \cap V(\varphi_t(\theta)) =$ $= 0$ for all $t \geq 0$. Hence $\theta \notin \Lambda^c$ because it doesn't satisfy the property of the claim. Hence $\Lambda \neq \emptyset$ i.e. $\Lambda = UM$. By the definition of Λ and Proposition II.1 this implies that M has no conjugate points. Moreover by (1) we have $\alpha(E^s_\theta, H(\theta)) \leq \sqrt{c}$ for all $\theta \in \Lambda$ i.e. for all $\theta \in UM$. The same argument replacing E^s by E^u shows that $\alpha(E^u_\theta, H(\theta)) \leq \sqrt{c}$ for all $\theta \in UM$. When M is compact there are no conjugate points (by Theorem B). Hence $\Lambda^c = \emptyset$ because II.2 the geodesic determined by an element $\theta \in \Lambda^c$ should contain conjugate points. Then $\Lambda = UM$ and the proof of Theorem A is completed as before.

Now let us prove Proposition II.1. Take symmetric linear maps $S(t): N(\varphi_t(\theta)) \circlearrowleft$ such that:

$$\text{graph } S(t) = (D\varphi_t)E \tag{2}$$

for all $0 \le t \le a$. Let $J(t)$ be a perpendicular Jacobi vector field along γ with $J(c) = 0$ and $\dot{J}(c) \ne 0$ for some $0 < c < a$. We want to prove that $J(t) \ne 0$ for all $0 < t < a$. For each $0 \le t \le a$ let $S(t): N(\varphi_t(\theta)) \circlearrowleft$ be a symmetric map whose graph is $(D\varphi_t)E$. Let $Y(t): N(\theta)) \to N(\varphi_t(\theta))$ be a family of isomorphisms satisfying $Y(c) = I$ and

$$\dot{Y}(t) = S(t)Y(t)$$

for $0 \le t \le a$. If $v \in E$, we can write

$$(D\varphi_t)v = (v(t), S(t)v(t))$$

where $v(t) \in N(\varphi_t(\theta))$. But, as we explained above, we have $S(t)v(t) = \dot{v}(t)$. Hence $v(t) = Y(t)w$, where $w = v(c)$, and then

$$(D\varphi_t)v = (Y(t)w, S(t)Y(t)w)$$

for all $0 \le t \le a$ and $J_1(t) = Y(t)w$ is a perpendicular Jacobi vector field. Hence, for all $0 \le t \le a$

$$\langle J(t), \dot{J}_1(t)\rangle - \langle \dot{J}(t), J_1(t)\rangle = \langle J(c), \dot{J}_1(c)\rangle -$$

$$- \langle \dot{J}(c), J_1(c)\rangle = -\langle \dot{J}(c), J_1(c)\rangle.$$

Then, for all $0 \le t \le a$

$$\langle J(t), S(t)Y(t)w\rangle - \langle \dot{J}(t), Y(t)w\rangle = -\langle \dot{J}(c), w\rangle.$$

Therefore

$$\langle Y^*(t)(S(t)J(t) - \dot{J}(t)), w\rangle = -\langle \dot{J}(c), w\rangle$$

for all $0 \le t \le a$ and $w \in N(\theta)$. This implies

$$\dot{J}(t) = S(t)J(t) + Y^*(t)^{-1}\, \dot{J}(c).$$

The solution of this equation with $J(c) = 0$ is

$$J(t) = Y(t)\int_c^t Y(s)^{-1}Y^*(s)^{-1}\dot{J}(c)ds.$$

Hence

$$\langle Y^{-1}(t)J(t),\dot{J}(c)\rangle = \int_c^t \|Y^*(s)^{-1}\dot{J}(c)\|^2 \, ds.$$

Then $J(t) \neq 0$ for all $0 \leq t \leq a$ different from c.

Finally let us prove Proposition II.2. We claim that under the hypothesis of II.2, there exists $\tau(c,\lambda,C) > 0$ such that $\gamma/[0,\tau]$ contains points conjugate to $\gamma(0)$ and there exists $t_o \in (0,\tau]$ such that:

$$(D\varphi_{t_o})E \cap V(\gamma(t_o),\dot{\gamma}(t_o)) \neq \{0\}.$$

Clearly this claim implies II.2 because it grants the existence of points in $\gamma/(0,+\infty)$ conjugate to $\gamma(0)$. Moreover, applying the claim to the geodesic $\gamma_1(t) = \gamma(t+t_o)$ and the space $(D\varphi_{t_o})E$, it follows the existence of $0 < t_1 \leq \tau$ such that

$$(D\varphi_{t_1})(D\varphi_{t_o})E \cap V(\gamma(t_o+t_1),\dot{\gamma}(t_o+t_1)) \neq \{0\}.$$

Continuing this method we obtain a sequence $0 < t_n < \tau$ such that denoting $s_n = t_o + \ldots + t_n$ we have

$$(D\varphi_{s_n})E \cap V(\gamma(s_n),\dot{\gamma}(s_n)) \neq \{0\}$$

thus completing the proof of II.2. To prove the claim it suffices to find $\tau_o = \tau_o(c,\lambda,C)$ such that there exists a point in $\gamma/(0,\tau_o)$ conjugate to $\gamma(0)$. Once this proved it follows that $\gamma(\epsilon,\tau_o+\epsilon)$ contains a point conjugate to $\gamma(\epsilon)$ if ϵ is small enough and then there exists $\epsilon \leq t \leq \tau_o + \epsilon$ such that $(D\varphi_t)E \cap V(\gamma(t),\dot{\gamma}(t)) \neq \{0\}$ because otherwise, by II.1, there would be no conjugate points in $\gamma/[\epsilon,\tau_o+\epsilon]$. Hence, taking, let us say, $\tau = \tau_o+1$ the claim is proved. To find $\tau_0 = \tau_0(c,\lambda,C)$ we shall use the following lemma due to Green [5].

Lemma II.3. Let M be a Riemannian manifold with curvature $\geq -c$, $c > 0$. Then there exists $A = A(c) > 0$ such that if $\gamma: [0,a] \to M$ is a geodesic arc without conjugate points and $J(t)$, $0 \leq t \leq a$ is a perpendicular Jacobi vector field on γ with $J(0) = 0$, then

$$\|\dot{J}(t)\| \leq A\|J(t)\|$$

for all $1 \leq t \leq a$.

Take a C^∞ function $\psi: \mathbb{R} \circlearrowleft$ satisfying $\psi(t) = 0$ if $|t| > 1$ and $\psi(0) = 1$. Set

$$B = \int_{-1}^{1} (|\psi''\psi| + c\psi^2)dt.$$

Define $D = ((c/3)+1+A)\exp A$ where A is given by the previous lemma. Take $T > 0$ such that

$$C\lambda^T < (DB)^{-1/2} \tag{4}$$

and let $\tau_0 = T+1$. We shall prove that $\gamma/(0,\tau_0)$ contains a point conjugate to $\gamma(0)$. Suppose that it doesn't. Let $J(t)$, $t \geq 0$, be a perpendicular Jacobi vector field on γ such that $J(0) = 0$, $\|\dot{J}(0)\| = 1$, $\dot{J}(0) \in E$. Since $\gamma(t)$ is not conjugate to $\gamma(0)$ for all $0 < t < \tau_0 = T+1$ there exists a perpendicular Jacobi vector field $\tilde{Z}(t)$ along γ with $\tilde{Z}(T) = 0$ and $\tilde{Z}(0) = \dot{J}(0)$. We have

$$1 = \langle \dot{J}(0), \tilde{Z}(0)\rangle = -\langle J(T), \dot{\tilde{Z}}(T)\rangle.$$

Define $Z(t) = \tilde{Z}(t)/\|\dot{\tilde{Z}}(T)\|$. Then $\|\dot{Z}(T)\| = 1$ and

$$\|Z(0)\| = \frac{\|\tilde{Z}(0)\|}{\|\dot{\tilde{Z}}(T)\|} = \frac{1}{\|\dot{\tilde{Z}}(T)\|} = -\frac{\langle J(T), \dot{\tilde{Z}}(T)\rangle}{\|\dot{\tilde{Z}}(T)\|} \leq$$

$$\leq \|J(T)\| \leq C\lambda^T.$$

Now consider the bilinear form I on the space Ω of piecewise C^2 perpendicular vector fields along $\gamma/[0,T+1]$ defined by

$$I(V,W) = -\int_0^{T+1} \langle \ddot{V}(t)+K(\gamma(t),\dot{\gamma}(t))V,W\rangle -$$

$$- \sum_i \langle (\Delta_i \dot{V})(t_i),W(t_i)\rangle$$

where $(\Delta_i \dot{V})(t_i)$ is the jump of $\dot{V}$ at the discontinuity t_i. Define $V \in \Omega$ as $V(t) = 0$ for $T \leq t \leq T+1$, $V(t) = Z(t)$ for $1 \leq t \leq T$ and $V(t) = t\Phi_1^t Z(1)$ for $0 \leq t \leq 1$, where $\Phi_1^t: T_{\gamma(1)}M \to T_{\gamma(t)}M$ denotes the parallel transport along γ. Define $W \in \Omega$ by $W(t) = \psi(t-T)\Phi_T^t \dot{Z}(T)$. Then

$$I(V,V) = -\int_0^1 t^2\langle K(\gamma(t),\dot{\gamma}(t))\Phi_1^t Z(1),\Phi_1^t Z(1)\rangle +$$

$$+ \langle Z(1)-\dot{Z}(1),Z(1)\rangle.$$

Using Green's lemma:

$$I(V,V) \leq \frac{1}{3} c\|Z(1)\|^2 + A\|Z(1)\|^2 + \|Z(1)\|^2.$$

But also from Green's Lemma and Gronwall's inequality we obtain:

$$\|Z(1)\| \leq \|Z(0)\| \exp A.$$

Hence

$$I(V,V) \leq (\tfrac{1}{3} c+A+1)(\exp A)\|Z(0)\|^2 = D\|Z(0)\|^2 .$$

Moreover

$$I(V,W) = -\|\dot{Z}(T)\| = 1$$

$$I(W,W) = -\int_{T-1}^{T+1} (\psi''\psi + \psi^2\langle K(\gamma(t),\dot{\gamma}(t))\Phi_T^t\dot{Z}(T),\Phi_T^t\dot{Z}(T)\rangle dt \leq$$

$$\leq \int_{-1}^{1} (|\psi''\psi| + c\psi^2)dt = B.$$

Therefore:

$$I(V+\lambda W,V+\lambda W) \leq D\|Z(0)\|^2 + 2\lambda + B\lambda^2 .$$

But

$$4-4DB\|Z(0)\|^2 > 0$$

because by (3) and (4)

$$DB\|Z(0)\|^2 \le DB(C\lambda^T)^2 < 1.$$

Hence $I(V+\lambda W, V+\lambda W) < 0$ for certain values of λ thus proving that the interval $(0,T+1)$ contains points conjugate to $\gamma(0)$.

To prove the corollary of Theorem A consider a perpendicular Jacobi vector field J along the geodesic γ with initial condition $(\gamma(0),\dot{\gamma}(0)) = \theta \in UM$ such that $J(0) = 0$. Let $U(t):N(\varphi_t(\theta)) \hookleftarrow$ be the symmetric map whose graph is $E^u_{\varphi_t(\theta)}$. Let $Y(t): N(\theta) \to N(\varphi_t(\theta))$ be a family of linear maps satisfying $Y(0) = I$ and

$$\dot{Y}(t) = U(t)Y(t) \tag{1}$$

for all $t \in \mathbb{R}$. Then if $v \in E^u_\theta$ is written as $v = (w,u)$ with $w \in N(\theta)$, $u \in N(\theta)$ we have

$$(D\varphi_t)v = (Y(t)w, U(t)Y(t)w)$$

for all t. Then

$$\|(D\varphi_t)v\|^2 = \|Y(t)w\|^2 + \|U(t)Y(t)w\|^2 \le$$
$$\le \|Y(t)w\|^2 + c\|Y(t)w\|^2 = (1+c)\|Y(t)w\|^2.$$

Moreover

$$\|(D\varphi_t)v\|^2 \ge C^2\lambda^{-2t}\|v\|^2 = C^2\lambda^{-2t}(\|w\|^2+\|u\|^2) \ge C^2\lambda^{-2t}\|w\|^2.$$

Hence

$$\|Y(t)w\| \ge C\lambda^{-t}(1+c)^{-1/2}\|w\|$$

for all $t \ge 0$ and w. This can be written as

$$\|Y(t)w\| \ge Ae^{at}\|w\| \tag{2}$$

for all $t \ge 0$, and w, for some $A > 0$, $a > 0$ depending only on

C, λ and c. Now define a perpendicular Jacobi vector field J_1 as $J_1(t) = Y(t)w$. Then

$$\langle J(t),\dot{J}_1(t)\rangle - \langle \dot{J}(t),J_1(t)\rangle =$$

$$= \langle J(0),\dot{J}_1(0)\rangle - \langle \dot{J}(0),J_1(0)\rangle = -\langle \dot{J}(0),J_1(0)$$

for all t. Then

$$\langle J(t),U(t)Y(t)w\rangle - \langle \dot{J}(t),Y(t)w\rangle = -\langle \dot{J}(0),w\rangle$$

for all t. Hence

$$\langle Y^*(t)(U(t)J(t)-\dot{J}(t)),w\rangle = -\langle \dot{J}(0),w\rangle$$

for all t and w. Then

$$\dot{J}(t) = U(t)J(t) + Y^*(t)^{-1}\,\dot{J}(0).$$

The solution of this equation with $J(0) = 0$ is given by

$$J(t) = Y(t)\int_0^t Y(s)^{-1}\,Y^*(s)^{-1}\,\dot{J}(0)ds.$$

Denote

$$u(t) = \int_0^t Y(s)^{-1}\,Y^*(s)^{-1}\,\dot{J}(0)ds.$$

Then

$$\langle u(t),\dot{J}(0)\rangle' = \langle \dot{u}(t),\dot{J}(0)\rangle = \|Y^*(t)^{-1}\,\dot{J}(0)\|^2.$$

But since Y satisfies equation (1) and $\|U(t)\| \le \sqrt{c}$ for all t, it follows that

$$\|Y(t)\| \le e^{t\sqrt{c}} \tag{3}$$

for all $t \ge 0$. Hence

$$\langle u(t),\dot{J}(0)\rangle \ge \|\dot{J}(0)\|^2\int_0^t e^{-2s\sqrt{c}}\,ds =$$

$$= \|\dot{J}(0)\|^2\frac{1}{2\sqrt{c}}(1-e^{-2t\sqrt{c}})ds.$$

Therefore

$$\|u(t)\| \geq \|\dot{J}(0)\|^{-1} \langle u(t), J(0)\rangle \geq \|\dot{J}(0)\| \frac{1}{2\sqrt{c}} (1-e^{-2t\sqrt{c}})$$

and then, by (2):

$$\|J(t)\| \geq A \frac{1}{2\sqrt{c}} (e^{at} - e^{(a-2\sqrt{c})t}) \|\dot{J}(0)\| .$$

Comparing (2) and (3) we obtain $\sqrt{c} > a$. Then $2\sqrt{c} - a > a$ and then, defining $\mu = 2\sqrt{c} - a$, we obtain that

$$\|J(t)\| \geq \|\dot{J}(0)\| A \frac{1}{\sqrt{c}} \sin h\, \mu\, t$$

for all $t \geq 0$.

III - Proof of Theorem B

If M is a Riemannian manifold and $\theta \in UM$ denote $\Lambda(\theta)$ the set of Lagrangian subspaces of $S(\theta)$. $\Lambda(\theta)$ has a natural manifold structure (see [1]). Denote $\Lambda(UM)$ the bundle on UM whose fiber on θ is $\Lambda(\theta)$. Let $\Lambda_k(\theta)$ be the set of Lagrangian subspaces $E \in \Lambda(\theta)$ such that $\dim E \cap V(\theta) = k$. Then $\Lambda_k(\theta)$ is a submanifold of $\Lambda(\theta)$ with codimension $k(k+1)/2$ ([1]). Then, denoting $\Lambda_k(UM)$ the set of elements $(\theta, E) \in \Lambda(UM)$ such that $E \in \Lambda_k(\theta)$, it follows that $\Lambda_k(UM)$ is a submanifold of $\Lambda(UM)$ with codimension $k(k+1)/2$. Define

$$\Gamma = \bigcup_{k \geq 2} \Lambda_k(UM)$$

and let Σ be the complement of Γ. Clearly Σ is an open and dense subset of $\Lambda(UM)$. Let S^1 be the unit circle of the complex plane. Define the function $m: \Sigma \to S^1$ as follows: if $(\theta, E) \in \Lambda_o(UM)$, take a symmetric map $L: N(\theta) \circlearrowleft$ such that graph $(L) = E$ and define

$$m(\theta, E) = \frac{1 - i\,\mathrm{tr}\, L}{1 + i\,\mathrm{tr}\, L} .$$

When $(\theta,E) \in \Lambda_1(UM)$, set

$$m(\theta,E) = -1.$$

Let us prove that m is continuous. Suppose that $(\theta,E) \in \Lambda_1(UM)$ and that $(\theta_n,E_n) \in \Lambda_0(UM)$ is a sequence converging to (θ,E). We can write $E_n = \text{graph}(L_n)$ where $L_n: N(\theta_n)\circlearrowleft$ is a symmetric linear map. We claim that there exists $C > 0$ such that for all n the number of eigenvalues of L_n (counted with multiplicity) larger in modulus than C is ≤ 1. If this property is false there exist integers $n_1 < n_2 < \ldots$, vectors x_j, y_j, and two real sequences $\{\lambda_j\}$ and $\{\mu_j\}$ such that $|\lambda_j| \to +\infty$, $|\mu_j| \to +\infty$ and:

$$\langle x_j,y_j\rangle = 0 \qquad \|x_j\| = \|y_j\| = 1$$

$$L_{n_j} x_j = \lambda_j x_j \qquad L_{n_j} y_j = \mu_j y_j$$

for all j. If S_j is the space spanned by $(x_j,\lambda_j x_j)$ and $(y_j,\mu_j y_j)$, the property $|\lambda_j| \to +\infty$ implies that the sequence $\{S_j\}$ converges to a two dimensional subspace of $V(\theta)$ and, since $S_j \subset E_{n_j}$, this two dimensional subspace is contained in E. This contradicts the property $E \in \Lambda_1(\theta)$ and proves the claim. Moreover the spectral radius of L_n goes to $+\infty$ when $n \to +\infty$ because otherwise, since the maps L_n are symmetric, we would have a sequence of integers $n_1 < n_2 < \ldots$ such that

$$\sup_j \|L_{n_j}\| < +\infty$$

and then E would be also a graph of a linear map $L: N(\theta)\circlearrowleft$ (obtained as the limit of a subsequence of $\{L_{n_j}\}$) thus implying $E \in \Lambda_0(\theta)$. Then every L_n has exactly one eigenvalue λ_n with $|\lambda_n| > C$ and $\lim|\lambda_n| = +\infty$. Hence $\lim_{n\to+\infty} |\text{tr } L_n| = +\infty$ and this implies $\lim_{n\to+\infty} m(\theta_n,E_n) = -1$.

Observe that m cannot be continuously extended to Γ because if $(\theta,E) \in \Gamma$ it is easy to see that given any $\lambda \in \mathbb{R}$ there exist, arbitrarily near to E, spaces $E' \in \Lambda_0(\theta)$ such that $E' = \text{graph}(L)$ and $\text{tr}\, L = \lambda$.

Now we shall define an index that to every continuous map $\gamma: S^1 \to \Lambda(UM)$ associates an integer $\text{Ind}(\gamma) \in Z$.

If $\gamma: S^1 \to \Sigma$ is a continuous map, define $\text{Ind}(\gamma)$ as the degree of $m \circ \gamma: S^1 \circlearrowleft$. Observe that if $\gamma_1: S^1 \to \Sigma$ and $\gamma_2: S^1 \to \Sigma$ are homotopic, then $\text{Ind}(\gamma_1) = \text{Ind}(\gamma_2)$. Given a continuous map $\gamma: S^1 \to \Lambda(UM)$ define $\text{Ind}(\gamma)$ as $\text{Ind}(\gamma) = \text{Ind}(\tilde{\gamma})$ where $\tilde{\gamma}: S^1 \to \Sigma$ is homotopic to γ. To show that this definition is correct we have to show that there exists $\tilde{\gamma}: S^1 \to \Sigma$ homotopic to γ and that $\text{Ind}(\tilde{\gamma})$ is independent of $\tilde{\gamma}$. The first property is proved taking $\tilde{\gamma}: S^1 \to \Sigma$ very near in the C^0 topology (in particular homotopic) to γ. This can be done because Γ is a union of submanifolds with codimensions ≥ 3. The independence property is proved observing that if $\hat{\gamma}: S^1 \to \Sigma$ is also homotopic to γ, then $\hat{\gamma}$ and $\tilde{\gamma}$ are homotopic and since Γ is a union of submanifolds with codimensions ≥ 3, the homotopy between $\hat{\gamma}$ and $\tilde{\gamma}$ can be taken avoiding Γ i.e. $\tilde{\gamma}$ and $\hat{\gamma}$ are homotopic as maps into Σ, and then, as we observed above, $\text{Ind}(\tilde{\gamma}) = \text{Ind}(\hat{\gamma})$.

Now suppose that there exists an invariant Lagrangian subbundle E on UM i.e. for all $\theta \in UM$ the fiber E_θ is Lagrangian and

$$(D\varphi_t)E_\theta = E_{\varphi_t(\theta)}$$

for all t. Given a continuous map $\alpha: S^1 \to UM$ we define $\hat{\alpha}: S^1 \to \Lambda(UM)$ by

$$\hat{\alpha}(t) = (\alpha(t), E_{\alpha(t)})$$

and we introduce an index $\text{ind}(\alpha)$ by

$$\operatorname{ind}(\alpha) = \operatorname{Ind}(\hat{\alpha}).$$

Every map $\alpha\colon S^1 \to UM$ can be written in the form $\alpha(t) = (\gamma(t), v(t))$ with $\gamma(t) \in M$ and $v(t) \in U_{\gamma(t)}M$. We shall denote α^* the map of S^1 into UM defined by $\alpha^*(t) = (\gamma(t), -v(t))$. Observe that α and α^* are always homotopic. Hence

$$\operatorname{ind}(\alpha^*) = \operatorname{ind}(\alpha). \tag{1}$$

We shall say that a continuous map $\alpha\colon S^1 \to UM$ is a pseudo geodesic if for all $s \in S^1$ where

$$E_{\alpha(s)} \cap V(\alpha(s)) \neq \{0\} \quad \text{or} \quad E_{\alpha^*(s)} \cap V(\alpha^*(s)) \neq \{0\} \tag{2}$$

there exists $\epsilon > 0$ and $k > 0$ such that

$$\varphi_{kt}(\alpha(s)) = \alpha(e^{it}s)$$

for all $-\epsilon \leq t \leq \epsilon$. In other words, writing α in the form $\alpha = (\gamma, v)$ as above, α is a pseudo geodesic if every $s \in S^1$ where (2) holds has a neighborhood V such that γ/V is a geodesic arc (parametrized by a multiple of arc lenght) and $v(t) = \dot{\gamma}(t)/\|\dot{\gamma}(t)\|$ for all $t \in V$.

The following is the key lemma in the proof of Theorem B.

<u>Lemma III.1</u>. If $\alpha\colon S^1 \to UM$ is a pseudo geodesic such that $E_{\alpha(s)} \cap V(\alpha(s)) \neq \{0\}$ for some $s \in S^1$, then $\operatorname{ind}(\alpha) > 0$.

Before proving III.1 we shall complete the proof of Theorem B. First we shall prove the following corollary of III.1.

<u>Corollary</u>. Every pseudo geodesic $\alpha\colon S^1 \to UM$ satisfies

$$E_{\alpha(t)} \cap V(\alpha(t)) = \{0\}$$

for all $t \in S^1$.

Proof: Suppose that $\alpha: S^1 \to UM$ is a pseudo geodesic and there exists $s \in S^1$ where $E_{\alpha(s)} \cap V(\alpha(s)) \neq \{0\}$ holds. By III.1, $\mathrm{ind}(\alpha) > 0$. Define $\bar{\alpha}: S^1 \to UM$ by $\bar{\alpha}(z) = \alpha(\bar{z})$. Then

$$\mathrm{ind}(\bar{\alpha}) = -\mathrm{ind}\,\alpha < 0 \tag{3}$$

because $\bar{\alpha}$ is just the composition of α with the map $S^1 \ni z \to \to \bar{z} \in S^1$ whose degree is -1. Write α in the form $\alpha(t) = = (\gamma(t), v(t))$ where $\gamma: S^1 \to M$ is a continuous map and $v(t) \in \in U_{\gamma(t)}M$ for all $t \in S^1$. Define $\beta: S^1 \to UM$ by

$$\beta(t) = (\gamma(\bar{t}), -v(\bar{t})).$$

This map is a pseudo geodesic because if $s \in S^1$ satisfies

$$E_{\beta(s)} \cap V(\beta(s)) \neq \{0\} \quad \text{or} \quad E_{\beta^*(s)} \cap V(\beta^*(s)) \neq \{0\}$$

then, from the definition of β follows that (2) is satisfied at $\bar{s}$ and then there exists a neighborhood V of s such that γ/V is a geodesic and $v(t) = \dot{\gamma}(t)/\|\dot{\gamma}(t)\|$ for t in V. But then the map $\bar{V} \ni t \to \gamma(\bar{t})$ is also a geodesic. Moreover, the property $v(t) = = \dot{\gamma}(t)/\|\gamma(t)\|$ for all $t \in V$ implies that the derivative of the map $V \ni t \to \gamma(\bar{t})$ is collinear with $-v(\bar{t})$ for all $t \in \bar{V}$. Hence β is a pseudo geodesic. But observe that $\beta = \bar{\alpha}^*$. Hence, using (1) and (2)

$$\mathrm{ind}(\beta) = \mathrm{ind}(\bar{\alpha}^*) = \mathrm{ind}(\bar{\alpha}) = -\mathrm{ind}(\alpha) < 0.$$

Since β is a pseudo geodesic this contradicts III.1 and proves the corollary.

The final step in the proof of Theorem B is to show that $E_\theta \cap V(\theta) = \{0\}$ for all $\theta \in UM$. By II.1 this implies that M has no conjugate points and completes the proof of Theorem B. Let us suppose that $E_\theta \cap V(\theta) \neq \{0\}$. Using this property we shall construct a pseudo geodesic $\alpha: S^1 \to UM$ with $\alpha(0) = \theta$. Since

$E_\theta \cap V(\theta) \neq \{0\}$ this pseudo geodesic contradicts the corollary of III.1 and shows that we must have $E_\theta \cap V(\theta) = \{0\}$ for all θ.

To construct α take $a > 0$ such that, denoting $(p_1, v_1) = \varphi_{-a}(\theta)$ and $(p_2, v_2) = \varphi_a(\theta)$, we have

$$E_{(p_1,v_1)} \cap V(p_1,v_1) = \{0\}, \quad E_{(p_1,-v_1)} \cap V(p_1,-v_1) = \{0\}$$

$$E_{(p_2,v_2)} \cap V(p_2,v_2) = \{0\}, \quad E_{(p_2,-v_2)} \cap V(p_2,-v_2) = \{0\}.$$

The existence of a is not immediate. It requires the following lemma, that will be proved after finishing the construction of α.

Lemma III.2. If $\theta \in UM$ and $E \subset S(\theta)$ is a Lagrangian subspace, then the set of $t \in \mathbb{R}$ where $(D\varphi_t)E \cap V(\varphi_t(\theta)) \neq \{0\}$ is discrete.

Apply this lemma to θ and E_θ. Then the set $S = \{t \in \mathbb{R} \mid E_{\varphi_t(\theta)} \cap V(\varphi_t(\theta)) \neq \{0\}\}$ is discrete. Moreover, if $\theta = (p,v)$ and we define $\xi = (p,-v)$, then, III.2 applied to ξ and E_ξ proves that the set $S' = \{t \in \mathbb{R} \mid E_{\varphi_t(\xi)} \cap V(\varphi_t(\xi)) \neq \{0\}\}$ is also discrete. Then, taking $a > 0$ such that a and $-a$ don't belong to $S \cup S'$, the desired properties are satisfied. Now take neighborhoods V_1, V_2 of (p_1,v_1) and (p_2,v_2) respectively such that

$$E_{(q,w)} \cap V(q,w) = \{0\} \tag{4}$$

and

$$E_{(q,-w)} \cap V(q,-w) = \{0\} \tag{5}$$

for all $(q,w) \in V_1 \cup V_2$. Since every point of UM is a non wandering point of φ (because the volume of UM is finite), we can take $\eta \in V_2$ and $T > 0$ such that $\varphi_T(\eta) \in V_1$. Now set $\omega_1 = \pi/4a$, $\omega_2 = \pi/T$ and take $\alpha\colon S^1 \to UM$ satisfying

$$\alpha(\exp \omega_1 ti) = \varphi_t(\theta)$$

for $-a \leq t \leq a$,

$$\alpha(\exp(\omega_2 t + \frac{\pi}{2})i) = \varphi_t(\eta)$$

for $0 \leq t \leq T$, and $\alpha(\text{esp } ti) \in V_2$ when $\pi/4 \leq t \leq \pi/2$ and $\alpha(\exp ti) \in V_1$ for $-\pi/2 \leq t \leq -\pi/4$. Obviously α satisfies the condition $\alpha(0) = \theta$. Moreover it is a pseudo geodesic because if at $s \in S^1$ condition (2) is satisfied, then by properties (4) and (5), the point $\alpha(s)$ cannot belong to $V_1 \cup V_2$. Hence the argument of s must be contained in the interval $(-\pi/4, \pi/4)$ or in the interval $(\pi/2, 3\pi/2)$. Since α restricted to these intervals is an arc of trajectory of the geodesic flow parametrized by a constant multiple of time, then the condition required in the definition of pseudo geodesics is satisfied.

To prove Lemma III.2 it suffices to show that if E is a Lagrangian subspace of $S(\theta)$ such that $E \cap V(\theta) \neq \{0\}$ there exists a neighborhood V of $t = 0$ such that $(D\varphi_t)E \cap V(\theta) = \{0\}$ for all $0 \neq t \in V$. Let $p: S(\theta) \to H(\theta)$ be the orthogonal projection. We claim that $p(E)$ is the orthogonal complement of $J_\theta(E \cap V(\theta))$ in $H(\theta)$. In fact

$$\dim p(E) + \dim J_\theta(E \cap V(\theta)) = \dim E - \dim(E \cap V(\theta)) + \\ + \dim(E \cap V(\theta)) = \dim E = \dim H(\theta).$$

Moreover, if $x \in p(E)$, we can write $x = y+z$ with $v \in E$ and $z \in E \cap V(\theta)$. Then, if $w \in E \cap V(\theta)$ we have $\langle y, J_\theta w\rangle = 0$ because y and w belong to the Lagrangian subspace E, and $\langle z, J_\theta w\rangle = 0$ because $z \in V(\theta)$ and $J_\theta w \in J_\theta(E \cap V(\theta)) \subset H(\theta)$. Hence $\langle x, J_\theta w\rangle = 0$ completing the proof of the claim. Take a basis $\{v_1, \ldots, v_n\}$ of $p(E)$ and let $p_t: S(\varphi_t(\theta)) \to H(\varphi_t(\theta))$ be the orthogonal projection. Then if t is near to 0 there exists a set of m linearly independent vectors $\{v_1(t), \ldots, v_m(t)\} \subset p_t((D\varphi_t)E)$.

Let $\{w_1,\dots,w_k\}$ be a basis of unitary vectors of $E \cap V(\theta)$. Then $k+m = \dim E = \dim H(\theta)$. Take perpendicular Jacobi vector fields $Y_1(t),\dots,Y_k(t)$, with $Y_i(0) = 0$ and $\dot{Y}_i(0) = J_\theta w_i$ for $1 \le i \le k$. Define $w_i(t) = Y_i(t)/\|Y_i(t)\|$ for $t \ne 0$. Then

$$\lim_{t\to 0} w_i(t) = \lim_{t\to 0} \frac{Y_i(t)/t}{\|Y_i(t)/t\|} = \frac{\dot{Y}_i(0)}{\|\dot{Y}_i(0)\|} = J_\theta w_i .$$

Moreover

$$p_t((D\varphi_t)w_i) = Y_i(t).$$

Hence

$$p_t((D\varphi_t)E) \supset \{v_1(t),\dots,v_m(t),w_1(t),\dots,w_k(t)\}.$$

When $|t|$ is small, $\{v_1(t),\dots,v_m(t),w_1(t),\dots,w_k(t)\}$ is near to $\{v_1,\dots,v_m,J_\theta w_1,\dots,J_\theta w_k\}$ that is a basis of $H(\theta)$ because $H(\theta) = p(E) \oplus J_\theta(E\cap V)$. Then $p_t((D\varphi_t)E) = H(\varphi_t(\theta))$ for $|t|$ small and $\ne 0$. Obviously $p_t((D\varphi_t)E) = H(\varphi_t(\theta))$ implies $(D\varphi_t)E \cap \cap V(\varphi_t(\theta)) = \{0\}$.

Finally, we shall prove Lemma III.1. Given a continuous map $\beta\colon S^1 \to \Lambda(UM)$ we shall denote $P(\beta,c) = \{t \mid (m\circ\beta)(t) = c\}$ and, if β is differentiable at $t \in S^1$, we shall denote $\sigma(\beta,t)$ the sign of the derivative $(m\circ\beta)'(t)$ i.e. $\sigma(t) = 1$ or -1 according to whether $(m\circ\beta)'(t)$ preserves or reverses orientation. Now let $\alpha\colon S^1 \to UM$ be a pseudo geodesic and define $\tilde{\alpha}\colon S^1 \to \Lambda(UM)$ by $\tilde{\alpha}(t) = (\alpha(t),E_{\alpha(t)})$. We have to prove that $\mathrm{Ind}(\tilde{\alpha}) > 0$. For this it suffices to show (as we explained above) that for all C^1 map $\beta\colon S^1 \to \Lambda(UM)$ that is a C^1 approximation of $\tilde{\alpha}$ (and then homotopic to $\tilde{\alpha}$) and satisfies $\beta(S^1) \cap \Gamma = \phi$ the degree of the map $m\circ\beta\colon S^1 \circlearrowleft$ (that is by definition $\mathrm{Ind}(\beta)$), is strictly positive. To prove this we shall show that there exists $\bar{c} > 0$ such that for all $c > \bar{c}$

$$\sum_{t\in P(\tilde{\alpha},c)} \sigma(\tilde{\alpha},t) > 0. \tag{6}$$

This implies that the degree of $m\circ\beta$ is >0 because this degree can be calculated as

$$\sum_{t\in P(\beta,c)} \sigma(\beta,t)$$

and an elementary application of the implicit function theorem shows that

$$\sum_{t\in P(\beta,c)} \sigma(\beta,t) \geq \sum_{t\in P(\tilde{\alpha},c)} \sigma(\beta,t).$$

Hence (6) suffices to prove III.1. To prove (6) we first transform $\tilde{\alpha}$ in a map of $[0,2\pi]$ in $\Lambda(UM)$ defining $\hat{\alpha}: [0,2\pi] \to \Lambda(UM)$ by $\hat{\alpha}(t) = \tilde{\alpha}(e^{it})$. Then (6) is equivalent to

$$\sum_{t\in P(\hat{\alpha},c)} \sigma(\hat{\alpha},t) > 0. \tag{7}$$

But since $m\circ\hat{\alpha}$ takes values in S^1, its derivative satisfies:

$$\frac{(m\circ\hat{\alpha})'(t)}{i(m\circ\hat{\alpha})(t)} \in \mathbb{R}$$

and

$$\sigma(\hat{\alpha},t) = \operatorname{sgn} \frac{(m\circ\hat{\alpha})'(t)}{i(m\circ\hat{\alpha})(t)}.$$

Hence (7) can be written as:

$$\sum_{t\in P(\hat{\alpha},c)} \operatorname{sgn} \frac{(m\circ\hat{\alpha})'(t)}{i(m\circ\hat{\alpha})(t)} > 0. \tag{8}$$

By Lemma III.2 the set of points $t \in [0,2\pi]$ where $(m\circ\hat{\alpha})(t) = -1$ is finite. Let $t_1,\ldots,t_r$ be this set. Without lose of generality we can assume it doesn't contain 0 or 2π. Each t_j has a neighborhood V_j such that $\hat{\alpha}/V_j$ can be written as

$$\hat{\alpha}(t) = (\alpha_j(t), E_{\alpha_j}(t))$$

where $\alpha_j: V_j \to UM$ is an arc of trajectory of the geodesic flow parametrized by a constant positive multiple of time i.e. there exists $k > 0$ such that

$$\alpha_j(t_j+s) = \varphi_{ks}(\alpha_j(t_j)).$$

Write, for $t \in V_j$

$$E_{\alpha_j(t)} = \text{graph } U_j(t)$$

where $U_j(t): N(\alpha_j(t)) \circlearrowleft$ is a linear symmetric map. Then, the invariance of the bundle E implies that U_j satisfies the Ricatti equation (see Section II):

$$k\dot{U}_j(t) + U_j^2(t) + K(t) = 0.$$

Now take $-1 < c < 0$ so near to -1 that

$$P(\hat{\alpha}, c) \subset \bigcup_{j=1}^{r} V_j.$$

Let t be a point in $P(\hat{\alpha}, c)$. Suppose that it belongs to V_j. Then

$$(m\circ\hat{\alpha})'(t) = \left(\frac{1-i\,\mathrm{tr}\, U_j}{1+i\,\mathrm{tr}\, U_j}\right)'(t) = -\frac{2i\,\mathrm{tr}\, \dot{U}_j(t)}{(1+i\,\mathrm{tr}\, U_j(t))^2}.$$

Then:

$$\frac{(m\circ\hat{\alpha})'(t)}{i(m\circ\hat{\alpha})(t)} = -\frac{2\,\mathrm{tr}\, \dot{U}_j(t)}{1+\mathrm{tr}^2 U_j(t)} = \frac{2}{k}\,\frac{\mathrm{tr}\, U_j^2(t) + \mathrm{tr}\, K(t)}{1 + \mathrm{tr}^2\, U_j(t)}.$$

Since U_j is symmetric we have

$$\mathrm{tr}\, U_j \le (n-1)^{1/2}\,(\mathrm{tr}\, U_j^2)^{1/2}.$$

Hence

$$\frac{(m\circ\hat{\alpha})'(t)}{i(m\circ\hat{\alpha})(t)} \ge \frac{2}{k}\,\frac{(n-1)^{-1}\,\mathrm{tr}^2\, U_j(t) + \mathrm{tr}\, K(t)}{1 + \mathrm{tr}^2\, U_j(t)}. \tag{9}$$

Moreover $(m\circ\hat{\alpha})(t) = c$ implies

$$\operatorname{tr} U_j(t) = -i\,\frac{1-c}{1+c}.$$

Therefore, if we take c very near to -1, the condition $(m\circ\alpha)(t) = c$ implies that $|\operatorname{tr} U_j(t)|$ is very large. Then the quotient at left becomes near to $\frac{2}{k}(n-1)^{-1}$. Hence, if c is taken close enough to -1, it follows that

$$\frac{(m\circ\hat{\alpha})'(t)}{i(m\circ\hat{\alpha})(t)} \geq \frac{1}{(n-1)k}$$

for all $t \in P(\hat{\alpha},c)$. This obviously implies (8).

REFERENCES

[1] D.V. Anosov, Geodesic flows on closed Riemannian manifolds with negative curvature, Proc. Steklow Inst. Math. 90 (1967). Amer. Math. Soc.

[2] V.I. Arnold, Une classe charactéristique intervenant dans les conditions de quantification, Appendix I in Théorie des perturbations et méthodes assymptotiques, V.P. Maslov, Dunod, Paris (1972).

[3] P. Eberlein, Geodesic flows on negative curved manifolds I, Ann. of Math. 95 (1972), 492-410.

[4] P. Eberlein, When is a geodesic flow of Anosov type? I, J. Differential Geometry 8, (1973), 437-463.

[5] L. Green, A theorem of E. Hopf, Michigan Math. J. 5 (1958), 31-34.

[6] J. Hadamard, Les surfaces a courbures apposées et leur lignes geodesiques, J. Math. Pures et appl. (1898), 27-73.

[7] M. Hirsch, C. Pugh, M. Shub, Invariant manifolds, Springer Verlag Lecture Notes in Mathematics 583 (1976).

[8] E. Hopf, Statistik der geodetischen linen in Mannigfaltigkeiten negativer Krummung, Ber. Vesh. Sochs. Akad. Wiss. Leipzig 91 (1939), 261-304.

[9] W. Klingenberg, Riemannian manifolds with geodesic flows of Anosov type, Ann. of Math. 99 (1974), 1-13.

Instituto de Matemática Pura e Aplicada (IMPA)
Estrada Dona Castorina 110
22460 Rio de Janeiro, RJ
Brazil

S PATTERSON & C ROBINSON[1]

Basins of sinks near homoclinic tangencies

§1. Introduction

When a one parameter family of diffeomorphisms in two dimensions creates a new homoclinic intersection for a dissipative saddle point, there results a cascade of sinks each of which arises from a saddle node bifurcation. In this paper we are concerned with the basin of attraction of these sinks. Besides the local information known about the basin of the sink given by the general theorem on the saddle node bifurcation, information about this basin and its boundary is obtained in a fixed neighborhood.

In fact each of the primary saddle node bifurcations takes place as part of the formation of a new horseshoe. The result about the basins of the sinks making up the primary cascade is obtained by studying one parameter families of area decreasing diffeomorphisms in dimension two which create horseshoes. For each horseshoe created, the two periodic points of lowest period are formed by a saddle node bifurcation. Just after this primary saddle node bifurcation in the creation of each horseshoe, the basin of attraction of the periodic sink is determined inside the rectangle associated with the

[1] Partially supported by grant from the National Science Foundation.

horseshoe. In order to apply to the horseshoes created near homoclinic tangencies, general sufficient conditions which determine the basin are developed in terms of the derivatives of the family of nonlinear maps creating the horseshoe. In addition to studying the basins just after the bifurcation, the evolution of the sink and the attracting sets spawned from them as the parameter varies is discussed.

In sections 2 and 3, we state and discuss the main theorems about the basin of the sink just after the primary saddle node bifurcation when creating a horseshoe. In section 2, it is discussed in terms of a family of quadratic maps, while section 3 gives the conditions needed on a more general family of nonlinear maps. The proof of these theorems is given in section 5. In section 4, we discuss how such families of maps arise from the creation of homoclinic tangencies, and in section 6 verify the general conditions of section 3 for these families arising from the creation of homoclinic tangencies.

§2. Main Result.

We want to consider the situation where an area decreasing map creates a horseshoe as the parameter varies. We take as the prototype of this situation the composition of a linear hyperbolic map, $A(x,y) = (\mu x, \lambda y)$, followed by a fixed family of quadratic maps, $Q(x,y) = (y, t-x-y^2)$:

$$F_t(x,y) = (\lambda y, t-\mu x-\lambda^2 y^2) \tag{2.1}$$

where $0 < \mu < 1 < \lambda$ and $\Delta = \mu\lambda < 1$.

This map is easily seen to be equivalent to the Henon family $\{H_{a,b}(x,y) = (y, b-ax-y^2): a > 0\}$ by the linear change of variables $T(x,y) = (\lambda^{-1}x, \lambda^{-2}y)$ and correspondence of parameters $a = \mu\lambda$ and

$b = \lambda^2 t$: $T^{-1} \circ F_t \circ T(x,y) = (y, \lambda^2 t - \mu\lambda x - y^2)$. See [6], [5], [2] and [7] for a discussion of this map.

The conditions for F_t to have a fixed point are given by the two equations $\lambda y = x$ and $t - \mu x - \lambda^2 y = y$. Defining two new constants by

$$\delta = \mu + \lambda^{-1} \quad \text{and} \quad t_0 = -\frac{1}{4}\delta^2,$$

it is easily checked that $t = t_0$ is the bifurcation parameter value where F_t changes from no fixed points for $t < t_0$ to two fixed points for $t > t_0$. Various calculations show the following facts where some of the calculations are easy and others are more involved.

(2.2.i) For $t < t_0 \equiv -\frac{1}{4}\delta^2$, there are no fixed points.

(2.2.ii) For $t = t_0$, there is exactly one fixed point which we label p_{t_0}. It has eigenvalue 1 and $\Delta = \lambda\mu$ and is a saddle node fixed points. Let E^c be the one dimensional space through the origin spanned by the eigenvector for the eigenvalue 1. Similarly let $E^s_{t_0}$ be the one dimensional space corresponding to the eigenvalue Δ.

(2.2.iii) For $t > t_0$, there are two fixed points which we label p_t and q_t. Letting p_t be the fixed point with smaller numerical values, it is a fixed saddle point for this entire range of parameters. Let E^u_t be the unstable line (through the origin) spanned by the eigenvector of modulus greater than one, and E^s_t the stable line.

(2.2.iv) For $t_0 < t < t_1 \equiv \frac{3}{4}\delta^2$, q_t is a sink with eigenvalues progressing from positive real to complex and finally too negative real as t increases from t_0 to t_1.

(2.2.v) At $t = t_1$, q_t has eigenvalues -1 and $-\lambda\mu$ and undergoes a period doubling bifurcation.

(2.2.vi) For $t > t_1$, q_t is a twisted saddle point with reflection in the unstable manifold (with a negative eigenvalue which is greater than one in absolute value) and F_t has a nearby sink of period two which will continue to period double as t increases.

Our concern is with the basin of attraction of the sink q_t for $t > t_0$ (and $t < t_1$). Our first and main theorem states that locally this basin is bounded by the stable manifold of p_t for $t > t_0$ but t near t_0. In particular, as t approaches t_0 from above the area of this basin does not go to zero but the basin has a definite size. More precisely we have the following

<u>Theorem I</u>. Let F_t be the family of area decreasing maps given (2.1). Then there exists $\varepsilon > 0$ such that for $t_0 < t < t_0 + \varepsilon$ the basin of the sink q_t contains an open disk D_t with the following properties: (i) the boundary of D_t is the union of two curves L_t and S_t, $\partial D_t = L_t \cup S_t$, with $L_t \cap S_t = \emptyset$, (ii) L_t is an open interval in the line through p_t in the direction of the unstable eigenvector with one end at p_t, $L_t \subset \{p_t\} + E_t^u$, (iii) S_t is a closed curve segment in $W^s(p_t, F_t)$ with one end at p_t (iv) $F_t(L_t) \subset D_t$, (v) $F_t(D_t) \subset D_t$, and (vi) $D_t \subset$ $\subset W^s(q_t, F_t)$.

The proof of Theorem I is given in section 5. A similar theorem holds for a more general family of area decreasing maps $\{F_t\}$ which create a horseshoe as the parameter varies. The specific requirements for $\{F_t\}$, (3.1)-(3.7), and the statement of the theorem, Theorem II, are given explicitly in section 3.

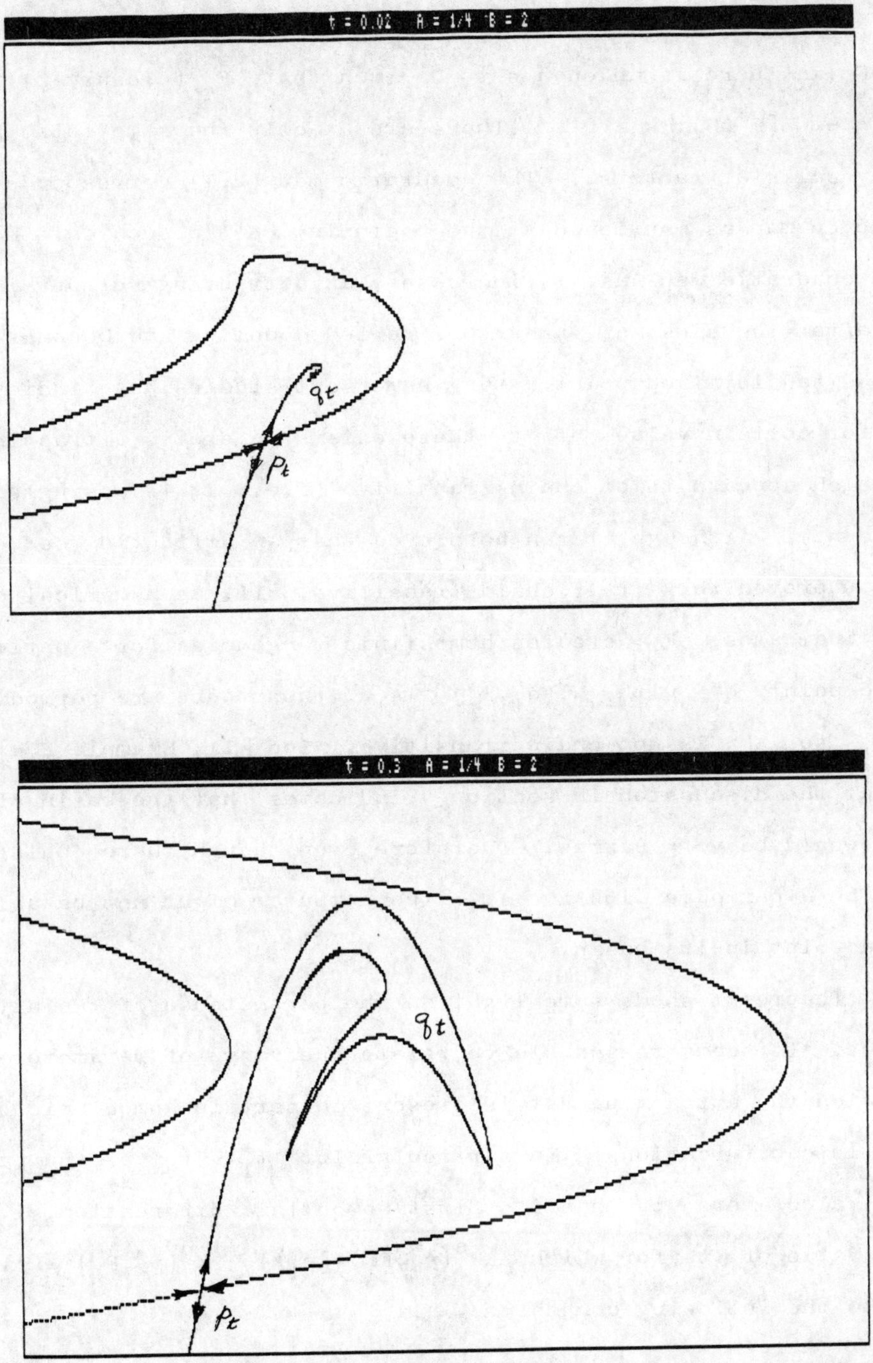

Figure 1

$W^s(p_t, F_t)$ and $W^u(p_t, F_t)$ for (2.1) with $\mu = \frac{1}{4}$, $\lambda = 2$, and $t - t_0 = 0.02$ and 0.3

We continue to discuss the development of the attractint set for larger values of t. Here we call A an <u>attracting set</u> for a map F if there is an open set U such that F (closure $U) \subset U$ and $A = \bigcap \{F^k(U): k \geq 0\}$. (These are exactly the same sets that Conley calls attractors.) The quadratic map (2.1) corresponds to the Hénon map as mentioned at the beginning of the section. For these quadratic maps as t increases, an attracting set persists long after the sink q_t undergoes period doubling and becomes a twisted saddle fixed point. Many numerical studies have indicated that for certain values of t there exists an exotic attracting set A_t which appears to be chain transitive (i.e.; it is a single chain component). Although it can be proved that an attracting set exists, it is unproved that it is chain transitive. If, as numerical evidence indicates, F_t creates homoclinic tangencies for a periodic saddle point q'_t with $W^u(q'_t,F_t) \subset A_t$, then there are periodic sinks in A_t so A_t is not chain transitive. See [11, Example 2.4], [10]. In fact the discussion in section 4 indicates that the basin of such sinks would be very narrow. Therefore even though there could possibly be a periodic sink in A_t, the computer might not be able to keep a point in its basin.

Theorem I sheds some light on the persistence of the attracting set. It seems reasonable to relate the range of parameter values for which the attracting set is present to certain homoclinic or heteroclinic bifurcations. In particular let $t_1^* > t_0$ be the parameter value where p_t has its first homoclinic bifurcation, i.e. first value of t for which $(W^s(p_t,F_t)-\{p_t\}) \cap W^u(p_t,F_t) \neq \emptyset$. We make the following conjecture.

<u>Conjecture I</u>: The attracting set which is initially $\{q_t\}$ persists with basin bounded locally by $W^s(p_t,F_t)$ for $t \in (t_0,t_1^*)$.

Now let t_2^* be the first parameter value for which $W^s(p_t,F_t)$ meets $W^u(q_t,F_t)$. It can be seen that $t_2^* > t_1^*$. Recall that $\Delta = \det(DF_t)$.

<u>Conjecture II</u>: For Δ small enough, the attracting set spawned by the initial saddle node bifurcation persists precisely for those t in $[t_0,t_2^*]$.

This conjecture is given plausibility by the fact that for small Δ, the period doubling sequence for q_t will be completed before $t = t_2^*$, so one might expect some transitivity along $W^u(q_t,F_t)$. Once a hook of $W^s(p_t,F_t)$ pushes through $W^u(q_t,F_t)$ orbits are allowed to "leak out of A_t" behind the hook. See [7] for a discussion of the order of the various bifurcations as Δ varies.

Many important questions remain concerning the evolution of A_t. Other recent research has determined results on the development of the basin and its boundary, [4], [8].

§3. Theorem for a general nonlinear family.

This section gives conditions on a family $\{F_t\}$ which are less restrictive than the quadratic form (2.1). Using these conditions a slightly modified version of the main theorem is stated.

The idea of the assumptions is that a saddle node fixed point p_0 is created at some $t = t_0$ whenever there is a rectangle B whose image $F_t(B)$ is shaped like a horseshoe and is pulled across B as t increases. If F_t decreases area in B, the basin of the sink created contains a region like D_t in Theorem I. The more specific assumptions used are now given using coordinates but obviously many other nonlinear maps would satisfy these conditions after a choice of local coordinates (cf. [12, p.846] and [11, Prop.3.3]).

(3.1) There are a rectangle $B = [x_1,x_2] \times [y_1,y_2]$ and a family of maps $\{F_t\}$ for the parameter $t_{-1} \le t \le t_1$ such that i) for each t both F_t and F_t^{-1} are defined on B so that the domain of F_t includes $B \cup B_t^-$ where $B_t^- = F_t^{-1}(B)$, ii) for each t, F_t is C^2, and iii) F_t varies continuously in the C^2 topology as t varies.

(3.2) The family preserves orientation and decreases area in $B \cup B_t^-$ for each t, i.e. $0 < \det DF_t(x,y) < 1$ for all (x,y) in $B \cup B_t^-$ and $t_{-1} \le t \le t_1$. (If the family reverses orientation and decreases area, $-1 < \det DF_t(x,y) < 0$, then related results can be shown but several changes need to be made in the statements as well as proofs.)

(3.3) For $t = t_{-1}$, the image of B misses B, i.e. for $t = t_{-1}$ $F_t(B) \cap B = \emptyset$.

(3.4) Let $t = t_1$ and fix any x in $[x_1,x_2]$. Then as y varies from y_1 to y_2, $F_{t_1}(x,y)$ starts outside B and below B (the second coordinate of $F_t(x,y)$ is less than y_1, $(F_{t_1})_2(x,y) < y_1$), crosses B and exists the top of B ($(F_{t_1})_2(x,y)$ becomes larger than y_2), reenters B, crosses B and exists the bottom ($(F_{t_1})_2(x,y)$ becomes less than y_1). In particular, $(F_{t_1})_2(x,y_j) < y_1$ for $j = 1,2$.

(3.5) $F_t(B)$ never intersects the sides of B for $t_{-1} \le t \le t_1$, $F_t \cap (\{x_j\} \times [y_1,y_2]) = \emptyset$ for $j = 1,2$. Also the first coordinate of $F_t^{-1}(x_j,y)$ is less than x_1 for $j = 1,2$, $(F^{-1})_1(x_j,y) < x_1$.

(3.6) For $j = 1,2$, $F_t(x,y_j)$ does not intersect B for $t_{-1} \le t \le t_1$ and in fact $(F_t)_2(x,y_j) < y_1$. (If follows that

$F_t^{-1}(B)$ does not intersect the top or bottom of B, $[x_1,x_2] \times \{y_j\}$.) Further assume $y_1 < (F_t^{-1})_2(x,y) < y_2$ for all (x,y) in $B \cup B_t^-$.

(3.7) Let $DF_t(x,y) = \binom{a \;\; b}{c \;\; d}$ where a, b, c, and d depend on (x,y) and t. Assume that for $t_{-1} \leq t \leq t_1$ and (x,y) in $B_t^- \cup B$ that

(i) $bc < 0$ and $|a| < \Delta = \det DF_t$.

(ii) $D^2F_{t2}(v,v) < 0$, where the second derivative of the second coordinate function is evaluated at (x,y) and acts as a bilinear map on the column vector $v = (d-1, -c)^{tr}$, i.e. $(-c)^2(\partial^2 F_{t2}/\partial y^2) + 2(-c)(d-1)(\partial^2 F_{t2}/\partial y \partial x) + (d-1)^2(\partial^2 F_{t2}/\partial x^2) < 0$.

(iii) $D^2F_{t2}(w,w) + (d/-b)D^2F_{t1}(w,w) < 0$ for $w = (b,1-a)^{tr}$ and

(iv) $D^2F_{t2}(w,w) + (a-\Delta/b)D^2F_{t1}(w,w) < 0$ for $w = (b,1-a)^{tr}$.

<u>Remarks:</u>

3.8.1) With assumptions (3.1-3.6) it is shown in [12] that F_t creates at some t_0, with $t_{-1} < t_0 < t_1$, a fixed point p_0 with the eigenvalues of $DF_{t_0}(p_0)$ being 1 and $\Delta_0 = \det DF_{t_0}(p_0)$ $(0 < \Delta_0 < 1)$. In section 5, it is shown that condition (3.7iv) insures the second derivative condition at the saddle node bifurcation is satisfied. It follows that for $t > t_0$ but near t_0, F_t has a fixed point sink q_t and a fixed point saddle p_t nearby.

3.8.2) Considering (3.7i), notice that for each fixed t and x^* and for y at which $F_{t2}(x^*,y)$ attains a maximum, that $d = 0$. Thus $\det DF_t = -bc > 0$ at these points. Thus the first condition is true in a neighborhood of these points and we assume it is true in all of $B \cup B_t^-$. We are imagining that F_t stretches in the y direction, $|\partial F_t/\partial y| > 1$, so it would follow that $|a| < |\partial F_t/\partial x|$ should be less than Δ.

3.8.3) We are considering the case where the image of B is bent downward. (Comparable changes in derivatives would treat the other cases.) Hence we are considering cases where $\partial d/\partial y = \partial^2 F_{t2}/\partial y^2 < 0$. In fact in the proof we need to use that this second derivative is negative enough to overcome the possible adverse effects of combinations of other second partial derivatives, hence assumptions (3.7ii-iv). Conditions (3.7ii) and (3.7iii) imply that the sets $\{F_{t2}(x,y)-y = 0\}$ and $\{(F_t^{-1})_1(x,y)-x = 0\}$ are convex to the left $(d^2x/dy^2 < 0)$ and downward $(d^2y/dx^2 < 0)$ respectively. As noticed above (3.7iv) insures the second derivative condition at the saddle node.

Theorem II. Let $\{F_t\}$ be a family of orientation preserving, area decreasing maps satisfying (3.1)-(3.7). Then there exist t_0 and $\varepsilon > 0$ such that $t_{-1} < t_0 < t_0+\varepsilon < t_1$ and for $t \in (t_0, t_0+\varepsilon)$ F_t has a fixed saddle point p_t and a fixed sink q_t such that the basin of q_t in $B \cup B_t^-$ is bounded by the stable manifold of p_t in that there is an open disk D_t with the following properties.

(i) $\partial D_t = L_t \cup S_t$

(ii) L_t is an open interval in $\{x_1\} \times [y_1, y_2]$

(iii) S_t is a closed curve segment in $W^s(p_t, F_t)$

(iv) $F_t(L_t) \subset D_t$

(v) $F_t(D_t) \subset D_t$ and

(vi) $D_t \subset W^s(q_t, F_t)$.

The proof is given in section 5.

§4. Creation of a horseshoe drom a homoclinic tangency.

The main theorem discusses the basin of the sink formed during the creation of a horseshoe. In this section we want to indicate how this occurs whenever a one parameter family of maps creates new homoclinic intersections.

Let $\{f_t\}$ with t in an interval $I \subset \mathbb{R}$ be a one parameter family of diffeomorphisms of the plane $\mathbb{R}^2$ or more generally a two manifold. Suppose that for t in I, f_t has a dissipative fixed point p_t: $f_t(P_t) = P_t$, the spectrum of the derivative, spec $Df(P_t) = \{\eta_t, \rho_t\} \subset \mathbb{R}$ with $|\eta_t| < 1$ and $|\rho_t| > 1$, and $|\det Df_t(P_t)| = |\eta_t \rho_t| < 1$. Suppose further that $\{f_t\}$ nondegenerately creates new homoclinic intersections for P_t at $t = t_0$ in the interior of I. This means that there are a coordinate system on $\mathbb{R}^2$, open one disks $\gamma_t^u \subset W^u(P_t, f_t)$ and $\gamma_t^s \subset W^s(P_t, f_t)$, and $\epsilon > 0$ such that (i) $\gamma_t^u \cap \gamma_t^s = \phi$ for $t \in (t_0 - \epsilon, t_0)$,
(ii) $\gamma_t^n \cap \gamma_t^s$ consists of one point $(x_1, 0)$ for $t = t_0$,
(iii) $\gamma_t^u \cap \gamma_t^s$ consists of two points for $t \in (t_0, t_0 + \epsilon)$, and
(iv) for $t = t_0$, γ_t^s is an interval in the x-axis containing $(x_1, 0)$ and γ_t^u is given by $y = a(x - x_1)^2$ + h.o.t. with $a \neq 0$.

This situation was studied by Gavrilov and Silnikov [3] who proved there was a cascade of sinks and later by S. Nehouse, [9], [10] who proved there were values of t for which f_t has infinitely many sinks. Also see C. Robinson [11] and [12]. To prove there is a cascade of sinks, it can be shown that for some N and $n \geq N$, one may use the hyperbolic estimates and careful scaling arguments to construct a box B_n above γ_t^s such that $f_t^n(B_n)$ is pulled across B_n to form a horseshoe as t increases, see e.g. [11, Sec. 5]. Further the boxes $\{B_n\}$ do not overlap with each other and they get closer to γ_t^s as $n \to \infty$. The following theorem shows that the results of the previous sections apply to this situation.

<u>Theorem III</u>. Let $\{f_t\}$ and B_n be as described above. Then there is an $N > 0$ such that if $n \geq N$ then $F_t = f_t^n$ and $B = B_n$ satisfy conditions (3.1)-(3.7). In particular there are values t_n and $t_n + \varepsilon_n$ such that for $t_n < t < t_n + \varepsilon_n$ there is a sink q_t^n and a saddle p_t^n for f_t in B_n. Further the basin of attraction for q_t^n contains a region in $B_n \cup F_t^{-1}(B_n)$ whose boundary is the stable manifold of a saddle p_t^n of f_t and one of the ends of B_n.

The proof is contained in section 6.

It is shown in [9] or [11] that the boxes B_n may be taken to be product boxes $B_n = I_n \times J_n$ with the length of I_n equal to $C_1 \rho^{-n}$ and the length of J_n equal to $C_2 \rho^{-2n}$. Thus the local basin of attraction for the sink q_t^n lies inside B_n and so has dimensions proportional to ρ^{-n} by ρ^{-2n}. Assuming $\rho = 2$, for $n \geq 32$ the width of the basin is about the accuracy of a computer, 2^{-64}. It is also clear that it would be very difficult to find one of these sinks even if its period was much lower than 32, say 15. In fact the sinks found usually have period closer to 7. Also, it is shown that there are parameter values t for which f_t has infinitely many sinks. However it seems very unlikely that there can be two sinks of relatively low period. The reason is that although there is a cascade of sinks q_t^n for $t = t_n$ of every period $n \geq N$ for some N, these sinks occur for different parameter values t, i.e. the intervals $(t_n, t_n + \varepsilon_n)$ do not overlap, [11]. The simultaneous sinks are caused by secondary saddle node bifurcations. This is related to the recent result of L. Tedeschini-Lalli/Yorke [13] which shows that the set of parameter values with infinitely many "Newhouse" sinks has measure zero. This situation should be further investigated but it might explain the difficulty in numerically finding simultaneous sinks. More indirectly, it might shed light on the question of whether the Hénon attractor is indeed chain transitive. See [6], [1], [5].

§5. Proof of Theorems I and II

We consider a general family of nonlinear maps which creates a horseshoe and satisfies conditions (3.1)-(3.7) or the family of quadratic maps satisfying (2.1). In the arguments where the proof is simple or even obvious for the case of quadratic maps, we indicate this so the reader can see why the result is reasonable and can skip the general proof if he or she so desires.

In both cases we construct a forward invariant open disk D_t which is contained in the basin of attraction of the sink q_t for the parameter t just larger than the saddle node bifurcation value t_0. In both cases part of the boundary of D_t is a compact curve segment in the stable manifold of the saddle point p_t, $W^s(p_t,F_t)$. However, the rest of the boundary of D_t, L_t, is chosen differently in the two cases. In each case L_t is a line segment over which D_t overflows under F_t^{-1}, or $F_t(L_t) \subset$ interior D_t. Clearly there are many choices for such a line segment. For the quadratic map (2.1) we chose $L_t \subset \{p_t\} + E_t^c$ for $t = t_0$ and $L_t \subset \{p_t\} + E_t^u$ for $t > t_0$, because this line is dynamically important in this case. (For $t=t_0$, F_t fixes the x-coordinate on this line.) There is also no natural choice of a rectangle B in this case. When F_t is assumed to satisfy (3.1)-(3.7) the above choice seems arbitrary. Furthermore the assumptions are given in terms of a rectangle B and so a more natural choice seems to be to choose L_t to lie in the left boundary of B, $L_t \subset \{x_1\} \times [y_1,y_2]$.

In the constructions we need to consider $\{F_t\}$ defined on $B \cup F_t^{-1}(B)$. More precisely, $B = [x_1,x_2] \times [y_1,y_2]$ and $F_t^{-1}(B)$ may extend both past $\{x_1\} \times [y_1,y_2]$ and $\{x_2\} \times [y_1,y_2]$. Since we are interested in the hook of $F_t^{-1}(B)$ which extends to the right of B but not the feet which extend to the left we define

$$B_t^* = (B \cup F_t^{-1}(B)) \cap \{(x,y): x \geq x_1\}.$$

For the statements about the quadratic family (2.1) or (5.1) below, B_t^* can be considered the region below $\{p_t\} + E_t^u$ for $t > t_0$ or $\{p_0\} + E^c$ for $t = t_0$.

We prove the theorems by first showing that both components of $\hat{W}^s(p_t,F_t) \equiv W^s(p_t,F_t) - \{p_t\}$ leave B_t^* exiting through the left side of B_t^*. (In the case of (2.1) we show that they cross the unstable line $\{p_t\} + E_t^u$.) We then take L_t to be the piece of $x = x_1$ cut off by the two components of $\hat{W}^s(p_t,F_t)$ and show $F_t(L_t) \subset \text{int } B_t^*$. Finally we show that for appropriate parameter values all of D_t lies in the basin of attraction of q_t.

We determine the location of $W^s(p_t,F_t)$ for $t > t_0$ by comparing it with the strong stable manifold of the saddle node p_t for $t = t_0$, the bifurcation value. We also label the strong stable manifold for $t = t_0$ by $W^s(p_t,F_t)$; the strong stable manifold is the set of points which tend exponentially fast toward the fixed point. (This manifold is often denoted by $W^{ss}(p_t,F_t)$.) The manifold $W^s(p_t,F_t)$ forms the boundary of the weak stable manifold, $W^{ws}(p_t,F_t)$, which is the set of all points which tend toward p_0 under forward iteration (but not necessarily at an exponential rate). Then for $t > t_0$ and t near t_0, the basin of the sink q_t, $W^s(q_t,F_t)$, is very near $W^{ws}(p_{t_0},F_{t_0})$ for $t = t_0$.

Before starting the constructions we want to make some simplifications in the equations for the map (2.1) and some choices for the general map satisfying (3.1)-(3.7). First, when considering (2.1) it is convenient to translate the bifurcation parameter so $t_0 = 0$ and the coordinates so p_0 is at the origin. Under such a translational change of coordinates, letting $\Delta = \lambda\mu$, and continuing to write F_t for the mpa in the new coordinates, we have

$$(5.1)\qquad \begin{aligned} F_t(x,y) &= (\lambda y,\ t - \mu x + y + \Delta y - \lambda^2 y^2) \\ F_t^{-1}(x,y) &= (\mu^{-1}t + x + \Delta^{-1}x - \mu^{-1}x^2 - \mu^{-1}y,\ \lambda^{-1}x). \end{aligned}$$

Turning to the more general family, certain orientation choices must be made. Here we make a particular choice and note that minor modifications in the statements as well as proof are necessary for the other choices. Our choices are given by

$$(5.2)\qquad \begin{aligned} &\text{i)}\ \ b > 0 \\ &\text{ii)}\ \ c < 0 \\ &\text{iii)}\ \ \partial d/\partial y < 0 \end{aligned}$$

for all (x,y) in B_t^* and $t_{-1} \le t \le t_1$. (Recall that $DF_t(x,y) = \begin{pmatrix} a & b \\ c & d \end{pmatrix}$ where $a = a(x,y,t)$ etc.)

As stated in section 3, assumptions (3.1)-(3.6) imply that $\{F_t\}$ creates a quasi-sink in B_t^*, [12]. In fact the index argument used and the continuity of eigenvalues allow us to conclude that there is a parameter $t = t_0$ such that F_t has a fixed point p_t with one eigenvalue equal to one. By a translation of the parameter values we can assume $t_0 = 0$. We prove below that the assumptions imply that p_0 is weakly attracting from above and thus is a saddle node bifurcation with a saddle p_t and a sink q_t for $t > 0$ small enough.

We focus primarily on the dynamics of F_0 so write $F = F_0$,

$$W^s(p_0) = W^s(p_0,F_0),\quad DF(p_0) = \begin{pmatrix} a_0 & b_0 \\ c_0 & d_0 \end{pmatrix},\quad \text{and}\quad DF^{-1}(p_0) = \begin{pmatrix} a_0^- & b_0^- \\ c_0^- & d_0^- \end{pmatrix}.$$

Let $\Delta_0 = \det DF(p_0)$, so $DF(p_0)$ has eigenvalues Δ_0 and 1, with column eigenvectors $v^s = (b_0,\Delta_0-a_0)^{tr}$ and $v^c = (b_0,1-a_0)^{tr}$ respectively. (Note in the case of (5.1), $v^s = (1,\mu)^{tr}$ and $v^c = (1,\lambda^{-1})^{tr}$.) By (3.7i) and (5.2), both vectors have positive slopes. Since $0 < \Delta_0 < 1$ (F decreases area), v^c has slope greater than v^s. The following lemma shows that p_0 satisfies the conditions for a saddle node which is weakly attracting from above.

<u>5.3 Lemma</u>. A family satisfying (3.1)-(3.7) or (5.1) undergoes a saddle node bifurcation at $t = t_0$ (taken so $t_0 = 0$) creating a saddle and a sink for $t > t_0$. Moreover, p_0 is weakly attracting from above $W^s(p_0)$.

<u>Proof</u>: These facts are more or less obvious for (5.1) and can be verified by a direct calculation of fixed points.

Turning to the general family, the conditions for a saddle node are usually expressed with the eigenspaces being the axes. Therefore we change coordinates so that v^s and v^c lie on the positive axes. Let V be the 2 by 2 matrix whose columns are v^s and v^c, $V = (v^s,v^c)$. Then $\det V > 0$ by the relative slopes. Let $H(u,v) = V^{-1}(p_0+uv^s+wv^c)$. We need $\partial^2H_2/\partial w^2 < 0$ and $\partial H_2/\partial t > 0$. But,

$$\frac{\partial^2 H^2}{\partial w^2} = (V^{-1})_2 D^2F(p_0)(v^c,v^c)$$

where $(V^{-1})_2$ is the second row of V^{-1}. Thus

$$(\det V)\frac{d^2H_2}{dw^2} = (a_0-\Delta_0,b_0)D^2F(p_0)(v^c,v^c)$$

$$= b_0D^2F_2(p_0)(v^c,v^c) + (a_0-\Delta_0)D^2F_1(p_0)(v^c,v^c)$$

and this is less than zero by (3.7iv). Moreover, letting t vary, keeping V^c and v^s fixed and therefore V fixed,

$$(\det V)\frac{\partial H_2}{\partial t} = (a_0-\Delta_0)\frac{\partial F_{t1}}{\partial t} + b_0\frac{\partial F_{t2}}{\partial t} > 0$$

by (3.7iv). //

Letting B^* be B_t^* for $t=t_o$, we now show that both of the components of $\hat{W}^s(p_o)=W^s(p_0-\{p_0\})$ exit B^* through its left boundary $\{x_1\}\times[y_1,y_2]$. We show that the left components go directly to this edge while the right component passes initially upward to the right only to turn and also leave B^* through its left boundary. Furthermore, while these components of $\hat{W}^s(p_0)$ may pass through $\{x_1\}\times[y_1,y_2]$ more than once, there is a finite distance along both directions which contains all the points of $W^s(p_0)$ whose entire forward orbit lies completely inside B^*. Note for the family (5.1) we ignore the left component of $\hat{W}^s(p_0)$ and show the right component goes up and crosses the center line E_0^c.

To prove these facts, we determine the curves (regions) where F and F^{-1} fix (increase) the x and y coordinates. Thus we define the four functions δ_x, δ_y, δ_x^-, and δ_y^- on B^* as follows:

$$\delta_x(x,y) = F_1(x,y)-x, \quad \delta_y(x,y) = F_2(x,y)-y$$

$$\delta_x^-(x,y) = (F^{-1})_1(x,y)-x, \quad \delta_y^-(x,y) = (F^{-1})_2(x,y)-y.$$

Now for convenience of notation we write $\{\delta_y > 0\}$ for $\{(x,y)\in B^* : \delta_y(x,y) > 0\}$ etc.

<u>5.4 Lemma</u>. a) The curve $\{\delta_y=0\}$ is tangent at p_0 to E^c and the region $\{\delta_y\geq 0\}$ is a convex set above and to the left of $\{p_0\}+E^c$. For the quadratic map (5.1), the curve $\{\delta_y=0\}$ is the parabola $x=\lambda y-\mu^{-1}\lambda^2y^2$.

b) The curve $\{\delta_x^-=0\}$ is tangent at p_0 to E^c and the region $\{\delta_x^-\geq 0\}$ is a convex set below and to the right of $\{p_0\}+E^c$. For (5.1), $\{\delta_x^-=0\}$ is the parabola $y=\lambda^{-1}x-x^2$.

c) The curves $\{\delta_x=0\}$ and $\{\delta_y^-=0\}$ both have positive slopes and lie strictly between $\{\delta_y=0\}$ and $\{\delta_x^-=0\}$ except at p_0, i.e. in the region $(\{\delta_y<0\}\cap\{\delta_x^-<0\})\cup\{p_0\}$. Also $\{\delta_x<0\}$ is below

$\{\delta_x = 0\}$, and $\{\delta_x^- > 0\}$ is below $\{\delta_y^- = 0\}$. For (5.1), $\{\delta_x = 0\} = \{\delta_y^- = 0\} = E^c$.

d) $\{\delta_y > 0\} \subset \{\delta_y^- < 0\}$, $\{\delta_y^- > 0\} \subset \{\delta_y < 0\}$, $\{\delta_x > 0\} \subset \{\delta_x^- < 0\}$, and $\{\delta_x^- > 0\} \subset \{\delta_x < 0\}$.

e) $\{\delta_y < 0\}$ and $\{\delta_x < 0\}$ are semi-invariant under F. The sets $\{\delta_x^- < 0\}$ and $\{\delta_y^- < 0\}$ are semi-invariant under F^{-1}. (This means $F\{\delta_y < 0\} \cap B^* \subset \{\delta_y < 0\}$ etc.)

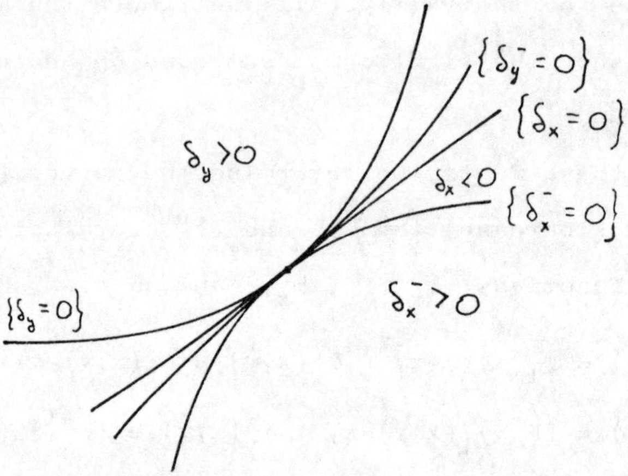

Figure 2

<u>Proof</u>: These facts for the family (5.1) are a direct calculation and are left to the reader. This and the proof of Lemma 5.3 are the main simplifications obtained by assuming F is given by (5.1).

For the general family of nonlinear maps satisfying (3.1)-(3.7) and (5.2) the argument is somewhat long and is as follows. Consider the curve $\{\delta_y = 0\}$. Since p_0 is a fixed point, $p_0 \in \{\delta_y = 0\}$. Now $\partial(\delta_y)/\partial x = (\partial/\partial x)(F_2(x,y)-y) = \partial F_2/\partial x = c(x,y,t_0)$ and by (5.1) $c < 0$. The implicit function theorem yields a function $h: [y_1, y_2] \to R$ such that $F_2(h(y),y)-y = 0$. Differentiating once gives

(5.5) $$DF_2(h(y),y)\begin{pmatrix}h'(y)\\1\end{pmatrix} - 1 = 0.$$

Thus $ch'(y) + d - 1 = 0$ and $h'(y) = (1-d)/c$. Thus $v = (h'(y),1)^{tr} = ((1-d)/c,1)^{tr}$ is a column vector field tangent to $\{\delta_y=0\}$. In particular, $v(p_0) = ((1-d_0)/c_0,1)^{tr}$ which is a multiple of V^c. Thus $\{\delta_y=0\}$ is tangent to E^c at p_0.

Differentiating (5.5) with respect to y yields

$$D^2F_2(v,v) + DF_2\begin{pmatrix}h''\\0\end{pmatrix} = 0,$$

so $h''(y) = -(1/c)D^2F_2(v,v) < 0$ by (3.7ii). We conclude that $\{\delta_y=0\}$ is given by a function $x = h(y)$ and in concave to the left. Thus $\{\delta_y=0\}$ lies strictly to the left of $\{p_0\} + E^c$ except at p_0. Furthermore, since $\partial(\delta y)/\partial x = c < 0$, $\{\delta y < 0\}$ $\{\delta y < 0\}$ lies to the left of $\{\delta_y = 0\}$ and 5.4a) is established.

The proof of 6.3b) is very similar. Here we are concerned with the curve $\{\delta_x^- = 0\} = \{(x,y): F_1^{-1}(x,y)-x = 0\}$ so the shape of this curve is controlled by putting conditions on the derivative of F^{-1}. However we stated the conditions in (3.7) in terms of derivatives of F so they would be easier to apply. Thus we need to translate these conditions. Let $G = F^{-1}$. The key relationship is of course that if $F(z) = z'$ then $DG(z')DF(z) = DF(z)DG(z') = id$. Now

$$DF(z) = \begin{pmatrix}a(z) & b(z)\\ c(z) & d(z)\end{pmatrix}, \qquad DG(z') = \begin{pmatrix}a^-(z') & b^-(z')\\ c^-(z') & d^-(z')\end{pmatrix}$$

so we have

$$\begin{pmatrix}a^-(z') & b^-(z')\\ c^-(z') & d^-(z')\end{pmatrix} = \frac{1}{\Delta(z)}\begin{pmatrix}d(z) & -b(z)\\ -c(z) & a(z)\end{pmatrix}.$$

Now consider $\{\delta_x^- = 0\}$. Since $\delta_x^-(p_0) = 0$ and $\partial(\delta_x^-)/\partial y =$ $= (\partial/\partial y)(G_1(x,y)-x) = b^- = -b/\Delta < 0$, the implicit function theorem implies there is a smooth curve through p_0 given by $y = g(x)$ such that $\delta_x^-(x,g(x)) = 0$. Furthermore, this curve may be extended until it leaves B^*. Differentiating yields

$$(5.6) \qquad DG_1(x,g(x)) \begin{pmatrix} 1 \\ g'(x) \end{pmatrix} = 1,$$

so $g'(x) = (1-a^-(z')/b^-(z')$ where $z' = (x,g(x))$. In particular, $(x_0,g(x_0)) = p_0$ so $g'(x_0) = (1-a_0^-)/b_0^-$. Thus $(1,g'(x_0))^{tr}$ is a vector in E^c and the curve is tangent to E^c at p_0. Differentiating (5.6) gives

$$(5.7) \qquad g''(x) = -(1/b^-(z'))D^2G_1(z')(w',w')$$

where $z' = (x,g(x))$ and $w' = (1,g'(x))^{tr}$.

Let $z = G(z')$ and $w = DG(z')w'$. We want to show w is a multiple of the vector given in (3.7iii). Notice that $z' = (x,g(x))$ is on $\delta_x'(z') = 0$ so $z = F^{-1}(z') = (x,G_2(x,g(x)))$. Differentiating with respect to x gives $w = DG(z')w' = (1,w_2)$. Next z' satisfies $\delta_x^-(z') = 0$ and $z = F^{-1}(z')$ so $\delta_x(z) = 0$. Differentiating $F_1 \circ G(x,g(x)) = x$ gives $DF_1(z)w = 1$ so $w_2 = (1-a)/b$. Thus w is a vector as in (3.7iii).

Next differentiating $DF(z)DG(z') = id$ and evaluating at (w',w'), we get

$$D^2F(z)(w,w) + DF(z)D^2G(z')(w',w') = 0.$$

Thus

$$D^2G(z')(w',w') = -DG(z')D^2F(z)(w,w)$$

and

$$D^2G_1(z')(w',w') = -DG_1(z')D^2F(z)(w,w)$$
$$= -a^-(z')D^2F_1(z)(w,w) - b^-(z')D^2F_2(z)(w,w).$$

Now from (5.7)

$$g''(x) = -(1/b^-(z'))D^2G_1(z')(w',w')$$
$$= \frac{a^-(z')}{b^-(z')} D^2F_1(z)(w,w) + D^2F_2(z)(w,w)$$
$$= \frac{d(z)}{b(z)} D^2F_1(z)(w,w) + D^2F_2(z)(w,w)$$

which is negative by (3.7iii). Thus we conclude that $\{\delta_x^- = 0\}$ is concave down.

Furthermore since $\partial(\delta_x^-)/\partial x = b^- = -b/\Delta < 0$, we conclude that $\{\delta_x^- > 0\}$ lies below $\{\delta_x^- = 0\}$ and 6.3b is established.

To prove (5.4c), notice that the curve $\{\delta_x=0\}$ has slope $(1-a)/b$ which is strictly positive. Similarly $\{\delta_y^- = 0\}$ has slope $(1-d^-)/c^- = (\Delta-a)/(-c)$ which is also positive.

To prove these curves lie between $\{\delta_x^- = 0\}$ and $\{\delta_y = 0\}$, first observe that $B^* \cap F\{\delta_x = 0\} = \{\delta_x^- = 0\}$. Next notice that since p_0 is the only fixed point for F, we have $\{\delta_x = 0\} \cap$ $\cap \{\delta_y = 0\} = \{p_0\}$. Thus a point $p = (x,y) \neq p_0$ which lies on $\{\delta_x = 0\}$ must either lie in i) $\{\delta_x^- \geq 0\}$, or ii) $\{\delta_y > 0\}$, or iii) $\{\delta_y < 0\} \cap \{\delta_x^- < 0\}$. In the first case $\delta_x^-(p) \geq 0$ so $\delta_y(p) < 0$ by parts 5.4a) and 5.4b). Since $\delta_y(p) < 0$ and $\delta_x(p) =$ $= 0$, $F(p)$ would lie strictly below $\{\delta_x^- = 0\}$. Since $F\{\delta_x = 0\} =$ $= \{\delta_x^- = 0\}$, $\delta_x^-(F(p)) = 0$. This contradiction shows case i) is impossible. Likewise in case ii), $\delta_y(p) > 0$ so p lies above $\{\delta_x^- = 0\}$ by parts 5.4a) and 5.4b). Then $\delta_y(p) > 0$ and $\delta_x(p) = 0$ so $F(p)$ would lie above p and so above $\{\delta_x^- = 0\}$. Again we obtain a contradiction showing that only case iii) is possible. This proves 5.4c).

Part 5.4d) follows directly from (5.4a), (5.4b), (5.4c), and the observation that if $\delta_y(z) > 0$ then $\delta_y^-(F(z)) < 0$ etc.

Finally, (5.4e) follows from this last observation and (5.4d). This completes the proof of Lemma 5.4. //

We can now discuss the position of $W^s(p_0)$. The left hand component of $\hat{W}^s(p_0)$ extends from p_0 locally into the region $\{\delta_x^- < 0\} \cap \{\delta_y^- < 0\}$. (This paragraph can be skipped for the proof of Theorem I for the family of quadratic maps (5.1).) This region is F^{-1} semi-invariant so the branch extends downward and to the left until it exits B^*. Since $\delta_y^- > 0$ on $[x_1, x_2] \times \{y_1\}$, by assumption (3.6), this component must exit through the left hand boundary $\{x_1\} \times [y_1, y_2]$ as required. Now since the stable manifolds vary continuously with t on compact sets, for small $\epsilon > 0$ and $t \in [t_0, t_0+\epsilon]$ the left hand component of $\hat{W}^s(p_t, F_t)$ exits B_t^* via $\{x_1\} \times [y_1, y_2]$, perhaps intersecting more than once before eventually extending away to the left. In any case we let z_t^* be the uppermost such point of intersection of the left hand component of $\hat{W}^s(p_t, F_t)$ with the left hand boundary (among the points of $\hat{W}^s(p_t, F_t)$ whose entire forward orbit lies inside B_t^*).

The determination of the position of the right hand component of $\hat{W}^s(p_t, F_t)$ is more delicate because it starts extending up and to the right, bends up and crosses $\{p_t\} + E_t^u$ (or $\{p_0\} + E^c$ for $t = 0$), and finally extends to the left and exits through the left boundary. See Devaney [2] for another such argument for a quadratic area preserving map. We introduce some notation for the right hand curve segment in $\hat{W}^s(p_t, F_t)$ of length r. Let $W_r^s(p_t, F_t)$ be all points in $W^s(p_t, F_t)$ whose distance from p_t along the stable manifold is less than or equal to r. (Notice that this is different from what is often also denoted by W_r^s: the set of point z such that the distance of all forward iterates, $f^j(z)$, from p_t is less than r within the total ambient space R^2.) Now let $\hat{W}_r^s(p_t, F_t) = W_r^s(p_t, F_t) - \{p_t\}$ and

$$S_t(r) = \text{closure of the right hand component of } \hat{W}_r^s(p_t, F_t).$$

By stable manifold theory, $S_t(r)$ is a closed 1-disk embedded in $W^s(p_t,F_t)$ with one end point at p_t and tangent to E^s_t at p_t. We may think of $S_t(r)$ as parametrically generating the right half of $W^s(p_t,F_t)$ as r increases. As many of the arguments consider the case $t = 0$, we write $S(r) = S_0(r)$.

To show that $S(r)$ bends up, we show that the curve $F^{-1}((x_0,x_2] \times \{y_0\})$ forms a barrier where $p_0 = (x_0,y_0)$ is the fixed point of F.

<u>5.8 Lemma</u>. Let $J = (x_0,x_2] \times \{y_0\}$ be the line extending to the right from the fixed point p_0. Then $F^{-1}(J)$ is a curve which extends from p_0 to the left end of B^* inside B^*. Further $F^{-1}(J) \cap \{z \in S(r): f^j(z) \in B^* \text{ for } j \geq 0\} = \phi$. Thus if $S(r)$ exits B^* it goes through the left end $\{x_1\} \times [y_1,y_2]$. (For the quadratic family (5.1), B^* is the region below and to the right of the center line E^c.)

<u>Proof</u>: The curve $\{\delta^-_y = 0\}$ has positive slope so $\delta^-_y(z) > 0$ for all points z on J. Thus $F^{-1}(J)$ lies above $y = y_0$. Also $F^{-1}(J)$ lies below $y = y_2$ by (3.6), and $(F^{-1})_1(x_2,y_0) < x_1$ by (3.5). Thus $F^{-1}(J)$ extends to the left end of B^* (or E^c for (5.1)). Let R be the region in B^* with lower boundary equal to the union of J and the left component of $\hat{W}^s(p_0)$. Then $F^{-1}(R) \cap B^*$ has $F^{-1}(J)$ as part of its boundary. The stable vector has positive slope, so for small r', $S(r')$ enters the interior of R. Thus $F^{-1}S(r') \supset S(r')$ lies in the interior of $F^{-1}(R)$. Since R is semi-invariant by F^{-1}, $F^{-1}(R) \cap B^*$ is also semi-invariant by F^{-1}. Thus $\bigcap\{F^{-j}S(r') \cap B^*: j \geq 0\} \supset \{z \in S(r): f^j(z) \in B^* \text{ for } j \geq 0\}$ is contained, and so the second set is contained in $F^{-1}(R) \cap B^*$. This proves the lemma. //

The following lemma shows that $S(r)$ extends to the left end of B^*.

<u>5.9 Lemma</u>. There exists $r_1 > 0$ such that

a) $S(r_1) \cap \partial B^* = S(r_1) \cap (\{x_1\} \times [y_1, y_2]) \neq \phi$ and

b) for $r > r_1$ $\{z \in S(r): f^j(z) \in B^*$ for all $j \geq 0\} \subset S(r_1)$

i.e. the only way $S(r) - S(r_1)$ intersects B^* is with points which leave B^* under forward iteration before they finally return to B^* forever.

<u>Proof</u>: By Lemma 5.8, the curve $F^{-1}((x_0, x_2] \times \{y_0\})$ forms a barrier to $S(r)$ and it can only exit the left end of B^* giving the first equality in a) for any r. It remains to show that $S(r)$ exits B^* for some r. Let I be a fundamental interval in $S(r)$ near p_0, $I = \text{closure}(S(r_0) - F(S(r_0)))$ for small r_0. By taking r_0 small, $I \subset \{\delta_x^- > 0\} \cap \{\delta_y^- > 0\}$. It will suffice to show that for each $z \in I$ there exists a positive N such that $\pi_1 F^{-N}(z) < x_1$, where $\pi_1(x,y) = x$. We first show that there is a positive integer K such that $F^{-K}(z) \in \{\delta_x^- < 0\}$. Suppose not, then the backward orbit $O_-(z) \subset \{\delta_x^- > 0\}$ and this is bounded because it cannot cross $F^{-1}((x_0, x_2] \times \{y_0\})$. Moreover since $\{\delta_x^- > 0\} \supset \{\delta_y^- > 0\}$ we have $O_-(z)$ converging monotonically in each coordinate to a fixed point to the right of p_0. This is impossible since p_0 is the only fixed point in B^* so we have established the existence of K. This completes the proof for (5.1) because $F^{-K-1}(z) \in \{\delta_x > 0\}$ which is above E^c.

Now consider the fate of $F^{-K}(z)$ under iteration of F^{-1}. Since $\{\delta_x^- < 0\}$ is F^{-1} semi-invariant, each successive F^{-1} iterate of $F^{-K}(z)$ moves monotonically to the left. Using the fact that $\{\delta_y^- = 0\}$ has positive slope and $F^{-1}((x_0, x_2] \times \{y_0\})$ is a barrier, these points must enter the region where $\{\delta_y^- < 0\}$. Then the region $\{\delta_y^- < 0\} \cap \{\delta_x^- < 0\}$ is F^{-1} semi-invariant. If the orbit does not exit B^* it must converge to a fixed point which must be p_0. Thus $F^j(z)$ would converge to p_0 from above which is forbidden by Lemma 5.3. This contradiction completes the proof of (5.9). //

We are now in a position to prove the theorem. Since $S(r_1)$ contains all the intersections of $S(r)$ with $\{x_1\} \times [y_1,y_2]$, there exists an $\epsilon > 0$ such that for each t in $[0,\epsilon)$ $S_t(r_1)$ contains all the intersections of $S_t(r)$ with $\{x_1\} \times [y_1,y_2]$ (in the sense given in Lemma 5.9b.) Let w_t^* be the lowest point of this intersection and S_t be the curve segment in $W^s(p_t,F_t)$ from z_t^* to w_t^*, i.e. both the right and left branches of $\hat{W}^s(p_t,F_t)$. Likewise let L_t be the open interval in $\{x_1\} \times [y_1,y_2]$ with end points z_t^* and w_t^*. (In the case of (5.1), L_t is the open interval in $\{p_t\} \times E_t^u$, or $\{p_0\} + E^c$ for $t = 0$, from p_t to w_t^*.) Let D_t be the open disk bounded by the Jordan curve $L_t \cup S_t$. Notice if tangencies of $S_t(r_1)$ and $\{x_1\} \times [y_1,y_2]$ occur, the point w_t^* and likewise z_t^* may only vary in a semi-continuous manner and so D_t may be only semi-continuous. Also note that D_t may extend beyond the left end of B^* so is not necessarily contained inside B^*.

We now show that these curve and D_t have all the properties stated in the theorem. Conditions (i)-(iii) of Theorems I or II are true by construction. To prove part (iv), let S_t' be given by $F_t^{-1}(S_t) = S_t'$. Since S_t has its ends at w_t^*, which is the lowest intersection of the top branch, and at z_t^*, which is the highest intersection of the bottom branch, S_t' can not have any intersection between $F_t^{-1}(S_t) \cap L_t = S_t' \cap L_t = \emptyset$. Thus $S_t \cap F_t(D_t) = \emptyset$. Also $L_t \subset \{\delta_x > 0\}$ so $F_t(L_t) \cap L_t = \emptyset$. Thus $F_t(L_t) \cap \partial D_t = \emptyset$. Since $F_t(L_t) \cap D_t \neq \emptyset$, $F_t(L_t) \subset \operatorname{int} D_t$ giving condition (iv). Next, since $F_t(L_t) \subset D_t$ and $F_t(S_t) \subset S_t \subset$ closure D_t, it follows that $F_t(D_t) \subset D_t$ giving condition (v). This leaves only condition (vi) to check which is the statement of the following lemma.

5.10 Lemma.

a) $D_0 \subset W^{ws}(p_0,F_0)$, the weak stable manifold of p_0, i.e. the set of all points z which tend toward p_0 under forward iteration of F_0 (not necessarily at an exponential rate).

b) For $\epsilon > 0$ small enough and $0 < t \leq \epsilon$, $D_t \subset W^s(q_t,F_t)$.

Proof: First we consider part (a) for which $t = 0$. By Lemma 5.3 the fixed point p_0 is weakly attracting from above; so there exists a neighborhood N of p_0 such that N_0 = upper component of $(N-W^s(p_0))$ is in the basin of p_0, $N_0 \subset W^{ws}(p_0)$. Now take z in D_0. Then z must lie in one of the following regions:
i) $\{\delta_x < 0\} = \{\delta_x < 0\} \cap \{\delta_y < 0\}$, ii) $\{\delta_x \geq 0\} \cap \{\delta_y < 0\}$, or iii) $\{\delta_y \geq 0\} = \{\delta_y \geq 0\} \cap \{\delta_x > 0\}$. In any of these cases we show there is a forward iterate, $F^k(z)$ for $k \geq 0$, for which $F^k(z)$ lies in $N_0 \subset W^{ws}(p_0)$. Since $F^k(z)$ is in the basin, z is also and the lemma is proved.

To show such a k exists consider case i). The region $\{\delta_x < 0\} \cap \{\delta_y < 0\}$ is F semi-invariant by Lemma 5.4e, so both coordinates of $F^j(e)$ are monotone. Because there are no fixed point outside N_0, $F^j(z)$ must enter N_0 giving case i). Considering case ii), then $F^j(z)$ must either enter N_0 or $\{\delta_x < 0\}$. However if it enters $\{\delta_x < 0\}$ then by case i) a further iterate enters N_0. Lastly for case iii), some iterate must enter the region $\{\delta_y < 0\}$ reducing it to either case i) or ii).

For part b, let N_t = upper component $\{N-W^s(p_t,F_t)\}$. By the local unfolding of the saddle node, for $\epsilon > 0$ small enough and for $0 < t \leq \epsilon$, $N_t \subset W^s(q_t,F_t)$. On the other hand, by compactness, there is a uniform $k > 0$ such that $F_0^k(D_0) \subset N_0$. Thus for

$\varepsilon > 0$ possibly smaller and $0 < t \leq \varepsilon$, it follows that $F_t^k(D_t) \subset$ $\subset N_t$. Thus $F_t^k(D_t) \subset W^s(q_t, F_t)$ and so D_t is also in $W^s(q_t, F_t)$. This completes the proof of the lemma and the theorem. //

§6. Verification of assumptions for homoclinic tangencies.

The essential ideas of the analysis is seen by taking f_t to be linear near P_t, $f_t(x,y) = (\eta x, \rho y)$ for all t. We assume for simplicity $1 > \eta_t \equiv \eta > 0$, $\rho_t \equiv \rho > 0$ and $P_t = 0 = (0,0)$ are independent of t. Also we take $t_0 = 0$. Since $W^u(0, f_0)$ has an orbit of tangency with $W^s(0, f_0)$, we scale these coordinates so the point $Q_0 = (1,0)$ and $Q_1 = (0,1)$ are two points in this orbit and let k be that positive integer such that $f_0^k(Q_1) = Q_0$.

The boxes B_n may be taken to be product boxes $B_n = I_n \times J_n$ where $I_n = [1 - C_1\rho^{-n}, 1 + C_1\rho^{-n}]$ and $J_n = [\rho^{-n+k} - C_2\rho^{-2n}, \rho^{-n+k} + C_2\rho^{-2n}]$. The interval J_n is centered at ρ^{-n+k} because it takes $n-k$ iterates to pass from B_n near Q_0 to near Q_1.

As a result of the linearization near P between Q_0 and Q_1, we have $f_t^{n-k}(x,y) = (\eta^{n-k}x, \rho^{n-k}y)$ from B_n to a neighborhood of Q_1. The map f_t^k from a neighborhood of Q_1 into a neighborhood of Q_0 is a fixed family of nonlinear maps independent of n. If we take a simple quadratic map we get a map equivalent to (2.1),

$$f_t^k(x,y) = (1 + \alpha(y-1), \beta x + \gamma t - \gamma(y-1)^2)$$

where $\alpha, \beta, \gamma = \pm 1$ embody the orientation choices. For our analysis we make the choices $\alpha = \gamma = 1$ and $\beta = -1$ so for (x,y) in B_n and $m = n-k$,

$$f_t^n(x,y) = f_t^k(\eta^m x, \rho^m y) = (\rho^m y, t - \eta^m x - (\rho^m y - 1)^2).$$

To get a map as in (2.1), let $\mu = \eta^m$, $\lambda = \rho^m$, and change the coordinates by $y' = y-\lambda^{-1}$, $x' = x-1$, and $t' = t + \mu + \lambda^{-1}$.

To treat the more general case, let $H = f_t^k$ from a neighborhood of Q_1 into a neighborhood of Q_0. Let $\mu = \eta^m$ and $\lambda = \rho^m$. Take coordinates in B_n near Q_0 by $x = x_0 + \lambda^{-1}u$ and $y = y_0 + \lambda^{-2}v$. Then

$$F(u,v) = \begin{pmatrix} \lambda H_1(\mu x_0+\mu\lambda^{-1}u,\ \lambda y_0+\lambda^{-1}v) - \lambda x_0 \\ \lambda^2 H_2(\mu x_0+\mu\lambda^{-1}u,\ \lambda y_0+\lambda^{-1}v) - \lambda^2 y_0 \end{pmatrix}.$$

Conditions (3.1)-(3.6) were checked in [11] for careful choices of x_0, y_0, and parameter range. If $DH = \begin{pmatrix} \alpha & \beta \\ \gamma & \delta \end{pmatrix}$ then

$$DF = \begin{pmatrix} \mu\alpha & \beta \\ \lambda\mu\gamma & \lambda\delta \end{pmatrix}.$$

For $m = n-k$ large enough $|a| = |\mu\alpha| = |\eta^m\alpha| < \Delta = \eta^m\rho^m \det DH$. At Q_1, $\delta = 0$ so $\det DH(Q_1) = -\gamma\rho > 0$ by assumption. Therefore at this point γ and ρ have opposite signs. For n large enough this will be the case at all points considered so c and b will have opposite signs. The vectors $v = (\lambda\delta-1,\ -\lambda\mu\gamma)^{tr} \approx (\lambda\delta, 0)^{tr}$ and $w = (\beta, 1-\mu\alpha)^{tr} \approx (\beta, 1)^{tr}$ for large n so $\mu \approx 0$ and $\lambda\mu \approx 0$.

$$D^2F_1 = \begin{pmatrix} \lambda^{-1}\mu^2\alpha_x & \mu\lambda^{-1}\beta_x \\ \mu\lambda^{-1}\beta_x & \lambda^{-1}\beta_y \end{pmatrix}$$

$$D^2F_2 = \begin{pmatrix} \mu^2\gamma_x & \mu\delta_x \\ \mu\delta_x & \delta_y \end{pmatrix}.$$

Then

$$\begin{aligned} D^2F_2(w,w) &+ (d/-b)D^2F_1(w,w) \\ &= [\delta_y + O(\mu)] + (\lambda\delta/-\beta)[\lambda^{-1}\beta_y + O(\lambda^{-1}\mu)] \\ &= \delta_y - (\delta/\beta)\beta_y + O(\mu). \end{aligned}$$

Since we are considering the case $\delta_y < 0$, and since $\delta \approx 0$ for large n, this term is negative. Similarly

$$D^2F_2(v,v) = \mu^2\lambda^2\delta_y(-\gamma)^2 + 2\gamma_y\,\delta(-\gamma) + \gamma_x\delta^2 + O(\lambda^{-1}))$$

is negative for δ near enough to zero. Lastly

$$\begin{aligned} bD^2F_2(w,w) &+ (a-\Delta)D^2F_1(w,w) \\ &= \beta[\delta_y + O(\mu)] + [\mu\alpha - \lambda\mu\ \det DH][\lambda^{-1}\beta_y + O(\lambda^{-1}\mu)] \\ &= \beta\delta_y + O(\mu) \end{aligned}$$

is negative for n large enough.

Finally letting $\Delta_0 = \det DH$,

$$\begin{aligned} \frac{\partial F_2}{\partial t} + \frac{a-\Delta}{b}\frac{\partial F_1}{\partial t} &= \lambda^2\frac{\partial H_2}{\partial t} + \lambda\frac{\mu\alpha - \lambda\mu\Delta_0}{\beta}\frac{\partial H_1}{\partial t} \\ &= \lambda^2\frac{\partial H_2}{\partial t} - (\mu\Delta_0/\beta)\frac{\partial H_1}{\partial t} + O(\mu\lambda^{-1}) \end{aligned}$$

which has the same sign as $\frac{\partial H_2}{\partial t}$ for μ and λ^{-1} small enough. This completes the checking of condition (3.7). //

References

[1] R. Devaney and Z. Nitechi, "Shift automorphisms in the Hénon mapping," Commun. Math. Phys. 67 (1979), pp. 137-146.

[2] R. Devaney, "Homoclinic bifurcations and the area-conserving Hénon mapping," J. Diff. Equat. 51 (1984), pp. 254-266.

[3] N.K. Gavrilov and L.P. Silnikov, "On the three dimensional dynamical systems close to a system with a structurally unstable homoclinic curve, I" Math. USSR Sbornik 17 (1972), pp. 467-485; II Math. USSR Sbornik 19 (1973), pp.139-156.

[4] C. Grebogi, E. Ott, J. Yorke, "Metamorphoses of basin boundaries," preprint University of Maryland, 1985.

[5] J. Guckenheimer and P. Holmes, Nonlinear Oscillations, Dynamical Systems, and Bifurcations of Vector Fields, Springer-Verlag, New York/Berlin/Heidelberg/Tokyo, 1983.

[6] M. Hénon, "A two-dimensional mapping with a strange attractor," Commun. Math. Phys. 50 (1976), pp. 69-77.

[7] P. Holmes and D. Whitley, "On the attracting set for Duffings's equation I, II," to appear.

[8] C. Jones, private communication.

[9] S. Newhouse, "The abundance of wild hyperbolic sets," Publ. Math. IHES 50 (1979), pp. 101-151.

[10] S. Newhouse, Lectures on dynamical systems, Progress in Math. vol. 8, Birkhäuser 1980, pp. 1-114.

[11] C. Robinson, "Bifurcation to infinitely many sinks," Commun. Math. Phys. 90 (1983), pp. 433-459.

[12] C. Robinson, "Cascade of sinks," Trans. Amer. Math. Sco. 288 (1985), pp. 841-849.

[13] L. Tedeschini-Lalli and J. Yorke, "How often do simple dynamical processes have infinitely many coexisting sinks," preprint University of Maryland, 1985.

S. Patterson
School of Mathematics
University of Minnessota
Minneapolis, Minnessota 55455
USA

C. Robinson
Department of Mathematics
Northwestern University
Evanston, IL 60201
USA

R ROUSSARIE

Weak and continuous equivalences for families on line diffeomorphisms

In problems of bifurcations, one may use several equivalence relations, more or less strong. For instance, the vector fields families may be compared up to the following relations:

a) Weak Equivalence: $X_\mu \sim Y_\mu \Leftrightarrow$ There exists a homeomorphic change of parameter $\varphi(\mu)$ such that, for each μ, X_μ is topologically equivalent to $Y_{\varphi(\mu)}$. (This means that there exists a homeomorphism h_μ, for each μ, sending orbits of X onto orbits of $Y_{\varphi(\mu)}$, in such a way that the time-orientation is preserved).

b) Continuous Equivalence: $X_\mu \underset{\text{cont}}{\sim} Y_\mu \Leftrightarrow$ the same definition as above with the following extra hypothesis: the equivalence homeomorphism h_μ depends continuously on μ.

In this article, I want to point out some differences between these two relations. It is known that the continuous equivalence may give rise to moduli for families which are weakly equivalent (see an example for one-parameter families of vector fields in $\mathbb{R}^3$, in the paper of J. BELOQUI, contained in this volume). Here, I want to describe a simpler occurence of this phenomenon of moduli. I limit myself to local families (deformations) of line diffeomorphisms, where the topological equivalence of vector fields in the above definitions is replaced by the topological conjugacy of diffeomorphisms. The term "local", means that one considers the equivalence relations for germs of families $F_\mu(x)$ at $(0,0) \in \mathbb{R}\times\mathbb{R}^k$, with $x \in \mathbb{R}$, $\mu \in \mathbb{R}^k$ and $F_o(0) = 0$. We consider specially the case where $DF_o(0) = 1$.

By suspension of families of line diffeomorphisms, one will derive results about families of vector fields in the plane.

I - <u>The results</u>

Let a 3-parameter local family of line diffeomorphisms F_μ. Generically (in the C^∞ topology), one can suppose that F_μ is differentiably equivalent to $F_\mu(x) \sim x + Q(x,\mu)(\nu_o(\mu)+\nu_1(\mu)x+\nu_2(\mu)x^2+x^4)$ where $Q(x,\mu)$ is C^∞ with $Q(0,0) > 0$ and $\frac{D(\nu_o,\nu_1,\nu_2)}{D(\mu_1,\mu_2,\mu_3)}(0) \neq 0$.
(One uses the preparation theorem to obtain this reduction).
The condition on the Jacobian of parameters, allows the substitution of $\mu = (\mu_1,\mu_2,\mu_3)$ by $\nu = (\nu_o,\nu_1,\nu_2)$. Then:

$$F_\mu \sim G_\nu = x + Q(x,\nu)(\nu_o+\nu_1x+\nu_2x^2+x^4)$$

(with another function $Q(x,\nu)$, again such that $Q(0,0) > 0$).
Now, the fixed points of G_ν are the same as the fixed points of $\tilde{G}_\nu(x) = x + \nu_o + \nu_1x + \nu_2x^2 + x^4$, and $\tilde{G}_\nu(x)-x$ and $G_\nu(x)-x$ have the same sign. So, the two families G_ν and $\tilde{G}_\nu$ are weakly equivalent (with $\varphi \equiv \mathrm{Id}$) and one can say:

<u>Proposition 1</u>. Every generic 3-parameter local family F_μ of line diffeomorphisms is weakly equivalent to the following family:

$$G_\nu(x) = \nu_o + (1+\nu_1)x + \nu_2x^2 + x^4.$$

Moreover, one can choose a differentiable change of parameter from μ to $\nu = (\nu_o,\nu_1,\nu_2)$.

It follows that the bifurcation set $\Sigma(F_\mu)$ of any generic 3-parameter family F_μ is diffeomorphic to the "swallow tail" set (discriminant set of the equation: $x^4 + \nu_2x^2 + \nu_1x + \nu_o = 0$).

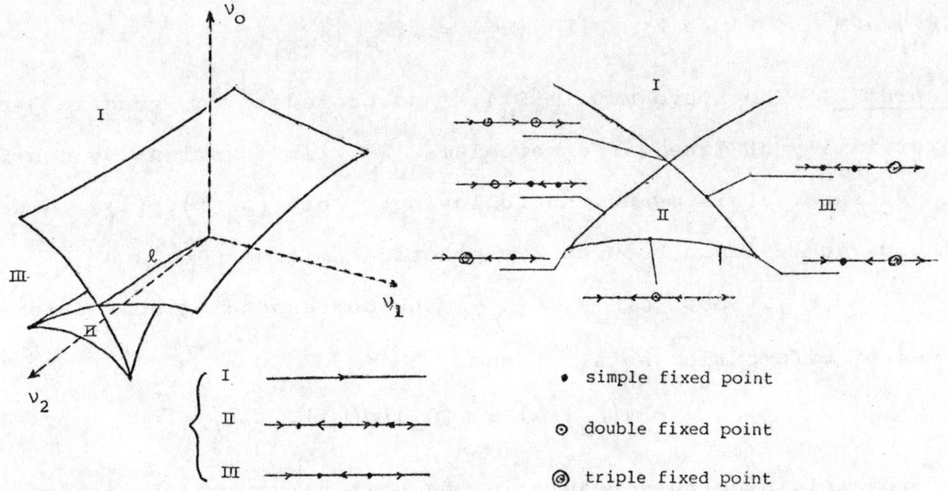

The Swallow Tail Set

Phase portraits in a plane $\{\nu_2 = C^t > 0\}$

Figure 1

Now, I recall the C^∞-invariant introduced by J. Mather for interval diffeomorphisms [M]. Let I, any closed interval in $\mathbb{R}$ and let: $S_I^\infty = \{f, C^\infty$ diffeomorphism of $I \mid f(x) > x, \ \forall\, x \in \text{int}(I)\}$. Then, J. Mather has defined a map:

$$\rho: S_I^\infty \to M^\infty = \text{Diff}_o^\infty(\pi^1)/\sim$$

where $\text{Diff}_o^\infty(\mathbb{T}^1)$ is the group of C^∞ diffeomorphisms of the circle $\mathbb{T}^1 = \mathbb{R}/\mathbb{Z}$, preserving 0 and orientation, and $\sim$ is the following equivalence relation: $h \sim h' \Leftrightarrow$ the graphs of h and h' differ by a translation of $\mathbb{T}^2 = \mathbb{R}^2/Z^2$; ρ is invariant by C^∞ conjugacy. I shall recall the construction of ρ in the next paragraph. One may also see [Y] for more details. For the moment, I return to the generic 3-parameter families of line diffeomorphisms $F_\mu(x)$. Let $\Sigma(F_\mu)$ its bifurcation set and $\ell(F_\mu)$ the line of self-intersection of $\Sigma(F_\mu)$. For each $\lambda \in \ell(F_\mu)$, the diffeomorphisms F_μ has two semi-stable fixed points p_μ, q_μ and we have: $F_\mu | I_\mu \in S_{I_\mu}^\infty$ with

$I_\mu = [p_\mu, q_\mu]$. So, we can associate to the family F_μ a map $\pi(F_\mu)$: $\ell(F_\mu) \to M^\infty$ defined by $\pi(F_\mu)(\mu_o) = \rho(F_{\mu_o} | I_{\mu_o})$.

<u>Theorem 2</u>. The above map $\pi(F_\mu)$, associated to any generic 3-parameter family of line diffeomorphisms F_μ, is invariant by continuous equivalence. This means the following: let $(h_\mu(x), \varphi(\mu))$ a continuous equivalence between two generic families G_μ and F_μ $(G_{\varphi(\mu)} \underset{h_\mu}{\sim} F_\mu)$. Let $\ell(F_\mu)$, $\ell(G_\mu)$ the corresponding self-intersection lines of bifurcation sets. Then

$$\pi(F_\mu)(\mu) = \pi(G_\mu)(\varphi(\mu)).$$

Before explaining this result in the next paragraph, I want to make some comments and to give some consequences.

1. Of course, it is possible to extend the existence of moduli for any generic k-parameter families of line diffeomorphisms, with $k \geq 3$.

2. The Mather map ρ is surjective. It follows easily from this that one can construct uncountably many generic, 3-parameter families, which will be non continuously equivalent.

3. I think that the map π doesn't define completely the type of the family, up to continuous equivalence (see the remark at the end of the article).

4. The above result may be translated to families of vector fields on the plane. For example, one may consider a generic 4-parameter family of Hopf-type [T]. Such a family may be reduced to an equation of the following form (here z is a complex coordinate in $R^2 \simeq C$):

$$\dot{z} = z(\mu_o + \mu_1|z|^2 + \mu_2|z|^4 + \mu_3|z|^6 \pm z^8) + O(|z|^{10})$$

$$\mu = (\mu_o, \mu_1, \mu_2, \mu_3) \in \mathbb{R}^4 \text{ is the parameter.}$$

The study of this differential equation (up to weak or continuous equivalence) reduce to the study of the family of Poincaré-map P_μ on the x-axis (up to weak or continuous equivalence). This family is a 4-parameter analogous of the generic 3-parameter family described above. It has a pair of semi-stable fixed points (corresponding to a pair of semi-stable closed orbits for the equation), along a 2-dimensional surface L in the parameter space (the equation of L is given by: $\mu_0 = p^2$, $\mu_1 = -ps$, $\mu_2 = s^2+2p$, $\mu_3 = -2s$, with $s,p > 0$, $s^2 > p$). We can define a map $\pi: L \to M^\infty$ by $\pi(\mu) = \rho(P_\mu)$, which is invariant by topological invariance of vector fields.

II - The proof.

1. We begin with some well-known properties of local diffeomorphisms on the line. Let $f(x) = x + \alpha x^k + \ldots$, $\alpha \neq 0$, $k > 2$, a germ of non-hyperbolic C^∞ diffeomorphism at $0 \in \mathbb{R}$. It is easy to show that, for any given k, every such diffeomorphism is topologically equivalent to $X(1)$, where $X(t)$ is the flow of $X = x^k \frac{\partial}{\partial x}$. If now, one looks to the different ways to conjugate f and $X(1)$, it is the same as to look to the different flows $U(t)$ in which one can embed f (f embeds in $U(t)$ if $U(1) = f$). Of course, given some $x_0 \neq 0$, one can choose for $U(t,x_0)$, $t \in [0,1]$, any homeomorphism from $[0,1]$ to $[x_0,f(x_0)]$.

2. One considers now a 1-parameter family of line diffeomorphisms.

Here, there is only a single generic 1-parameter family up to continuous equivalence: the saddle-node bifurcation $f_\mu(x) = x+\mu+x^2$. But now the possibilities for embeddings in flows are quite different. In some sense, there is only one way (for $\mu=0$) to make such an embedding. This rigidity phenomenon was discovered by Newhouse, Palis and Takens and presented in [N.P.T.]; J. Palis and myself

have considered again this question in [P.R.] to give some topological interpretation and formulation of this result. Because I need it in what follows, I recall this topological formulation.
We call <u>regular family of flows</u> $X_\mu(t)$, any 1-parameter family of topological flows on $\mathbb{R}$, continuous in x, t, μ, with parameter $\mu \in \mathbb{R}^+ = [0,+\infty)$ and the following properties:

a) The flows X_μ have only one stationary point at $(0,0) \in \mathbb{R}^+ \times \mathbb{R}$ ($X_0(t,0) \equiv 0$); for $(\mu,x) \neq (0,0)$ and $t > 0$, $X_\mu(t,x) > x$.

b) There exists two topological segments ℓ, r (homeomorphic to $\mathbb{R}^+$), with end points on the line $\mu = 0$, respectively on the left and the right of 0, transversal to the lines $\{\mu = \text{constant}\}$ and such that the time-function $t(\mu)$ (time needed to go from ℓ to r), is a strictly decreasing in μ.

Now, we look to families of homeomorphisms on the line, f_μ, $\mu \in R^+$ which embed in at least one regular family of flows ($\exists\ X_\mu(t)$, such that $X_\mu(1) = f_\mu$; I shall call it a <u>regular embedding</u> of f_μ). The topological formulation of the rigidity phenomenon in [P.R.], is like this: <u>let</u> $X_\mu(t)$ <u>and</u> $\tilde{X}_\mu(t)$ <u>two regular embeddings of</u> f_μ. <u>Then</u> $X_0(t) \equiv \tilde{X}_0(t)$.
We may say the same thing in a different way: if f_μ embeds in a regular $X_\mu(t)$ and $\tilde{f}_\mu$ embeds in a regular $\tilde{X}_\mu(t)$ and if $(h_\mu,\varphi(\mu))$ is a continuous equivalence of f_μ with $\tilde{f}_\mu$, then h_0 is a conjugacy between $X_0(t)$ and $\tilde{X}_0(t)$.

Next, we consider the particular case of generic C^∞ 1-parameter family f_μ of saddle-node type. It is another part of the results of [N.P.T.], that f_μ embeds in some regular family $X_\mu(t)$, such that $X_0(t)$ is the unique C^∞-flow in which f_0 embeds. So, in the case of saddle-node bifurcations, we have: <u>Given a continuous equivalence</u> $(h_\mu,\varphi(\mu))$ <u>between two saddle-node bifurcations</u> f_μ <u>and</u>

$\tilde{f}_\mu$, <u>then</u> h_o <u>is a conjugacy between the unique</u> C^∞ <u>flows in which</u> f_o <u>and</u> $\tilde{f}_o$ <u>embed</u>.

3. Consider now a family of C^∞ line diffeomorphisms f_μ, $\mu \in \mathbb{R}^+$, which has two distinct saddle-node bifurcations for $\mu = 0$ at points $p,q \in \mathbb{R}$ $(p < q)$. We call it a "double saddle-node bifurcation". To the diffeomorphism f_o, we can associate the Mather invariant $\rho(f_o) \in M^\infty$. I recall now the construction of $\rho(f_o)$:

- there exists a unique C^∞ flow X_t such that $f_o = X_1$ on $[p,q[$.
- there exists a unique C^∞ flow X^t such that $f_o = X^1$ on $]p,q]$.

Choose $a \in]p,q[$; we define $\varphi_a \in \mathrm{Diff}_o(\mathbb{T}^1)$ by:

$$X_t(a) = X^{\varphi_a(t)}(a).$$

If we choose another point $\bar{a} \in]p,q[$, we can write $\bar{a} = X_T(a)$ for some T and we have $\varphi_{\bar{a}}(t) = \varphi_a(t+T) - \varphi_a(T)$.
So, the class of φ_a in M^∞ depends only on f_o. This class is $\rho(f_o)$ by definition.

Suppose now that $(h_\mu, \varphi(\mu))$ is a continuous equivalence between two double saddle-node bifurcations f_μ and $\tilde{f}_\mu$. Let X_t, X^t the flows associated as above to f_o, $\tilde{X}_t$, $\tilde{X}^t$ the flows associated to $\tilde{f}_o$. We know that h_o conjugates X_t and $\tilde{X}_t$, X^t and $\tilde{X}^t$. And so, if we choose some $a \in]p,q[$ to define φ_a, we have $\tilde{\varphi}_{h_o(a)} = \varphi_a$ and $\rho(f_o) = \rho(\tilde{f}_o)$.

We see that the invariant $\rho(f_o)$, defined as a C^∞-invariant becomes a C^o-<u>invariant for families</u>. Next, using the topological formulation recalled above, it is easy to extend this result to any topological family f_μ, f_o being C^∞, which admits a double regular embedding.

4. We return now to the generic 3-parameter families. Suppose that F_ν, $\tilde{F}_\nu$ are two such families, and that $(h_\nu, \varphi(\nu))$ is a continuous

equivalence between them. Let $\ell, \tilde{\ell}$ the self-intersection lines of their respective bifurcation sets $\Sigma(F_\nu)$ and $\Sigma(\tilde{F}_\nu)$. Choose a differentiable segment σ, transversal to $\Sigma(F_\nu)$ and entering by some point $\nu_o \in \ell$ in the region I (see the figure 2).

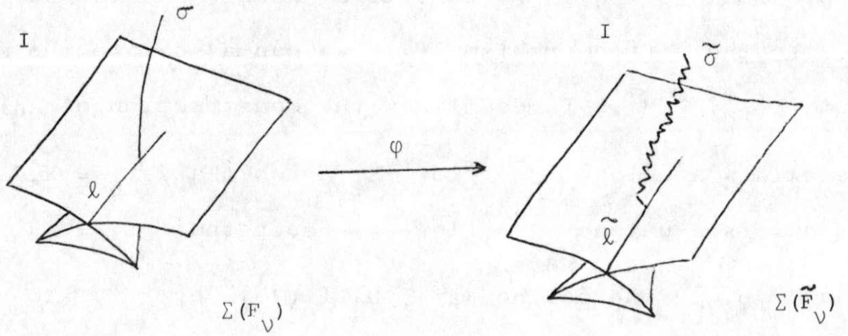

Figure 2

Then F_ν, above σ, is a double C^∞ saddle-node bifurcation. Let $\tilde{\sigma} = \varphi(\sigma)$. The continuous equivalence $(h_\nu, \varphi(\nu))$ induces a continuous equivalence between $F_\nu|\sigma$ and $\tilde{F}_\nu|\tilde{\sigma}$. Now, the last family is a continuous family (C^∞ for each value of the parameter), with a double regular embedding (because the property of regular embedding is preserved by continuous equivalence). So, it follows from the sub paragraph 3 above, that $\rho(\tilde{F}_{\varphi(\nu_o)}) = \rho(F_{\nu_o})$, which is the desired formula for Theorem 2.

Final remark: It may be asked if the above map $\pi: \ell \to M^\infty$ gives complete invariant for the continuous equivalence. The answer is no. In fact, it is easy to define a second invariant which may be non trivial even when π is trivial (for example, when F_ν embeds in a family of flows).

Look again to a double C^∞ saddle-node bifurcation f_μ. Choose three fundamental domains for f_μ: D_1, on the left of p, D_2 between p and q, D_3 on the right of q. For each $\mu > 0$, let $n(\mu)$ the number of iterations of f_μ to go from D_1 to D_2 and $m(\mu)$ the number of iterations to go from D_2 to D_3.

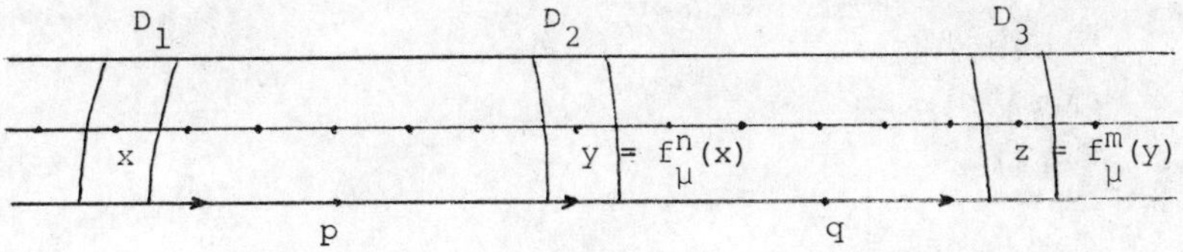

Figure 3

Then, it is easy to show that the limit of $\frac{\text{Log } n(\mu)}{\text{Log } m(\mu)}$ exists for $\mu \to 0$ and is a topological invariant.

This new invariant may be used to defined a second map $\pi' : \ell \to \mathbb{R}$, for each generic 3-parameter family of line diffeomorphisms, which will be invariant by continuous equivalence. One may conjecture that (π, π') is a complete set of invariants for the continuous equivalence of generic 3-parameter families of line diffeomorphisms.

REFERENCES

[M] J. MATHER: Commutators of diffeomorphisms, Comm. Math. Helv., vol. 48, (1973), 195-233.

[N.P.T.] S. NEWHOUSE, J. PALIS and F. TAKENS: Bifurcations and stability of families of diffeomorphisms. Publ. Math. IHES, nº 57, p.5-72.

[P.R.] J. PALIS and R. ROUSSARIE: Topological invariants as number of translation. Lect. Notes in Math. 1125 (1984), 64-86.

[T] F. TAKENS: Unfoldings of certain singularities of vector fields. Generalized Hopf bifurcations. Journal of Diff. Eq. 14 (1973), 476-493.

[Y] J.Ch. YOCCOZ: Centralisateurs et conjugaison différentiable des difféomorphismes du cercle (Thèse, Université d'Orsay) (1985).

Département de Mathématiques
Université de Dijon. UER-MIPC
Laboratoire de Topologie
U.A. nº 755 du CNRS
21000 - DIJON - FRANCE

P SAD

Central eigenvalues of complex saddle–nodes

One of the most important results in the classification of singularities of differential equations describes all complex saddle-nodes up to analytic equivalences. Roughly speaking, several copies of the normal form $x^{k+1}dy - y(1+\lambda x^k)dx = 0$, where $(x,y) \in \mathbb{C}^2$, $k \in \mathbb{N}$ and $\lambda \in \mathbb{C}$, are glued together along vertical sectors by transformations which preserve its solutions; all complex saddle-nodes are obtained in this way (see [1]). When these glueings are composed starting at some one-dimensional transversal to the x-plane, the resulting diffeomorphism has a fixed point only if there exists a central invariant manifold. Such a diffeomorphism is called the central holonomy and obviously is an analytic invariant for the differential equation. In this note we intend to show that in some sense it is also an invariant for the much weaker notion of topological equivalence.

In order to be precise, consider the set of complex saddle-nodes $x^{k+1}dy - y(1+\lambda x^k+xA(x,y))dx = 0$, where A is holomorphic for $|x|+|y|$ small and $|A(x,y)| = \mathcal{O}(|(x,y)|^{k+1})$; all these equations have $y = 0$ as the <u>central manifold</u>. We will be here interested in the subset S_k of those equations whose central holonomies have eigenvalues outside the unit circle (these numbers will be refered to as the <u>central eigenvalues</u>).

Theorem. Let μ and μ' be the central eigenvalues of two topologically equivalent saddle-nodes in S_k. Then:

1) $\mathrm{Arg}\,\frac{\mu}{|\mu|} = \mathrm{Arg}\,\frac{\mu'}{|\mu'|} \mod \mathbb{Z}$.

2) If $\frac{1}{2\pi}\mathrm{Arg}\,\frac{\mu}{|\mu|} \notin \mathbb{Q}$, we have also $|\mu| = |\mu'|$.

As usually, topological equivalence means a homeomorphism preserving solutions of the differential equations seen as foliations; we assume that in our case the orientations of the central manifolds are preserved as well.

We are grateful to J. Martinet for explanations involving Proposition 2.1 below in the case where no central manifold exists.

§1. Blowing-up the Central Holonomy

For simplicity the differential equations we will work with are of the type

(1) $\quad \omega = x^2dy - y(1+xA(x,y))dx = 0$

where $(x,y) \in \mathbb{C}^2$ and $A(x,y)$ is holomorphic in a polydisc around $(0,0) \in \mathbb{C}^2$.

There exists a formal change of coordinates

(2) $\quad \varphi(x,y) = (x, y + \sum_{n\geq 1} a_n(y)x^n)$, all $a_n(y)$ analytic for $|y| \leq R$, $R > 0$ which transforms (1) into its normal form

(3) $\quad \omega_\lambda = x^2dy - y(1+\lambda x)dx = 0, \quad \lambda \in \mathbb{C}$.

Solutions of $\omega_\lambda = 0$ are written as

(4) $\quad y = C\,x^\lambda \exp(-\frac{1}{x})$, C varying in $\mathbb{C}$.

The complex number $\underline{\lambda}$ is associated to the holonomy of the solution $y = 0$, as explained below.

A basic fact is that for any open sector U in the x-plane of opening at most 2π there exists a convergent change of coordinates (defined in some polydisc around $(0,0) \in \mathbb{C}$)

(5) $\quad \varphi_U(x,y) = (x, \varphi_U(x,y))$

which transforms (1) into (3) and is assymptotic to the formal difeomorphism in (2) (see [1]).

Both equations (1) and (3) have $y = 0$ as a solution - the central manifold. An easy calculation shows that their holonomies (called hereafter central holonomies) are respectively of the type $y \mapsto e^{2\pi i\lambda}y$ and $y \mapsto e^{2\pi i\lambda}y + y\xi(y)$, where $\xi(y)$ is analytic in a neighborhood of $0 \in \mathbb{C}$ and $\xi(0) = 0$. We intend now to blow up the central holonomy of (1) using the foliation $\omega = 0$.

For the moment we consider $\omega_\lambda = 0$. Let $\Sigma_u = \{(u,y);\ y \in \mathbb{C}\}$, $\Sigma_u(r) = \{(u,y);\ |y| \leq r\}$, for each $u \in S^1$ and $P(y) = \mu y(1+Q(y))$ be an analytic map seen as a map from $\Sigma_{-1}(r)$ to Σ_{-1}, $Q(0) = 0$. The foliation $\omega_\lambda = 0$ intersects each real three dimensional plane $T_u = \{(x,y) \in \mathbb{C}^2;\ x = ru,\ r \in \mathbb{R}\}$, $u \in S^1$, along the curves

(6) $\quad y(r) = yr^\lambda \exp(-\frac{1}{ru} + \frac{1}{u}).$

It will be useful to keep in mind that

(7) $\quad \lim_{r\to 0} |y(r)| = \infty \quad \text{if} \quad \frac{\pi}{2} < \operatorname{Arg} u < \frac{3\pi}{2}$

$\quad\quad \lim_{r\to 0} |y(r)| = 0 \quad \text{if} \quad -\frac{\pi}{2} < \operatorname{Arg} u < \frac{\pi}{2}$

If $\Gamma = \{(x,y);\ x \in \mathbb{C};\ |y| = 1\}$ one may think of transporting P to Γ along the curves of (6) for $u = -1$. To $(-1,y) \in \Sigma_{-1}$ we associate $(r_y, \hat{y}) \in \Gamma$ such that

(8) $\quad \hat{y} = [(r_y)^\lambda \exp(\frac{1}{r_y} - 1)]y, \qquad |\hat{y}| = 1.$

This mapping takes Σ_{-1} into a real two dimensional cylinder $\hat{\Sigma}_{-1} \subset \Gamma$; it lifts $P: \Sigma_{-1}(r) \to \Sigma_{-1}$ to $\hat{P}: \hat{\Sigma}_{-1}(r) \to \hat{\Sigma}_{-1}$ according to the diagram

$$\begin{array}{ccc} \hat{\Sigma}_{-1}(r) & \xrightarrow{\hat{P}} & \hat{\Sigma}_{-1} \\ \uparrow & \circlearrowright & \uparrow \\ \Sigma_{-1}(r) & \xrightarrow[P]{} & \Sigma_{-1} \end{array}$$

We think of $\hat{P}$ as a blowing-up of P because $\hat{P}$ extends itself to the circle $x = 0$, $|y| = 1$ as the linear rotation $y \longmapsto \frac{\mu}{|\mu|} \cdot y$; this is the content of the next lemma.

<u>Lemma 1.1</u>. $\lim\limits_{|y| \to 0} \frac{\hat{P}(\hat{y})}{\hat{y}} = \frac{\mu}{|\mu|}$.

<u>Proof</u>:

1 - This is a straightforward computation:

$$\frac{\hat{P}(\hat{y})}{\hat{y}} = \frac{[r_{P(y)}^{\lambda} \exp(\frac{1}{r_{P(y)}} - 1)]P(y)}{[r_y^{\lambda} \exp(\frac{1}{r_y} - 1)]y} = \frac{P(y)}{y} \cdot \left(\frac{r_{P(y)}}{r_y}\right)^{\lambda} \exp\left(\frac{1}{r_{P(y)}} - \frac{1}{r_y}\right).$$

Putting $\left(\frac{r_{P(y)}}{r_y}\right)^{\lambda} = a(y)\eta(y)$, with $|\eta(y)| = 1$, we get

$a(y)\exp(\frac{1}{r_{P(y)}} - \frac{1}{r_y}) = \left|\frac{y}{P(y)}\right|$ (because $|\hat{y}| = |\hat{P}(\hat{y})| = 1$) so that

$$\lim_{|y| \to 0} a(y)\exp\left(\frac{1}{r_{P(y)}} - \frac{1}{r_y}\right) = \frac{1}{|\mu|}.$$

2 - From (8) we have:

$$\mathrm{Log}|\hat{y}| = \mathrm{Log}|y| + \Re e\left(\lambda \,\mathrm{Log}\, r_y + \frac{1}{r_y} - 1\right)$$

$$\Rightarrow r_y \,\mathrm{Log}|y| + \Re e(\lambda r_y \,\mathrm{Log}\, r_y + 1 - r_y) = 0 \Rightarrow$$

(9) $$\lim_{|y| \to 0} r_y \,\mathrm{Log}|y| = -1.$$

Hence $\lim\limits_{|y|\to 0} \dfrac{r_{P(y)}\ \mathrm{Log}|P(y)|}{r_y\ \mathrm{Log}|y|} = \lim\limits_{|y|\to 0} \dfrac{r_{P(y)}}{r_y} \dfrac{\mathrm{Log}|P(y)|}{\mathrm{Log}|y|}$

$$= \left(\lim_{|y|\to 0} \frac{r_{P(y)}}{r_y}\right) \lim_{|y|\to 0} \frac{\mathrm{Log}|\mu y(1+Q(y))|}{\mathrm{Log}|y|} = 1,$$

so that $\lim\limits_{|y|\to 0} \dfrac{r_{P(y)}}{r_y} = 1.$

In particular $\lim\limits_{|y|\to 0} a(y) = 1$. This implies $\lim\limits_{|y|\to 0} \exp\left(\dfrac{1}{r_{P(y)}} - \dfrac{1}{r_y}\right) =$

$= \dfrac{1}{|\mu|}$, so at last $\lim\limits_{|y|\to 0} \dfrac{\hat{P}(\hat{y})}{\hat{y}} = \dfrac{\mu}{|\mu|}$. ■

Now we consider again equation (1) and its central holonomy H defined in $\Sigma_{-1}(r)$ for $r > 0$ small. In the same way as before we may lift it to $\hat{H}$ acting on $\hat{\Sigma}_{-1}(r)$, this time using the curves in T_{-1} defined by $\omega = 0$. Assume that $H(y) = \mu y(1+Q(y))$ for $\mu = e^{2\pi i\lambda}$, $Q(0) = 0$ and Q holomorphic.

<u>Proposition 1.1</u>. $\hat{H}$ extends continuously to the base of $\hat{\Sigma}_{-1}(r)$ as the linear rotation $y \longmapsto \dfrac{\mu}{|\mu|} y$.

<u>Proof</u>:

1 - Simply apply Lemma 1 to $P = \varphi_U \circ H \circ \varphi_U^{-1}$, where φ_U comes from (5) and U is, for example, the sector $\frac{\pi}{2} < \mathrm{Arg}\ x < \frac{3\pi}{2}$, and use that φ_U is tangent to the identity along $\{0\}\times\mathbb{C}$.

<u>Remarks</u>:

1 - In fact the above statements make sense for Σ_{-R_1} and $\Gamma = \{(x,y);\ x \in \mathbb{C};\ |y| = R_2\}$ where R_1 and R_2 are sufficiently small.

2 - One should notice that the solutions of (1) leave a trace in Γ given by

$$(10)\quad \begin{cases} \dot{y}(t) = R_2 e^{it} \\ \dot{x}(t) = \dfrac{ix^2}{1+\lambda x^k + xA(x, R_2 e^{it})} . \end{cases}$$

Inside Γ are placed the cylinders $\hat{\Sigma}_u$. As long as $\Re e\, u < 0$ $\hat{\Sigma}_u$ is transverse to (6) for $|x|$ small (depending on $u \in S^1$). The holonomies H_u's defined in small disks of Σ_u lift to $\hat{H}_u$, this time using the planes T_u instead of T_{-1}. Finally, the flow of (10) induces a conjugation between $\hat{H}_{-1}$ and $\hat{H}_u$, which restricts to the identity along $x = 0$, $|y| = R_2$.

§2. Domains for the Central Holonomies

Let us consider now the central holonomies H_{ru} defined in sections Σ_{ru}, $u \in S^1$, and analyze how the sizes of their domains behave as $r \to 0$. We shall see that $u = -1$ is completely singular in this analysis.

Remember that equation (1) can be "normalized" in the sectors $U^+: -\frac{\pi}{2} < \text{Arg}\, u < \frac{3\pi}{2}$ and $U^-: -\frac{3\pi}{2} < \text{Arg}\, u < \frac{\pi}{2}$ by holomorphic maps φ_{U^+} and φ_{U^-} which are assymptotic to the formal difeomorphism (2). Besides, $\varphi_{U^-} \circ \varphi_{U^+}^{-1}$ and $\varphi_{U^+} \circ \varphi_{U^-}^{-1}$ leave the equation (3) invariant and are both assymptotic to the identity as $|x| \to 0$. Since equation (1) has a central manifold it must be

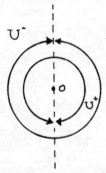

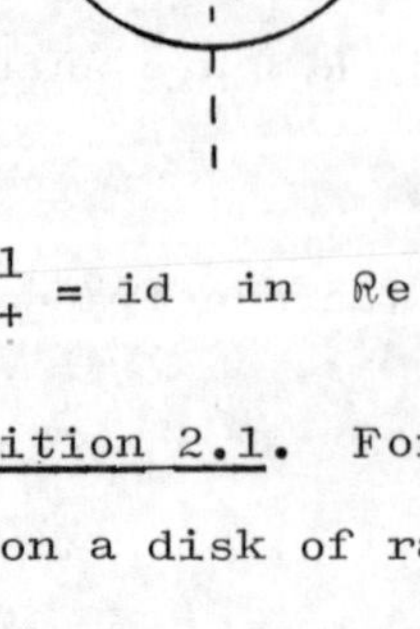

$\varphi_{U^-} \circ \varphi_{U^+}^{-1} = \text{id}$ in $\Re e\, u > 0$ (see [1], pg. 111).

<u>Proposition 2.1</u>. For each $u \in S^1$ the holonomy H_{ru} is not defined on a disk of radius independent of $r \in \mathbb{R}_+$ unless $u = -1$.

Proof:

1 - Fix $u = u_o$, $-\frac{\pi}{2} < \text{Arg } u \leq \frac{\pi}{2}$ and consider $\varphi_{U^-} \circ H_{ru_o} \circ \varphi_{U^+}^{-1} = P_{ru_o}$.

Then $P_{ru_o}(y)$, $y \in \Sigma_{ru_o}$, is obtained as follows: first follow (4) along $|x| = r$ until reaching Σ_{-ru_o}; apply $\varphi_{U^-} \circ \varphi_{U^+}^{-1}$; follow again (4) along $|x| = r$ until reaching Σ_{ru_o} and finally apply $\varphi_{U^+} \circ \varphi_{U^-}^{-1}$ (which is the identity!). Now $\varphi_{U^-} \circ \varphi_{U^+}^{-1}$ distorts sizes by a bounded ammount, so we can do all computations for the normalized system (3). Moving from Σ_{ru_o} to Σ_{-ru_o} changes a radius η to $|(-1)^{\lambda} [\exp - \frac{1}{r}(u_o v - u_o)]|\eta$, where $v \in S^1$ and $0 \leq \text{Arg } v \leq \pi$; as $r \to 0$, this goes to ∞ because $-\frac{\pi}{2} < \text{Arg } u_o \leq \frac{\pi}{2}$.

2 - So it is enough to prove the proposition when $\frac{\pi}{2} < \text{Arg } u_o \leq \frac{3\pi}{2}$.

With the obvious changes, the holonomy H_{ru_o} is described as above; by the same argument, we are left with the case $\pi \leq \text{Arg } u_o \leq \frac{3\pi}{2}$ only. In order to handle it, we complete one turn going from Σ_{ru_o} to Σ_{-ru_o} and coming back to Σ_{ru_o}. The radius η changes to

$$|1^{\lambda}|[\exp \Re e \frac{2u_o}{r}][\exp \Re e(\frac{1}{rvu_o} - \frac{1}{ru_o})] = |1^{\lambda}|[\exp \Re e \frac{u_o}{r}(1+v)]\eta.$$

This keeps bounded as $r \to 0$ only if $u_o = -1$. ■

Corollary 2.1. Assume that a sequence $(x_n, y_n) \in \mathbb{C}^2$ with $|x_n| \to 0$ and $|y_n| \geq \eta > 0$ satisfies the following property: for each $n \in \mathbb{N}$, there exists a path $(x_n(t), y_n(t))$ starting at (x_n, y_n) and contained in the leaves of $\omega = 0$ such that $y_n(t) \xrightarrow[t\to\infty]{} 0$ but $|x_n(t)| \geq \delta_n > 0$. Then $\frac{x_n}{|x_n|} \xrightarrow[n\to\infty]{} -1$.

This follows from the fact that $(x_n(t), y_n(t))$ approaches $y = 0$ keeping a positive distance from $x = 0$ only if it turns around infinitely many times around the y-axis. Proposition 2.1 yields immediately the corollary.

§3. Proof of the Theorem

From now on all normal forms (3) are supposed to satisfy $\operatorname{Im} \lambda > 0$.

Let $\omega = 0$ and $\omega' = 0$ be differential equations of type (1), and $h: (\mathbb{C}^2, 0) \to (\mathbb{C}^2, 0)$ a topological equivalence between them. Without loss of generality one may suppose that:

1) $h(\bigcup_{u \in S^1} \Sigma_{\epsilon u}(\delta)) \subset \bigcup_{u \in S^1} \Sigma_{\epsilon u}$, for some $\epsilon > 0$, $\delta > 0$.

2) $h(\Gamma_{\epsilon'}(\delta')) \subset \Gamma_{\epsilon'}$, for some $\epsilon' > 0$, $\delta' > 0$, where

$\Gamma_{\epsilon'}(\delta') = \bigcup_{v \in S^1} \{(x,y) \in \mathbb{C}^2;\ |x| \leq \delta' \text{ and } y = \epsilon' v\}$.

3) $h(\Sigma_{-\epsilon}(\delta)) \subset \Sigma_{-\epsilon}$.

All these properties hold true after submitting $\underline{h}$ to isotopies along the leaves of $\omega' = 0$. We will add another property to this list.

4) $h(\widehat{\Sigma_{-\epsilon}(\delta)}) \subset \hat{\Sigma}_{-\epsilon}$.

To explain why we can do so let us consider again the flow given in equation (10) for $\omega' = 0$. This flow is always transverse to the cylinders $\hat{\Sigma}_{\epsilon u}(\delta)$ as long as $\Re e\, u < 0$ and $\delta > 0$ is small (depending on $u \in S^1$). Since $h(\hat{\Sigma}_{-\epsilon}(\delta))$ is "assymptotically tangent" to $\hat{\Sigma}_{-\epsilon}$ as follows from Corollary 2.1, a small isotopy of $\underline{h}$ along the solutions of (10) for $\omega' = 0$ will carry $h(\hat{\Sigma}_{-\epsilon}(\delta))$ over $\hat{\Sigma}_{-\epsilon}$.

The central holonomies $H: \Sigma_{-\epsilon}(\delta) \to \Sigma_{-\epsilon}$ and $H': \Sigma_{-\epsilon}(\delta) \to \Sigma_{-\epsilon}$ associated to $\omega = 0$ and $\omega' = 0$ are now related by the formula $H' \circ h = h \circ H$. The next lemma shows that the same relation holds true for their blowing-up's.

Lemma 3.1. $\hat{H}'\circ h = h\circ\hat{H}$.

Proof:

1 - First of all we claim that there exists $k \in \mathbb{N}$ such that $h(\hat{z}) = \widehat{(H')^k(h(z))}$ (same notation as §1). Writing $\hat{z} = \varphi(z)$ and $\hat{z}' = \varphi'(z')$ in connection to equations $\omega = 0$ and $\omega' = 0$, the equality above becomes $h\circ\varphi = \varphi'\circ(H')^k\circ h$ for some $k \in \mathbb{N}$. This follows from the fact that $\psi = (\varphi')^{-1}\circ h\circ\varphi\circ h^{-1}$ is continuous and z, $\psi(z)$ belong to the same orbit of H', which is conjugated to a linear contraction.

2 - Therefore one can write:

$$h\hat{H}(\varphi(z)) = h\varphi(H(z)) = \varphi'(H')^k h(H(z)) = \varphi'(H')^{k+1}(h(z)) =$$
$$= \varphi' H'[(H')^k(h(z))] = \hat{H}'\varphi'[(H')^k(h(z))] = \hat{H}'\varphi'[(\varphi')^{-1}h(\varphi(z))]$$
$$= \hat{H}'h(\varphi(z)).$$ So that $h\hat{H} = \hat{H}'h$. ∎

Call $\mu = dH(0)$ and $\mu' = dH'(0)$.

Corollary 3.1. $\dfrac{\mu}{|\mu|} = \dfrac{\mu'}{|\mu'|}$.

Proof: 1. From Lemma 3.1, h conjugates the extensions of $\hat{H}$ and $\hat{H}'$ to the case of $\hat{\Sigma}_{-\varepsilon}(\delta)$ which are linear rotations $\hat{z} \mapsto \frac{\mu}{|\mu|}\hat{z}$, and $\hat{z}' \mapsto \frac{\mu'}{|\mu'|}\hat{z}'$ respectively (Proposition 1.1). Hence $\frac{\mu}{|\mu|} = \frac{\mu'}{|\mu'|}$. ∎

In order to prove the second part of the theorem, we assume from now on that $\lambda_1 \notin \mathbb{Q}$ (here we are writing $\lambda = \lambda_1 + i\lambda_2$ and $\lambda' = \lambda_1' + i\lambda_2'$; Corollary 3.1 says that $e^{2\pi i\lambda_1} = e^{2\pi i\lambda_1'}$). It follows that $h(0,y) = (0,\alpha y)$ for $(0,y) \in \Gamma_{\varepsilon'}$ (that is, h has to be linear at the base of $\hat{\Sigma}_{-\varepsilon} \subset \Gamma_{\varepsilon'}$).

Take a sequence $(n_k) \in \mathbb{N}$ such that $\lim_{n_k\to\infty} e^{2\pi i\lambda_1 n_k}$ exists.

Given $y_0 \in \Sigma_{-\varepsilon}$ and $y_n = H^n(y_0)$, $n \in \mathbb{N}$, let us choose a subsequence (m_k) of (n_k) in order to have $y = \lim_{m_k \to \infty} \hat{y}_{m_k}$, where $(0,y) \in \Gamma_{\varepsilon'}$. If we denote by $\theta(z) = \frac{z}{|z|}$, we find from (8):

$$\theta(\hat{y}_{n_k}) = \exp[i(\lambda_2 \text{ Log } r_{n_k} + 2\pi\lambda_1 n_k)] \cdot \theta(\frac{y_{n_k}}{\mu^{n_k}})$$

(11)

$$\theta(\hat{y}'_{n_k}) = \exp[i(\lambda'_2 \text{ Log } r'_{n_k} + 2\pi\lambda'_1 n_k)] \cdot \theta(\frac{y'_{n_k}}{\mu'^{n_k}})$$

where $y'_n = h(y_n)$, $n \in \mathbb{N}$. Now let L and L' be the analytic linearizations of H and H', i.e., $L(H(z)) = \mu z$ and $L'(H'(z)) = \mu' z$ and $L(0) = L'(0) = 1$. These linearizations may be found from the formulae (see [2])

(12) $$L(z) = \lim_{n\to\infty} \frac{H^n(z)}{\mu^n} \quad \text{and} \quad L'(z) = \lim_{n\to\infty} \frac{H'^n(z)}{\mu'^n}.$$

Hence (11) and (12) imply that

(13) $$\alpha = \lim_{m_k\to\infty} \exp[i\ \lambda_2 \text{ Log } r_{m_k} - \lambda'_2 \text{ Log } r'_{n_k}] \cdot \theta(\frac{L(y_o)}{L'(y_o)}).$$

Since this is true for any subsequence (m_k) of (n_k), after applying (9) we arrive at following conclusion ((9) implies $\lim_{n_k\to\infty} 2\pi\ \lambda_2\ n_k\ r_{n_k} = \lim_{n_k\to\infty} 2\pi\ \lambda'_2\ n_k\ r'_{n_k} = 1$).

(14) $\lim_{n_k\to\infty} \exp[i(\lambda'_2-\lambda_2)\text{Log } n_k] = b$ exists and depends only on $y_o \in \Sigma_{-\varepsilon}$, provided that $(e^{2\pi i \lambda_1 n_k})$ converges as $n_k \to \infty$.

This is not possible; in fact, changing (n_k) by (mn_k), where $m \in \mathbb{N}$, we see that (14) implies

(15) $$\lim_{n_k\to\infty} \exp[i(\lambda'_2-\lambda_2)\text{Log } mn_k] = b.$$

Hence $\exp[i(\lambda'_2-\lambda_2)\text{Log } m] = 1 \quad \forall\ m \in \mathbb{N}$ and therefore $\lambda'_2 = \lambda_2$. ∎

REFERENCES

[1] J. Martinet and J.-P. Ramis, Problèmes de Modules pour des Équations Différentielles Non-Linéaires du Premier Ordre, Publications Mathématiques IHES nº 55,

[2] P. Sad, Dinâmica das Funções Racionais, Colóquio Brasileiro de Matemática, 1983.

Instituto de Matemática Pura e Aplicada (IMPA)
Estrada Dona Castorina 110
22460 Rio de Janeiro, RJ
Brazil

F TAKENS

Transitions from periodic to strange attractors in constrained equations

1. Introduction

We shall deal with bifurcations in dynamical systems in which simple attractors (attracting fixed points or periodic attractors) change into complicated or chaotic attractors. From the technical point of view we deal mainly with one-parameter families of continuous and piecewise C^1 endomorphisms of $\mathbb{R}$ (for a detailed description see below). Such piecewise C^1 endomorphisms, as well as smooth endomorphisms, arise in dynamical systems in which there are "different time scales" in the sense that certain parts of the dynamics are much faster than others; this is formalized in the definition of constrained equation below.

The piecewise C^1 endomorphisms we consider are continuous maps $f: \mathbb{R} \to \mathbb{R}$ which are, when restricted to $(-\infty,0]$ and when restricted to $[0,+\infty)$, C^1; we call these maps PC^1 endomorphisms. Differentiable and generic one-parameter families of such endomorphisms are defined in the obvious way using the C^1 topology on both $(-\infty,0]$ and $[0,\infty)$; since we only study local properties one may take the weak or the strong Whitney topology. We study one-parameter families f_u, $\mu \in \mathbb{R}$, of such PC^1 endomorphisms with:

- for $\mu = 0$ and $x = 0$: $f_\mu(x) = 0$, $\frac{\partial}{\partial\mu} f_\mu(x) > 0$;
- for $\mu = 0$ and $x = 0$, $\partial_\ell f_\mu(x) \in (0,1)$ and $\partial_r f_\mu(x) \in (-\infty,-1)$;

where ∂_ℓ, ∂_r denotes the left, respectively right derivative with respect to x. These left and right derivatives are often denoted by σ and $-\lambda$: $\sigma = \partial_\ell f_\mu(x)\big|_{(\mu,x)=(0,0)}$ and $\lambda = -\partial_r f_\mu(x)\big|_{(\mu,x)=(0,0)}$. This may seem a somewhat special situation, however:

- if we agree to study PC^1 endomorphisms and their bifurcations in generic one-parameter families then the only new local bifurcations we find are those where 0 is a fixed point;
- for bifurcations in one-parameter families of PC^1 endomorphisms where 0 is a fixed point, $\frac{\partial}{\partial\mu} f_\mu(0) \neq 0$ is generically satisfied; our condition that this should be positive is just to fix the direction of the μ-axis;
- in the same way, $\partial_\ell f_\mu(x) \neq -1,0,1$ and $\partial_r f_\mu(x) \neq -1,0,1$ for $(\mu,x) = (0,0)$ are generic conditions - the fact that we singled out $\sigma \in (0,1)$, $\lambda > 1$ is because it is the only complicated and interesting case (except for the analogous case with l(eft) and r(ight) interchanged).

Under the above hypothesis we shall prove in section 2 that:

- for $\mu < 0$, $|\mu|$ small, f_μ has an attracting fixed point near 0; there are no local bifurcations near zero;
- there are open subsets $B_2, B_3, \ldots, B_\infty \subset (1,\infty)\times(0,1)$ with $\bigcup_i \bar{B}_i = [1,\infty)\times[0,1]$ so that when $(\lambda,\sigma) \in B_i$, then for $i < \infty$, f_μ, for $\mu > 0$, μ small, has a periodic attractor of period i and is locally structurally stable;

  for $i = \infty$, f_μ, for $\mu > 0$, μ small, has near 0 a chaotic "attractor".

We shall give explicit formulas for the sets B_i.

One should compare this with known results on one-parameter families of diffeomorphisms, vector fields, or (smooth) interval endomorphisms where a point attractor changes to a chaotic attractor. In such generic one-parameter families with only local bifurcations of attractors, this always, i.e., in all known cases, involves infinitely many bifurcations like period doubling sequences [4]. For general information on bifurcation to chaos see e.g. [5].

Next we indicate here, and make precise in section 3, how these PC^1 endomorphisms as well as smooth endomorphisms occur as return or Poincaré maps in "flows with different time scales". For this we need to define so called "constrained equations" or "equations lents-rapides". For more general definitions see [1] and [8]; here we give a definition without any attempt to generality.

Consider on R^3 a differential equation of the form

$$\dot{x} = X(x,y,z)$$

$$\dot{y} = Y(x,y,z)$$

$$\varepsilon\dot{z} = -\frac{\partial V}{\partial z}(x,y,z).$$

The corresponding constrained equation "describes" the limiting behaviour of solutions of this equation when ε tends to zero. To be more formal, a <u>constrained equation</u> is given by C^∞ functions (V,X,Y) as above. In order to define a solution or an integral curve of a constrained equation we need the following definitions:

$$\pi(x,y,z) = (x,y);$$

$$S_V = \{(x,y,z) \mid \frac{\partial V}{\partial z}(x,y,z) = 0\};$$

$$S_V^{min} = \{(x,y,z) \mid \frac{\partial V}{\partial z}(x,y,z) = 0,\ \frac{\partial^2 V}{\partial z^2}(x,y,z) \geq 0\}.$$

A map $\gamma: (\alpha,\beta) \to S_V^{min}$ is a solution of the constrained equation (V,X,Y) if:

(i) $\pi\gamma$ is continuous;

(ii) for each $t_o \in (\alpha,\beta)$, $\gamma(t_o) = \lim_{t \searrow t_o} \gamma(t)$;

(iii) $\partial_r(\pi\gamma(t)) = (X(\gamma(t)), Y(\gamma(t)))$;

(iv) for each $t_o \in (\alpha,\beta)$ $\gamma^-(t_o) = \lim_{t \nearrow t_o} \gamma(t)$ exists;

(v) V is strictly decreasing along some curve from $\gamma^-(t_o)$ to $\gamma(t_o)$ in $\pi^{-1}(\gamma(t_o))$ whenever $\gamma(t_o) \neq \gamma^-(t_o)$.

It is known that for generic V, in most points of S_V, S_V projects diffeomorphically on the (x,y)-plane; in the other points of S_V one has either a cusp of a fold. We shall not deal with cusp points in this paper; at a fold point S_V has the shape indicated in the figure below. The lines of fold points are in the boundary of S_V^{min}. Solutions which arrive at a fold point have in general a discontinuity (to a lower V level of S_V^{min}).

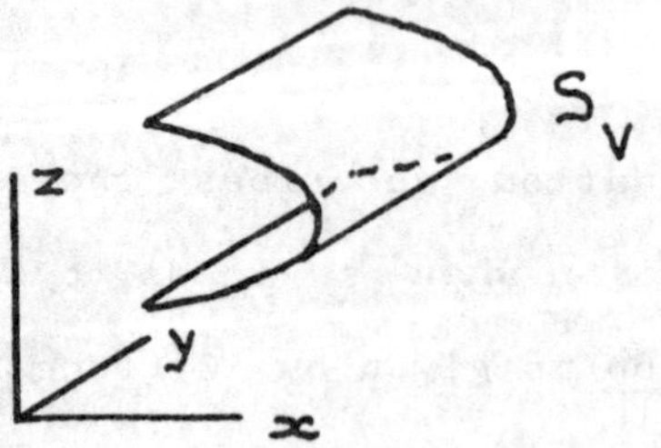

For these discontinuities to exist, even for solutions to exist one has to make some extra assumptions like: V bounded below and for each (x,y), $S_V \cap \pi^{-1}(x,y)$ finite, see [8]. We shall always assume these extra assumptions satisfied.

Just as with flows in the plane, one can describe examples by indicating the phase portrait (or part of it). In points of S_V^{min} where it projects to the (x,y)-plane as a local diffeomorphism, the constrained equation is equivalent with a 2-dimensional flow. We draw the projections of solutions on the (x,y)-plane. The corresponding vector field is in principle multi valued, the number of values changing in the (projections of) fold lines.

In our first example we show how arbitrary interval endomorphisms can appear as return map of a constrained equation. The shape of the fold line is the graph (turned upside down) of the map of

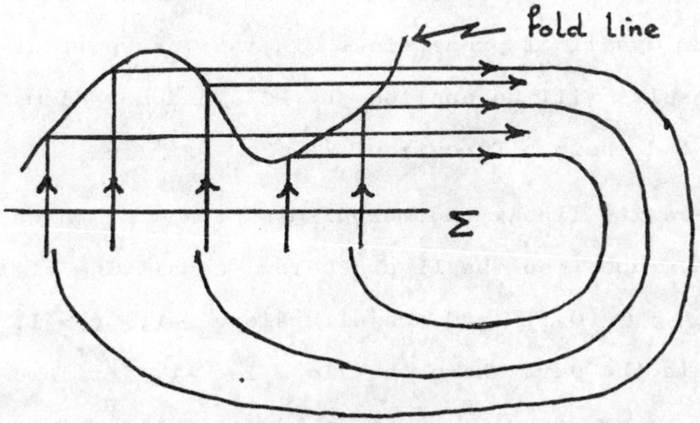

first return to the crossection Σ.

In the second example we show how PC^1 endomorphisms arrise.

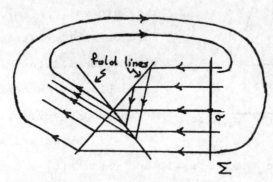

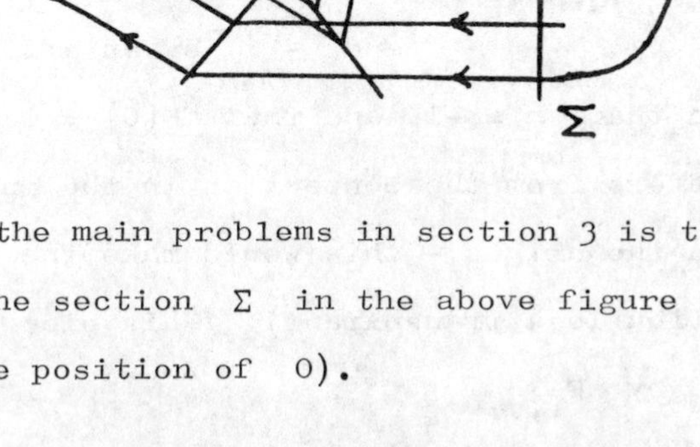

In fact one of the main problems in section 3 is to show that the return map of the section Σ in the above figure is indeed PC^1 (with σ in the position of 0).

So the results of this paper show new transitions from periodic to chaotic attractors which occur in generic one-parameter families of flows with different time scales as formalized above. In section 4 we point to some connections with related work.

2. Piecewise linear and piecewise differentiable endomorphisms

We consider first a class of piecewise linear endomorphisms and study their dynamical properties. In the last part of this section the results will be applied to PC^1 endomorphisms (see the introduction) and their bifurcations.

The piecewise linear endomorphisms $F: \mathbb{R} \to \mathbb{R}$ which we shall analyse have a graph (see the figure below) consisting of two lines, one with slope $\sigma \in (0,1)$ and one with slope $-\lambda$, $\lambda > 1$. F has a fixed point in the part where the slope is $-\lambda$. It is clear that,

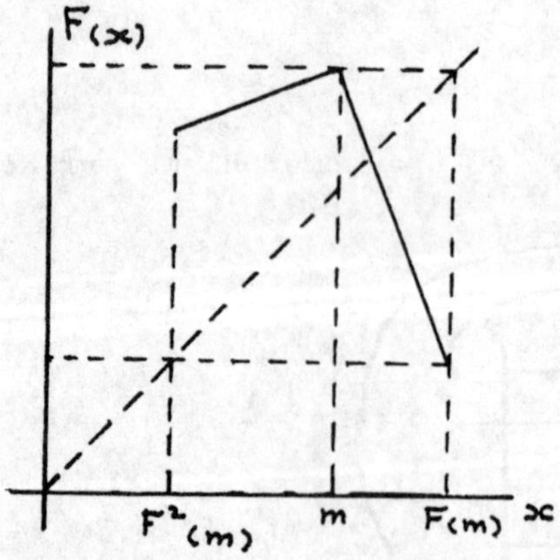

if m denotes the point where the derivative of F changes, the ω-limit of each point will be in the interval $[F^2(m),F(m)]$. Therefore we shall restrict our attention to that interval. The map F is completely determined (up to an affine conjugacy) by the slopes σ and $-\lambda$; by an affine transformation we can arrange that $m = -1$ and that $F(0) = 0$ (note that this convention differs from the convention in the introduction where we had m in the origin - this would make the computations in the present section less transparant). This piecewise linea map is also denoted by $F_{\lambda,\sigma}$.

In the next figure we indicate some of the consequences of the above conventions. Also, as indicated, we define $L = [-\lambda^2, -1]$

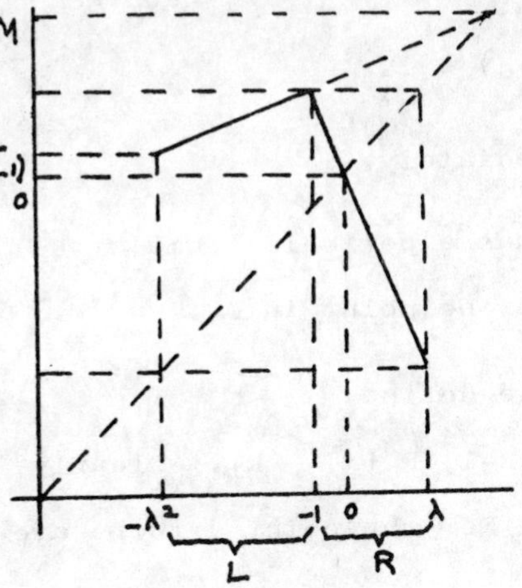

and $R = [-1,\lambda]$; (M,M) is the point on the diagonal which lies on the prolongation of the line in the graph of F with slope σ. It follows from a simple calculation that

$$M = \frac{\sigma+\lambda}{1-\sigma}.$$

Lemma 1. For $k \geq 2$ there is a point with k successive iterates in L, i.e., a point x such that $x, F(x), \ldots, F^{k-1}(x) \in L$, if and only if $\lambda \geq \lambda_k(\sigma)$ with

$$\lambda_k(\sigma) = \frac{\sigma^{-k+1} - \sigma}{1 - \sigma}.$$

(note that $\lambda_{k+1}(\sigma) > \lambda_k(\sigma)$). For $k = 1$ the above holds trivially.

Proof: If there is a point with k successive iterates in L, then this certainly holds for the point $-\lambda^2$. This is equivalent with

$$\sigma^{k-1}(M+\lambda^2) \geq M+1.$$

This inequality is equivalent with the following ones

$$\sigma^{k-1} \left(\frac{\sigma+\lambda}{1-\sigma} + \lambda^2\right) \geq \frac{\sigma+\lambda}{1-\sigma} + 1$$

$$\sigma^{k-1}(\sigma+\lambda+\lambda^2-\lambda^2\sigma) \geq \sigma + \lambda + 1 - \sigma$$

$$\sigma^{k-1}(\sigma+\lambda+\lambda^2-\lambda^2\sigma) \,/\, (1+\lambda) \geq 1$$

$$\sigma^{k-1}(\lambda+\sigma-\lambda\sigma) - 1 \geq 0 \qquad (*k)$$

We shall refer to this last inequality as (*k). For the

present proof we observe that the left hand side of (*k) is an increasing function of λ, so (*k) is equivalent with $\lambda \geq \lambda_k(\sigma)$ where $\lambda_k(\sigma)$ satisfies

$$\sigma^{k-1}(\lambda_k(\sigma)+\sigma-\lambda_k(\sigma)\cdot\sigma) = 1.$$

From this the lemma follows immediately.

<u>Lemma 2</u>. Whenever $\lambda \geq \lambda_k(\sigma)$, $F_{\lambda,\sigma}$ has a periodic orbit with period $k+1$ with k points in L and one point in R.

<u>Proof</u>: Writing again F for $F_{\lambda,\sigma}$, we define $[a,b] = \{x \in L \mid F^i(x) \in L,\ i=0,1,\dots,k-1,\ F^k(x) \in R\}$. Clearly $F^{k-1}(b) = -1$, so that $F^{k+1}(b) = -\lambda^2$. For the point a there are two possibilities:

$a = -\lambda^2$; in that case F^{k+1} clearly has a fixed point in $[a,b]$;

$a > -\lambda^2$; then $F^k(a) = -1$ and $F^{k+1}(a) = \lambda$ so that $F^{k+1}(b) < a < b < F^{k+1}(a)$ and F^{k+1} has a fixed point in $[a,b]$.

In both cases this fixed point of F^{k+1} in $[a,b]$ is on the required periodic orbit.

<u>Lemma 3</u>. $F_{\lambda,\sigma}$ has a periodic attractor with period $k+1$, $k \geq 1$, having k points in L and one point in R, if and only if

$$\lambda_k(\sigma) < \lambda < \bar{\lambda}_k(\sigma),$$

with $\bar{\lambda}_k(\sigma) = \sigma^{-k}$. Moreover $\bar{\lambda}_k(\sigma) > \lambda_k(\sigma)$ if and only if $\sigma < \sigma_k$, where $\sigma_k \in (0,1)$ is a solution of $\sigma^{-1}+\sigma^k-2 = 0$; $\sigma_{k+1} < \sigma_k$ and $\lim_{k\to\infty} \sigma_k = \frac{1}{2}$.

<u>Proof</u>: If -1 is not on the periodic orbit of period $k+1$, then dearly $\sigma^k\cdot\lambda < 1$ is equivalent with this periodic orbit being an attractor. If -1 belongs to the periodic orbit, then $\lambda = \lambda_k(\sigma)$, $k \geq 2$, and $F^{k+1}_{\lambda,\sigma}$ has in any of the corresponding fixed points in

L a left derivative $-\sigma^k\cdot\lambda$ and a right derivative $\sigma^k\cdot\lambda^2$. For such a periodic point to be an attractor, we need $\sigma^k\cdot\lambda^2 < 1$. But on the contrary we have

$$\sigma^k\cdot(\lambda_k(\sigma))^2 = \frac{\sigma^k(\sigma^{-k+1}-\sigma)^2}{(1-\sigma)^2} = \frac{\sigma^{-k+2}(1-\sigma^k)^2}{(1-\sigma)^2} \geq 1.$$

Because the left hand side of (*k) is an increasing function of $\lambda, \lambda_k(\sigma) < \bar{\lambda}_k(\sigma)$ is equivalent with (*k), in which $\bar{\lambda}_k(\sigma)$ is substituted for λ, or with

$$\sigma^{-1} + \sigma^k - 2 > 0.$$

For $k = 1$ and any $\sigma \in (0,1)$ this inequality holds. For $k \geq 2$ we consider the function $g_k(\sigma) = \sigma^{-1}+\sigma^k-2$ for $\sigma \in (0,1]$. We have

- $g_k(1) = 0$;
- $g_k'(\sigma) = -\sigma^{-2} + k\cdot\sigma^{k-1}$ is monotonously increasing;
- $g_k'(1) > 0$;
- $\lim_{\sigma\to 0} g_k(\sigma) = \infty$, $\lim_{\sigma\to 0} g_k'(\sigma) = -\infty$.

This implies that the graph of g_k is as indicated below. Consequently $g_k(\sigma) = 0$ has a unique solution in $(0,1)$ which we denote by σ_k; g_k is positive on $(0,\sigma_k)$. Also it is easy to see that $g_k(\sigma_{k+1}) > 0$, so $\sigma_{k+1} < \sigma_k$. For $\sigma \in (0,1)$ we have $\lim_{k\to\infty} g_k(\sigma) = \sigma^{-1} - 2$, hence $\lim_{k\to\infty} \sigma_k = \frac{1}{2}$.

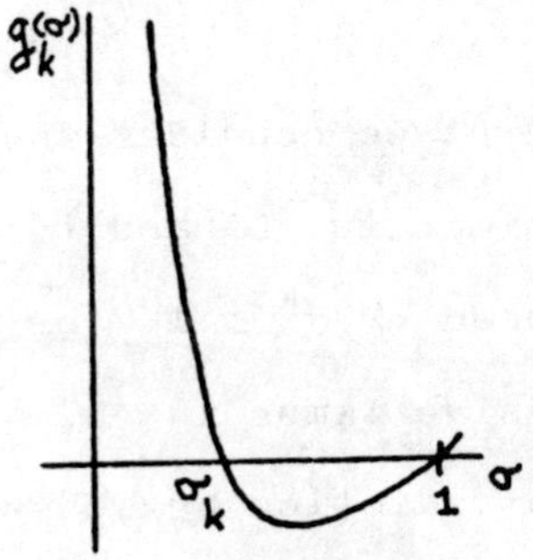

<u>Definition 4</u>. We define the open subsets B_i, $i \geq 2$, (see the introduction) of $(1,\infty)\times(0,1)$ by

$$B_i = \{(\lambda,\sigma) \mid \lambda_{i-1}(\sigma) < \lambda < \bar{\lambda}_{i-1}(\sigma),\ \sigma \in (0,1)\}$$

for $2 \leq i < \infty$ and $B_\infty = (1,\infty)\times(0,1) - \bigcup_{i=2,3,\ldots} \bar{B}_i$.

Lemma 5. For $i \neq j$, $B_i \cap B_j = \phi$.

Proof: Due to the previous results it is enough to show that $\bar{\lambda}_k(\sigma) < \lambda_{k+1}(\sigma)$ for all $\sigma \in (0,1)$. In order to prove this we substitute $\bar{\lambda}_k(\sigma) = \sigma^{-k}$ for λ in (*k+1) and show that it is never satisfied:

$$\sigma^k(\sigma^{-k}+\sigma-\sigma^{-k+1}) - 1 \geq 0 \quad \text{or}$$

$$1 + \sigma^{k+1} - \sigma - 1 \geq 0 \quad \text{or}$$

$$\sigma^{k+1} - \sigma \geq 0.$$

This last inequality is never satisfied for $\sigma \in (0,1)$.

From the present results we know that the B_i's must be as indicated in the figure below.

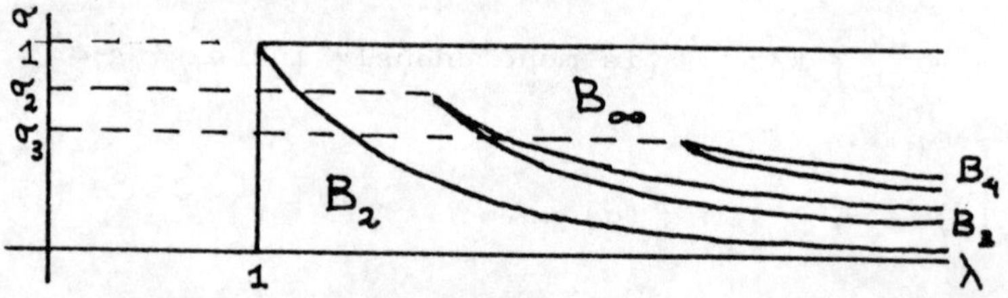

Lemma 6. Let $x \in L$ be a point such that $x, F(x), \ldots, F^{k-1}(x) \in L$ and $F^k(x), F^{k+1}(x) \in R$ and let $F^i(x) \neq -1$ for $i = 0, \ldots, k+1$. Then $(F^{k+2})'(x) > 1$.

Proof: We prove this by contradiction. First we consider the case $k \geq 2$. For an orbit segment which is k times in L and twice in R to have a non-expanding derivative, we need $\lambda^2\sigma^k \leq 1$ or $\lambda \leq \sigma^{-k/2}$. For an orbit to stay k successive times in L, we need (*k) to hold. We get the required contradiction by substituting $\sigma^{-k/2}$ for λ in (*k) and showing that it is never satisfied:

$$\sigma^{k-1}(\sigma^{-k/2}+\sigma-\sigma^{-k/2+1}) - 1 \geq 0 \quad \text{or}$$

$$\sigma^{k/2-1} + \sigma^{k} - \sigma^{k/2} - 1 \geq 0.$$

This is never satisfied for $k \geq 2$ and $\sigma \in (0,1)$ because:

$$\sigma^{k} - \sigma^{k/2} < 0$$

$$\sigma^{k/2-1} - 1 \leq 0.$$

For $k = 1$, we observe that an orbit segment passing successively through L, R, and R is only possible if

$$\lambda - \sigma(\lambda^{2}-1) \leq \lambda^{-1}.$$

In order to see this, we observe that $[-1,\lambda^{-1}] \subset R$ is the interval of those points which are mapped by F into R. In order to have an orbit segment of the required type we need $F(L) \cap [-1,\lambda^{-1}] \neq \phi$ from which the above inequality follows immediately. Observe that the left hand side of this inequality is a decreasing function of σ. For an orbit segment of the above type to have a non-expanding derivative, we need $\sigma\lambda^{2} \leq 1$ or $\sigma \leq \lambda^{-2}$. We obtain the required contradiction by substituting λ^{-2} for σ in the above inequality and showing that it is never satisfied:

$$\lambda - \lambda^{-2}(\lambda^{2}-1) \leq \lambda^{-1} \quad \text{or}$$

$$\lambda - 1 - \lambda^{-1} + \lambda^{-2} \leq 0 \quad \text{or}$$

$$(\lambda-1)(1-\lambda^{-2}) \leq 0.$$

This last inequality is never satisfied for $\lambda > 1$.

<u>Corollary</u>. A consequence of Lemma 6 is that for $(\lambda,\sigma) \in B_{\infty}$, there are constants $N \in \mathbb{N}$ and $c > 1$ such that for any $x \in \mathbb{R}$, $n \geq N$, $|(F^{n}_{\lambda,\sigma})'(x)| \geq c^{n}$. This means that F cannot have periodic attractors. Since the interval $[F^{2}(-1),F(-1)]$ still contains all ω-limit sects, this interval must contain one or more chaotic

attractors. The above statement about the norm of the derivative of $F^n_{\lambda,\sigma}$ can be interpreted as sensitive dependence of the dynamics on initial conditions. See also section 4.

<u>Lemma 7</u>. For $(\lambda,\sigma) \in B_i$, i finite, $F^j_{\lambda,\sigma}(-1)$ approaches the periodic attractor as j tends to infinite.

<u>Proof</u>: We use the idea of Lemma 2. We show that $[a,b]$ as defined there, is mapped into itself by $F^i_{\lambda,\sigma}$. Since $(\lambda,\sigma) \in B_i$, the derivative of $F^i_{\lambda,\sigma}$ is smaller than one in norm on $[a,b]$. This means that the second alternative, namely $F^i_{\lambda,\sigma}(b) < a < b < F^i_{\lambda,\sigma}(a)$ cannot occur. Hence $F^i_{\lambda,\sigma}(b) = a$ and consequently $F^i_{\lambda,\sigma}[a,b] \subset [a,b]$. The lemma now follows from the fact that $F^2_{\lambda,\sigma}(-1) = a$.

<u>Corollary</u>. For $(\lambda,\sigma) \in B_i$, i finite, and $F = F_{\lambda,\sigma}$, we define $Z \subset [F^2(-1),F(-1)]$ as the set of points whose ω-limit is <u>not</u> the periodic attractor. Z is clearly closed (and non-empty: $0 \in Z$); $F(Z) = F^{-1}(Z) = Z$. From the above lemma it follows that $-1 \notin Z$ so that F is smooth on a neighbourhood of Z. By Lemma 6 then, Z is a hyperbolic set with expanding derivative. Z cannot have interior points: if Z would contain a non-trivial interval J, then J as well as $F^j(J)$ would be contained in Z and hence not contain 0. But then $F^j|J$ is monotonous and, for n big, arbitrarily expanding. Hence Z would contain arbitrarily big intervals which is absurd. Since F, restricted to a neighbourhood of Z is C^2, it follows from [2] that the Lebesgue measure of Z is zero. Hence, apart from the periodic attractor, there are no other attractors. $F|[F^2(-1),F(-1)]$ is structurally stable.

<u>Extension to piecewise C^1 endomorphisms</u>. Let $f: \mathbb{R} \to \mathbb{R}$ be a PC^1 endomorphism as defined in the introduction, i.e., f is continuous and both $f|(-\infty,0]$ and $f|[0,\infty)$ are C^1. We assume that $f(0) > 0$,

that $f' < -1$ on $[0,\infty)$, and that $f' \in (0,1)$ on $(-\infty,0]$. In this case the ω-limit of each point $x \in \mathbb{R}$ is contained in $[f^2(0),f(0)]$; from now on we restrict our attention to this interval. If the derivative of f is constant on the intervals $[f^2(0),0]$ and $[0,f(0)]$ then the previous theory applies. It is however also simple to verify that if, for some i, each pair (λ,σ), for which there are points $x \in [0,f(0)]$ and $x' \in [f^2(0),0]$ such that $f'(x) = -\lambda$ and $f'(x') = \sigma$, is contained in B_i, then also the above statements about the dynamics of $F_{\lambda,\sigma}$, with $(\lambda,\sigma) \in B_i$, hold for f. Roughly speaking this means that f need not be piecewise linear: if its derivative is almost constant on the intervals $[f^2(0),0]$ and $[0,f(0)]$ the conclusions are the same.

<u>Local bifurcations</u>. Let $f_\mu : \mathbb{R} \to \mathbb{R}$ be a smooth one-parameter family of PC^1 endomorphisms, i.e., $f_\mu(x)$ depends continuously on (μ,x) and its restrictions to $\mathbb{R}\times(-\infty,0]$ and to $\mathbb{R}\times[0,\infty)$ are C^1. As announced in the introduction we assume that

- for $\mu = 0$ and $x = 0$: $f_\mu(x) = 0$, $\frac{\partial}{\partial\mu} f_\mu(x) > 0$;
- for $\mu = 0$ and $x = 0$ $\partial_\ell f_\mu(x) = \sigma \in (0,1)$, $-\partial_r f_\mu(x) = \lambda > 1$.

With these assumptions it follows easily that for small negative values of μ, f_μ has only a point attractor near zero - there are no local bifurcations. For small positive values of μ, the interval $[f_\mu^2(0),f_\mu(0)]$ is small, so if $(\lambda,\sigma) \in B_i$, there is a positive μ_o such that whenever $0 < \mu < \mu_o$, $x \in [f_\mu^2(0),0]$, $x' \in [0,f_\mu(0)]$, we have

$$(-f'_\mu(x'),\ f'_\mu(x)) \in B_i .$$

This means that $f_\mu|[f_\mu^2(0),f_\mu(0)]$, $0 < \mu < \mu_o$, has all the dynamical properties derived for the corresponding piecewise linear maps F. In detail:

Theorem. For f_μ, λ, σ as above and $(\lambda,\sigma) \in B_i$, there is a positive μ_o such that for $0 < \mu < \mu_o$:

- if i is finite, f has a periodic attractor near 0; the ω-limit of every point in $[f_\mu^2(0), f_\mu(0)]$, with the exception of a set of points with Lebesgue measure zero, is this periodic attractor; $f_\mu | [f_\mu^2(0), f_\mu(0)]$ is structurally stable;
- if $i = \infty$, f_μ has no periodic attractors in $[f_\mu^2(0), f_\mu(0)]$: the derivative of f_μ^n wherever defined on this interval is expanding for n sufficiently big; since this interval is positively invariant, it has to contain "chaotic attractors".

3. Transition maps.

In this section we deal with constrained equations as defined in the introduction. We are especially interested in the following situation (see also the figure below): p, q, and r are points in S_V; p and q are fold points and r is an interior point of S_V^{min}; $\pi(p) = \pi(q) = \pi(r)$; V is decreasing from p to q and from q to r; the projections of the fold lines meet transversally. We assume that the "slow vector field" (which is given by (X,Y)) in the points p, q, and r is as indicated in the next figure. Let $\gamma: (-\varepsilon, +\varepsilon) \to S_V^{min}$ be a solution with $\gamma(0) = r$ and $\lim_{t \nearrow 0} \gamma(t) = p$. Let Σ_- and Σ_+ be two sections in S_V^{min} intersecting $\gamma(-\varepsilon, +\varepsilon)$ transversally in a point of $\gamma(-\varepsilon, 0)$,

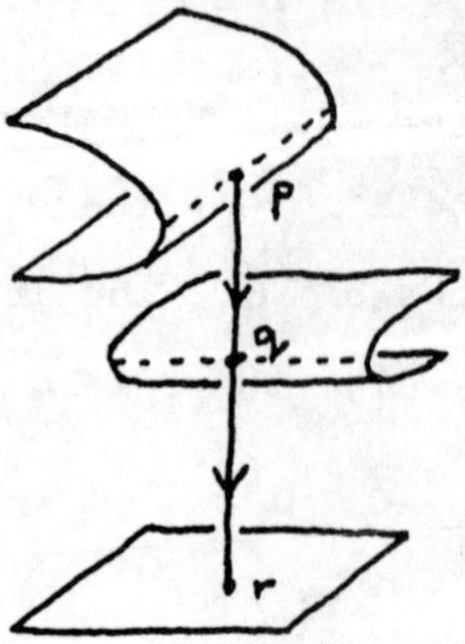

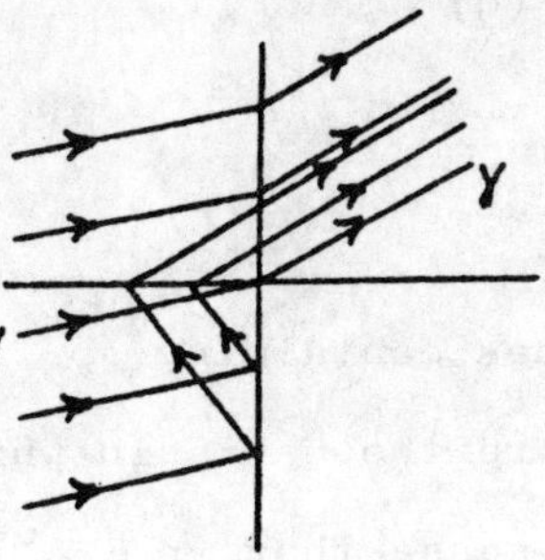

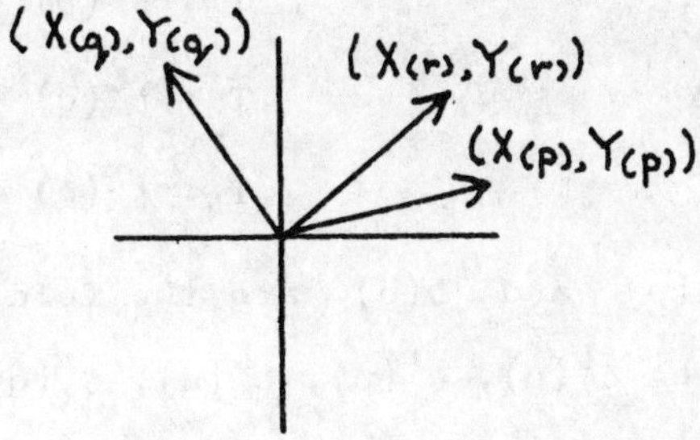

respectively $\gamma(0,\varepsilon)$; the intersections of $\gamma(-\varepsilon,+\varepsilon)$ with Σ_-, Σ_+ are denoted by σ_-, σ_+ respectively. The <u>transition map</u> $T: \Sigma_- \to \Sigma_+$ is defined (at least in a neighbourhood of σ_-) by: $T(x)$ is the first point of intersection with Σ_+ of the positive solution, of the constrained equation, starting in x.

<u>Proposition</u>. In the above situation, T is continuous and piecewise C^1 in the sense that T is C^1 on Σ_-^{ℓ} and Σ_-^r, where Σ_-^{ℓ} and Σ_-^r are the closed (i.e., including σ_-) intervals in which Σ_- is devided by σ_-.

<u>Corollary</u>. As a consequence of this proposition there is an open set of one-parameter families of constrained equations having Poincaré (or return) maps which are, up to local conjugacy, PC^1 endomorphisms which bifurcate as described in section 2. See also the last figure of the introduction.

<u>Proof</u>: In the same way as the transition map $T: \Sigma_- \to \Sigma_+$ was defined, we can also define

$$T_1 \colon \Sigma_- \to \ell(p)$$

$$T_2 \colon \ell^\ell(p) \to \ell^\ell(q)$$

$$T_3 \colon \ell^\ell(q) \to \Sigma_+$$

$$T_4 \colon \ell^r(q) \to \Sigma_+$$

where $\ell(p)$ and $\ell(q)$ are the fold lines containing p and q, and where $\ell^\ell(p)$, $\ell^r(p)$, $\ell^\ell(q)$, $\ell^r(q)$ are the parts in which these lines are divided by p and q in the sense that $p \in \ell^\ell(p)$, $p \in \ell^r(p)$, $q \in \ell^\ell(q)$, and $q \in \ell^r(q)$. The projections in the (x,y)-plane of all these lines are indicated in the next figure. We have to show that $T_1,\ldots,T_4$ are C^1. In a neighbourhood of r,

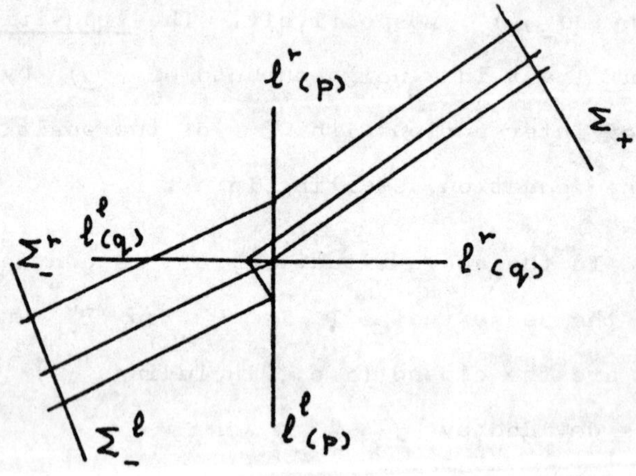

our constrained equation is equivalent with a smooth flow on a surface. From this one easily sees that T_3 and T_4 are C^∞.

Next we prove that T_1 is C^∞. Observe that coordinate transformations of the form

$$\tilde{x} = \tilde{x}(x,y)$$
$$\tilde{y} = \tilde{y}(x,y)$$
$$\tilde{z} = \tilde{z}(x,y,z)$$

transform one constrained equation into another. With such a coordinate transformation we can arrange that $p = (0,0,0)$, and that near p $S_V = \{x=z^2\}$ and $S_V^{min} = S_V \cap \{z \geq 0\}$. For the existence of such coordinates see e.g. [3]. As coordinates on S_V we use y and z. On S_V^{min} (except on the fold line) we define the <u>induced vector field</u>: its integral curves are just the solution (near p) of the constrained equation. For these solutions we have

$$\dot{x} = X(x,y,z)$$
$$\dot{y} = Y(x,y,z)$$
$$x = z^2,$$

so

$$2z\dot{z} = X(z^2,y,z).$$

This means that the induced vector field is

$$Y(z^2,y,z)\,\frac{\partial}{\partial y} + \frac{1}{2z}\cdot X(z^2,y,z)\,\frac{\partial}{\partial z}.$$

The <u>reduced vector field</u> is obtained by multiplying the induced vector field by z (and extending its domain of definition to all of S_V (in a neighbourhood of p). This reduced vector field

$$z\cdot Y(z^2,y,z)\,\frac{\partial}{\partial y} + \frac{1}{2}\cdot X(z^2,y,z)\,\frac{\partial}{\partial z}$$

is C^∞; its integral curves are, up to parametrization, solution curves of the constrained equation. This implies that T_1 is C^∞.

Finally we prove that T_2 is C^1. As in the previous case we choose adapted coordinates but now for a neighbourhood of q. So we assume that $q = (0,0,0)$, $S_V = \{x=z^2\}$, and that $S_V^{min} = S_V \cap \{z \geq 0\}$. We also assume that the projection of $\ell(p)$ on the (x,y)-plane is (locally) equal to the x-axis. In these coordinates $X(0) \neq 0$ and $Y(0) \neq 0$. As above we construct the reduced vector field

$$z\cdot Y(z^2,y,z)\,\frac{\partial}{\partial y} + \frac{1}{2}\cdot X(z^2,y,z)\,\frac{\partial}{\partial z}.$$

Using y and z as coordinates in S_V, we observe that the z-axis is tangent to the integral curve of the reduced vector field through 0, but has no higher order contact with this integral curve. This means that the transition map $\mathfrak{J}$ from the positive z-axis to the y-axis (this transition map can be defined with integral curves of the reduced vector field or with solutions of the constrained equation) has the form $\mathfrak{J}(z) = z^2 \cdot t(z)$ with $t(0) \neq 0$. See also the figure in which the situation (z- and y-axis and integral curves of the reduced vector field) in S_V is indicated. Let S denote the projection of the positive z-axis, as a curve in S_V^{min}, on the x-axis, so $S(z) = z^2$. The transition map T_2 is equal to $\mathfrak{J} \cdot S^{-1}$ (except for the projection of $\ell(p)$ on the x-axis which is a diffeomorphism). This means that, in the present coordinates $T_2(x) = x \cdot t(\sqrt{x})$; this is C^1 but in general not C^2. This completes the proof of our proposition.

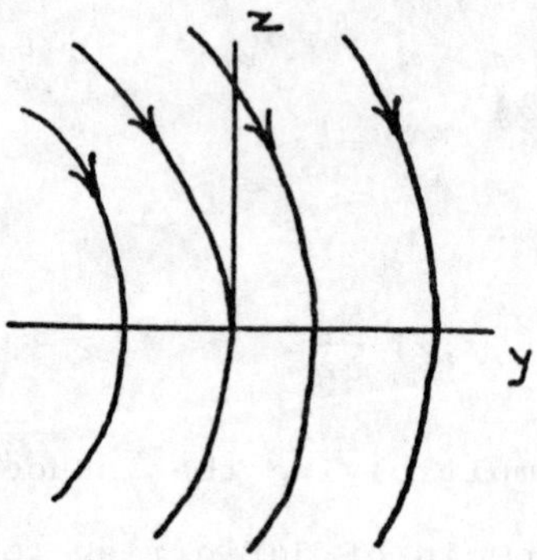

4. Some final remarks.

1. In the introduction we motivated our definition of a solution of a constrained equation by considering the system

$$\dot{x} = X(x,y,z)$$

$$\dot{y} = Y(x,y,z)$$

$$\varepsilon \dot{z} = -\frac{\partial V}{\partial z}(x,y,z).$$

However it is not obvious that, for ε sufficiently small, the solutions of the above system are close to the solutions of the constrained equation defined by (V,X,Y). In fact this is in general not the case. This problem was considered by Benoit in his thesis [1]; it turned out that for the examples in this paper the solutions of the constrained equations are good approximations of the solutions of the corresponding differential equations when ε is small.

2. In [9] I made some observations about the symbolic dynamics of the piecewise linear dynamical systems which we also consider here. It is clear that for $(\lambda,\sigma) \in B_\infty$, the set of inverse images of the "maximum" (or "-1" in the notation in the first part of section 2) namely

$$\bigcup_{i \in \mathbb{N}} ((F_{\lambda,\sigma})^i)^{-1}(-1)$$

is dense in $[F^2_{\lambda,\sigma}(-1), F_{\lambda,\sigma}(-1)]$. This implies that the dynamics of $F_{\lambda,\sigma}$ restricted to this last interval is completely determined, up to topological conjugacy, by the ordered structure of the set of these inverse images of the maximum. This is exactly what is described by the symbolic dynamics. So from [9] it follows that for $(\lambda,\sigma) \in B_\infty$ the dynamics is completely described by one real invariant.

3. We did not give here a precise definition of a "chaotic attractor". In fact we did not prove that there is an attractor at all in the sense of a closed set K, containing a dense orbit and having a positively invariant neighbourhood U whose largest invariant set is K. However with the more liberal definition of attractor in [6] there is always an attractor in $[F^2(-1), F(-1)]$ because it attracts all orbits. If $(\lambda,\sigma) \in B_\infty$ this attractor is certainly not periodic. We can only prove that it has sensitive dependence on initial conditions in a special sense: for nearby points, x, x' the distance

between $F^{i_o}(x)$ and $F^{i_o}(x')$ increases as long as $F^i(x)$ and $F^i(x')$ are on the same side of the maximum -1 for all $0 \le i < i_o$. In other words the distance increases until the orbits can be distinguished by the methods os symbolic dynamics. I do not know whether for all $(\lambda,\sigma) \in B_\infty$ there is an orbit of $F_{\lambda,\sigma}$ which is dense in $[F^2_{\lambda,\sigma}(-1), F_{\lambda,\sigma}(-1)]$. The numerical simulations, of which the results are shown below, indicate that this might be the case.

4. The constrained equations which we considered in this paper had two slow variables (x and y) and one fast variable (z). It is easy to generalize the definition to arbitrary numbers of slow and fast variables. In the case of more than one fast variable our examples remain, i.e., there is still an open set of one-parameter families of constrained equations in which a periodic attractor bifurcates to a chaotic - or a subharmonic - attractor as described. If one changes the number of slow variables the situation is more complicated: with only one slow variable there are (generically) no chaotic attractors. For more than two slow variables we don't even know what the analogue of our present example would be. The situation here is different from "ordinary" dynamical systems like diffeomorphisms and vector fields. For these ordinary dynamical systems any low dimensional example has a corresponding high dimensional example: one obtains the high dimensional example by putting the low dimensional example in a normally contracting invariant manifold; by making the normal contraction sufficiently strong the invariant manifold and its differentiability are persistent under small perturbations. This means that all phenomena which occur generically in low dimensional (ordinary) dynamical systems also occur generically in high dimensional dynamical systems, e.g. see [7]. For dynamical systems given by endomorphisms or by constrained equations,

a corresponding principle is not known. It seems to me a basic problem to clarify this situation.

5. In the two figures below we show the results of numerical simulations of the dynamics of $F_{\lambda,\sigma}$: in the first diagram for $\sigma = \frac{1}{2}$ while λ varies from 1 to 20, in the second diagram for $\lambda = 50$ while σ varies from 0 to $\frac{1}{2}$. These were obtained by iterating $F_{\lambda,\sigma}$ 100,000 times while moving the parameter linearly with the number of iterates (in the first case λ from 1 to 20 in the second case σ from 0 to $\frac{1}{2}$) and by plotting after each iteration the point (x,y), with x proportional to the changing parameter and y proportional to the computed value of $F_{\lambda,\sigma}$ (this value was rescaled so as to make the interval $[F^2_{\lambda,\sigma}(-1), F_{\lambda,\sigma}(-1)]$ independent of λ and σ). In this way we get the attracting set of $F_{\lambda,\sigma}$ as function of the changing parameter. In both cases the succession of periodic regions and chaotic regions and the periods in the periodic regions show clearly the structure corresponding to the location of $B_2, B_3, \ldots, B_\infty$ as determined in section 2.

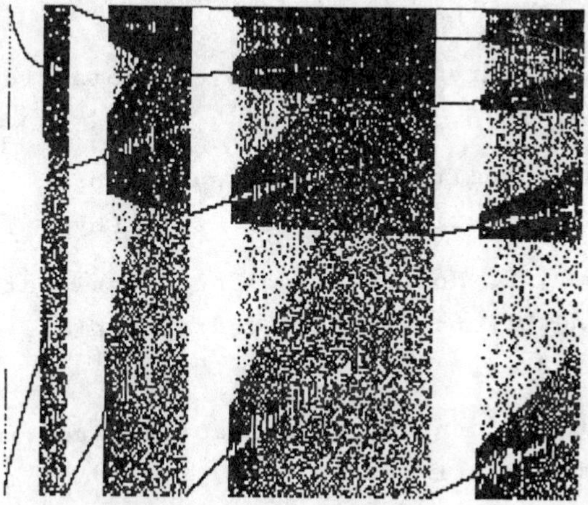

$\sigma = \frac{1}{2}$, λ from 1 to 20;

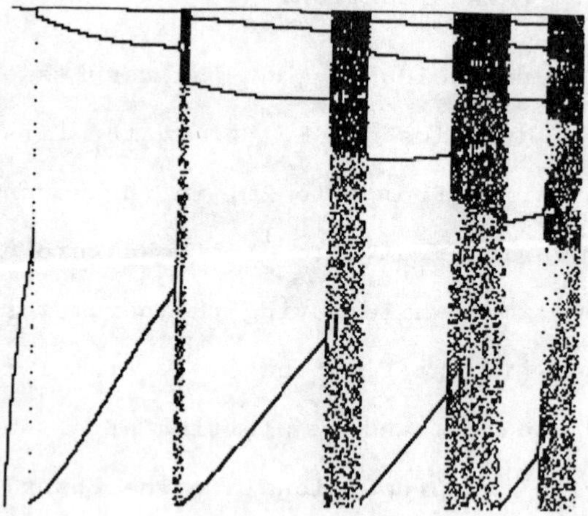

$\lambda = 50, \quad \sigma$ from 0 to $\frac{1}{2}$

References

[1] E. Benoit, Systèmes lents-rapides dans $\mathbb{R}^3$ et leur canards, Astérisque, 109-110, (1983), 159-191.

[2] R. Bowen, Equilibrium states and the ergodic theory of Anosov diffeomorphisms, Lecture Notes in Math. 470, (1975), Springer-Verlag, Berlin.

[3] Th. Bröcker, Differentiable germs and catastrophes, Cambr. Univ. Press, 1975.

[4] M. Feigenbaum, Qualitative universality of a class of non-linear transformations, Journ. Stat. Phys. 19, (1978), 25.

[5] J. Guckenheimer, P. Holmes, Nonlinear oscillations, dynamical systems, and bifurcations of vector fields, Springer-Verlag, New York, 1983.

[6] J. Milnor, On the concept of attractor, preprint Institute for Advanced Study, Princeton, 1984.

[7] J. Palis, F. Takens, Topological equivalence of normally hyperbolic dynamical systems, Topology 16, (1977), 335-345.

[8] F. Takens, Constrained equations; a study of implicit differential equations and their discontinuous solutions, in Structural stability, the theory of catastrophes, and applications in the sciences, P. Hilton (ed.), Lecture Notes in Math. 525, (1976), Springer-Verlag, Berlin.

[9] F. Takens, Implicit differential equations; some open problems, in Singularités d'applications differentiable, O. Burlet, F. Ronga (eds.), Lecture Notes in Math. 535, (1976), Springer-Verlag, Berlin.

Mathematatish Instituut
Rijksuniversiteit Groningen
G.P.O. 800
Groningen - The Netherlands